Gene Control

Gene Control

David S. Latchman

Garland Science
Taylor & Francis Group
NEW YORK AND LONDON

Garland Science
Vice President: Denise Schanck
Editor: Elizabeth Owen
Editorial Assistant: David Borrowdale
Production Editor: Karin Henderson
Illustrator: Matthew McClements, Blink Studio Ltd
Cover Design: Andrew Magee
Copyeditor: Nik Prowse
Typesetting: Georgina Lucas
Proofreader: Sally Livitt
Indexer: Liza Furnival

David S. Latchman obtained his PhD from Cambridge University and a DSc from London University. He is currently Master of Birkbeck University of London and Professor of Genetics at Birkbeck and University College London. Prior to this he held appointments as Dean of the Institute of Child Health and Director of the Windeyer Institute of Medical Sciences at University College London. He is currently Chairman of London Higher (the umbrella organization which represents all London's universities) and serves on the Mayor of London's Promote London Council and the London Skills and Employment Board.

ISBN 978-0-8153-6513-6

Library of Congress Cataloging-in-Publication Data

Latchman, David S.
    Gene control / David S. Latchman.
        p. ; cm.
    Includes bibliographical references and index.
    ISBN 978-0-8153-6513-6
    1. Genetic regulation. I. Title.
    [DNLM: 1. Gene Expression Regulation--physiology. 2. Transcription, Genetic--physiology. QU 475 L351g 2010]
    QH450.L38 2010
    572.8'65--dc22

                        2009046632

Published by Garland Science, Taylor & Francis Group, LLC, an informa business,
270 Madison Avenue, New York NY 10016, USA,
and 2 Park Square, Milton Park, Abingdon, OX14 4RN, UK.

Printed in the United States of America

15 14 13 12 11 10 9 8 7 6 5 4 3 2 1

Visit our web site at http://garlandscience.com

Mixed Sources
Product group from well-managed forests, controlled sources and recycled wood or fibre
www.fsc.org   Cert no. SW-COC-002985
© 1996 Forest Stewardship Council

Cover image shows the crystal structure of a translation termination complex formed with release factor RF2. The image contains RF2 (yellow), P-site tRNA (orange), E-site tRNA (red), mRNA (green), 16S rRNA (cyan), 23S and 5S rRNA (gray), 30S proteins (blue), and 50S proteins (magenta). Courtesy of Harry Noller and John Donohue, University of California, Santa Cruz.

To my mother and in memory of my father

Preface

Gene Control is a significant expansion and reorganization of my previous book *Gene Regulation: A Eukaryotic Perspective* which went through five editions between 1990–2005. Gene control is of vital importance in both normal development and the proper functioning of the adult organism as well as playing a critical role in the development of different human diseases. This book will therefore continue to be of value to students, scientists, and clinicians interested in the topic of gene control.

The first edition of *Gene Regulation* was based on the fact that at that time sufficient information had become available to justify a book which dealt with the topic of eukaryotic gene regulation in its own right rather than simply as an adjunct to the much better understood prokaryotic systems.

Since that time information on eukaryotic gene regulation has continued to accumulate at an ever-increasing rate. Topics such as the modification of histones, which were mentioned only briefly in the first edition, have emerged as being of central importance. Similarly, other aspects of gene regulation were completely unknown at that time, such as the key role of small inhibitory RNAs. These topics and many others have been progressively incorporated into the subsequent editions, each of which was significantly revised and updated.

However, in preparing a sixth edition it was clear that a more radical revision to the work was required. The continuing major advances in our understanding of gene regulation now allow both a mechanistic understanding of the processes involved and an analysis of how they operate in specific biological systems.

To reflect this, the much expanded and revised work now has a new title, *Gene Control*, and is organized into two parts. The first part provides a detailed mechanistic analysis of the processes involved in controlling gene expression. After an introductory chapter, three pairs of chapters cover the fundamental processes involved in gene regulation. In each pair of chapters, the first chapter deals with the basic process itself and the second deals with the manner in which it is involved in regulating gene expression. Thus, Chapters 2 and 3 are about chromatin structure and its role in gene regulation, Chapters 4 and 5 cover the process of transcription itself and the manner in which it is regulated, and Chapters 6 and 7 explain post-transcriptional processes and their role in the regulation of gene expression.

In contrast, the second part of the book deals with specific biological processes and the role played by gene control in their regulation. Thus, Chapter 8 is on cellular signaling processes, Chapter 9 covers the regulation of gene expression in development, and Chapter 10 discusses the key role played by gene-regulatory processes in the specification of individual differentiated cell types. Finally, Chapters 11 and 12 cover the alterations in gene expression which can cause specific human diseases. Thus, Chapter 11 discusses the role of gene regulation in cancer while Chapter 12 deals with gene regulation in inherited and infectious diseases of humans, as well as discussing the manner in which advances in our understanding of gene regulatory processes may lead to improved therapies for human diseases.

Finally, I should like to thank the staff of Garland Science who have produced this book so efficiently, particularly Elizabeth Owen who originally suggested the restructuring/expansion of this book, as well as being a valuable source of advice throughout its preparation, and David Borrowdale who supervised the preparation of the illustrations most effectively. As always, I am indebted to Maruschka Malacos who has coped most efficiently with the need to introduce considerable new material as well as to rearrange and modify existing material in the preparation of this work.

David S. Latchman

Acknowledgments

In writing this book I have benefited greatly from the advice of many geneticists, cell biologists, and biochemists. I would like to thank the following for their suggestions in preparing this edition.

David Elliott (Newcastle University, UK); Maureen Ferran (Rochester Institute of Technology, USA); Tom Geoghegan (University of Louisville, USA); Michael W. King (Indiana State University, USA); Olga Makarova (University of Leicester, UK); Nick Plant (University of Surrey, UK); Ava Udvadia (University of Wisconsin-Madison, USA); Ian Wood (University of Leeds, UK).

I would also like to thank all those colleagues who have given permission for material from their papers to be reproduced in this book and have provided images suitable for reproduction.

Contents in Brief

Contents in Detail

Levels of Gene Control

1

INTRODUCTION

The evidence that eukaryotic gene expression must be a highly regulated process is available to anyone visiting a butcher's shop. The various parts of the mammalian body on display differ dramatically in appearance, ranging from the muscular legs and hind quarters to the soft tissues of the kidneys and liver. However, all these diverse types of tissues arose from a single cell, the fertilized egg or zygote, raising the question of how this diversity is achieved.

It is clear that regulatory processes must exist which allow different cells to form in different places in an orderly manner during embryonic development and to maintain their differences in the adult organism. Moreover, cells need to be able to alter their pattern of gene expression in response to changes such as alterations in the levels of nutrients, growth factors, hormones, etc. It is the aim of this book to consider the processes regulating tissue-specific gene expression in eukaryotes and the manner in which they produce these differences in the nature and function of different tissues and cell types.

1.1 PROTEIN CONTENT VARIES AMONG DIFFERENT CELL TYPES

The fundamental dogma of molecular biology is that DNA produces RNA, which in turn produces proteins. The genetic information in the DNA specifying particular functions is converted into an RNA copy which is then translated into protein. The action of the protein then produces the **phenotype**, be it the presence of a functional globin protein transporting oxygen in the blood or the activity of a proteinaceous enzyme capable of producing the pigment causing the appearance of brown rather than blue eyes. If the differences in the appearance of tissues described above are indeed caused by differences in gene expression in different tissues, they should be produced by differences in the proteins present in these tissues and the cell types of which they are composed. Such differences can be detected both by specific methods which study the expression of one particular protein and by general methods aimed at studying the expression of all proteins in a given tissue.

Specific methods can be used to study the expression of individual proteins in tissues and cells

A simple means of determining whether differences in protein composition exist in different tissues or cell types involves investigating the expression of individual known proteins in such tissues using a specific **antibody** to that protein. Such antibodies can be used in conjunction with the one-dimensional polyacrylamide gel electrophoresis technique to investigate the expression of a particular protein in different tissues. In this technique, proteins are denatured by treatment with the detergent sodium dodecyl sulfate (SDS) and then subjected to electrophoresis in a polyacrylamide gel which separates them according to their size.

Subsequently, in a technique known as **Western blotting**, the gel-separated proteins are transferred to a nitrocellulose filter which is incubated with the antibody. The antibody recognizes and binds specifically to the protein against which it is directed. This protein will be present at a particular position on the filter, dependent on how far it moved in the electrophoresis step and hence on its size. The binding of the antibody is then visualized by an enzymatic or fluorescence-based detection procedure. If a tissue contains the protein of interest, a band will be observed in the track containing total protein from that tissue and the intensity of the band observed will provide a measure of the amount of protein present in the tissue. If none of the particular protein is present in a given tissue, no band will form (**Figure 1.1**).

This method allows the presence or absence of a specific protein in a particular tissue to be assessed using one-dimensional gel electrophoresis without the complicating effect of other unrelated proteins of similar size, since these will fail to bind the antibody. Similarly, quantitative differences in the expression of a particular protein in different tissues can be detected on the basis of differences in the intensity of the band obtained when extracts prepared from the different tissues are used in this procedure.

The specific reaction of a protein with an antibody can also be used directly to investigate its expression within a particular tissue. In this method, known as **immunohistochemistry**, thin sections of the tissue of interest are reacted with the antibody which binds to those cell types expressing the protein. As before, the position of the antibody is visualized by an enzymatic detection procedure or more usually by the use of a fluorescent dye which can be seen in a microscope when appropriate filters are used (**Figure 1.2**). This method can be used not only to provide information about the tissues expressing a particular protein but, since individual cells can be examined, it also allows detection of the specific cell types within the tissue which are expressing the protein.

The results of experiments using these specific methods to study the expression of particular proteins indicate that while some proteins are present at similar abundance in all tissues, the abundance of other proteins varies between different tissues and a large number are specific to one or a few tissues or cell types. Thus, the differences between different tissues in appearance and function are correlated with qualitative and quantitative differences in protein composition, as assayed using antibodies to individual proteins.

General methods can be used for studying the overall protein composition of tissues and cells

As well as examining the expression of individual proteins in different tissues, it is also possible to use more general methods to compare the overall

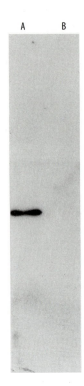

Figure 1.1
Western blot with an antibody to guinea-pig casein kinase showing the presence of the protein in lactating mammary gland (track A) but not in liver (track B). Courtesy of Alison Moore, from Moore A, Boulton AP, Heid HW et al. (1985) *Eur. J. Biochem.* 152, 729–737. With permission from Wiley-Liss, Inc., a subsidiary of John Wiley & Sons, Inc.

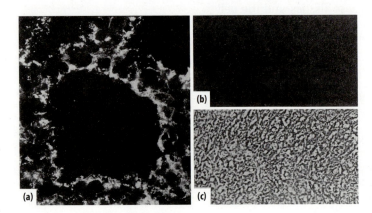

Figure 1.2
Use of the antibody to casein kinase to show the presence of the protein (bright areas) in frozen sections of lactating mammary gland (a) but not in liver (b). (c) A phase-contrast photomicrograph of the liver section, confirming that the lack of staining with the antibody in (b) is not due to the absence of liver cells in the sample. Courtesy of Alison Moore, from Moore A, Boulton AP, Heid HW et al. (1985) *Eur. J. Biochem.* 152, 729–737. With permission from Wiley-Liss, Inc., a subsidiary of John Wiley & Sons, Inc.

protein composition of different tissues. All the proteins present in a tissue can be subjected to one-dimensional gel electrophoresis and examined by using a stain which reacts with all proteins. This method allows the visualization of all cellular proteins separated according to their size on the gel, rather than using Western blotting with a particular antibody to focus on a specific protein. However, this method is of relatively limited use to investigate variations in the overall protein content of different tissues. This is because of the very large number of proteins in the cell and the limited resolution of the technique which separates proteins solely on the basis of their size. For example, two entirely different proteins in two tissues may be scored as being the same protein simply on the basis of a similarity in size.

A more detailed investigation of the protein composition of different tissues can be achieved by two-dimensional gel electrophoresis. In this procedure (**Figure 1.3**) proteins are first separated on the basis of differences in their charge, in a technique known as isoelectric focusing. The separated proteins, still in the first gel, are subsequently layered on top of an SDS-polyacrylamide gel and separated by electrophoresis according to their size. Thus, a protein moves to a position determined both by its size and its charge. The much greater resolution of this method allows a number of differences in the protein composition of particular tissues to be identified. Some spots or proteins are found in only one or a few tissues and not in many others while others are found at dramatically different abundance in different tissues (**Figure 1.4**).

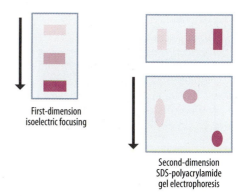

First-dimension
isoelectric focusing

Second-dimension
SDS-polyacrylamide
gel electrophoresis

Figure 1.3
Two-dimensional gel electrophoresis.

(a)

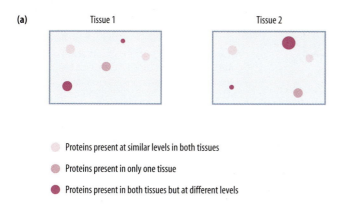

Tissue 1   Tissue 2

Proteins present at similar levels in both tissues

Proteins present in only one tissue

Proteins present in both tissues but at different levels

(b)

Basic   Stable pH gradient   Acidic

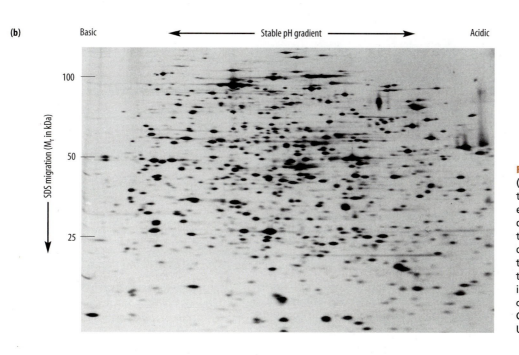

Figure 1.4
(a) Schematic results of two-dimensional gel electrophoresis allowing the detection of proteins specific to one tissue or expressed at different levels in different tissues. (b) Actual result showing the separation of all the proteins in an *Escherichia coli* bacterial cell. M_r, molecular weight. Courtesy of Patrick O'Farrell, University of California.

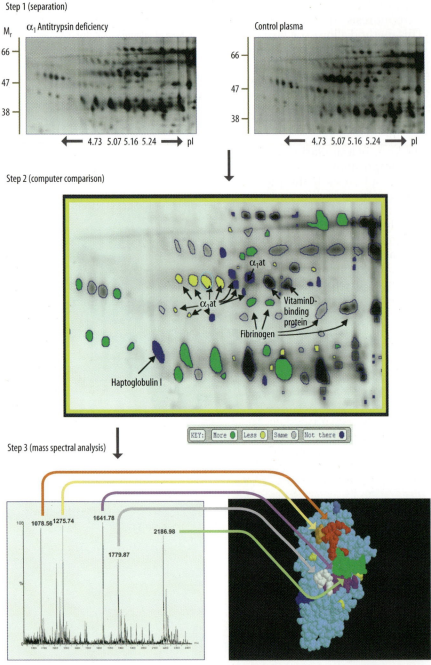

Step 1 (separation)

Step 2 (computer comparison)

Step 3 (mass spectral analysis)

Figure 1.5
Analysis of differences between two protein samples by two-dimensional gel electrophoresis to separate the proteins (step 1), computer comparison of the two gels to identify proteins which differ between the two samples (step 2) followed by excision of a protein spot which is altered, and its identification by mass-spectral analysis of the peptides produced by trypsin digestion (step 3). Courtesy of Kevin Mills & Bryan Winchester, University College London.

The different appearances of different tissues are indeed paralleled therefore by both qualitative and quantitative differences in the proteins present in each tissue. It should be noted however, that some proteins can be shown by two-dimensional gel electrophoresis to be present at similar levels in virtually all tissues. Presumably such so-called housekeeping proteins are involved in basic metabolic processes common to all cell types.

Early experiments using two-dimensional gel electrophoresis simply involved examining the patterns of spots generated by different tissues and describing which spots were present or absent in specific tissues or showed different intensities in different tissues. More recently however, methods have been developed which allow the protein responsible for forming a spot of particular interest to be identified and characterized (**Figure 1.5**). These methods have allowed the development of a new field of study, known as **proteomics**, in which the power of two-dimensional gels to separate a wide range of proteins is combined with the ability to study specific proteins individually.

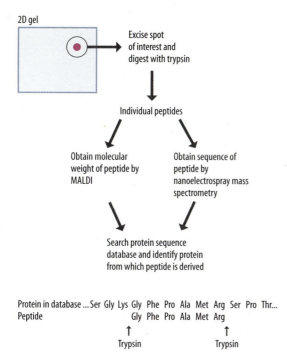

Figure 1.6
Use of mass spectrometry to determine the molecular weight and amino acid sequence of a particular protein spot on a two-dimensional (2D) gel. The information obtained in this way can be used to search databases of known protein sequences and identify the protein producing the spot.

In these methods, individual spots of interest are excised from the two-dimensional gel and digested into their constituent peptides using the **proteolytic enzyme**, trypsin. The individual peptides from a single protein spot are then analyzed by **mass spectrometry**. This can involve both matrix-assisted laser desorption/ionization (**MALDI**), which allows the molecular weight of the peptide to be determined, and nanoelectrospray mass spectrometry, which allows the amino acid sequence of a particular peptide to be obtained (**Figure 1.6**). Often these two techniques are performed sequentially in a tandem mass spectrometer which first determines the molecular weight of a peptide produced by trypsin digestion and then fragments it further, allowing its sequence to be determined.

This peptide molecular weight and sequence information can then be used to search data bases of known protein sequences which have been obtained either by direct protein sequencing or more frequently predicted from the ever-expanding amount of DNA sequence information. As trypsin always cleaves proteins after a lysine or arginine residue, it is possible to align the peptide sequence against that of a known protein and hence identify the protein responsible for the original spot (see Figure 1.6).

In a more recent development of this technique, the two-dimensional gel electrophoresis step is eliminated. Instead, the initial sample containing a variety of different proteins is first digested with a **protease**, such as trypsin. The resulting highly complex mixture of peptides is then separated by liquid **chromatography** into different fractions. Each of these fractions still contains a mixture of different peptides from a number of different proteins but each fraction is obviously less complex than the original mixture prior to fractionation. The individual fractions can therefore be analyzed by repeated cycles of mass spectrometry to determine the molecular weight and sequence of individual peptides, as described above (**Figure 1.7**). In turn, this will allow identification of proteins and their corresponding peptides which are present at different levels in different biological samples.

Proteomic methods can therefore be used to characterize individual proteins which show interesting patterns of expression in different tissues or in response to a specific stimulus or whose expression is altered in a specific disease. Hence they combine the power to look at a wide range of proteins with the ability to identify an individual protein. In terms of gene regulation, the use of these proteomic methods reinforces the conclusion from the other studies described above, that the protein content of different tissues and cell types is both qualitatively and quantitatively different.

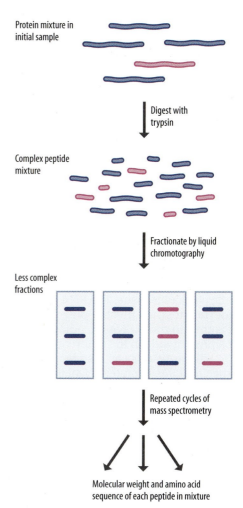

Figure 1.7
Complex biological samples containing many different proteins can be digested with trypsin and the resulting peptides fractionated by liquid chromatography. The different fractions obtained in this way contain less complex peptide mixtures, allowing the molecular weight and amino acid sequence of each individual peptide in the mixture to be determined by repeated cycles of mass spectrometry.

Figure 1.8
Northern-blot hybridization using a probe specific for the α-fetoprotein mRNA. The RNA is detectable in the embryonic yolk sac sample (track A) but not in the adult liver sample (track B). From Latchman DS, Brzeski H, Lovell-Badge R & Evans MJ (1984) *Biochim. Biophys. Acta*. 783, 130–136. With permission of Elsevier.

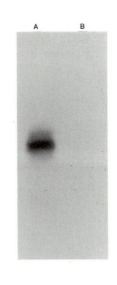

1.2 THE mRNA CONTENT VARIES AMONG DIFFERENT CELL TYPES

Proteins are produced by the translation of specific messenger RNA (mRNA) molecules on the **ribosome** (see Section 6.5). Having established that quantitative and qualitative differences exist in the protein composition of different tissues, it is necessary to ask whether such differences are paralleled by tissue-specific differences in the abundance of their corresponding mRNAs. Although it seems likely that differences in the mRNA populations of different tissues do indeed underlie the observed differences in proteins, it is formally possible that all tissues have the same mRNA species and that production of different proteins is controlled by regulating which of these are selected by the ribosome to be translated into protein.

As with the study of proteins, both specific and general techniques exist for studying the mRNAs expressed in a given tissue.

Specific methods can be used to study the expression of individual mRNAs in different tissues and cells

A number of different specific methods exist which can be used to detect and quantify one specific mRNA, using a cloned DNA probe derived from its corresponding gene. In one such method, **Northern blotting**, the RNA extracted from a particular tissue is electrophoresed on an agarose gel, transferred to a nitrocellulose filter, and hybridized to a radioactive probe derived from the gene encoding the mRNA of interest. The presence of the RNA in a particular tissue will result in hybridization of the radioactive probe and the visualization of a band on **autoradiography**, the intensity of the band being dependent on the amount of RNA present (**Figure 1.8**). Although such experiments do detect the RNA encoding some proteins, such as actin or tubulin, in all tissues, very many others are found only in one particular tissue. The RNA for the globin protein is found only in **reticulocytes**, that for myosin only in muscle, while (in the example shown in Figure 1.8) the mRNA encoding the fetal protein α-fetoprotein is shown to be present only in the embryonic yolk sac and not in the adult liver.

Similar results can also be obtained using other methods for detecting specific RNAs in material isolated from different tissues. In particular, many studies on the expression of individual mRNAs now use the quantitative **reverse transcriptase**/polymerase chain reaction (RT/PCR) technique. This method relies on the use of the **polymerase chain reaction** (**PCR**) to specifically amplify small amounts of a particular nucleic acid sequence (**Figure 1.9**). As PCR can only be carried out with DNA and not with RNA, the RNA is first converted into a **complementary DNA** copy (**cDNA**) by the enzyme reverse transcriptase. Subsequently, short oligonucleotide **primers**, which specifically hybridize to the DNA derived from the RNA of interest, are used to direct its replication by a DNA **polymerase** enzyme in a PCR reaction (see Figure 1.9). The resulting double-stranded DNA is then melted into its

Figure 1.9
In the reverse transcriptase/polymerase chain reaction (RT/PCR) method, mRNA is converted into a complementary DNA (cDNA) copy by the enzyme reverse transcriptase. The cDNA derived from an individual mRNA is then specifically amplified by hybridization with complementary primers and DNA synthesis by a DNA polymerase enzyme. Repeated cycles of amplification result in the production of a large amount of PCR product from a small amount of a specific cDNA/mRNA.

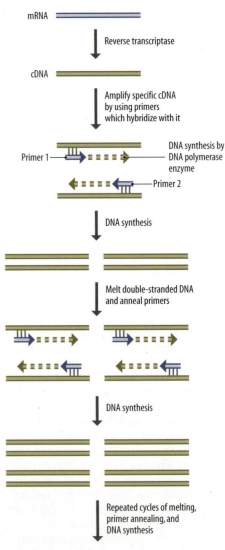

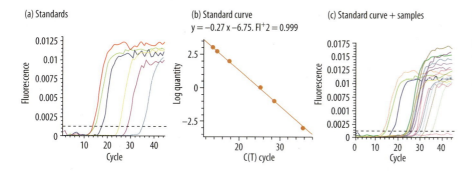

(a) Standards

(b) Standard curve
$y = -0.27 x - 6.75.$ Fl$^+$2 = 0.999

(c) Standard curve + samples

Figure 1.10
Use of RT/PCR to quantify the amount of
p53 mRNA in different samples. A number
of different standards containing known
amounts of p53 mRNA were subjected to
RT/PCR and graphs prepared of the amount
of PCR product generated by different
numbers of amplification cycles in each
case (a). This allows the preparation of
a standard curve (b). In turn this allows
the quantity of p53 mRNA in samples
where this is unknown to be calculated
by comparing it to the standards, since
the samples fall within the range of
the standards, validating the use of
the standard curve for quantification
of unknown samples (c). Courtesy of
Vishwanie Budhram-Mahadeo, University
College London.

single-stranded components, allowing further annealing of the primers and another round of DNA synthesis.

Repeated cycles of amplification, involving melting, primer annealing, and DNA synthesis, result in a small amount of a specific mRNA/cDNA being amplified to produce a large amount of product. This method therefore allows the analysis of very small amounts of mRNA which could not be analyzed by Northern blotting. Moreover, by comparing the amount of product produced from different samples at specific numbers of amplification cycles, it is possible to compare the amount of a specific mRNA in each of the samples.

In addition, it is possible to produce standard curves showing the amount of PCR product produced at different amplification cycles starting with known quantities of a particular RNA. This allows the amount of that RNA to be precisely quantified in samples where this is not yet known. This is achieved by using a standard curve relating the amount of the mRNA present to the amount of PCR product generated at different numbers of amplification cycles (**Figure 1.10**).

As with protein studies, methods studying expression in RNA isolated from particular tissues can be supplemented by methods allowing direct visualization of the RNA in particular cell types. In such a method, known as *in situ* **hybridization**, a fluorescent probe specific for the RNA to be detected is hybridized to a section of the tissue of interest in which cellular morphology has been maintained. Visualization of the position at which the fluorescent probe has bound (**Figure 1.11**) not only allows an assessment of whether the particular tissue is expressing the RNA of interest but also of which individual cell types within the tissue are responsible for such expression. This technique has been used to show that the expression of some mRNAs is confined to only one cell type, paralleling the expression of the corresponding proteins by that cell type (see Figure 1.11).

General methods can be used to study the overall population of mRNAs expressed in different tissues and cells

As in the case of proteins, methods for studying individual RNAs can be supplemented by methods for studying RNA populations. Methods have

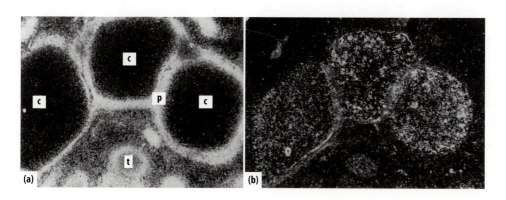

(a)   (b)

Figure 1.11
Localization of the RNA for type I
collagen (a) and type II collagen (b)
in 10-day chick embryo leg by *in situ*
hybridization. Note the different
distributions of the bright areas
produced by binding of each probe
to its specific mRNA. c, cartilage;
p, perichondrium; t, tendons.
Courtesy of Paul Brickell, from
Devlin CJ, Brickell PM, Taylor ER
et al. (1988) *Development* 103,
111–118. With permission from the
Company of Biologists Ltd.

Figure 1.12
Gene-chip analysis of mRNA expression patterns in different tissues. Fluorescent sequences are prepared from the total mRNA present in each tissue. Hybridization of these fluorescent sequences to the gene chip containing many different gene sequences will reveal the expression level of each gene in each tissue. The figure shows this for genes which are not expressed in a particular tissue (pale ovals), genes which are expressed at low level (mid-colored ovals), and genes which are expressed at high level (darkest ovals).

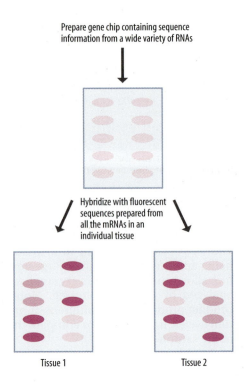

been developed that can combine the ability to look at variation in the total RNA population of different tissues with the ability to look specifically at the variation in a specific mRNA. By analogy with the proteomic methods which achieve this at the protein level (see Section 1.1), these methods are often referred to as **transcriptomic** methods.

In these methods, so-called gene chips are prepared in which DNA sequences derived from thousands of different mRNAs are laid out in a regular array at a density of over 10,000 different DNA sequences per square centimeter (**Figure 1.12**). This is achieved either by spotting out cDNA clones derived from the different mRNAs at high density on a glass slide or by synthesizing short oligonucleotides derived from each mRNA *in situ* on the gene chip.

Once this has been achieved, the gene chips can be hybridized with fluorescent sequences prepared from the total mRNA present in different tissues and the pattern of hybridization observed (see Figure 1.12). Where a specific mRNA is present in a tissue, it will be present in the fluorescent sequences and a signal will be obtained on the gene chip with the strength of the signal being proportional to the amount of the RNA which is present. Similarly, if the mRNA is absent then it will be absent from the probe and no signal will be obtained. The very large amount of DNA sequence information which is now available allows large numbers of DNA sequences derived from individual genes to be included on the chips, taking advantage of the ability to spot out many different DNA sequences onto a very small chip. Hence, the expression of virtually all the mRNAs in a cell can now be compared in different tissues and cell types or in response to specific stimuli (**Figure 1.13**).

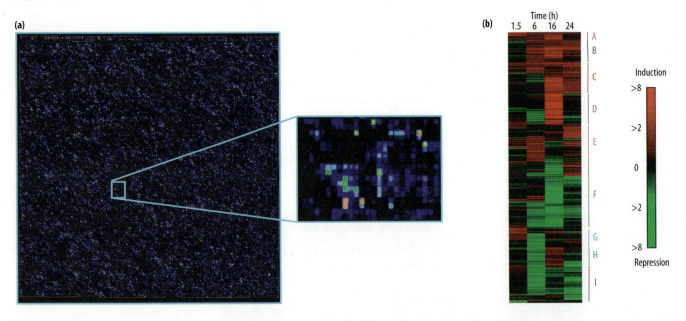

Figure 1.13
Gene-chip analysis of gene expression patterns using the Affymetrix gene chip system. Panel (a) shows a single gene chip hybridized to labeled RNA with a close-up of a specific region. Panel (b) shows the changes in expression of different genes during the activation of lung fibroblasts. Red and green show up-regulation and down-regulation, respectively. Letters represent groups of genes showing similar patterns of expression over time. Courtesy of Rachel Chambers & Mike Hubank, University College London.

This very powerful method allows the expression profile of individual tissues and cell types to be obtained and compared. The use of this method has provided further support for the conclusion that the mRNA populations of different tissues show both qualitative and quantitative differences.

Both methods which look at individual mRNAs and transcriptomic methods clearly demonstrate that the qualitative and quantitative variation in protein composition between different tissues and cell types is paralleled by a similar variation in the nature of the mRNA species present in different tissues. Indeed, in a study of a wide range of **yeast** proteins and mRNAs, the abundance of an individual protein, determined by two-dimensional gel electrophoresis, generally correlated well with the abundance of the corresponding mRNA, determined by a gene-chip approach. A similar correlation has also been observed in a survey of human liver mRNAs and proteins. Thus, the production of different proteins by different tissues is primarily regulated by controlling the population of mRNAs which are present in each tissue rather than by selecting which mRNAs are translated by the ribosome, although some examples of such regulated translation exist (see Section 7.5).

1.3 THE DNA CONTENT OF DIFFERENT CELL TYPES IS GENERALLY THE SAME

Having established that the RNA and protein content of different cell types shows considerable variation, the fundamental dogma of molecular biology, in which DNA makes RNA which in turn makes protein, directs us to consider whether such variation is caused by differences in the DNA present in each tissue. In theory, DNA corresponding to an RNA that is required in one particular tissue only might be discarded in all other tissues (**Figure 1.14a**). Alternatively, it might be activated in the tissue where RNA was required by a selective increase in its copy number in the **genome** via an amplification event (**Figure 1.14b**) or by some rearrangement of the DNA necessary for its activation (**Figure 1.14c**).

It is clear however, that in general these mechanisms do not operate to regulate gene expression. The DNA content of different tissues and cell types is normally the same, and differential expression to produce different mRNA populations in different cell types is the primary control point regulating gene expression (**Figure 1.14d**). As with the studies of protein (see Section 1.1) and RNA (see Section 1.2) content, the evidence for this comes both from methods studying individual genes and from methods studying the entire genome in different cell types. These will be discussed in turn.

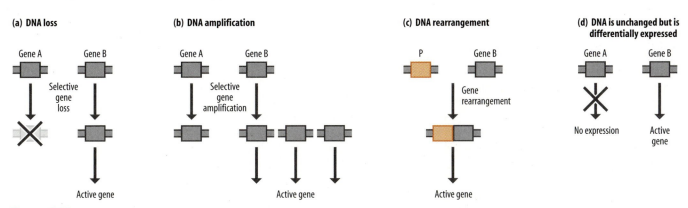

Figure 1.14
Potential mechanisms by which change at the DNA level could be involved in the activation of a specific gene (gene B) in a specific cell type. In the DNA-loss mechanism (a), genes not required in a particular cell type are deleted with the remaining genes being active. In the DNA-amplification mechanism (b), genes needed in a particular cell type are selectively amplified, resulting in high-level expression of the amplified gene. In the DNA-rearrangement mechanism (c) the genes required in a particular cell type undergo a DNA rearrangement which, for example, links the gene to promoter (P) elements required for its transcription. Panel (d) shows the correct mechanism in which DNA content does not change but the expression of individual genes is regulated in different tissues, with gene B but not gene A being expressed in the cell type/tissue illustrated.

Methods Box 1.1
SOUTHERN BLOTTING (FIGURE 1.15)

- Digest DNA with a restriction endonuclease which cuts it at specific sequences.
- Electrophorese on an agarose gel.
- Denature DNA and transfer DNA in gel to nitrocellulose filter.
- Hybridize filter with radiolabeled probe for gene of interest.
- Analyze structure of a specific gene based on patterns of hybridization obtained when DNA is cut with different restriction endonucleases.

Specific methods can be used to study individual genes in different tissues and cells

In the same way that individual proteins and RNAs can be studied by Western blotting and Northern blotting, respectively, the structure of individual genes in different cell types can be studied by **Southern blotting**. This is the original blotting technique, first described by Ed Southern in 1975, with the other techniques being subsequently named on the basis of the different points of the compass! (**Methods Box 1.1**, **Figure 1.15**).

Southern blotting can therefore be used to compare the structure of a particular gene in tissues where it is expressed with that in tissues where it is not expressed. When this is done, in the vast majority of cases no difference is detected (**Figure 1.16**). Specific DNA bands produced by a particular gene are not lost in tissues where the gene is not expressed, as expected on a DNA-loss model. Similarly, the bands do not become more intense in expressing tissues, as expected in a DNA-amplification model, and there is no evidence of different-sized DNA bands appearing in expressing tissues, as would be expected in a DNA-rearrangement model.

General methods can be used to study the total DNA in different tissues and cells

The conclusions based on Southern blotting of individual genes can be progressively extended as more and more genes and the entire human genome are analyzed by DNA sequencing using DNA isolated from different tissues and cell types. These whole-genome results support those based on individual genes. In general, there is no evidence for DNA sequence changes when specific genes or chromosomal regions are compared in different cell types. These techniques would detect even single-base changes and therefore represent a clear proof that the DNA content of different tissues and cell types is generally the same.

These conclusions from molecular biology experiments such as Southern blotting and DNA sequencing are complemented by functional studies which show that a whole new organism can be produced from a single differentiated cell. This would be difficult or impossible if specific DNA changes occur during the process of cell **differentiation**. For example, if the genes required for cell types such as muscle and brain are lost in the development of the intestine, it would not be possible for a differentiated intestinal cell to give rise to all the differentiated cell types in the adult organism, including muscle and brain. In fact, such totipotency of differentiated cells has been observed in both animals and plants.

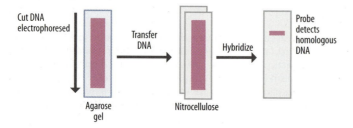

Figure 1.15
Procedure for Southern-blot analysis, involving electrophoresis of DNA which has been cut with a restriction enzyme, its transfer to a nitrocellulose filter, and hybridization of the filter with a specific DNA probe for the gene of interest.

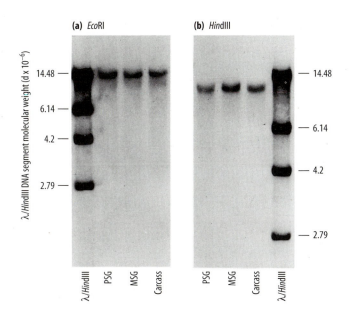

Figure 1.16
Southern blot of DNA prepared from the posterior silk gland (PSG), the middle silk gland (MSG), or the carcass of the silk moth, *Bombyx mori,* with a probe specific for the fibroin gene. Note the identical size and intensity of the band produced by *Eco*RI or *Hind*III digestion in each tissue, although the gene is only expressed in the posterior silk gland. Courtesy of Ronald Manning, from Manning RF & Gage LP (1978) *J. Biol. Chem.* 253, 2044–2052. With permission from The American Society for Biochemistry and Molecular Biology.

In plants, the production of a whole organism from a single differentiated cell has been achieved in several species, including both the carrot and tobacco plants. In the carrot, for example, regeneration can occur from a single differentiated phloem cell, which forms part of the tubing system through which nutrients are transported in the plant. If a piece of root tissue is placed in culture (**Figure 1.17**), single **quiescent** cells of the phloem can be stimulated to grow and divide and an undifferentiated callus-type tissue forms. The disorganized cell mass can be maintained in culture indefinitely but if the medium is suitably supplemented at various stages embryonic development will occur and eventually result in a fully functional flowering plant containing all the differentiated cell types and tissues normally present. The plant that forms is fertile and cannot be distinguished from a plant produced by normal biological processes.

The ability to regenerate a functional plant from fully differentiated cells eliminates the possibility that genes required in other cell types are lost in the course of plant development. It is noteworthy, however, that one

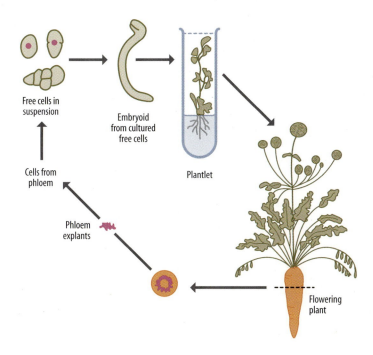

Figure 1.17
Scheme for the production of a fertile carrot plant from a single differentiated cell of the adult plant phloem, by growth in culture as a free cell suspension which develops into an embryo and then into an adult plant. From Steward FC, Mapes MO, Kent AE & Holsten RD (1964) *Science* 143, 20–27. With permission from The American Association for the Advancement of Science.

Figure 1.18
Nuclear transplantation in Amphibia. The introduction of a donor nucleus from a differentiated cell into a recipient egg whose nucleus has been destroyed by irradiation with ultraviolet (UV) light can result in an adult frog with the genetic characteristics of the donor nucleus. From Gurdon JB (1974), Gene Expression. With permission from Oxford University Press.

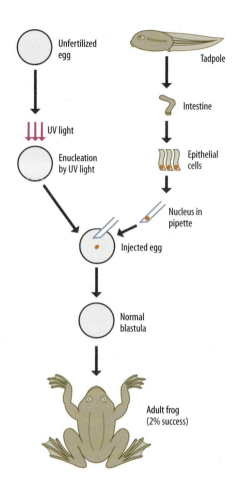

differentiated cell type does not transmute directly into another; rather, a transitional state of undifferentiated proliferating cells serves as an intermediate. Although differentiation does not apparently involve permanent irreversible changes in the DNA, it does appear to be relatively stable and, although reversible, requires an intermediate stage for changes to occur. This semi-stability of cellular differentiation will be discussed further in Chapter 3.

Although no complex animal has been regenerated by culturing a single differentiated cell in the manner used for plants, other techniques have been used to show that differentiated cell nuclei are capable of giving rise to very many different cell types. These experiments involve the use of **nuclear transplantation** (**Figure 1.18**). In this technique the nucleus of an unfertilized frog egg is destroyed, either surgically or by irradiation with ultraviolet light, and a donor nucleus from a differentiated cell of a genetically distinguishable strain of frog is implanted. Development is then allowed to proceed to test whether, in the environment of the egg cytoplasm, the genetic information in the nucleus of the differentiated cell can produce an adult frog.

When the donor nucleus is derived, for example, from differentiated intestinal epithelial cells of a tadpole, an adult frog is indeed produced in a small proportion of cases (about 2%) and development to at least a normal swimming tadpole occurs in about 20% of cases with the resulting organisms having the genetic characteristics of the donor nucleus. Such successful development supported by the nucleus of a differentiated frog cell is not unique to intestinal cells and has been achieved with the nuclei of other cell types, such as the skin cells of an adult frog. Although these experiments are technically difficult and have a high failure rate, it is therefore possible for the nuclei of differentiated cells to produce organisms containing a range of different cell types and which are perfectly normal and fertile.

These experiments were carried out in Amphibia over 30 years ago. However, it is only within the last few years that similar experiments have been successfully performed in mammals. In 1997 Wilmut and co-workers reported experiments in which a donor nucleus derived from a mammary gland of a 6-year-old sheep was introduced into an unfertilized egg whose own nucleus had been destroyed. In this manner, they produced an adult sheep (Dolly) which, as in Gurdon's experiments with frogs, had the genetic characteristics of the sheep from which the mammary gland cell nucleus had been derived (**Figure 1.19c**). Similar "cloning" experiments have now been successfully carried out in several other mammalian species including mice (**Figures 1.19a** and **b**) and pigs.

Although these studies have provoked considerable ethical discussion on the potential cloning of humans, their relevance for the area of gene regulation is that they indicate that in mammals, as in Amphibia, a nucleus from an adult differentiated cell is capable of supporting embryonic development and producing the full range of cell types present in an adult organism.

These experiments show that the nuclei of individual differentiated animal and plant cells are **totipotent** and can regenerate an adult organism. These functional studies therefore reinforce molecular methods in indicating that irreversible changes to the DNA do not occur as a general mechanism for gene control during differentiation. This indicates that the model of differential expression of the DNA to produce different mRNA populations in different tissues (see Figure 1.14d) is the correct one. The mechanism of such differential expression is discussed in Section 1.4

(a)

(b)

(c)

Exceptional cases do exist in which changes to the DNA occur in specific tissues or cell types

Although DNA changes are not a general mechanism of gene control, they do occur in particular specialized cases. We will now discuss one case each of DNA loss, DNA amplification, and DNA rearrangement to indicate particularly why such cases appear to be due to the specialized requirements of the particular situation.

(a) DNA loss

The best-known case of DNA loss is in the development of the mammalian red blood cell, a highly specialized cell containing large amounts of the blood pigment hemoglobin which functions in the transport of oxygen in the blood. During the differentiation of red blood cells the region of the cell containing the nucleus is pinched off surrounded by a region of the cell membrane and is eventually destroyed (**Figure 1.20**).

The resulting cell, known as the reticulocyte, is entirely anucleate and completely lacks DNA. However, it continues synthesis of large amounts of globin (the protein part of the hemoglobin molecule), as well as small amounts of other reticulocyte proteins, by repeated translation of mRNA molecules produced in red blood cell precursors prior to the loss of the nucleus. Such RNA molecules must therefore be highly stable and resistant to degradation. Eventually, however, other cytoplasmic components (including ribosomes) are lost, protein synthesis ceases, and the cell assumes the characteristic structure of the **erythrocyte** which is essentially a bag full of oxygen-transporting hemoglobin molecules.

Although this process offers an example of DNA loss during differentiation, it is clearly a highly specialized case in which the loss of the nucleus is primarily intended to allow the cell to fill up with hemoglobin and to assume a shape facilitating oxygen uptake, rather than as a means of gene control. The genes for globin and other reticulocyte proteins are not selectively retained but are lost with the rest of the nuclear DNA (**Figure 1.21**). The tissue-specific pattern of protein synthesis observed in the reticulocyte is therefore achieved not by selective gene loss but by the processes which, earlier in development, resulted in high-level transcription of the globin genes and the production of long-lived, stable globin mRNA molecules.

(b) DNA amplification

In the case of red blood cells, DNA loss represents a response to the specialized role of the cell rather than a mechanism of gene control. A similar conclusion can be reached in cases of gene amplification. These occur in situations where a very high level of mRNA/protein production is required in a short time such that a single copy of a gene could not produce enough mRNA/protein.

One such case involves the genes encoding the eggshell or chorion proteins in *Drosophila melanogaster*. The chorion genes are selectively amplified (up to 64 times) in the DNA of the cells which surround the egg follicle, allowing the synthesis in these cells of the large amounts of chorion mRNA and protein needed to construct the eggshell. Such amplification can be observed readily in a Southern blot experiment using a recombinant DNA probe derived from one of the chorion genes (**Figure 1.22**).

Although this amplification event is part of normal embryonic development, it appears to be a response to a very specialized set of circumstances, requiring a novel means of gene regulation. Eggshell construction in

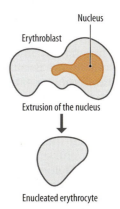

Figure 1.20
Extrusion of the nucleus from an erythroblast, resulting in an anucleate erythrocyte.

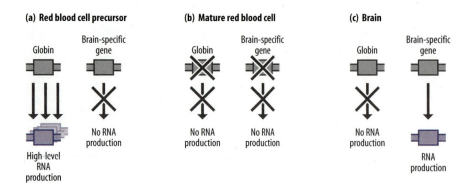

Figure 1.21
In red blood cell precursors the globin gene is highly expressed and a brain-specific gene would not be expressed (a). In mature blood cells, all genes are lost so that no new RNA is produced (b). In the brain, the brain-specific gene would be expressed, but the globin gene would not be expressed (c).

Drosophila occurs over a very short period (about 5 h) and necessitates the synthesis of mRNA for the chorion proteins at very high rates, too high to be achieved by transcription of a single unique gene.

It is this combination of high-level synthesis and a very short period to achieve it which necessitates the use of gene amplification. The very short time period makes it impossible to produce the high levels of protein required by use of a highly stable mRNA, as occurs in the globin case (see above) where high levels of protein are required but over a much longer time period.

(c) DNA rearrangement

When mammals are exposed to foreign bacteria or viruses, their **immune systems** respond by synthesizing specific antibodies directed against the proteins of these organisms with the aim of neutralizing the harmful effects of the infection. The requirement for the synthesis of diverse kinds of antibody by the mammalian immune system is obviously vast. The body must be able to defend itself, for example, against a bewildering variety of challenges from infectious organisms by the production of specific antibodies against their proteins.

These antibodies or **immunoglobulins** are produced by the covalent association of two identical **heavy chains** and two identical **light chains** to produce a functional molecule. The combination of specific heavy and light chains produces the specificity of the antibody molecule. Each immunoglobulin chain contains, in addition to a constant region (which is relatively similar in different antibody molecules), a highly **variable region** which differs widely in amino acid sequence in different antibodies. It is this variable region that actually interacts with the **antigen** and determines the specificity of the antibody molecule.

As any heavy chain can associate with any light chain and as approximately 1 million types of antibody specificities can be produced, at least 1000 genes encoding different heavy chains and a similar number encoding the light chains would be required. Copy-number studies have shown, however, that the **germ line** does not contain anything like this number of intact immunoglobulin genes and no specific amplification events have been detected in the DNA of antibody-producing **B cells**.

Rather, functional immunoglobulin genes are created by DNA rearrangements in the B-cell lineage. The germ-line DNA contains a large number of tandemly repeated DNA segments encoding different variable regions which are separated by over 100 kb from a much smaller number of DNA segments encoding the constant region of the molecule and a joining region (**Figure 1.23**). This organization is maintained in most somatic cell types, but in each individual B cell a unique DNA rearrangement event occurs by which one specific variable region is brought together with one joining/constant region by deletion of the intervening DNA. In this manner, a different functional gene, containing specific variable and constant regions, is produced in each individual B cell.

It is clear that the primary role of the rearrangement of the immunoglobulin genes in B cells is the generation of the diversity of antibodies required to cope with the vast number of different possible antigens rather than to produce the high-level expression of the immunoglobulin genes which

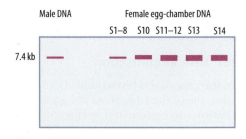

Figure 1.22
Amplification of the chorion genes in the ovarian follicle cells of *Drosophila melanogaster*. Note the dramatic increase in the intensity of hybridization of the chorion gene specific probe to the DNA samples prepared from Stage (S) 10 to Stage 14 egg chambers compared to that seen in Stages 1–8 or in male DNA. From Spradling AC & Mahowald AP (1980) *Proc. Natl. Acad. Sci. USA* 77, 1096–1100. With permission from Allan Spradling, Carnegie Institution of Washington.

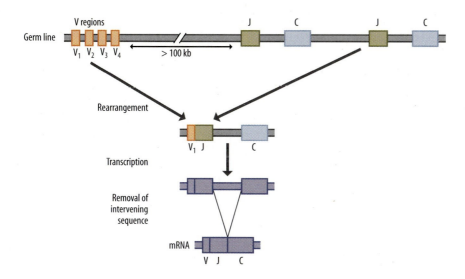

Figure 1.23
Rearrangement of the gene encoding the γ light chain of immunoglobulin results in the linkage of one specific variable region (V) to a specific joining region (J) with removal of the intervening DNA. The gene is then transcribed into RNA and the region between the joining (J) and constant regions (C) is removed by RNA splicing to create a functional mRNA.

occurs in such cells. Nonetheless, high-level immunoglobulin gene expression does occur as a consequence of such rearrangements. The DNA encoding the variable regions of the antibody molecule is closely linked to the **promoter** elements that direct the transcription of the gene. Such elements show no activity in the unrearranged DNA of non-B cells, but become fully active only when the variable region is joined to the constant region. In the case of the immunoglobulin heavy-chain genes, this is because the intervening sequence between the joining and constant regions of these genes contains an **enhancer** element (see Section 4.4) which, although not a promoter itself, greatly increases the activity of the promoter adjacent to the variable-region element. Rearrangement not only brings the promoter element into a position where it can produce a functional mRNA containing variable, joining, and constant regions but also facilitates the high-level transcription of this gene by juxtaposing promoter and enhancer elements that are separated by over 100 kb of DNA before the rearrangement event.

The element of gene control in this process is, however, secondary to the need to produce diversity by rearrangement. If the immunoglobulin enhancer is linked artificially close to the immunoglobulin promoter and introduced into a variety of cells, it will activate transcription from the promoter only in B cells and not in other cell types. Thus, the enhancer is active only in B cells and would not activate immunoglobulin transcription even if functional immunoglobulin genes containing closely linked variable, joining, and constant segments were present in all other tissues. The immunoglobulin genes are therefore regulated by tissue-specific activator sequences in much the same way as other genes, but such activation has been made to depend on the occurrence of a DNA rearrangement whose primary role lies elsewhere.

It is clear from the examples of DNA changes which we have discussed that such cases represent special situations whose particular requirements have necessitated the use of unusual mechanisms. In general, the DNA of different cell types is quantitatively and qualitatively identical. Given that we have established previously that the mRNA content of different cell types can vary dramatically, it is now necessary to investigate how such differences in mRNA content can be produced from the similar DNA present in such cell types.

1.4 TRANSCRIPTIONAL OR POST-TRANSCRIPTIONAL CONTROL?

In theory, different mRNA populations could be produced from the same DNA in different tissues or cell types by two mechanisms (**Figure 1.24**). Firstly, control process could operate to decide which genes were transcribed to produce the **primary RNA transcript** within the nucleus. In such

(a) Transcriptional control

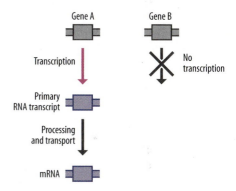

(b) Post-transcriptional control

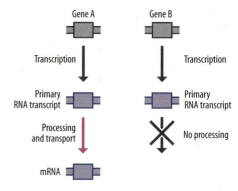

Figure 1.24
Gene control could operate at the level of transcription by regulating which genes are transcribed into the primary RNA transcript (a). Alternatively, in post-transcriptional control, all genes would be transcribed, with gene regulation determining which primary RNA transcripts were processed to mature mRNA (b).

a transcriptional control mechanism, subsequent post-transcriptional processes (such as the removal of intervening sequences (**introns**) from the primary RNA transcript and its transport to the cytoplasm; see Chapter 6) would follow automatically (**Figure 1.24a**). Alternatively, it is possible that all genes would be transcribed in all tissues with gene control operating at the post-transcriptional level by deciding which primary RNA transcripts should be processed and transported to the cytoplasm for translation into protein (**Figure 1.24b**).

Although there is evidence that **post-transcriptional control** does operate in a number of cases (see Chapter 7), the predominant mechanism for regulating gene expression in **eukaryotes** operates at the level of transcription, controlling which genes are transcribed into a primary RNA transcript in different tissues or cell types. The evidence for such regulation of gene transcription comes from several types of study which will be considered in turn.

Studies of nuclear RNA suggest that gene transcription is regulated

If regulation of gene expression takes place at the level of transcription, the differences in the cytoplasmic levels of particular mRNAs between different tissues should be paralleled by similar differences in the levels of these RNAs within the nuclei of different tissues. In contrast, regulatory processes in which a gene was transcribed in all tissues and the resulting transcript either spliced or transported to the cytoplasm in a minority of tissues would result in cases where differences in mRNA content occurred without any corresponding difference in the nuclear RNA (**Figure 1.25**). A study of the level of particular RNA species in the nuclear RNA of individual tissues or cell types serves therefore as an initial test to distinguish transcriptional and post-transcriptional regulation.

The earliest studies in this area focused on the highly abundant RNA species produced during terminal differentiation which could be readily studied simply because of their abundance. As with cytoplasmic mRNA, the RNA in the nucleus derived from a particular gene can be studied by Northern blotting (see Section 1.2). Moreover, this method allows the separation of different species within nuclear RNA by size and their visualization by hybridization to an appropriate probe.

In the case of genes with many intervening sequences (introns), such as that encoding the egg protein ovalbumin, many different RNA species can be observed in the nucleus by Northern blotting. These include a large RNA which is the primary transcript, a series of intermediate-sized RNAs from which some of the intervening sequences (introns), have still to be removed and the fully processed RNA which is the same size as the cytoplasmic mRNA and is about to be transported to the cytoplasm (**Figure 1.26**) (see Chapter 6 for a detailed account of the post-transcriptional processes which convert the primary RNA transcript into a mature mRNA).

The identification of such potential precursors of the mature mRNA allowed a study of their expression in tissues either producing or not producing ovalbumin mRNA and protein. Cytoplasmic ovalbumin mRNA and protein are present only in the oviduct following stimulation with estrogen and disappear when estrogen is withdrawn. Similarly, the mRNA and protein are absent in other tissues, such as the liver, and cannot be induced by treatment with the hormone in these tissues. Studies of the distribution of both the fully processed nuclear RNA species and the larger precursors showed that these species could only be detected in the nuclear RNA of the oviduct following estrogen stimulation and were absent in the liver nuclear RNA or in

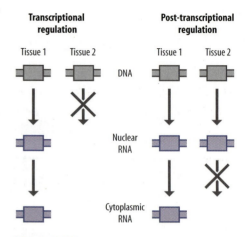

Transcriptional regulation   **Post-transcriptional regulation**

Tissue 1   Tissue 2   Tissue 1   Tissue 2

DNA

Nuclear RNA

Cytoplasmic RNA

Figure 1.25
Consequences of transcriptional or post-transcriptional regulation on the level of a specific nuclear RNA in a tissue that expresses the corresponding cytoplasmic mRNA (tissue 1) and one that does not (tissue 2).

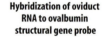
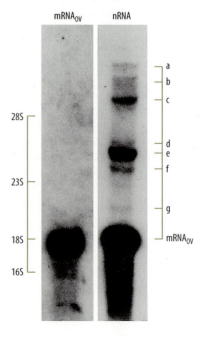

Hybridization of oviduct RNA to ovalbumin structural gene probe

mRNA$_{OV}$ nRNA

28S
23S
18S
16S

a
b
c
d
e
f
g
mRNA$_{OV}$

Figure 1.26
Northern blot showing that unspliced and partially spliced precursors (a–g) to the ovalbumin mRNA (mRNA$_{OV}$) are detectable in the nuclear RNA (nRNA) of estrogen-stimulated oviduct tissue. Courtesy of Bert O'Malley, from Roop DR, Nordstrom JL, Tsai SY et al. (1978) *Cell* 15, 671–685. With permission from Elsevier.

Hybridization of various RNA preparations to ovalbumin structural gene probe

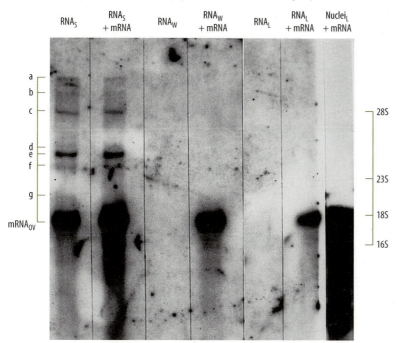

Figure 1.27
Northern blot showing that the nuclear
precursors to ovalbumin mRNA (mRNA$_{ov}$;
seen in Figure 1.26) are detectable in the
nuclear RNA of estrogen-stimulated oviduct
(RNA$_S$) but are absent in the nuclear RNA
of estrogen-withdrawn oviduct (RNA$_W$)
and of liver (RNA$_L$). Moreover, ovalbumin
mRNA mixed with withdrawn oviduct
(RNA$_W$+mRNA) or liver (RNA$_L$+mRNA)
nuclear RNA is not degraded, showing
that the absence of RNA for ovalbumin in
these nuclear RNAs is not due to a nuclease
specifically degrading the ovalbumin RNA.
Courtesy of Bert O'Malley, from Roop DR,
Nordstrom JL, Tsai SY et al. (1978) *Cell* 15,
671–685. With permission from Elsevier.

oviduct nuclear RNA following estrogen withdrawal (**Figure 1.27**). The distribution of these precursors in the nucleus exactly parallels that of the cytoplasmic mRNA, a finding entirely consistent with the transcriptional induction of the ovalbumin gene in the oviduct in response to estrogen.

These early studies of very abundant mRNAs, such as ovalbumin, have now been abundantly supplemented by many others measuring the nuclear RNA levels of other specific genes whose expression changes in particular situations such as those encoding a number of highly abundant mRNAs present only in the soya bean embryo or that encoding the developmentally regulated mammalian liver protein, α-fetoprotein. In general, such studies have led to the conclusion that in most cases alterations in specific mRNA levels in the cytoplasm are accompanied by parallel changes in the levels of the corresponding nuclear RNA species.

It appears from these studies that in mammals and other higher eukaryotes the regulation of transcription, resulting in parallel changes in nuclear and cytoplasmic RNA levels, is the primary means of regulating gene expression. However, the studies described so far suffer from the defect that they measure only steady-state levels of specific nuclear RNAs. It could be argued that the gene is being transcribed in the non-expressing tissue but that the transcript is degraded within the nucleus at such a rate that it cannot be detected in assays of steady-state RNA levels (**Figure 1.28**). This possibility

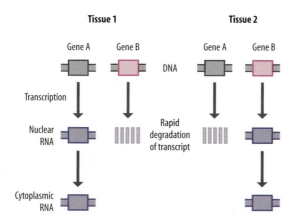

Figure 1.28
Model in which lack of expression of genes
in particular tissues is caused by rapid
degradation of their RNA transcripts.

Methods Box 1.2
MEASUREMENT OF TRANSCRIPTION RATES

(a) Pulse labeling to measure transcription (Figure 1.29)
- Add radioactively labeled nucleotides to cells.
- After 5–10 min harvest cells and isolate RNA (including radioactively labeled newly synthesized RNA).
- Hybridize labeled RNA to dot blot containing DNA from genes the transcription of which is being measured.

(b) Nuclear run-on assay to measure transcription (Figure 1.30)
- Isolate nuclei from cells.
- Add radioactively labeled nucleotide to the isolated nuclei.
- After 1–2 h isolate RNA from nuclei (including labeled RNA produced by transcribing polymerase "running on" to the end of the gene).
- Hybridize labeled RNA to dot blot containing DNA from genes the transcription of which is being measured.

necessitates the direct measurement of gene transcription itself in the different tissues in order to establish unequivocally the existence of transcriptional regulation. Methods to do this have been devised and these will now be discussed (**Methods Box 1.2**).

Pulse labeling studies directly demonstrate transcriptional control

The synthesis of RNA from DNA by the enzyme **RNA polymerase** involves the incorporation of ribonucleotides into an RNA chain. Therefore the synthesis of any particular RNA can be measured by adding a radioactive ribonucleotide (usually uridine labeled with tritium) to the cells and measuring how much radioactivity is incorporated into RNA specific for the gene of interest. Clearly, if the degradation mechanisms above do exist, they will given time degrade the radioactive RNA molecule produced in this way and drastically reduce the amount of labeled RNA detected. In order to prevent this, the rate of transcription is measured by exposing cells briefly to the labeled uridine in a process referred to as **pulse labeling**. The labeled uridine is incorporated into nascent RNA chains that are being made at this time and even before a complete transcript has had time to form the cells are lysed and total RNA is isolated from them. This RNA, which contains labeled partial transcripts from genes active in the tissue used, is then hybridized to a piece of DNA derived from the gene of interest, the number of radioactive counts that bind providing a measure of the incorporation of labeled precursor into its corresponding RNA (**Figure 1.29**) (**Methods Box 1.2a**).

This method provides the most direct means of measuring transcription and has been used, for example, to show that the induction of globin production which occurs in Friend erythroleukemia cells in response to treatment with dimethyl sulfoxide is mediated by increased transcription of the globin gene.

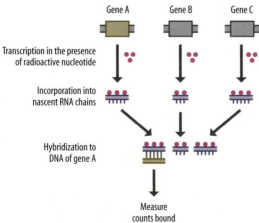

Figure 1.29
Pulse-labeling assay to assess the transcription rate of a specific gene (gene A) by measuring the amount of radioactivity (dots) incorporated into nascent transcripts.

Although pulse labeling provides a very direct measure of transcription rates, the requirement to use very short labeling times to minimize any effects of RNA degradation limits its applicability. In the globin case, it was possible to measure the amount of radioactivity incorporated into globin RNA in the very brief labeling times used (5 or 10 min) only because of the enormous abundance of globin RNA and the very high rate of transcription of the globin gene. With other RNA species, the rates of transcription are insufficient to provide measurable incorporation of label in the short pulse time. More label will of course be incorporated if longer pulse times are used but such pulse times allow the possibility of RNA turnover and are therefore subject to the same objections as the measurement of stable RNA levels.

Although pulse labeling can be used to establish unequivocally that transcriptional control is responsible for the massive synthesis of the highly abundant RNA species present in terminally differentiated cells, it cannot be used to demonstrate the generality of transcriptional control processes and in particular their applicability to RNAs which, although regulated in different tissues, never become highly abundant. However, another method exists which, although less direct than pulse labeling, is more sensitive and can be applied to a wider variety of cases including non-highly abundant mRNAs. This method is discussed below.

Nuclear run-on assays allow transcriptional control to be demonstrated for a wide range of genes

The primary limitation on the sensitivity of pulse labeling is the existence within the cell of a large pool of non-radioactive ribonucleotides, which are normally used by the cell to synthesize RNA. When labeled ribonucleotide is added to the cell, it is considerably diluted in this pool of unlabeled precursor. The amount of label incorporated into RNA in the labeling period is therefore very small, since the majority of ribonucleotides incorporated are unlabeled. The sensitivity of this method is thus reduced severely, resulting in its observed applicability only to genes with very high rates of transcription.

Interestingly, however, although transcription takes place in the nucleus, most of the pool of precursor ribonucleotides is present in the cytoplasm. It is possible by removing the cytoplasm and isolating nuclei to remove much of the pool of unlabeled ribonucleotide. When this occurs the RNA polymerase ceases transcribing due to lack of ribonucleotide but remains associated with the DNA (**Figure 1.30**). If labeled ribonucleotides are then added directly to the isolated nuclei in a test tube, the RNA polymerase resumes transcribing and runs on to the end of the gene incorporating labeled ribonucleotide into the RNA transcript. Because the labeled ribonucleotides are not diluted in the unlabeled cytoplasmic pool, considerably more label is incorporated into any particular RNA transcript than is observed in pulse-labeling experiments. The label incorporated into any particular transcript is detected by hybridization to its corresponding DNA exactly as in pulse-labeling experiments (**Methods Box 1.2b**).

This method, which is known as a nuclear run-on assay, is therefore much more widely applicable than pulse labeling and can be used to quantify the transcription of genes that are never transcribed at levels detectable by pulse labeling. Moreover, very many studies have now established that the RNA synthesized by isolated nuclei in the test tube is similar to that made by intact

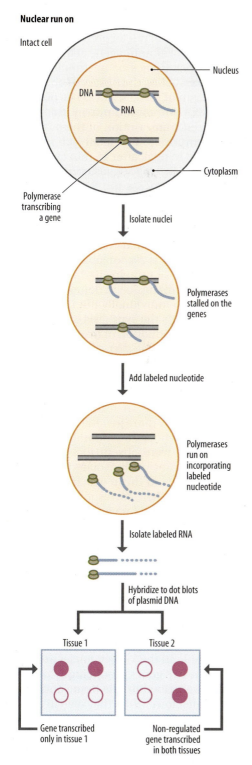

Nuclear run on

Intact cell

Nucleus

DNA

RNA

Cytoplasm

Polymerase transcribing a gene

Isolate nuclei

Polymerases stalled on the genes

Add labeled nucleotide

Polymerases run on incorporating labeled nucleotide

Isolate labeled RNA

Hybridize to dot blots of plasmid DNA

Tissue 1   Tissue 2

Gene transcribed only in tissue 1

Non-regulated gene transcribed in both tissues

Figure 1.30

Nuclear run-on assay. Following isolation of nuclei, RNA polymerase ceases transcribing due to lack of ribonucleotides but remains associated with the DNA. When radioactively labeled ribonucleotide is added, the polymerase resumes transcription and runs on to the end of the gene incorporating the labeled ribonucleotide into the RNA transcript. The labeled RNA made in this way in different tissues can then be used to probe filters containing DNA from genes whose transcription is to be measured in the different tissues.

whole cells and that the method is therefore not only sensitive but also provides an accurate measure of transcription free from artefact.

In initial studies, nuclear run-on assays were used to measure the transcription of highly abundant RNA species. For example, nuclei isolated from the erythrocytes of adult chickens (which unlike the equivalent cells in mammals do not lose their nuclei) were shown to transcribe the gene encoding the adult β-globin protein. In contrast, nuclei prepared from similar cells isolated from embryonic chickens failed to transcribe this gene and instead transcribed the gene encoding the form of β-globin made in the embryo. Such a finding indicates that the developmentally regulated production of different forms of globin protein is under transcriptional control.

Similarly, the presence of ovalbumin-specific RNA only in the nuclear RNA of hormonally stimulated oviduct tissue described earlier, is paralleled by the ability of nuclei prepared from stimulated oviduct cell nuclei to transcribe the ovalbumin gene at high levels in run-on assays. In contrast, nuclei from other tissues or unstimulated oviduct cells failed to transcribe this gene, paralleling the observed absence of ovalbumin RNA in the nuclei and cytoplasm of these tissues. The observed increase in nuclear and cytoplasmic RNA for ovalbumin in response to estrogen is indeed caused by increased transcription of the ovalbumin gene. Interestingly, in this study the incorporation of label into ovalbumin RNA in the run-on assay was observed to peak after 15 min and did not decrease at longer labeling times of up to 1 h. This suggested that unlike intact cells isolated nuclei do not degrade or process the RNA that they synthesize and allowed the use of longer labeling times, further increasing the sensitivity of this technique (see Methods Box 1.2 and compare section a with section b).

This increased sensitivity has allowed the use of nuclear run-on assays to demonstrate that transcriptional control of many genes encoding specific proteins is responsible for the previously observed differences in the levels of these proteins in different situations. Cases where such transcriptional control has been demonstrated in this manner are far too numerous to mention individually but involve a range of different tissues and organisms, such as the synthesis of α-fetoprotein in fetal but not adult mammalian liver, production of insulin in the mammalian pancreas, the expression of the *D. melanogaster* yolk protein genes only in ovarian follicle cells, the expression of aggregation-stage-specific genes in the slime mold *Dictyostelium discoideum*, and the expression of soya bean seed proteins such as glycinin in embryonic but not adult tissues.

These studies on individual genes for particular proteins have been supplemented by more general studies which, for example, examined 12 different genes whose corresponding mRNAs were present in mouse liver cytoplasm but were absent in brain cytoplasm. These included both genes encoding previously isolated liver-specific proteins, such as albumin or transferrin, and those which had been isolated simply on the basis of the presence of their corresponding cytoplasmic RNA in the liver and not in other tissues and for which the protein product had not yet been identified. Measurement of the transcription rate of these genes in nuclei isolated from brain or liver showed that such transcription was detectable only in the liver nuclei (Figure 1.31), indicating that the difference in cytoplasmic mRNA level was produced by a corresponding difference in gene transcription.

In these experiments, the rate of transcription of the 12 genes was also measured in nuclei prepared from kidney tissue. In this tissue mRNAs corresponding to two of the genes were present at a considerably lower level than that observed in the liver, whereas RNA corresponding to the other 10 genes was undetectable. As with the brain nuclei, the level of transcription detectable in the kidney nuclei exactly paralleled the level of RNA present. Only the two genes producing cytoplasmic mRNA in the kidney were detectably transcribed in this tissue and the level of transcription of these genes was much lower than that seen in the liver nuclei.

These studies on a large number of different liver RNAs of different abundances, when taken together with the studies of many genes encoding specific proteins, indicate that for genes expressed in one or a few cell types

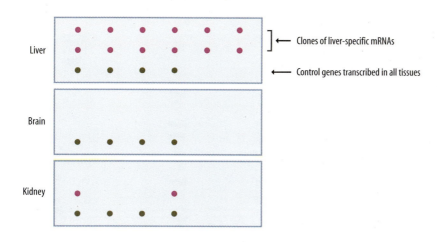

Liver — Clones of liver-specific mRNAs

— Control genes transcribed in all tissues

Brain

Kidney

Figure 1.31
Nuclear run-on assay to measure the transcription rates of genes encoding liver-specific mRNAs and of control genes expressed in all tissues. Note that the liver-specific genes are not transcribed at all in brain, whereas in the kidney the only two of these genes to be transcribed are those that are known to produce a low level of mRNA in the kidney.

increased levels of RNA and protein in a particular cell type are brought about primarily by increases in gene transcription.

Polytene chromosomes provide further evidence for transcriptional control

Although pulse labeling and nuclear run-on studies of many genes have conclusively established the existence of transcriptional regulation, it is necessary to discuss another means of demonstrating such regulation in which increased transcription can be directly visualized. The chromosomal DNA in the salivary glands of *Drosophila* is amplified many times, resulting in giant **polytene chromosomes** which contain approximately 1000 DNA molecules (**Figure 1.32**). All of the DNA in the *Drosophila* genome is amplified in this process and there is no preferential amplification of the genes which are expressed in the salivary gland. Thus, this example does not affect our general conclusion that changes in the DNA do not constitute a widespread method of gene control (see Section 1.3).

However, the very large size of the polytene chromosomes allows the visualization of events which cannot be observed in normal size chromosomes. Polytene chromosomes exhibit along their length areas known as puffs, in which the DNA has decondensed, resulting in the expansion of the chromosome (**Figure 1.33**). If cells are allowed to incorporate labeled ribonucleotides into RNA and the resulting RNA is then hybridized back to the polytene

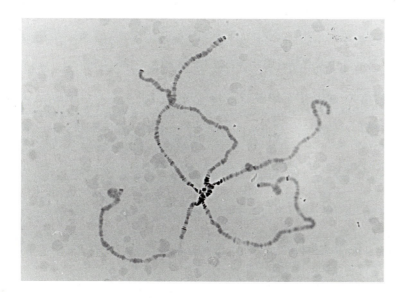

Figure 1.32
Polytene chromosomes from the salivary gland of *D. melanogaster.* Courtesy of Michael Ashburner, University of Cambridge.

chromosomes, it localizes primarily to the positions of the puffs. Hence these puffs represent sites of intense transcriptional activity which because of the large size of the polytene chromosomes can be visualized directly.

Most interestingly, many conditions which in *Drosophila* result in the production of new proteins, such as exposure to elevated temperature (heat shock) or treatment with the steroid hormone ecdysone, also result in the production of new puffs at specific sites on the polytene chromosomes, each treatment producing a different specific pattern of puffs.

This suggests that these sites contain the genes encoding the proteins whose synthesis is increased by the treatment and that this increased synthesis is mediated via increased transcription of these genes which can be visualized in the puffs. In the case of ecdysone treatment this has been confirmed directly by showing that the radioactive RNA synthesized immediately after ecdysone treatment hybridizes strongly to the ecdysone-induced puffs but not to a puff which regresses upon hormone treatment. In contrast, RNA prepared from cells prior to ecdysone treatment hybridizes only to the hormone-repressed puff and not to the hormone-induced puffs (**Figure 1.34**).

Similarly, RNA labeled after heat shock hybridizes intensely to puff 87C, which appears following exposure to elevated temperature and is now known to contain the gene encoding the 70 kDa heat-shock protein (hsp70), which is the major protein made in *Drosophila* following heat shock (see Sections 4.3 and 8.1 for further discussion of this gene and its induction by elevated temperature).

The large size of polytene chromosomes allows a direct visualization of the transcriptional process and indicates that, as in other situations, gene activity in the salivary gland is regulated at the level of transcription.

Transcriptional control can operate at the level of chromatin structure and at the level of production of the primary RNA transcript

The experiments described above indicate therefore that the process of transcription represents a key target for gene-control processes. However, such regulation of transcription could operate at two different levels. As will be discussed in Chapter 2, the DNA is packaged with proteins to form a structure known as **chromatin**. Transcriptional regulation could therefore be achieved by altering the chromatin structure so that constitutively active regulatory molecules could gain access and switch on transcription (**Figure 1.35a**).

Alternatively, the chromatin structure could be accessible at all times with the activation of transcriptional regulatory proteins inducing transcription to produce the primary RNA transcript (**Figure 1.35b**). In fact it appears that both these mechanisms are frequently used together in particular cases and a move to a more open chromatin structure is followed by the activation of a transcriptional regulatory protein (**Figure 1.35c**).

For this reason, the structure of chromatin will be discussed in Chapter 2 and its role in the regulation of gene transcription will be considered in Chapter 3. Similarly, the actual process of gene transcription by RNA polymerases will be discussed in Chapter 4 and its regulation considered in Chapter 5. In addition, because a number of examples of post-transcriptional control do exist, post-transcriptional events will be discussed in Chapter 6 and their regulation considered in Chapter 7.

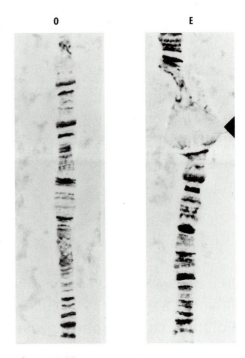

Figure 1.33
Transcriptionally active puff (arrowhead) in a polytene chromosome of *D. melanogaster*. The puff appears in response to treatment with the steroid hormone ecdysone (E) and is not present prior to hormone treatment (O). Courtesy of Michael Ashburner, University of Cambridge.

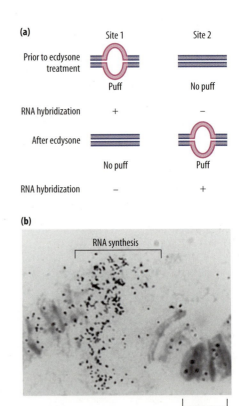

Figure 1.34
(a) Schematic diagram illustrating that the newly synthesized RNA made following ecdysone stimulation can be labeled with [³H]uridine and shown to hybridize to the puffs that form following ecdysone treatment. Conversely, a puff that regresses after hormone treatment binds only the labeled RNA synthesized before addition of the hormone. (b) Autoradiograph showing hybridization of [³H]uridine-labeled RNA to a newly induced puff. (b) Courtesy of Jose Bonner, Indiana University.

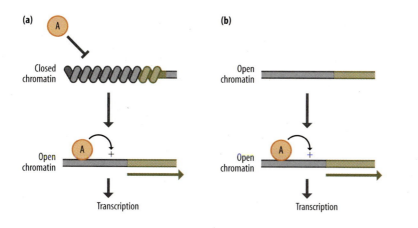

1.5 SMALL RNAs AND THE REGULATION OF GENE EXPRESSION

So far in this chapter we have considered gene control in terms of the central dogma of molecular biology, namely that DNA produces RNA which in turn produces proteins. Of course, it has been known for some time that some cellular RNAs have a direct functional role rather than being translated into protein. These include the U RNAs (rich in uridine) which are involved in the process by which the primary transcript is processed to produce the mature mRNA (see Section 6.3) and the ribosomal and transfer RNAs which play a key role in the translation of the mRNA into protein (see Section 6.6).

Recently, however, it has become clear than small RNAs between 20 and 30 bases in length not only function at the RNA level but play a key role in inhibiting the expression of protein-coding genes by interacting with specific mRNAs. As these small RNAs inhibit gene expression at more than one level, they will be introduced in this section. Their roles in regulating chromatin structure and in post-transcriptional gene regulation at the level of mRNA turnover or mRNA translation will be discussed in Chapters 3 (Section 3.4) and 7 (Section 7.6) respectively (**Figure 1.36**).

In this section, we will discuss in turn two major classes of small RNA which regulate cellular gene expression. These are the **microRNAs (miRNAs)** and the short interfering RNAs (**siRNAs**) which differ in their mechanism of production and in their biological roles. A third class of small RNAs, piRNAs are expressed in germ cells where they appear to function to inhibit the activity of mobile genetic elements (transposons) rather than in the regulation of cellular gene expression. Therefore they will not be discussed further.

Figure 1.35
Transcriptional control could operate by changes to the structure of chromatin allowing access to constitutively active proteins which activate transcription (A) (a). Alternatively, the chromatin could be accessible at all times with transcription controlled by the presence or absence of activator proteins (A) (b). In many cases however, these two mechanisms of transcriptional regulation are combined (c).

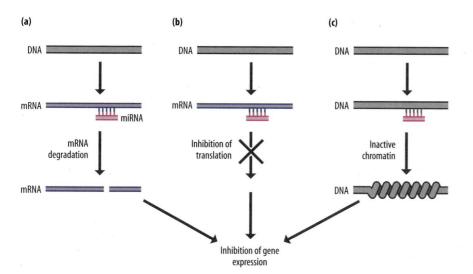

Figure 1.36
Small RNAs can inhibit gene expression by inducing mRNA degradation (a), blocking mRNA translation (b), or inducing the reorganization of their target gene into an inactive chromatin structure (c). miRNA, microRNA.

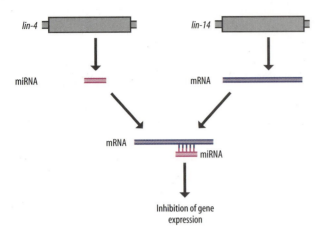

Figure 1.37
In the nematode, the *lin-4* gene produces a 22-nucleotide miRNA which binds to the *lin-14* mRNA and inhibits its translation.

miRNAs are processed from a single-stranded precursor which folds to form a double-stranded hairpin loop

The first miRNA to be discovered was identified by genetic experiments in a nematode worm. It was shown that inactivation of the *lin-4* gene by **mutation** had the same effect as a **gain-of-function mutation** in the *lin-14* gene, indicating that the two genes were likely to be antagonistic to one another. Moreover, it was shown subsequently that the *lin-4* gene produces a 22-**nucleotide** miRNA which can base pair with the *lin-14* gene mRNA. This led to the suggestion that base-pairing of the *lin-4* miRNA to the *lin-14* gene mRNA inhibits *lin-14* gene expression (**Figure 1.37**). This idea was subsequently shown to be correct since, as discussed in Section 7.6, binding of the *lin-4* miRNA to the *lin-14* mRNA blocks the translation of the mRNA into the corresponding protein.

Interestingly, this interaction plays a key role in nematode development, with *lin-4* being expressed at the second larval stage in development when it blocks *lin-14* expression. In contrast, *lin-4* is not expressed at the first larval stage, so allowing *lin-14* mRNA to be translated at the appropriate point in embryonic development.

Following this initial discovery of an miRNA in 1993, there has been an explosion of information on miRNAs. A large number of miRNAs have been discovered in a range of organisms from human and flies to plants and worms and shown to play a key role in specific biological events. For example, humans have been shown to express over 1000 different miRNAs. Moreover, binding of an miRNA to an mRNA can occur on the basis of only partial complementarity between their sequences, with only some bases in the miRNA forming base pairs with the mRNA. This allows an individual miRNA to bind to multiple mRNAs which have different target sequences capable of partially base pairing with the miRNA. In turn, this allows one individual miRNA to inhibit the expression of many different target genes (**Figure 1.38**).

By this means, miRNAs play a key role in regulating gene expression in a wide range of cell types in many different organisms and a number of

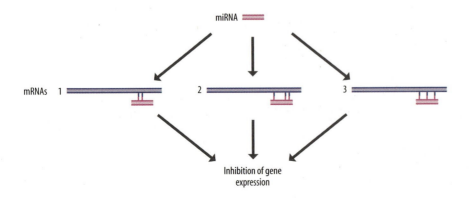

Figure 1.38
A single miRNA can bind to a number of target mRNAs and inhibit gene expression. 1, 2 and 3 indicate different mRNAs. Note that this binding only requires partial complementarity between the miRNA and a target mRNA.

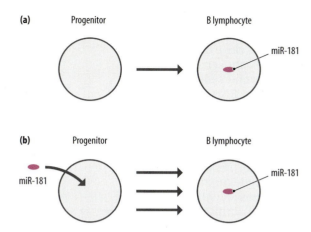

Figure 1.39
In the mouse, the miR-181 miRNA is expressed specifically in B lymphocytes (a). Its artificial expression in progenitor cells induces increased production of B lymphocytes, indicating its functional importance (b).

examples will be discussed in subsequent chapters. A particularly dramatic example is provided by the mouse miRNA, miR-181. This miRNA is expressed specifically in **B lymphocytes** but not in their progenitor cells. However, its artificial over-expression in the progenitor cells is sufficient to induce their differentiation to mature B lymphocytes (**Figure 1.39**), suggesting that this microRNA is likely to play a key role in the normal differentiation of these cells.

The key role of miRNAs in regulating gene expression in a range of organisms focuses attention on the manner in which they are synthesized. Interestingly, in animals miRNAs are transcribed as larger precursor RNAs, known as pri-miRNAs, with transcription being carried out by RNA polymerase II, the same polymerase which transcribes the genes encoding mRNAs (see Section 4.1). As with mRNA precursors, the pri-miRNA is modified at its 5′ end by **capping** and at its 3′ end by polyadenylation (see Sections 6.1 and 6.2 for an account of these post-transcriptional processes).

Unlike mRNA precursors, however, the pri-miRNA folds to form a partially double-stranded structure, with a hairpin loop in which some complementary bases are paired to one another (**Figure 1.40a**). This loop binds a complex containing a protein known as Drosha. Following binding, Drosha cleaves the pri-miRNA to generate a pre-miRNA of approximately 70 bp containing the double-stranded region (**Figure 1.40b**).

All these events take place in the nucleus of the cell and are followed by the transport of the pre-miRNA to the cytoplasm. Here it is cleaved again by the Dicer protein (**Figure 1.40c**). This removes the single-stranded loop of the pre-miRNA, leaving two single-stranded RNA molecules base-paired to one another (**Figure 1.40d**). One of these single strands is then degraded, leaving the other which represents the mature miRNA free to bind to its target mRNAs (**Figure 1.40e**).

Many siRNAs are processed from a double-stranded precursor

In contrast to miRNAs, many siRNAs are processed from a double-stranded RNA rather than from a single-stranded RNA which has formed a hairpin loop. The production of siRNA was initially characterized as a defense mechanism against invading viruses, which often produce double-stranded RNAs as part of their life cycle. In response to this, the same Dicer protein as is involved in miRNA processing binds to the double-stranded RNA. Dicer then cuts the double-stranded RNA into smaller double-stranded RNA molecules. Following separation of the two strands of these smaller RNAs, one of the strands forms the siRNA. It can then bind to viral mRNA which is identical in sequence to the original double-stranded RNA and promote its degradation, thereby inhibiting viral gene expression (**Figure 1.41**).

As well as functioning as a cellular defense mechanism against viral infection, the existence of siRNAs also explains a phenomenon observed in some transgenic experiments involving the introduction of specific genes into animals or plants. In some such experiments, it was noted that not only

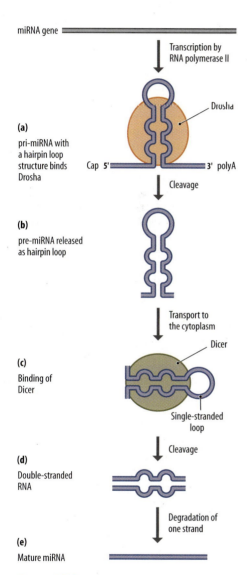

Figure 1.40
Processing of miRNA precursors. The pri-miRNA precursor is transcribed by RNA polymerase II and forms a hairpin loop structure which binds the Drosha protein (a). Cleavage by Drosha generates a pre-miRNA containing the hairpin loop which is transported to the cytoplasm (b). Subsequently, the pre-miRNA binds the Dicer protein (c) which cleaves off the single-stranded loop. This releases double-stranded RNA (d), one strand of which is degraded and the remaining strand constituting the mature miRNA (e). polyA, polyadenylated RNA.

was the introduced transgene not expressed but the endogenous copy of the same gene was also specifically silenced.

This effect occurs when the transgene inserts into the genome close to a cellular promoter which transcribes the transgene in an antisense orientation. This **antisense RNA** transcript can bind to the normal sense RNA transcript of the transgene and produce a double-stranded RNA. In turn, this double-stranded RNA will bind Dicer, which will cut it up to produce siRNAs. Such siRNAs can bind both to the mRNA of the transgene and to mRNA derived from the corresponding endogenous gene, inhibiting their expression (**Figure 1.42**).

As well as silencing the expression of invading viruses and artificially introduced transgenes, siRNA is also involved in regulating endogenous cellular genes in normal rather than artificial situations and this has recently been demonstrated in mammalian cells. This can involve the transcription of a **pseudogene** which is homologous to the functional gene but cannot encode a protein due, for example, to it containing **stop codons**, which would terminate translation prematurely (see Section 6.6). In these cases, the pseudogene is transcribed to produce an antisense transcript which can hybridize to the protein-encoding sense transcript of the functional gene. This double-stranded RNA is evidently a substrate for the production of siRNA which can then bind to further copies of the functional RNA and inhibit its expression (**Figure 1.43**).

Interestingly, it appears that expression of the mammalian gene encoding histone deacetylase 1 (Hdac1) is regulated in this manner, with an Hdac1 pseudogene being transcribed in the opposite direction to the functional *Hdac1* gene. This is of particular interest since, as will be discussed in Chapters 2 (Section 2.3) and 3 (Section 3.3) **acetylation**/deacetylation of histones plays a key role in the regulation of chromatin structure and therefore in transcriptional regulation. This example therefore illustrates the manner in which the siRNA system can potentially regulate the expression of a gene encoding a protein which itself has a gene-regulatory role.

As will be discussed further in subsequent chapters, both miRNAs and siRNAs play key roles in the regulation of gene expression. Indeed, mice lacking the gene for Dicer die early in embryonic development, indicating the key role of these pathways, both of which involve the Dicer protein in producing the functional mature small RNA.

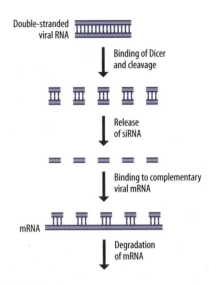

Figure 1.41
Double-stranded viral RNA binds the Dicer protein which cleaves the RNA into small double-stranded siRNAs. Following release of single-stranded siRNA, it can bind to the complementary viral mRNA and induce its degradation, thereby inhibiting viral gene expression.

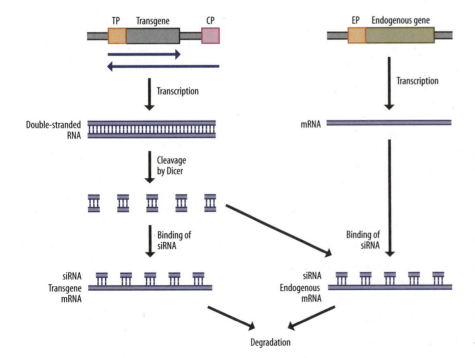

Figure 1.42
In cases where a transgene inserts close to a downstream cellular promoter (CP), an antisense transcript will be produced. This can bind the normal transcript produced from the transgene promoter (TP) to produce a double-stranded RNA. The double-stranded RNA will be cleaved by Dicer to produce siRNA, which can bind to both the transgene mRNA and to the mRNA from the corresponding endogenous gene. This results in degradation of both mRNAs, inhibiting the expression of both the transgene and the corresponding endogenous gene.

CONCLUSIONS

The central dogma of molecular biology is that DNA makes RNA and RNA makes protein. In this chapter we have seen that the DNA content of different tissues and cell types is generally the same. However, the RNA and protein contents of different tissues and cell types are both qualitatively and quantitatively different. This indicates that gene control must operate to produce different mRNA and thus different protein populations in different tissues and cell types from the same **genomic DNA**. A variety of evidence indicates that the major gene control point is at the level of transcription. Regulatory processes therefore control which genes are transcribed into the primary RNA transcript with processing of this transcript to produce the mRNA following subsequently.

In turn, such transcriptional regulation appears to operate both at the level of chromatin structure in opening up the gene to allow access to proteins involved in transcription and also at the level of the actual process of gene transcription by RNA polymerase II. Accordingly, the structure of chromatin and its role in gene control are discussed in Chapters 2 and 3, respectively. Similarly, the process of transcription itself and its regulation are discussed in Chapters 4 and 5, respectively.

Although transcriptional control is the major form of gene control in eukaryotes, post-transcriptional gene regulation also occurs, often as a significant supplement to transcriptional control. Chapter 6 therefore discusses the processes which modify the primary transcript to produce the mRNA as well as the translation of the mRNA into protein, whereas Chapter 7 discusses the role of these processes in gene regulation. Interestingly, the recent explosion in information on the small inhibitory RNAs (miRNA and siRNA) (see Section 1.4) indicates that they can inhibit gene expression both by regulating chromatin structure and by regulating post-transcriptional processes such as RNA turnover and translation into protein. These effects are discussed in Chapters 3 (Section 3.4) and 7 (Section 7.6), respectively.

Evidently, the fundamental processes involved in gene regulation, which will be discussed in Chapters 2–7, play a vital role in embryonic development and the functioning of the adult organism. The role of gene regulation in the cellular response to external signals, in embryonic development itself, and in the differentiation and functioning of specific cell types is therefore discussed in Chapters 8–10.

Finally, in view of the complexity of gene-regulatory processes and their vital role, it is not surprising that they can go wrong and that this has damaging effects on the organism. Chapters 11 and 12 therefore discuss the role of gene-control processes in cancer and other human diseases, respectively, as well as the manner in which the manipulation of gene-regulatory processes may ultimately be of therapeutic benefit.

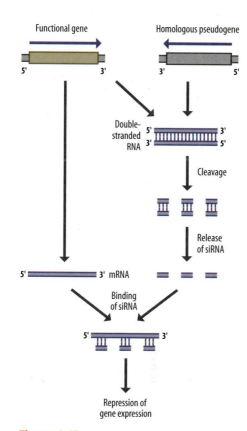

Figure 1.43
Transcription of a functional gene in the sense direction and of a homologous pseudogene in the antisense direction produces a double-stranded RNA. This is cleaved to produce siRNA which binds to the transcript of the functional gene and inhibits gene expression.

KEY CONCEPTS

- The central dogma of molecular biology is that DNA produces RNA which in turn produces proteins.

- Different tissues and cell types of the same organism show differences in the different proteins which are present and in their relative abundance.

- Similarly, different tissues and cell types show differences in the RNAs which are present and in their relative abundance.

- Unlike the RNA and protein content, the DNA of different tissues and cell types is generally the same in a specific organism.

- Gene control must therefore operate to produce different RNA populations in different cell types from the same DNA.

- The major point of gene control is at the level of transcription regulating which genes are transcribed to produce a primary RNA transcript in different cell types.

- Small RNAs play a key role in gene regulation, acting to regulate both transcriptional and post-transcriptional processes.

FURTHER READING

1.1   Protein content varies among different cell types

Gershoni JM & Palade GE (1983) Protein blotting: principles and applications. *Anal. Biochem.* 131, 1–15.

Kislinger T, Cox B, Kannan A et al. (2006) Global survey of organ and organelle protein expression in mouse: combined proteomic and transcriptomic profiling. *Cell* 125, 173–186.

O'Farrell PH (1975) High resolution two-dimensional electrophoresis of proteins. *J. Biol. Chem.* 250, 4007–4021.

Ong SE & Mann M (2005) Mass spectrometry-based proteomics turns quantitative. *Nat. Chem. Biol.* 1, 252–262.

Pandey A & Mann M (2000) Proteomics to study genes and genomes. *Nature* 405, 837–846.

1.2   The mRNA content varies among different cell types

Lockhart DJ & Winzeler EA (2000) Genomics, gene expression and DNA arrays. *Nature* 405, 827–836.

Thomas PS (1980) Hybridization of denatured RNA and small DNA fragments transferred to nitrocellulose. *Proc. Natl Acad. Sci. USA* 77, 5201–5205.

1.3   The DNA content of different cell types is generally the same

Claycomb JM & Orr-Weaver TL (2005) Developmental gene amplification: insights into DNA replication and gene expression. *Trends Genet.* 21, 149–162.

Gurdon JB (1968) Transplanted nuclei and cell differentiation. *Sci. Am.* 219, 24–35.

Gurdon JB & Melton DA (2008) Nuclear reprogramming in cells. *Science* 322, 1811–1815.

Southern EM (1975) Detection of specific sequences among DNA fragments separated by gel electrophoresis. *J. Mol. Biol.* 98, 503–517.

Steward FC (1970) From cultured cells to whole plants: the induction and control of their growth and morphogenesis. *Proc. R. Soc. Lond. B Biol. Sci.* 175, 1–30.

Wilmut I, Schnieke AE, McWhis J, Kind AJ & Campbell KHS (1997) Viable offspring derived from fetal and adult mammalian cells. *Nature* 385, 810–813.

1.4   Transcriptional or post-transcriptional control?

Derman E, Krauter K, Walling L et al. (1981) Transcriptional control in the production of liver specific mRNAs. *Cell* 23, 731–739.

Roop DR, Nordstrom JL, Tsai S-Y, Tsai M-J & O'Malley BW (1978) Transcription of structural and intervening sequences in the ovalbumin gene and identification of potential ovalbumin mRNA precursors. *Cell* 15, 671–685.

1.5   Small RNAs and the regulation of gene expression

Carthew RW & Sontheimer EJ (2009) Origins and mechanisms of miRNAs and siRNAs. *Cell* 136, 642–655.

Ghildiyal M. & Zamore PD (2009) Small silencing RNAs: an expanding universe. *Nat. Rev. Genet.* 10, 94–108.

Golden DE, Gerbasi VR & Sontheimer EJ (2008) An inside job for siRNAs. *Mol. Cell* 31, 309–312.

Grosshans H & Filipowicz W (2008) Molecular biology: the expanding world of small RNAs. *Nature* 451, 414–416.

Kim VN, Han J & Siomi MC (2009) Biogenesis of small RNAs in animals. *Nat. Rev. Mol. Cell Biol.* 10, 126–139.

Neilson JR & Sharp PA (2008) Small RNA regulators of gene expression. *Cell* 134, 899–902.

Okamura K & Lai EC (2008) Endogenous small interfering RNAs in animals. *Nat. Rev. Mol. Cell Biol.* 9, 673–678.

Sasidharan R & Gerstein M (2008) Protein fossils live on as RNA. *Nature* 453, 729–731.

Siomi H & Siomi MC (2009) On the road to reading the RNA-interference code. *Nature* 457, 396–404.

Winter J, Jung S, Keller S, Gregory RI & Diederichs S (2009) Many roads to maturity: microRNA biogenesis pathways and their regulation. *Nat. Cell Biol.* 11, 228–234.

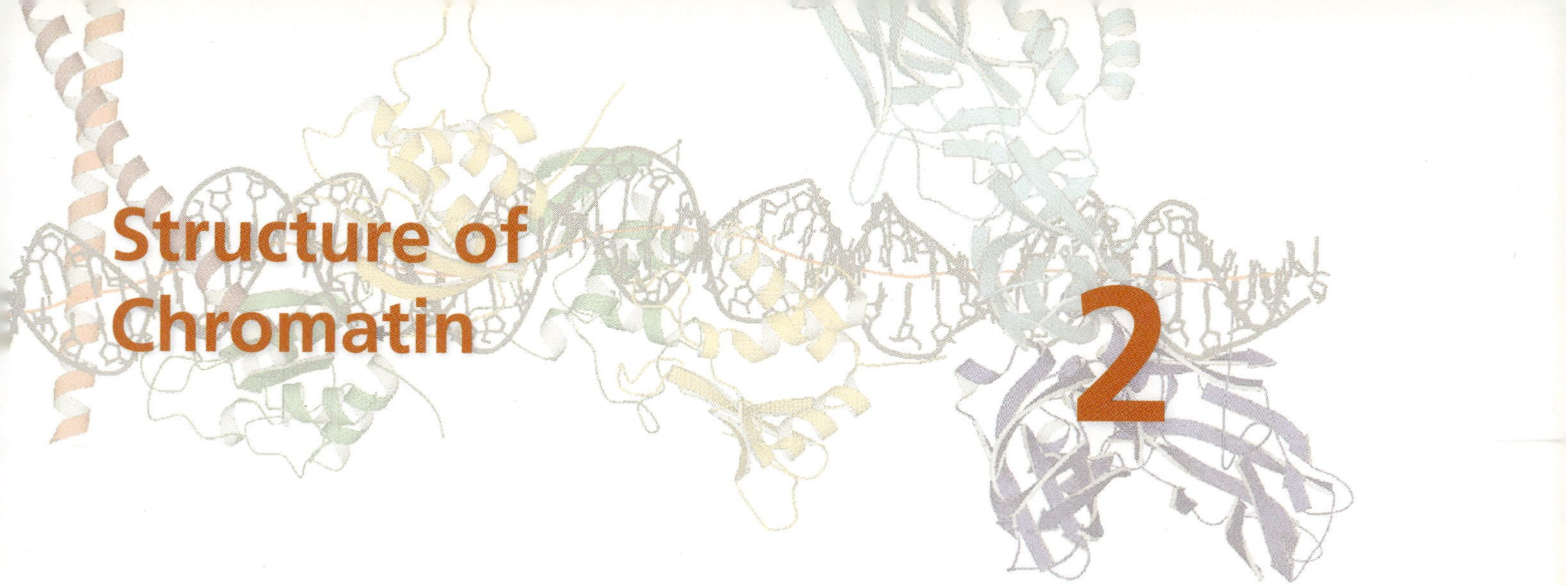

Structure of Chromatin

2

INTRODUCTION

In the previous chapter, we discussed the evidence indicating that the primary control of eukaryotic gene expression lies at the level of transcription with regulatory processes determining which genes are transcribed to produce the primary RNA transcript. To understand gene-control processes it is therefore necessary to investigate the mechanisms responsible for such transcriptional control.

Regulation of transcription in eukaryotes is much more complex than in prokaryotes

The fact that regulation at transcription is also responsible for the control of gene expression in bacteria suggests that insights into these procedures obtained in these much simpler organisms may be applicable to higher organisms. In bacteria, transcription of many genes such as those in the *lac* operon is controlled by negative regulatory mechanisms in which the presence of **repressor** molecules inhibits gene transcription. Inactivation of such repressors in response to a specific signal therefore leads to transcription of the gene.

It is possible therefore that regulation of transcription in eukaryotes might similarly occur by means of a protein which is present in all tissues and binds to the promoter region of a particular gene preventing its expression. In one particular cell type or in response to a particular signal such as elevated temperature, this protein would be inactivated either directly (**Figure 2.1a**) or by binding of another factor (**Figure 2.1b**), and would no longer bind to the gene.

However, most eukaryotic genes are inactive in most tissues and become active only in one tissue or in response to a particular signal. This suggests that it may be more economical to have a system in which the gene is constitutively inactive in most tissues without any repressor being required. Activation of the gene would then require a particular activating factor binding to its promoter. Positive regulatory mechanisms of this type are used to control the expression of some genes in bacteria in response to specific signals.

Such activation mechanisms would involve the specific expression pattern of the gene being controlled by the presence of a positively acting factor only in the expressing cell type (**Figure 2.2a**) or alternatively by the factor's requirement for a co-factor such as a steroid hormone to convert it to an active form (**Figure 2.2b**).

Based on the known mechanisms of gene regulation in bacteria, it is possible therefore to produce models of how gene regulation might operate in eukaryotes. Indeed, a considerable amount of evidence indicates that many cases of gene regulation do use activation type mechanisms of the type illustrated in Figure 2.2 in which the gene is not transcribed except in

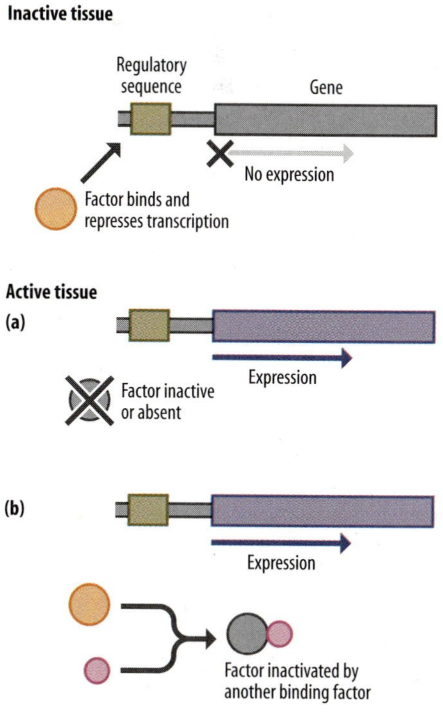

Figure 2.1
Model for the activation of gene expression in a particular tissue by inactivation (a) or binding out (b) of a repressor present in all tissues.

the presence of an activating factor. For example, as will be discussed in Sections 5.1 and 8.1, the effects of glucocorticoid and other steroid hormones on gene expression are mediated by the binding of the steroid to a receptor protein. This activated complex then binds to particular sequences upstream of steroid-responsive genes and activates their transcription.

Even in the case of steroid hormones, however, such mechanisms cannot account entirely for the regulation of gene expression. In the chicken, administration of the steroid hormone estrogen results in the transcriptional activation in oviduct tissue of the gene encoding ovalbumin, as discussed in Section 1.4. In the liver of the same organism, however, estrogen treatment has no effect on the ovalbumin gene but instead results in the activation of a completely different gene encoding the protein vitellogenin. Such tissue-specific differences in the response to a particular treatment are, of course, entirely absent in single-celled bacteria and cannot be explained simply on the basis of the activation of a single DNA-binding protein by the hormone (**Figure 2.3**).

An understanding of eukaryotic gene regulation will therefore require a knowledge of the ways in which relatively short-term regulatory processes mediated by the binding of proteins to specific DNA sequences interact with much longer-term regulatory processes, which establish and maintain the differences between particular tissues and also control their response to treatment with effectors such as steroids. These differences involve packaging DNA with proteins to form the structure known as chromatin.

Hence, eukaryotic gene regulation requires both long-term processes which regulate chromatin structure and more short-term processes which actually result in the activation of gene transcription. Only the second type of process is found in bacteria where regulation of chromatin structure does not occur. The structure of chromatin will therefore be discussed in this chapter, whereas the long-term control processes which regulate chromatin structure will be discussed in Chapter 3. The process of transcription itself and its control by regulatory factors is discussed in Chapters 4 and 5, respectively.

2.1 COMMITMENT TO THE DIFFERENTIATED STATE AND ITS STABILITY

Cells can remain committed to a particular differentiated state even in the absence of its phenotypic characteristics

It is evident that the existence of different tissues and cell types in higher eukaryotes requires mechanisms that establish and maintain such differences, and that these mechanisms must be stable in the long term. In general, tissues or cells of one type do not change spontaneously into another cell type and cells must be capable therefore of maintaining their differentiated phenotype throughout their lifetimes. Indeed, the long-term control processes that achieve this effect must regulate not only the ability to maintain the differentiated state more or less indefinitely but also the observed ability of cells to remember their particular cell type, even under conditions when they cannot express the characteristic phenotype of that cell type. For example, it is possible to regulate the behavior of cartilage-producing cells in culture by changing the medium in which they are placed. In one medium the cells are capable of expressing their differentiated phenotype and produce cartilage-forming colonies synthesizing an extracellular matrix containing chondroitin sulfate. By contrast, in a different medium favoring rapid division, the cells do not form such colonies and instead divide rapidly and lose all the specific characteristics of cartilage cells, becoming indistinguishable from undifferentiated fibroblast-like cells in appearance. Nonetheless, even after 20 generations in this rapid growth medium, the cells can resume the appearance of cartilage cells and synthesis of chondroitin sulfate if returned to the appropriate medium. This process does not

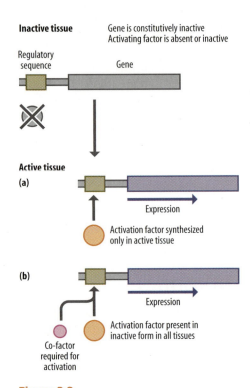

Figure 2.2
Model for the activation of gene expression in a particular tissue by an activator present only in that tissue (a) or activated by a co-factor (b) only in that tissue.

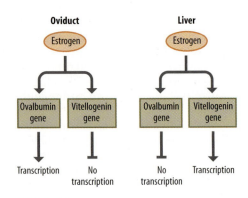

Figure 2.3
Estrogen stimulates the transcription of the ovalbumin gene in the oviduct but stimulates the transcription of the vitellogenin gene in the liver.

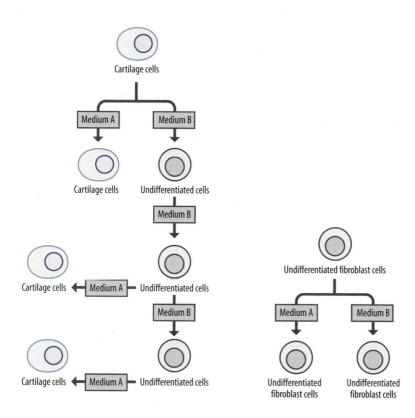

Figure 2.4
Cartilage cells cultured in medium A maintain their differentiated state but revert to an undifferentiated state in medium B. However, even after many generations in medium B, the cells can re-differentiate to cartilage cells if returned to medium A. In contrast, undifferentiated fibroblast cells, which have never had a cartilage phenotype, remain undifferentiated in medium A or medium B.

occur if other cell types or undifferentiated **fibroblast** cells are placed in an identical medium. The cartilage cells are therefore capable not only of maintaining their differentiated phenotype in a particular medium supplying appropriate signals, but also of remembering that phenotype in the absence of such signals and returning to the correct differentiated state when placed in the appropriate medium (**Figure 2.4**).

Cells can become committed to a particular differentiated state prior to actual phenotypic differentiation

The experiments described above suggest that cells belong to a particular lineage and that mechanisms exist to maintain the **commitment** of cells to a particular lineage, even in the absence of the characteristics of the differentiated phenotype. Considerable evidence exists to suggest that cells become committed to a particular differentiated stage or lineage well before they express any features characteristic of that lineage and that such commitment can be maintained through many cell generations.

Perhaps the most dramatic example of this effect occurs in the fruit fly, *Drosophila melanogaster*. In this organism, the larva contains many discs consisting of undifferentiated cells located at intervals along the length of the body and indistinguishable from each other in appearance. Eventually, these **imaginal discs** will form the structures of the adult, the most anterior pair producing the antennae and others producing the wings, legs, etc. To do this, however, the discs must pass through the intermediate pupal state where they receive the appropriate signals to differentiate into the adult structures. If they are removed from the larva and placed directly in the body cavity of the adult they will remain in an undifferentiated state because the signals inducing differentiation will not have been received (**Figure 2.5**).

This process can be continued for many generations. The disc removed from the adult can be split in two, one half being used to propagate the cells by being placed directly in another adult and the other half being tested to see what it will produce when placed in a larva that is allowed to proceed to an adult through the pupal state. In this experiment, a disc which would

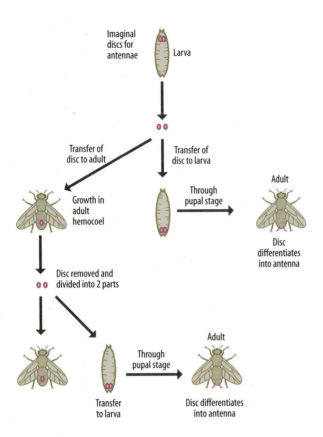

Figure 2.5
The use of larval imaginal discs to demonstrate the stability of the committed state. The disc cells maintain their commitment to produce a specific adult structure even after prolonged growth in an undifferentiated state in successive adults. From Gurdon, JB (1974) *Gene Expression*. With permission from Oxford University Press.

have given rise to an antenna can still do so when placed in a larva after many generations of passage through adult organisms in an undifferentiated state. This is the case even if the disc is placed at a position within the larva very different from that normally occupied by the imaginal disc producing the antenna. The cells of the imaginal disc within the normal larva must therefore have undergone a commitment event to the eventual production of a particular differentiated state, which they can maintain for many generations, prior to having ever expressed any phenotypic features characteristic of that differentiated state (see Figure 2.5).

These examples of the stability of the differentiated state and commitment to it imply the existence of long-term regulatory processes capable of maintaining such stability. However, there are cases where such stability breaks down; for example, following nuclear transplantation (see Section 1.3), indicating that such processes, although stable, are not irreversible.

Indeed, even in the imaginal discs of *Drosophila* the stability of the committed state can be lost by culture for long periods in adults with the discs giving rise, when eventually placed in the larva, to tissues other than the one intended originally. This breakdown of the committed state is not a random process but proceeds in a highly reproducible manner. Wing cells are the first abnormal cells produced by a disc that should produce an eye, with other cell types being produced only subsequently. Similarly, a disc that should produce a leg never produces genitalia, although it can produce other cell types. The various transitions that can occur in this system are summarized in **Figure 2.6**. Interestingly, these various transitions are precisely those observed in the **homeotic mutations** in *Drosophila* (see Sections 5.1 and 9.2) which convert one adult body part into another. Each of these transitions is likely therefore to be controlled by a specific gene whose activity changes when commitment breaks down.

These examples and the other experiments discussed in Section 1.3 eliminate irreversible mechanisms such as DNA loss as a means of explaining the process of commitment to the differentiated state. Rather, the semi-stability of this process and its propagation through many cell generations is due to the establishment of a particular pattern of association of the DNA with specific proteins. The structure formed by DNA and its associated

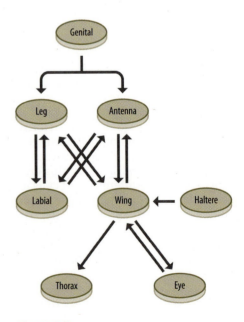

Figure 2.6
Structures that arise from specific imaginal discs of *Drosophila* following the breakdown of commitment. From Cove DJ (1971) *Genetics*. With permission from Cambridge University Press.

TABLE 2.1 THE HISTONES			
HISTONE	TYPE	MOLECULAR WEIGHT (Da)	MOLAR RATIO
H1	Lysine-rich	23,000	1
H2A	Slightly lysine-rich	13,960	2
H2B	Slightly lysine-rich	13,744	2
H3	Arginine-rich	15,342	2
H4	Arginine-rich	11,282	2

proteins is known as chromatin. An understanding of how long-term gene regulation is achieved therefore requires a knowledge of the structure of chromatin.

2.2 THE NUCLEOSOME

The nucleosome is the basic unit of chromatin structure

If the DNA in a single human individual were to exist as an extended linear molecule, it would have a length of 5×10^{10} km and would extend 100 times the distance from the Earth to the Sun. Even the DNA in a single cell would extend for 2 m if it was not compacted in some way. Clearly therefore, the DNA must be compacted to fit in the nucleus of the cell. This is achieved by folding the DNA in a complex with specific nuclear proteins into the structure known as chromatin. Of central importance in this process are the five types of histone protein (Table 2.1), which have a high proportion of the positively charged amino acids lysine and arginine that neutralizes the net negative charge on the DNA and allows folding to occur.

The basic unit of this folded structure is the **nucleosome**, in which approximately 200 bp of DNA are associated with a **histone octamer** containing two molecules of each of the four core histones, H2A, H2B, H3, and H4 (Figure 2.7). In this structure approximately 146 bp of DNA make almost two full turns around the histone octamer, and the remainder serves as linker DNA joining one nucleosome to another. As expected from the key role of the histones in nucleosome structure, these proteins are highly conserved in evolution. For example, the histone H4 of a cow differs by only two amino acids from that of a garden pea, whereas the histone H3 molecules exhibit only four amino acid differences in these two species.

Each of the four core histones has a similar structure, with an N-terminal tail and a helical region which forms three α-helices separated by two loops (Figure 2.8). The α-helical region forms a specific structure, known as the histone fold (Figure 2.9a) which in turn allows individual histones to associate with one another (Figure 2.9b).

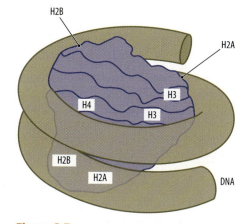

Figure 2.7
Structure of DNA and the core histones within a single nucleosome.

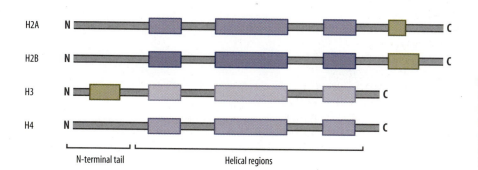

Figure 2.8
Structure of the four core histones, H2A, H2B, H3, and H4, showing the N-terminal tail and the three main α-helical regions (purple) which form the histone fold. Green boxes indicate other α-helical regions found in individual histones.

The detailed arrangement of the histones and the DNA within the nucleosome has been confirmed by high-resolution structural analysis using **X-ray crystallography**. In the nucleosomal structure, the histones form H2A–H2B and H3–H4 heterodimers via the histone fold. These heterodimers then associate to form the histone octamer with the DNA wrapped around the surface of the octamer (**Figure 2.10**). Interestingly, the N-terminal tails of the histones extend beyond the surface of the nucleosome particle and are likely to be involved in interactions between nucleosomes. This is of particular interest since these N-terminal histone tails are subject to specific modifications which are known to alter chromatin structure (see Section 2.3). These modifications may thus act by affecting the nucleosome–nucleosome interactions which are mediated via these regions of the histone molecule.

In this structure, the linker DNA is more accessible than the highly protected DNA tightly wrapped around the octamer and is therefore preferentially cleaved when chromatin is digested with small amounts of the DNA-digesting enzyme, micrococcal **nuclease**. If DNA is isolated following such mild digestion of chromatin, it produces a ladder of DNA fragments in multiples of 200 bp on gel electrophoresis, representing the results of cleavage in some but not all linker regions (**Figure 2.11a**, **track T**).

This can be correlated with the properties of nucleosomes isolated from the digested chromatin. A partially digested chromatin preparation can be fractionated into individual nucleosomes associated with 200 bases of DNA (**Figure 2.11a**, **track D**), dinucleosomes associated with 400 bases of DNA (**Figure 2.11a**, **track C**), and so on. The individual mononucleosomes, dinucleosomes, or larger complexes in each of these fractions can be observed readily in the electron microscope (**Figure 2.11b**). These experiments therefore provide direct evidence for the organization of DNA into nucleosomes within the cell.

This organization of DNA into nucleosomes can be directly visualized in the electron microscope when chromatin is isolated under conditions of low ionic strength. This yields a fiber which has a diameter of 10 nm and appears as **beads on a string** in the electron microscope, with nucleosomes joined by visible linker DNA (**Figure 2.12**). This beads-on-a-string structure constitutes the first stage in the packaging of DNA.

(a)

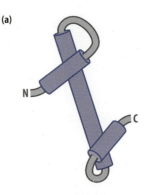

(b)

Figure 2.9
The histone fold formed by the three α-helical regions of the core histones (a). The histone fold regions of two histone molecules allow them to associate to form a heterodimer (b).

(a)

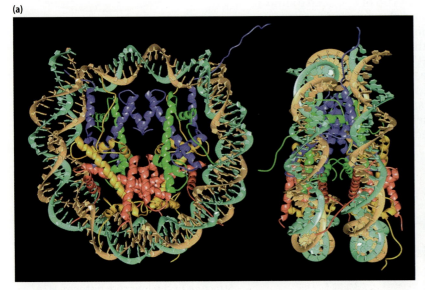

(b)

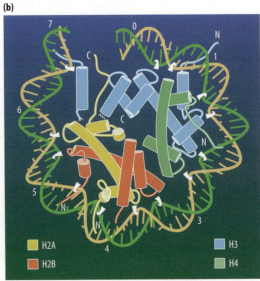

Figure 2.10
Two views of the structure of the nucleosome (a) and a schematic diagram indicating the positions of the different histones (b). C and N indicate the C- and N-termini. (a) Courtesy of Tim Richmond, from Luger K, Mäder AW, Richmond RK et al. (1997) *Nature* 389, 251–260. (b) Courtesy of Daniela Rhodes, from Rhodes D (1997) *Nature* 389, 231–232. Both with permission from Macmillan Publishers Ltd.

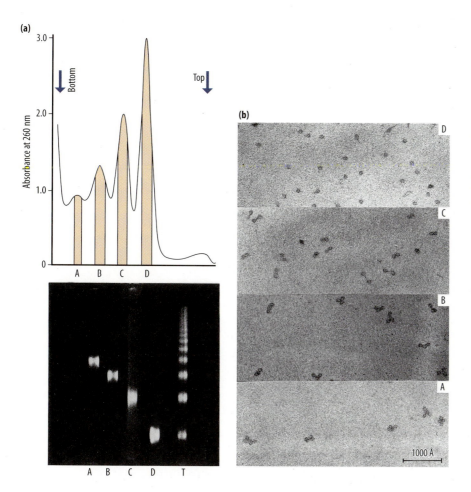

Figure 2.11
(a) Separation of mononucleosomes (D), dinucleosomes (C), trinucleosomes (B), and tetranucleosomes (A) by sucrose gradient centrifugation. The upper panel shows the peaks of absorbance produced by the individual fractions of the gradient; the lower panel shows the DNA associated with each fraction separated by gel electrophoresis. Track T shows the DNA ladder produced from a preparation containing all the individual nucleosome fractions. (b) Electron microscopic analysis of the fractions separated in panel (a). Mononucleosomes are clearly visible in fraction D, dinucleosomes in fraction C, and so on. Courtesy of John Finch, from Finch JT, Noll M & Kornberg RD (1975) *Proc. Natl Acad. Sci. USA* 72, 3320–3322.

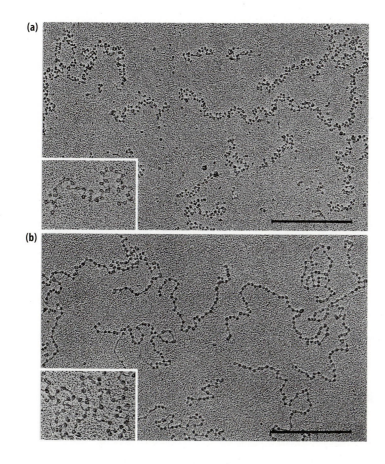

Figure 2.12
Beads-on-a-string structure of chromatin visualized in the electron microscope in the presence (a) or absence (b) of histone H1. The bars indicate 0.5 µm. Courtesy of F Thoma, from Thoma F, Koller T & Klug A (1979) *J. Cell Biol.* 83, 403–427. With permission from Rockefeller University Press.

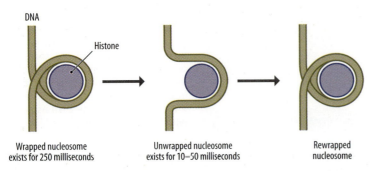

Figure 2.13
Dynamic nature of the nucleosome in which the DNA is continuously unwrapping and rewrapping itself around the histone octamer.

Nucleosome structure or position can be altered by chromatin-remodeling processes

The nucleosome is a highly dynamic structure. The DNA is continuously unwrapping from the histone octamer, remaining exposed for a short period and then rewrapping around the histones (**Figure 2.13**). This dynamic nature of the nucleosome is of considerable importance in the process of chromatin remodeling in which specific protein complexes hydrolyze ATP and use the energy to displace the nucleosome or modify its structure. Several multi-protein **chromatin-remodeling complexes** exist in eukaryotic cells and these are classified into several families, the best characterized of which are the **SWI/SNF** family and the ISWI family.

Each chromatin-remodeling complex contains an **ATPase** component which actually hydrolyzes ATP to generate the energy for the remodeling process. Such remodeling can have several outcomes (**Figure 2.14**). Firstly, it can involve a change in the structure of the nucleosome so that the DNA is more exposed (**Figure 2.14a**). Secondly, it can involve the displacement of the nucleosome so that it moves along the DNA (**Figure 2.14b**). Finally, it can involve the complete displacement of the nucleosome so that it is lost from the DNA molecule (**Figure 2.14c**). These changes make specific regions of DNA more accessible to specific regulatory molecules and their key role in gene control will be discussed further in Section 3.5.

Interestingly, as well as altering the association of DNA with the histone octamer, chromatin-remodeling complexes can also promote the exchange

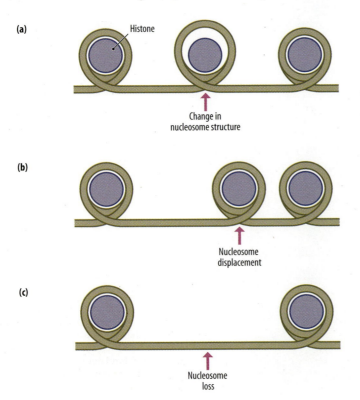

Figure 2.14
Chromatin remodeling can involve a change in nucleosome structure (a), the displacement of a nucleosome along the DNA (b), or the complete removal of a nucleosome (c).

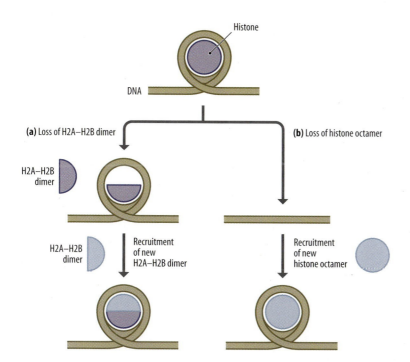

Figure 2.15
Chromatin remodeling can involve loss of a histone H2A–H2B dimer, leaving a partial histone octamer which then binds a new H2A–H2B dimer (light blue) (a). Alternatively it can involve the loss of the entire histone octamer leaving nucleosome-free DNA, followed by the recruitment of a new histone octamer (light blue) (b).

of histone molecules within a histone octamer so that pre-existing histone molecules are replaced with new ones (**Figure 2.15**). This can involve the loss of a histone H2A–H2B heterodimer and its replacement by another H2A–H2B dimer (**Figure 2.15a**). Alternatively, it can involve the loss of the complete histone octamer (consisting of two molecules of H2A, H2B, H3, and H4) followed by replacement with a new octamer of the core histones (**Figure 2.15b**).

This loss and subsequent replacement of all or part of the histone octamer would be somewhat pointless if like was always replaced with like. However, as discussed in the next section, the histones are subject to various **post-translational modifications** which alter their properties. Moreover, variant forms of the histones also exist. Exchange of histones catalyzed by remodeling complexes can therefore have significant effects in terms of gene regulation.

2.3 HISTONE MODIFICATIONS AND HISTONE VARIANTS

Histones are subject to a variety of post-translational modifications

Each of the core histones (H2A, H2B, H3, H4) is subject to a variety of post-translational modifications. These modifications target particularly the N-terminal portion of the histones which extend beyond the surface of the nucleosome particle and can therefore affect the interaction of the modified histones with other regulatory proteins. Alternatively, they can affect the interaction of the histones with the DNA itself or with other histone molecules in adjacent nucleosomes (**Figure 2.16**).

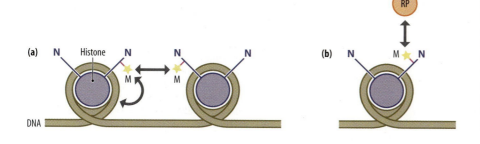

Figure 2.16
The N terminal (N) region of the histones extends outwards from the histone octamer in the nucleosome. Post-translational modifications (M) of this region can therefore affect the interaction of histones with adjacent nucleosomes or with the DNA (a). Alternatively, such modifications could affect the binding of a regulatory protein (RP) (b).

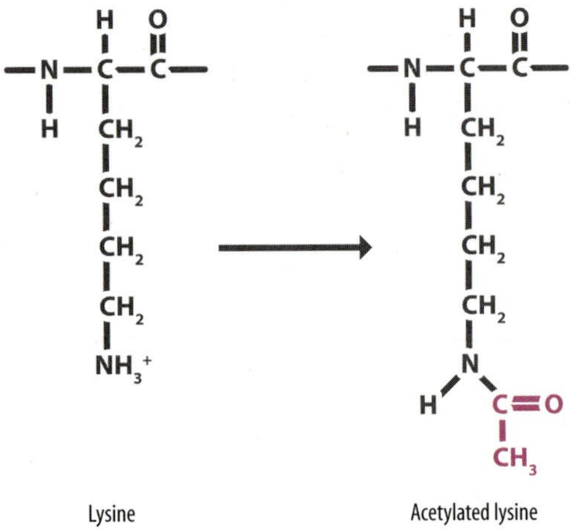

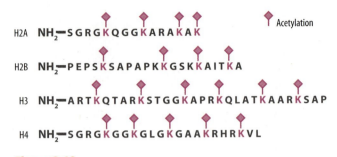

Figure 2.18
N-terminal sequence of the four core histones showing the lysine (K) amino acids which can be modified by acetylation.

Figure 2.17
Modification of a lysine amino acid by addition of an acetyl residue.

In many cases, individual histone molecules will show multiple different modifications. As described below, this leads to the idea of a "**histone code**" in which the overall pattern of modification of individual histone molecules plays a key role in the regulation of chromatin structure. The different post-translational modifications of the histones will be discussed in turn.

(a) Acetylation

In the process of acetylation, the free amino group on specific lysine residues is modified by one of the hydrogen atoms being substituted by an acetyl group ($COCH_3$). This reduces the net positive charge on the histone molecule (**Figure 2.17**). Such acetylation modifies multiple specific lysine residues at the N-terminus of each of the four core histones (**Figure 2.18**). Thus, each of the histones can exist in multiple forms with none, one, or multiple acetylated lysine residues.

The acetylation of histones is catalyzed by enzymes known as **histone acetyltransferases** (HATs). These are part of multi-protein complexes containing a variety of proteins including the catalytic subunit which actually acetylates the histones. A variety of HATs exist and are classified into three major families: the Gcn5 N-acetyltransferases (GNATs), the **p300/CBP** family, and the MYST family (named for its founder members Morf, Ybf2, Sas2, and Tip60).

Interestingly, the founder member of the GNAT family, the yeast protein Gcn5, had previously been characterized as a protein involved in transcription before its HAT activity was characterized. Similarly, p300/CBP were defined as key transcriptional **co-activator** proteins before their HAT activity was characterized (see Section 5.2 for further discussion of Gcn5 and p300/CBP). These findings are consistent with the data discussed in Section 3.3 which indicate the importance of histone acetylation in the control of chromatin structure and therefore in the regulation of transcription.

The action of HAT enzymes is opposed by **histone deacetylase** enzymes (HDACs), which remove the acetyl group from the lysine amino acids, with the balance of activity controlling the level of histone acetylation (**Figure 2.19**). As with HATs, the HDACs are multi-protein complexes with well-characterized examples including the Sin3 complex and the NuRD complex.

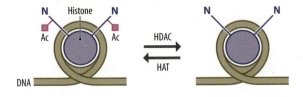

Figure 2.19
The level of acetylation (Ac) at the N-terminus (N) of the histones is controlled by the balance between histone acetyltransferase enzymes (HAT), which acetylate histones, and histone deacetylase enzymes (HDAC), which deacetylate histones.

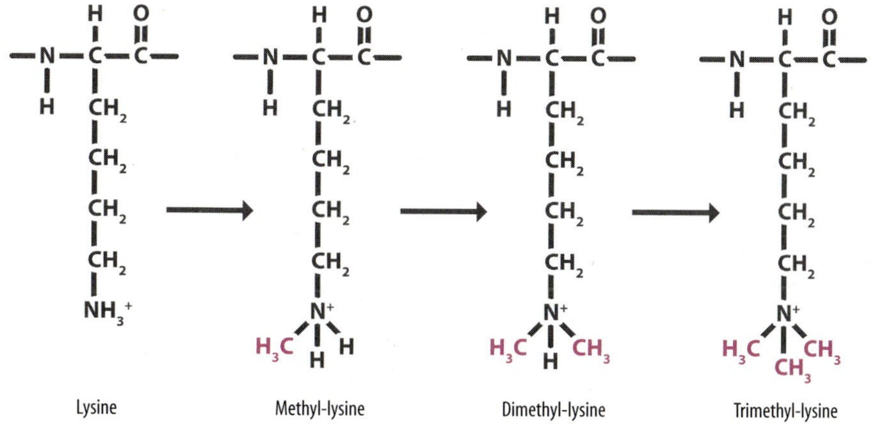

Lysine   Methyl-lysine   Dimethyl-lysine   Trimethyl-lysine

Figure 2.20
Modification of a lysine amino acid by addition of one, two, or three methyl residues. Compare with Figure 2.17.

Moreover, as with HATs, HDAC components had previously been shown to be involved in transcription even before their HDAC activity had been defined. For example, HDAC1, the first mammalian HDAC to be characterized, was shown to be homologous to Rpd3, a yeast protein identified on the basis that it repressed transcription.

(b) Methylation

As with acetylation, methylation targets lysine residues at the N-terminus of the histone molecule. However, in the methylation process, rather than replacing a hydrogen atom with an acetyl group, one, two, or three hydrogen atoms are replaced with methyl (CH_3) groups to form methyl-lysine, dimethyl-lysine or trimethyl-lysine (**Figure 2.20**). In addition, unlike acetylation, methylation can also target arginine residues. Moreover, unlike acetylation, methylation does not reduce the net positive charge on these basic amino acids.

Methylation can occur at multiple lysine or arginine residues at the N-terminus of each of the four histones H2A, H2B, H3, and H4. The residues in histones H3 and H4 which can be methylated are shown in **Figure 2.21**. As with acetylation (also shown in Figure 2.21), none, one or multiple specific residues can be methylated in any individual histone molecule.

Figure 2.21
N- terminal sequence of histones H3 and H4 showing the residues which can be methylated and comparing this with those that can be acetylated. The regions of histone H3 shown in Figure 2.22 and 2.23 are indicated by the solid and dashed underlines, respectively.

H3 NH₂—ARTKQTARKSTGGKAPRKQLATKAARKSAP

H4 NH₂—SGRGKGGKGLGKGAAKRHRKVL

◆ Acetylation
● Methylation

Interestingly, as seen in Figure 2.21 some lysine residues can be modified by either methylation or acetylation. As such modifications target the same atoms of the lysine molecule (compare Figures 2.17 and 2.20) they are mutually exclusive and such residues can either be acetylated or methylated (or unmodified) in any individual histone molecule (**Figure 2.22**).

In addition to such alternative modification of individual amino acids, modifications at one site often influence modifications at other sites. For example, methylation of histone H3 at position 9 in the N-terminus not only prevents acetylation at this position but also inhibits acetylation at position 14. Conversely, demethylation at position 9 is frequently accompanied by acetylation at position 14 (**Figure 2.23**).

Figure 2.22
Short region of histone H3 (solid underline in Figure 2.21) showing the lysine (K) at position 9 which can be acetylated or methylated but not both.

Ac
TARKST ⟷ TARKST ⟶ TARKST
9   Acetylation 9 Methylation 9
Me

Methylation of histones is catalyzed by specific histone methylase enzymes which methylate either arginine or lysine residues. It had been thought that such methylation was irreversible, with methylated histones ultimately being degraded or exchanged for non-methylated molecules. In the last few years, however, it has become clear that as with acetyl groups (see above) **methyl groups** can be removed from histones and enzymes which catalyze this reaction have been identified.

The demethylation event can take place via different mechanisms depending on the enzyme involved. For example, in the case of methylated lysine residues, the LSD1 enzyme catalyzes an amine oxidase reaction in which hydrogen atoms are removed and transferred to the FAD co-factor. This produces an imine intermediate, which then undergoes a deimination to produce unmethylated lysine and formaldehyde (CH_2O; **Figure 2.24a**). The LSD1 enzyme can remove the methyl groups from monomethyl and dimethyl-lysine but not from trimethyl-lysine.

In contrast, the JmjC demethylase enzymes act on trimethyl-lysine, as well as the monomethyl and dimethyl forms using a different mechanism to that employed by LSD1. This mechanism involves an oxidative demethylation reaction which releases succinate and CO_2. This produces a hydroxymethyl intermediate which then releases formaldehyde to leave a demethylated lysine (**Figure 2.24b**).

Although these enzymes use different mechanisms involving either deimination or demethylation, the result is the same and demonstrates that as with acetylation histone methylation is a reversible process.

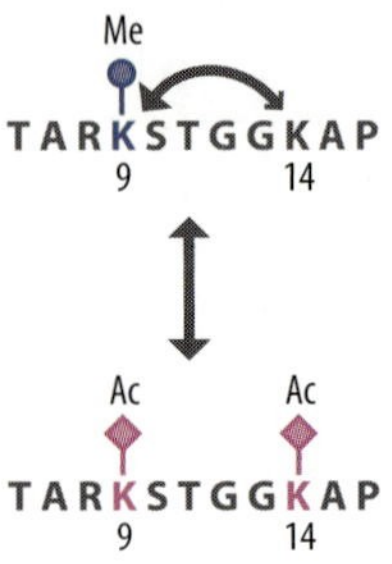

(c) Ubiquitination and sumoylation

As well as being modified by acetylation or methylation of their free amino group, lysine residues can also be modified by ubiquitination of this amino group. Unlike other modifications, however, ubiquitination does not involve

Figure 2.23
Short region of histone H3 (dashed underline in Figure 2.21) showing the lysines at positions 9 and 14. Interaction occurs between the modifications at the two positions so that methylation at position 9 inhibits acetylation at position 14, as well as preventing acetylation at position 9 itself.

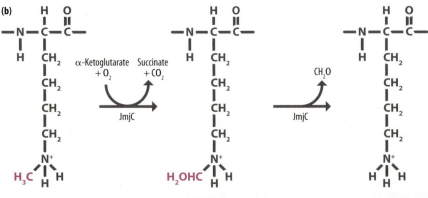

Figure 2.24
Mechanisms by which a methyl group can be removed from a lysine amino acid. (a) The enzyme LSD1 catalyzes an amine oxidase reaction in which two hydrogen atoms are transferred to the FAD co-factor producing an imine intermediate from monomethyl-lysine. The intermediate then undergoes hydrolysis to release formaldehyde (CH_2O) and unmethylated lysine. (b) The JmjC enzymes catalyze an oxidative demethylation reaction producing a hydroxymethyl-lysine intermediate with release of succinate and CO_2. The intermediate then releases formaldehyde (CH_2O) to generate unmethylated lysine.

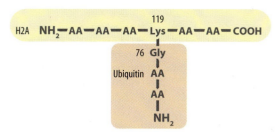

Figure 2.25
Linkage of ubiquitin to histone H2A. The C-terminal amino acid of ubiquitin (76) links to the lysine at position 119 of H2A. The amino acid backbone of the molecule is indicated by AA.

the addition of a small chemical group such as a methyl or acetyl residue. Rather, it involves the linkage of a lysine residue in the histone to the free amino group of the 76 amino acid protein **ubiquitin** to form a branched molecule (**Figure 2.25**).

Unlike acetylation or methylation, ubiquitination only occurs on H2A and H2B. Moreover, it targets only one lysine molecule in each histone, amino acid 119 in H2A and 120 in H2B. Hence, the ubiquitin molecule is located some distance away from the methyl and acetyl groups which as discussed above are located predominantly near the N-terminus of the histone molecules.

Although in some other proteins the addition of ubiquitin targets a protein for degradation, this is not the case for ubiquitinated H2A and H2B which remain stable proteins. As with acetylation, however, ubiquitination reduces the positive charge on the histone molecules since it removes the positive charge on the modified lysine residue. In addition, the ubiquitin protein itself contains a number of negatively charged amino acids which can neutralize the positively charged amino acids in the histone molecule.

As with the other modifications discussed so far, ubiquitination of histones can affect the modification of histones at other positions. Interestingly, ubiquitination of histone H2B actually promotes the methylation of a different histone, H3, at the lysines located at positions 4 and 79. The other cases we have discussed so far, involve a **histone modification** promoting further modifications of the same histone molecule (**Figure 2.26a**) whereas this example indicates that a histone modification can promote the modification of a different histone (**Figure 2.26b**).

As well as being modified by addition of ubiquitin, lysine residues on histones can also be modified by addition of small ubiquitin-related modifier (**SUMO**), which as its name suggests is a small protein related to ubiquitin. Interestingly, it appears that sumoylation of histones can lead to the recruitment of HDAC enzymes. This results in deacetylation of the sumoylated histone, providing another link between two different histone modifications (**Figure 2.27**).

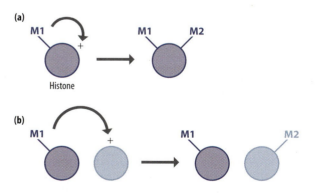

Figure 2.26
Post-translational modification of a histone molecule (M1) can stimulate further modification (M2) of the same histone (a) or a different histone (b).

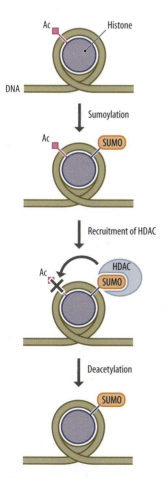

Figure 2.27
Addition of SUMO to a histone molecule can recruit a HDAC enzyme leading to deacetylation of the histone.

(d) Phosphorylation

Unlike other histone modifications discussed so far, **phosphorylation** of histones targets serine or threonine rather than lysine residues. Phosphorylation of histones represents an aspect of the very widespread regulation of cellular function by the phosphorylation of these residues in a number of different proteins (see Section 8.2 for discussion of the regulation of **transcription factor** activity by phosphorylation).

As with acetylation and methylation, phosphorylation occurs for all four core histones (H2A, H2B, H3, and H4) and occurs at the N-terminus of the protein close to the acetylated and methylated lysine residues. The pattern of phosphorylation, methylation, and acetylation at the N-terminus of histones H3 and H4 is illustrated in **Figure 2.28**.

The close association of these various modifications results in interaction between them. For example, phosphorylation of histone H3 on serine 10 is associated with acetylation of the adjacent lysine 14. Indeed, the demethylation of position 9 in histone H3 described above (see Figure 2.23) is accompanied not only by acetylation of lysines 9 and 14 but also by the phosphorylation of serine 10 (**Figure 2.29**).

Since the added phosphate molecule is negatively charged, phosphorylation – like acetylation or ubiquitination – will partially neutralize the net positive charge of the modified histone molecules and therefore affect their ability to interact with the negatively charged DNA or with other histones in adjacent nucleosomes (**Figure 2.30a**).

In addition to such charge effects, however, it is also likely that the post-translational modifications of the histones affect their interaction with regulatory proteins. Indeed, the linkage between the different modifications described throughout this section has led to the idea of a "histone code" in which individual regulatory proteins may recognize multiple modifications on histone molecules (**Figure 2.30b**). The findings discussed in this section therefore indicate the features of histone modifications which allow them to affect the interaction of the histones with DNA and other histones or with regulatory proteins, as suggested at the beginning of this section (compare Figures 2.16 and 2.30).

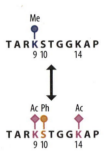

Figure 2.29
Short region of histone H3 (dashed underline in Figure 2.28) showing the lysines at positions 9 and 14, as well as the serine at position 10. Demethylation of lysine 9 promotes phosphorylation of serine 10, as well as acetylation of lysines 9 and 14.

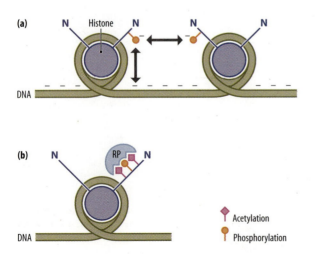

Figure 2.30
Histone modifications, such as phosphorylation, could modulate chromatin structure by altering the net charge on the histone molecule. This could affect its interaction with other histones or with the negatively charged DNA, for example by negative charges repelling one another (arrows) (a). In addition, however, associated modifications, such as acetylation and phosphorylation, could constitute a histone code which is recognized by regulatory proteins (RP) (b). Compare with Figure 2.16.

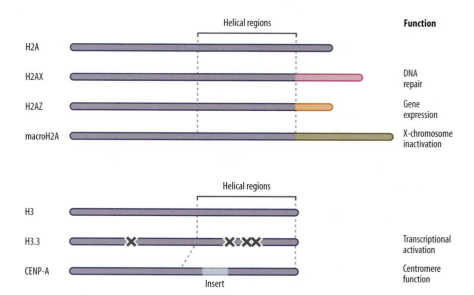

Figure 2.31
Comparison of the standard forms of histones H2A or H3 with some of their different variants. Note the different C-termini of the H2A variants. The position of the four amino acids which differ in H3.3 compared to histone H3 is indicated, as is the extra region present in CENP-A. The special function of each of the variants is also indicated.

Histone variants are encoded by distinct genes to those encoding the standard histone isoforms

The standard isoforms of the four core histones (H2A, H2B, H3, and H4), whether modified by post-translational modifications or not, constitute the great bulk of the histones in the cell. However, in all cases except H4, minor variant forms of the histones exist which show a number of amino acid differences from the standard forms of the histones. The degree of similarity between variant and the standard form of the histone varies (**Figure 2.31**). For example, the histone H3 variant CENP-A shows only 46% identity with histone H3, whereas another H3 variant, H3.3, differs by only four amino acids from H3, making it 96% identical. The variant forms of the histones are encoded by different genes from those encoding the standard forms of the histones and are less well conserved evolutionarily than the standard forms.

Nonetheless, these different **histone variants** have important functions, in particular chromosomal processes where **nucleosome remodeling** (see Section 2.2 and Figure 2.15) results in their replacing one or more of the standard histones. For example, the H3 variant CENP-A plays a key role in organizing chromatin structure at the **centromere**, the specialized region by which the chromosome attaches to the spindle during cell division (see Section 2.5).

Similarly, the H2A variant H2AX is associated with regions where DNA damage needs to be repaired. H2AX has a C-terminal region which is not found in H2A (see Figure 2.31) and this region contains a serine residue that can be phosphorylated. It appears that phosphorylation of this serine acts as a marker of damaged DNA and recruits the enzymes required for DNA repair (**Figure 2.32**). In agreement with this, loss of H2AX leads to increased chromosomal rearrangements due to the failure to repair damaged DNA. This example therefore combines the use of a variant histone with its modification by phosphorylation.

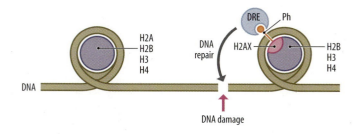

Figure 2.32
In regions of DNA damage, the histone variant H2AX replaces H2A. H2AX can be phosphorylated (Ph) on a serine residue not found in H2A. This recruits DNA repair enzymes (DRE) which repair the DNA.

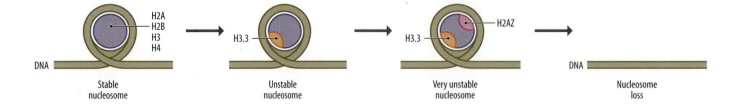

In terms of our primary interest in gene control, the most interesting histone variants are the H2A variant H2AZ and the H3 variant H3.3. Both these variants appear to localize to regions where genes are being actively transcribed and may therefore play a role in producing a more open chromatin structure. Nucleosomes containing either H2AZ or H3.3 in place of H2A or H3, respectively, appear to be more unstable than standard nucleosomes. This instability is enhanced in nucleosomes containing both H2AZ and H3.3, suggesting that their presence may result in loss of nucleosomes from specific regions of DNA, allowing transcriptional regulatory proteins to bind (**Figure 2.33**).

Figure 2.33
The replacement of histone H3 with the variant H3.3 and of histone H2A with the variant H2AZ makes the nucleosome progressively more unstable and may produce loss of nucleosomes from regions of DNA which bind transcriptional regulatory proteins.

2.4 THE 30 nm CHROMATIN FIBER

The 30 nm fiber represents a further compaction of the beads-on-a-string structure

Although the beads-on-a-string structure discussed in Section 2.2 produces a compaction of the DNA by winding it around the histone octamer, further compaction is still required. Indeed, when carefully extracted from cells using buffers with the same salt concentration as is found in the cell, chromatin is seen as a fiber of approximately 30 nm in diameter. This is a shorter, more compact structure than the 10 nm-diameter beads-on-a-string form which is isolated under conditions of low ionic strength (see Section 2.2). This indicates that the nucleosomes have associated more closely together to form a shorter, but thicker fiber.

Initial studies of the structure of the 30 nm fiber suggested that it had a **solenoid** structure (**Figure 2.34**). In this structure, the linear array of nucleosomes in the beads on a string is thrown into a helical structure in which nucleosomes are stacked above one another, producing a more compact structure. As this structure is a single helix with only one starting point, it is known as a one-start helix.

More recently, however, studies of the crystal structure of a tetranucleosome (four successive nucleosomes linked by DNA) have suggested an alternative structure, which is known as the two-start helix (**Figure 2.35**). In this model, the nucleosomes form a **zigzag ribbon** structure, with alternative nucleosomes being in different stands of the ribbon. The two strands of the ribbon are then wound into a double-helical ribbon structure (**Figure 2.36**). This has two strands of nucleosomes wound around one another which is why it is known as a two-start helix.

Although the density of the 30 nm fiber itself makes it difficult to determine its structure exactly, it appears that the two-start model is more likely to be correct. As well as being based on structural studies of the tetranucleosome, it also agrees with cross-linking studies on arrays of 12 nucleosomes. These suggested that the 30 nm fiber contains two rows of six nucleosomes rather than a single stack of 12 nucleosomes.

Figure 2.34
The solenoid or one-start helix model for the structure of the 30 nm chromatin fiber. From McGhee JD, Nickol JM, Felsenfeld G & Rau DC (1983) *Cell* 33, 831–841. With permission from Elsevier.

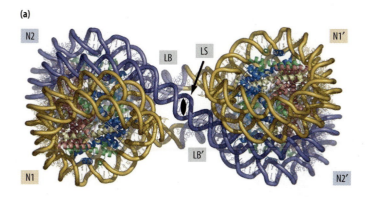

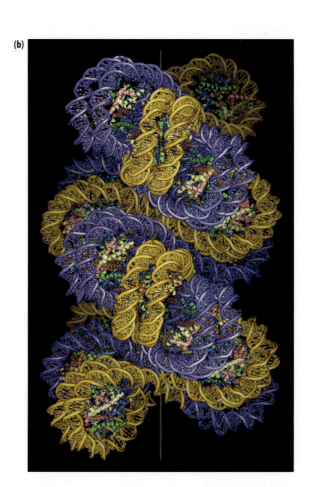

Figure 2.35
Two-start helix model of the 30 nm fiber. Panel (a) shows the crystal structure of a tetranucleosome. Panel (b) shows the model for the chromatin fiber derived by stacking individual four nucleosome structures (as shown in a) on top of one another. N, nucleosome. Courtesy of Tim Richmond, from Schalch T, Duda S, Sargent DF & Richmond TJ (2005) *Nature* 436, 138-141. With permission from Macmillan Publishers Ltd.

Histone H1 and post-translational modifications of the other histones are involved in the formation of the 30 nm fiber

Much of the DNA in the cell appears to exist in the 30 nm-fiber structure. Thus, an understanding of the factors involved in producing this structure from the 10 nm beads-on-a-string structure is of considerable importance. Moreover, as will be discussed in Section 3.1, DNA which is about to be transcribed adopts the beads-on-a-string structure. The factors which control the interconversion of the 10 and 30 nm fibers are of critical importance therefore in terms of both chromatin structure and gene control.

It appears that the histone modifications discussed in Section 2.3 play a critical role in the transition between the 30 nm fiber and the beads-on-a-string structure. As noted above, the N-terminal ends of the core histones extend as tails from the nucleosome particle, allowing them to promote the

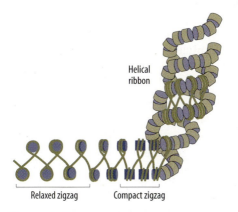

Figure 2.36
Schematic diagram showing how a zig-zag arrangement of nucleosomes can be compacted and converted into a helical ribbon. Courtesy of JB Rattner, from Woodcock CL, Frado LL & Rattner JB (1984) *J. Cell Biol.* 99, 42–52. With permission from Rockefeller University Press.

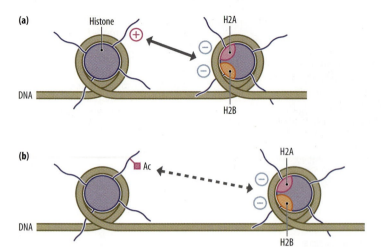

(a)

Histone

H2A

DNA

H2B

(b)

H2A

Ac

DNA

H2B

Figure 2.37
A positively charged region at the N-terminus of histone H4 interacts with a negatively charged region on histones H2A and H2B in the adjacent nucleosome and therefore promotes closer packing of the nucleosomes (a). Acetylation (Ac) of lysine 16 in histone H4 reduces the positive charge of this region and therefore promotes a more open chromatin structure (b).

interactions between adjacent nucleosomes which are required to form higher-order chromatin structures. It has been shown, for example, that the positively charged region at the N-terminus of histone H4, and particularly the lysine at position 16, interacts with a negatively charged cluster of seven amino acids on the surface of histone H2A–H2B in the adjacent nucleosome. This has the effect of pulling adjacent nucleosomes together (**Figure 2.37a**).

As noted in Section 2.3, lysine 16 of histone H4 is a target for modification by acetylation. This reduces the positive charge on this region of H4 and potentially weakens the interaction with the negatively charged region of H2A–H2B in the adjacent nucleosome and so promotes a more open chromatin structure (**Figure 2.37b**).

Histone H1 also plays a critical role in promoting the more tightly packed chromatin structure in the 30 nm fiber. Indeed as illustrated in Figure 2.12, when visualized in the electron microscope chromatin appears more tightly packed in the presence of histone H1 than in its absence. As indicated in Table 2.1, histone H1 is present in the cell in half the amount of each of the four core histones and it is not part of the histone octamer. Rather, it is present at the level of one molecule per nucleosome. It is thought that the central globular region of histone H1 contacts the linker DNA as it exits from the nucleosome and changes its path to form a more compact structure (**Figure 2.38**). This has the effect of sealing the two turns which the DNA makes around the octamer of the core histones and appears to be crucial for the formation of the 30 nm fiber.

In agreement with this idea, histone H1 is enriched in regions of DNA which are not being transcribed whereas it is depleted in regions where transcription is occurring and which have a less tightly packed chromatin structure. Interestingly, it has been shown that a reciprocal relationship exists between histone H1 and the poly(ADP-ribose) polymerase-1 (PARP-1), with PARP-1 being enriched in regions where H1 is depleted and vice versa. This correlates with studies on the pS2 gene, which is inducible by the steroid hormone estrogen. Induction of this gene by estrogen involves the displacement of histone H1 by PARP-1 and topoisomerase IIβ. This results in a transient double-stranded cut being made in the DNA by topisomerase IIβ. In turn this allows the DNA to unwind and alter its chromatin structure so that transcription can occur, with the break being resealed once the chromatin structure has altered (**Figure 2.39**).

It is clear, therefore, that both histone H1 and post-translational modifications of the core histones play key roles in the transition between the

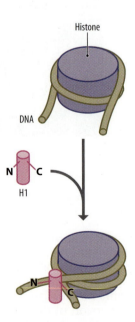

Histone

DNA

N   C

H1

N

C

Figure 2.38
The globular region of histone H1 (pink) interacts with the linker DNA as it exits the nucleosome and changes its path to produce a more compact structure.

30 nm fiber and the 10 nm beads-on-a-string structure, with this transition being required for transcription to occur.

Interestingly, in mature chicken erythrocytes, histone H1 is replaced by histone H5, which produces a very highly compact chromatin structure from which no transcription occurs. This reduces the portion of the cell occupied by the nucleus and allows the maximum space in the erythrocyte for its primary function, namely carrying oxygen. It therefore fulfils the same purpose as the total loss of the nucleus which occurs in mature mammalian erythrocytes (see Section 1.3).

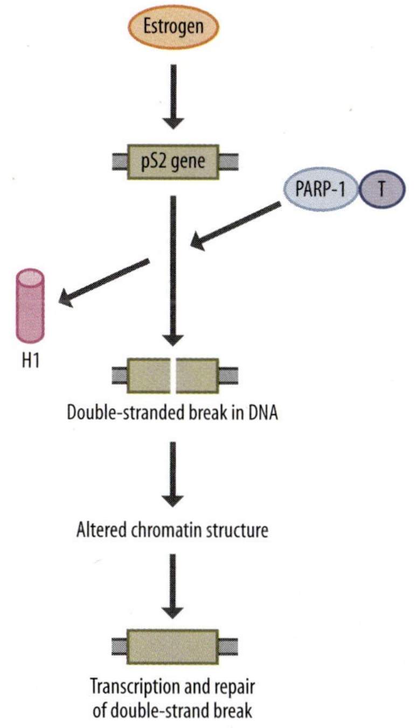

2.5 STRUCTURAL AND FUNCTIONAL DOMAINS IN CHROMATIN

The 30 nm fiber is further compacted by looping

Although the 30 nm fiber is the fundamental structure packaging most of the DNA in the cell, it appears not to exist as a simple linear structure. Rather, it is thrown into a series of loops of different sizes but normally each containing between 50,000 and 200,000 bp of DNA (**Figure 2.40**). As in the conversion of the 10 nm beads-on-a-string structure into the 30 nm fiber, this looped structure is shorter and thicker than the 30 nm fiber.

These loops are attached to the **nuclear matrix**, a network of RNA and protein fibrils which extends throughout the entire nucleus. Attachment is mediated via specific AT-rich regions of DNA, known as **matrix-attachment regions** (MARs) or scaffold-attachment regions (SARs), which are located at the base of the DNA loops (**Figure 2.41**).

Each of these loops therefore constitutes a structural domain of the chromatin. In some but not all cases these structural domains appear to correspond to functional domains in which a particular loop of DNA moves from the 30 nm-fiber structure to the beads-on-a-string structure, prior to the onset of transcription itself (**Figure 2.42**). As will be discussed in Section 4.1, such looping also allows DNA that is about to be active to extend outwards from the bulk of the chromatin, allowing it to localize to specific regions of the nucleus where transcription actually occurs (see Figure 4.21).

Specific functional domains of chromatin thus exist in which chromatin structure is altered in a particular situation to allow transcription to occur. The existence of such functional domains, whether or not they correspond to structural domains or not, raises the question of how their boundaries are defined, allowing two regions of chromatin adjacent to one another to have two distinct structures; that is, the beads-on-a-string structure and the 30 nm-fiber structure.

Figure 2.39
Estrogen-mediated activation of the pS2 gene involves displacement of histone H1 by PARP-1 and topisomerase IIβ (T). This results in a transient double-stranded break being produced in the DNA, which allows it to unwind and alter its chromatin structure. In turn this facilitates transcription of the pS2 gene.

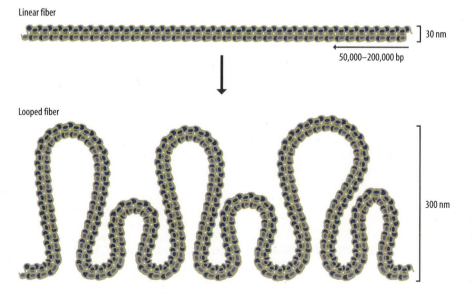

Linear fiber

30 nm

50,000–200,000 bp

Looped fiber

300 nm

Figure 2.40
The 30 nm fiber can be thrown into a series of loops, each of approximately 50,000–200,000 bp in size, producing a looped fiber with a diameter of approximately 300 nm.

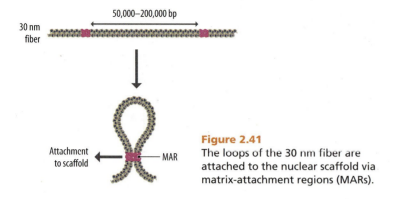

30 nm fiber

50,000–200,000 bp

Attachment to scaffold ← MAR

Figure 2.41
The loops of the 30 nm fiber are attached to the nuclear scaffold via matrix-attachment regions (MARs).

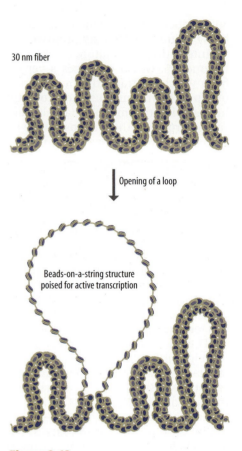

30 nm fiber

Opening of a loop

Beads-on-a-string structure poised for active transcription

Figure 2.42
An individual loop can alter its structure from that of the 30 nm fiber to the beads-on-a-string structure, allowing transcription to occur.

Two types of DNA element which mediate these boundary effects will be discussed in turn.

Locus-control regions regulate the chromatin structure of a large region of DNA

When specific genes are introduced into fertilized mouse eggs and used to create transgenic mice, the introduced genes are frequently expressed at very low levels and this expression is not increased when multiple copies of the genes are introduced. Similarly, identical copies of the introduced gene which integrate into the host chromosomes at different positions are expressed at very different levels, suggesting that gene activity is being influenced by adjacent chromosomal regions. This effect occurs even when the gene is introduced together with its adjacent regulatory elements. This has led to the concept that some DNA sequences, which are necessary for high-level gene expression independent of the position of the gene in the genome, are absent from the introduced gene.

This idea has been supported by studies in the mammalian β-globin gene cluster, which contains a non-functional pseudogene and four functional β-globin-like genes that are expressed successively with α-globin in erythroid cells during the development of the organism. Thus, ε-globin is expressed in the early embryo, the two γ-globin genes (which differ at only a single position where Gγ has a glycine and Aγ has an alanine) in the later embryo, and β-globin (together with a small amount of δ-globin) in the adult (**Figure 2.43**).

A region located 10–20 kb upstream of the β-globin genes has been shown to confer high-level, position-independent expression when linked to a single globin gene and introduced into transgenic mice. Moreover, this element acts in a tissue-specific manner since its presence allows the globin gene to be expressed in the correct pattern with high-level expression in erythroid cells and not in other cell types.

Most importantly this element is not only required for expression of the β-globin genes in transgenic mice but also plays a role in the natural expression of the globin genes. Its deletion in human individuals leads to a lack of expression of any of the genes in the cluster, resulting in a lethal disease known as Hispanic thalassemia in which no functional hemoglobin is produced (see Figure 2.43). This occurs even in cases where all the genes together with their promoter and enhancer regions remain intact.

The role of this element in stimulating activity of all the genes in the β-globin cluster has led to its being called a **locus-control region** (LCR). Following the original definition of an LCR in the β-globin cluster, similar LCR elements have been identified in other genes including the α-globin cluster, the major histocompatibility locus, and the CD2 and lysozyme

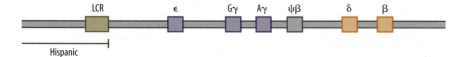

LCR   ε   Gγ   Aγ   ψβ   δ   β

Hispanic

Figure 2.43
Organization of the β-globin gene locus showing the position of the locus-control region (LCR) (green). The functional genes (purple and orange) encoding epsilon-globin (ε), the two forms of gamma-globin (Gγ and Aγ) delta globin (δ) and beta globin (β), and the non-functional β-globin-like pseudogene (gray) are indicated. The extent of the deletion in patients with Hispanic thalassemia is indicated by the line. Note that this deletion removes the LCR but leaves the genes themselves intact.

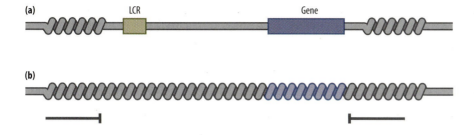

(a)   LCR   Gene

(b)

Figure 2.44
An inserted gene containing an LCR will be organized into an open chromatin conformation compared to flanking DNA (a). In contrast an inserted gene lacking the LCR will be subject to repression by adjacent regions which direct its organization into a closed chromatin organization (b).

genes. It is clear that LCRs constitute an important element essential for the correct regulation of gene expression.

It is likely that LCRs function by directing a structural change in a region of chromatin into a more open chromatin structure in specific cell types. In the β-globin case, this would occur in erythroid cells allowing the β-globin-like genes to be successively expressed in embryonic development and in the adult. In agreement with this, in many cases regions with LCR activity do not affect gene activity when introduced transiently into cells under conditions where the exogenous DNA is not packaged into chromatin but can do so when the same gene construct integrates into the host chromosome.

The presence of the LCR would render the adjacent DNA capable of being expressed in a position-independent manner, allowing high-level tissue-specific expression of the gene in transgenic mice regardless of the position in the genome into which it integrated. In contrast a gene lacking this sequence would lack this stimulatory action and would be subject to the influence of adjacent regulatory elements which might inhibit its expression (**Figure 2.44**). In agreement with this idea, a transgene containing the CD2 LCR is not inactivated when inserted into a tightly packed region of the DNA known as **heterochromatin** (see below), whereas this does occur when the same gene is inserted into this region without the LCR.

The link between LCRs and chromatin structure has been supported by detailed studies of the β-globin gene cluster at different stages of erythroid development. It has been shown that in early erythroid development the cluster forms a looped structure in which the active γ-globin genes are brought into close proximity with the LCR. In contrast, later in erythroid development at the stage when β-globin itself is expressed, the loop structure changes so that it is now the active β-globin and δ-globin genes which are associated with the LCR (**Figure 2.45**). This has led to the idea of an active chromatin hub at which the regulatory elements in the LCR are brought together with regulatory elements of the individual gene itself to ensure that each gene is actively transcribed at the appropriate stage of development.

A similar change in looping pattern, controlled by an LCR, is seen during **T-lymphocyte** development and involves a cluster of genes encoding **interleukins** 4, 5, and 13 and their associated LCR. In early T-cell development prior to the expression of any of these genes, the chromatin that contains them is actually associated with the chromatin region containing the **interferon**-γ gene, which is also not expressed at this stage. This is an inter-chromosomal association, since the interferon-γ gene is actually on a different chromosome to that containing the interleukin gene cluster (**Figure 2.46**). When **T cells** differentiate to produce cells expressing interleukins 4, 5, and 13, this inter-chromosomal association is lost. Instead, the chromatin in the interleukin gene region is organized into a series of small loops which allow the interleukin genes to be expressed (see Figure 2.46).

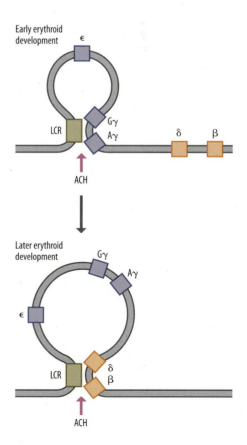

Early erythroid development

ε

LCR   Gγ   Aγ   δ   β

ACH

Later erythroid development

Gγ   Aγ

ε

LCR   δ   β

ACH

Figure 2.45
Organization of an active chromatin hub (ACH) in the β-globin gene cluster. At each stage of erythroid development, the looping of the cluster brings together the genes being transcribed with the LCR. Early in erythroid development the two γ-globin genes (Aγ and Gγ) are close to the LCR and are transcribed, whereas later in development this association with the LCR involves the β and δ globin genes, which are transcribed at this stage.

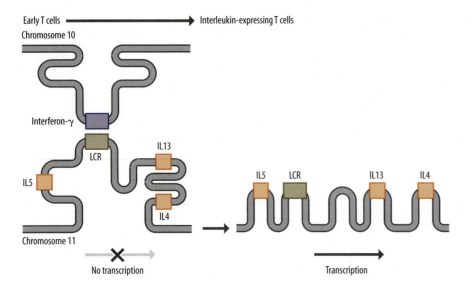

Early T cells ——————→ Interleukin-expressing T cells

Chromosome 10

Interferon-γ

LCR

IL5

IL13

IL4

Chromosome 11

No transcription

IL5 LCR IL13 IL4

Transcription

Figure 2.46
Early in T-cell development, the interleukin gene cluster on chromosome 11 (IL4, IL5, and IL13), together with their LCR, is associated with the region on chromosome 10 containing the interferon-γ gene, and none of the genes is transcribed. When interleukin-expressing T cells are produced, this inter-chromosomal link is lost and the interleukin gene cluster is reorganized into a series of smaller loops, which allows transcription to occur.

In this case, therefore, activation of one group of genes involves the loss of an inter-chromosomal association with a gene which remains transcriptionally inactive. However, as will be discussed in Section 4.1, transcriptional activation can also result in the creation of inter-chromosomal associations between active genes on different chromosomes, which co-localize to regions of the nucleus where transcription occurs.

The organization of the interleukin genes into the structure of small loops required for transcription involves the SATB1 protein which binds at a number of sites within the interleukin gene cluster. Interestingly, SATB1 is known to bind at MARs (see above) and play a key role in looping the chromatin and anchoring the loops at their base to the nuclear matrix. A number of other LCRs also contain MAR sequences. It is clear therefore that the LCR plays a key role in directing looping of the chromatin into structural and functional domains attached to the nuclear scaffold with the chromatin structure of each domain being regulated as a separate unit. In agreement with this idea, the MAR sequences contained within the LCR of the immunoglobulin gene have been shown to enhance the accessibility of the adjacent chromatin in B lymphocytes but not in other cell types.

Insulators block the inappropriate spread of particular chromatin structures

The existence of elements such as LCRs which can alter the chromatin structure of large regions of the chromosome leads to the question of how the action of such elements is confined to the appropriate gene or genes. Some mechanism must exist to prevent the effect of LCRs from spreading to other genes in adjacent regions of the chromosome and producing an inappropriate pattern of gene expression. Similarly, as will be discussed in Section 4.4, many genes contain enhancer elements which are able to act over long distances and which must also have their activity limited to specific regions of DNA so as to prevent inappropriate effects.

This problem is solved by the existence of **insulator** sequences which as their name implies block the spread of LCR or enhancer activity so that it is confined to the appropriate gene or genes and other adjacent genes are "insulated" from these effects. As shown in **Figure 2.47**, the region containing the LCR and the β-globin genes is flanked by two insulator sequences known as HS4 sequences which prevent the LCR inappropriately activating the flanking genes encoding the folate receptor and the odorant receptor gene (see Figure 2.47).

As with LCRs, insulator sequences can also allow position independent expression of an inserted gene in transgenic mice. Thus, they will prevent the spread of a closed chromatin structure from an adjacent region,

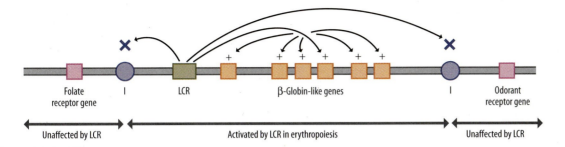

Folate receptor gene   I   LCR   β-Globin-like genes   I   Odorant receptor gene

Unaffected by LCR   Activated by LCR in erythropoiesis   Unaffected by LCR

allowing a transgene for example to be expressed regardless of the site in the genome where it has inserted.

LCRs and insulators share the ability of preventing the structure of a particular DNA region from being inappropriately influenced by that of adjacent regions. Moreover, as is the case with LCRs, some MARs can function as insulators linking the structural domains in the looped structure of chromatin with functional domains having a specific chromatin structure. Clearly, however, LCRs and insulators differ in that LCRs are involved in promoting a particular chromatin structure in a specific region of DNA whereas insulators block the spread of a particular structure from or to adjacent DNA regions (see Figure 2.47).

Heterochromatin is a very tightly packed form of chromatin

As well as preventing the inappropriate spread of open chromatin structure such as that produced by LCRs, insulators can also block the inappropriate spread of highly compact chromatin structures (**Figure 2.48**). For example, insulators can mark the boundary between the bulk of chromatin in the cell which is known as **euchromatin** and more compacted regions known as heterochromatin (**Figure 2.48a**). These regions were originally defined on the basis that heterochromatin is more condensed than euchromatin during the **interphase** stage when cells are not dividing and therefore stains more intensely with specific stains used in light microscopy.

Euchromatin contains most of the DNA in the cell and is packaged in the looped form of the 30 nm fiber. Euchromatin contains the regions of DNA which will be transcribed in specific cell types or under particular conditions. As noted above and described in more detail in Section 3.1, when particular regions of euchromatin are about to be transcribed they move into the more open beads-on-a-string structure, allowing transcription to occur.

In contrast, approximately 10% of the cell is packaged in a more compact form known as heterochromatin which is incompatible with transcription. Where an insulator is lost due to deletion or a chromosome **translocation**, the more compact structure of heterochromatin can spread into the adjacent euchromatin. This will silence the genes in the adjoining region, preventing them from being expressed at the appropriate place or time (**Figure 2.48b**).

Interestingly, when this occurs in the early *Drosophila* embryo the heterochromatin randomly spreads to a different extent depending on cell type, resulting in different numbers of adjacent genes being silenced (**Figure 2.49**). As these cells proliferate in embryonic development, they maintain this pattern of gene silencing, resulting in a phenomenon known as **position-effect** variegation, in which different groups of cells in the adult show different patterns of expressed and non-expressed genes (see Figure 2.49).

Although the precise structure of heterochromatin is uncertain, it appears to involve a more extensive looped structure compared to that found in euchromatin, with much tighter packing of the loops. Several non-histone proteins play a key role in the assembly of heterochromatin and these include heterochromatin protein one (**HP1**) and members of the **polycomb** group of proteins. Interestingly, different forms of heterochromatin appear to exist which contain different combinations of these proteins. For example, studies in *Drosophila* have identified regions of heterochromatin containing HP1 and polycomb proteins, either HP1 or polycomb alone, or neither protein

Figure 2.47
Insulator elements (I) flanking the β-globin gene cluster prevent the LCR from stimulating the expression of adjacent genes and confine its effect to the β-globin genes.

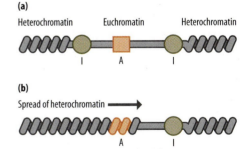

(a)
Heterochromatin   Euchromatin   Heterochromatin

I   A   I

(b)
Spread of heterochromatin ⟶

A   I

Figure 2.48
Insulator (I) elements can block the spread of the highly compact structure of heterochromatin into adjacent areas of euchromatin (a). Loss of an insulator results in spread of heterochromatin into adjacent areas and will prevent expression of a gene in this region (A) due to its being packaged into heterochromatin (b).

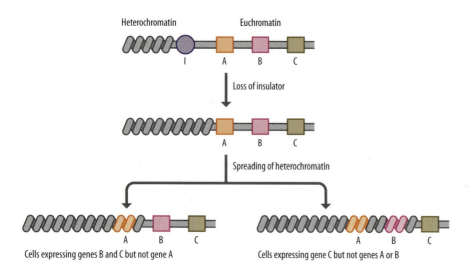

Figure 2.49
Spreading of heterochromatin due to loss of an insulator (I) can occur to a different extent in different cells in embryonic development, resulting in groups of cells with different patterns of expressed and non-expressed genes.

(see Sections 3.3 and 3.4 for further discussion of the role of HP1 and polycomb proteins in the regulation of chromatin structure).

The chromosome is the visible result of chromatin compaction

During the interphase stage when cells are not dividing, the bulk of chromatin is in the euchromatin structure, with some regions being more tightly packaged in heterochromatin. The heterochromatin regions including the centromere where DNA will attach to the mitotic spindle during cell division and the **telomeres** at the ends of the DNA molecule (**Figure 2.50**). As noted in Section 2.3, the histone H3 variant CENP-A plays a key role in the formation of centromeric heterochromatin.

As cells prepare for cell division, the entire DNA molecule undergoes further compaction, in a process involving proteins known as **condensins**, to form the chromosomes. Due to their very compact nature, the chromosomes are visible in the light microscope (Figure 2.50 and **Figure 2.51**). This compact structure is incompatible with gene expression but facilitates the accurate partitioning of the replicated DNA to the daughter cells produced by cell division.

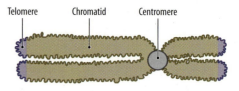

Figure 2.50
Structure of a chromosome during cell division to show the areas of heterochromatin at the centromere and the telomeres. The DNA has already replicated, so the chromosome therefore contains two sister chromatids, which will separate during cell division. Note that during interphase between cell divisions, the bulk of the DNA has a less compact structure, known as euchromatin, whereas the centromeres and telomeres maintain the more compact heterochromatin structure.

CONCLUSIONS

The extended form of DNA in the double-helical structure would be vastly too large to fit into the cell or even the organism. In this chapter we have seen how the DNA undergoes a series of compactions in association with specific proteins to form the looped structure of interphase chromatin and, ultimately, the chromosomes which are visible during cell division (**Figure 2.52**). This allows the double-helical DNA of a single cell, which would have a length of 2 m if fully extended, to fit within the nucleus which has a diameter of 6 μm.

Most importantly, however, this progressive hierarchy in chromatin structure is also used in the regulation of gene expression. For example, genes which are active or potentially active in a particular cell type move into the more open beads-on-a-string structure from the 30 nm fiber characteristic of non-active genes. The information on chromatin structure described in this chapter is therefore essential to our understanding of its role in the regulation of gene transcription, which is discussed in the next chapter.

Figure 2.51
Human chromosome spread stained with Giemsa stain. The upper panel shows a chromosome spread, and in the lower panel the chromosomes have been arranged in order of their chromosome number. Courtesy of AT Sumner, MRC Human Genetics Unit.

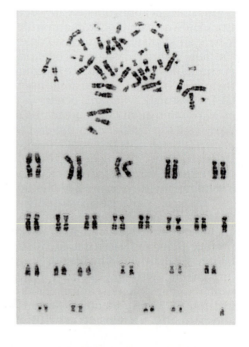

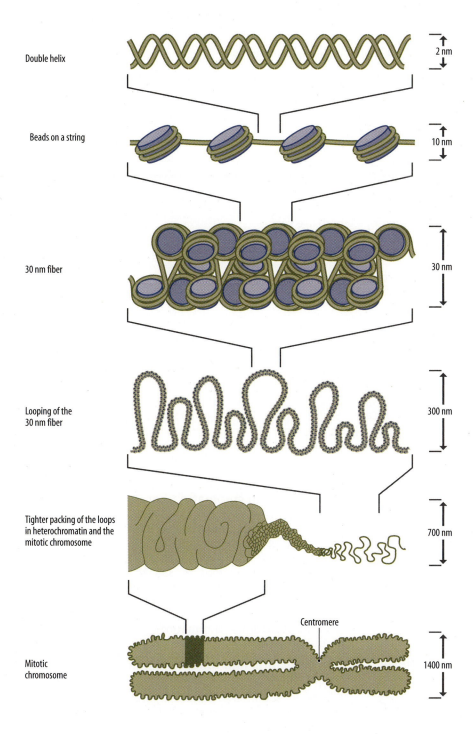

Double helix

Beads on a string

30 nm fiber

Looping of the
30 nm fiber

Tighter packing of the loops
in heterochromatin and the
mitotic chromosome

Mitotic
chromosome

2 nm

10 nm

30 nm

300 nm

700 nm

1400 nm

Centromere

Figure 2.52
Stages in the compaction of chromatin from the DNA double-helix to the mitotic chromosome.

KEY CONCEPTS

- In eukaryotes, short-term gene-control processes need to interact with long-term processes which determine the commitment of cells to become or remain a particular differentiated cell type.

- Such long-term control processes involve the manner in which particular regions of DNA are packaged with proteins into a structure known as chromatin.

- Packaging DNA into chromatin compacts it, so allowing it to fit within the nucleus, but packaging is also used to regulate gene expression.

- The negatively charged DNA associates with positively charged histone proteins to allow such compaction to occur.

- The first level of compaction involves association of the DNA with the core histones H2A, H2B, H3, and H4 to form a structure known as the nucleosome.

- This produces a 10 nm fiber with a beads-on-a-string structure of nucleosomes and DNA.

- The histones can be modified by various post-translational modifications and variant forms also exist which play different roles.

- Further compaction, involving histone H1 and regulation of histone modification, produces a 30 nm fiber with a helical structure of nucleosomes.

- This 30 nm fiber is further compacted by forming loops which are very closely packed in heterochromatin and in the **mitotic chromosome**, producing the most compact form of chromatin.

- LCRs regulate the chromatin structure of a region of DNA.

- Insulator sequences prevent the inappropriate spread of chromatin structure to adjacent DNA regions.

FURTHER READING

2.1   Commitment to the differentiated state and its stability

Coon HG (1966) Clonal stability and phenotypic expression of chick cartilage cells *in vitro*. *Proc. Natl Acad. Sci. USA* 55, 66–73.

Hadorn E (1968) Transdetermination in cells. *Sci. Am.* 219, 110–120.

2.2   The nucleosome

Khorasanizadeh S (2004) The nucleosome: from genomic organization to genomic regulation. *Cell* 116, 259–272.

Li B, Carey M & Workman JL (2007) The role of chromatin during transcription. *Cell* 128, 707–719.

Richmond TJ & Davey CA (2004) The structure of DNA in the nucleosome core. *Nature* 423, 145–150.

Saha A, Wittmeyer J & Cairns BR (2006) Chromatin remodelling: the industrial revolution of DNA around histones. *Nat. Rev. Mol. Cell Biol.* 7, 437–447.

2.3   Histone modifications and histone variants

Jin J, Cai Y, Li B et al. (2005) In and out: histone variant exchange in chromatin. *Trends Biochem. Sci.* 30, 680–687.

Klose RJ & Zhang Y (2007) Regulation of histone methylation by demethylimination and demethylation. *Nat. Rev. Mol. Cell Biol.* 8, 307–318.

Kouzarides T (2007) Chromatin modifications and their function. *Cell* 128, 693–705.

Lee KK & Workman JL (2007) Histone acetyltransferase complexes: one size doesn't fit all. *Nat. Rev. Mol. Cell Biol.* 8, 284–295.

Sims III RJ & Reinberg D (2008) Is there a code embedded in proteins that is based on post-translational modifications? *Nat. Rev. Mol. Cell Biol.* 9, 815–820.

Suganuma T & Workman JL (2008) Crosstalk among histone modifications. *Cell* 135, 604–607.

2.4   The 30 nm chromatin fiber

Lis JT & Kraus WL (2006) Promoter cleavage: a topoIIβ and PARP-1 collaboration. *Cell* 125, 1225–1227.

Robinson PJ & Rhodes D (2006) Structure of the '30 nm' chromatin fibre: a key role for the linker histone. *Curr. Opin. Struct. Biol.* 16, 336–343.

Tremethick DJ (2007) Higher-order structures of chromatin: the elusive 30 nm fiber. *Cell* 128, 651–654.

Woodcock CL, Frado LL & Rattner JB (1984) The higher-order structure of chromatin: evidence for a helical ribbon arrangement. *J. Cell. Biol.* 99, 42–52.

2.5   Structural and functional domains in chromatin

Allshire RC & Karpen GH (2008) Epigenetic regulation of centromeric chromatin: old dogs, new tricks? *Nat. Rev. Genet.* 9, 923–937.

Bushey AM, Dorman ER & Corces VG (2008) Chromatin insulators: regulatory mechanisms and epigenetic inheritance. *Mol. Cell* 32, 1–9.

Dean A (2006) On a chromosome far, far away: LCRs and gene expression. *Trends Genet.* 22, 38–45.

Gaszner M & Felsenfeld G (2006) Insulators: exploiting transcriptional and epigenetic mechanisms. *Nat. Rev. Genet.* 7, 703–713.

Gondor A & Ohlsson R (2006) Transcription in the loop. *Nat. Genet.* 38, 1229–1230.

Peters JM, Tedeschi A & Schmitz J (2008) The cohesin complex and its roles in chromosome biology. *Genes Dev.* 22, 3089–3114.

Role of Chromatin Structure in Gene Control

3

INTRODUCTION

As discussed in Chapter 1, the processes which control gene expression during cellular differentiation cannot be irreversible since it is possible for the DNA genome of a differentiated cell to give rise to an entire organism containing a wide variety of different cell types. This can occur, for example, when differentiated plant cells are cultured under certain conditions or when the nucleus of differentiated animal cells is transplanted into an **oocyte** (see Section 1.3).

Similarly, under certain conditions one type of differentiated cell can give rise to another differentiated cell type. This is seen, for example, in Amphibia in a process known as Wolffian lens regeneration or **transdifferentiation**. If the lens of the eye is surgically removed in a frog, the adjacent iris cells surrounding the lens lose their differentiated characteristics, proliferate, and then differentiate into lens cells to replace those which have been lost. The new lens cells are indistinguishable from the original lens cells in their pattern of gene expression, for example expressing high levels of proteins such as crystallins, which are the major lens proteins.

Gene-regulatory processes cannot therefore be irreversible and this eliminates models such as DNA loss (see Section 1.3). Nonetheless, such processes must be stable. In general one type of differentiated cell does not spontaneously change into a different cell type. Similarly, in many cases where differentiated cells undergo cell division, their differentiated characteristics are inherited by the resulting daughter cells.

As well as explaining the stability of the differentiated state, these long-term gene regulatory processes must also explain the experiments discussed in Chapter 2 (Section 2.1), which indicate that cells undergo a process of commitment to a particular differentiated state before actually differentiating. Moreover, they can maintain such commitment even when placed under conditions where they lose the phenotypic features of the differentiated state (see Section 2.1).

The stable but not irreversible nature of long-term regulatory processes has led to the idea that they involve changes in the structure of chromatin. These changes would facilitate transcription of specific genes and would occur prior to the actual process of gene transcription. Rather than involving irreversible changes to the DNA sequence itself, this would involve stable but reversible modifications of the DNA or of chromatin proteins, such as histones with which it is associated (**Figure 3.1**).

These reversible modifications are referred to as epigenetic changes. *Epi* is Greek for "on" and these changes are superimposed on the DNA rather than representing genetic changes in the DNA itself. Unlike genetic changes, epigenetic changes are potentially reversible, accounting for the stable but not irreversible nature of long-term gene control processes. As we shall discuss later in this chapter, changes in these epigenetic modifications often require a cell division to replicate the DNA which then allows alterations in

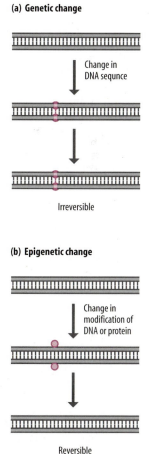

(a) Genetic change

Change in DNA sequnce

Irreversible

(b) Epigenetic change

Change in modification of DNA or protein

Reversible

Figure 3.1
Genetic changes to the sequence of the DNA are likely to be irreversible (a) whereas epigenetic changes modifying the DNA, or the proteins associated with it, are potentially reversible (b).

the modification of the DNA itself or the proteins associated with it. This requirement relates well to the observation, discussed in Section 1.3, that cultured differentiated phloem cells from plants can dedifferentiate and then proliferate prior to giving rise to all the different cells of the plant. Similarly, as noted above, in Wolffian lens regeneration, differentiated iris cells do not change directly into lens cells. Rather they first dedifferentiate and proliferate prior to differentiating into lens cells.

The structure of chromatin discussed in Chapter 2 is a key target for regulatory processes. These processes produce epigenetic modifications of the DNA itself and/or the proteins associated with it, which in turn alter chromatin structure. This chapter will discuss the role of changes in chromatin structure in the regulation of gene transcription and the epigenetic modifications which are involved in producing these changes.

3.1 CHANGES IN CHROMATIN STRUCTURE IN ACTIVE OR POTENTIALLY ACTIVE GENES

Active DNA is organized in a nucleosomal structure

As discussed in Section 2.2, the bulk of cellular DNA is associated with histone molecules in a nucleosomal structure. This leads to the obvious question of whether genes which are either being transcribed or are about to be transcribed in a particular tissue are also organized in this manner or whether they exist as naked, nucleosome-free DNA.

Two main lines of evidence suggest that actively transcribed genes are still organized into nucleosomes. First, if DNA that is being transcribed is examined in the electron microscope, in most cases the characteristic beads-on-a-string structure (see Section 2.2) is observed, with nucleosomes visible both behind and in front of the RNA polymerase molecules transcribing the gene (**Figure 3.2**). Although this structure may break down in genes which are being highly actively transcribed, such as occurs for the genes encoding **ribosomal RNA** during **oogenesis**, it is maintained in most transcribed genes.

Secondly, if DNA organized into nucleosomes is isolated as a ladder of characteristically sized fragments following mild digestion with micrococcal nuclease (see Section 2.2), the DNA from active genes is found in these fragments in the same proportion as in total DNA. Similarly, no enrichment or depletion in the amount of DNA from a particular gene found in these fragments is observed when the chromatin isolated from a tissue actively transcribing the gene is analyzed and compared to chromatin isolated from a tissue that does not transcribe it. The ovalbumin gene, for example, is found in nucleosome-sized fragments of DNA in these experiments regardless of whether chromatin from hormonally stimulated oviduct tissue or

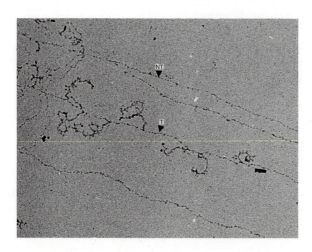

Figure 3.2
Electron micrograph of chromatin from a *Drosophila* embryo. Note the identical beads-on-a-string structure of the chromatin that is not being transcribed (NT) and the chromatin that is being transcribed (T) into the readily visible ribonucleoprotein fibrils. Courtesy of OL Miller, from McKnight SL, Bustin M, Miller OL Jr (1978) *Cold Spring Harb. Symp. Quant. Biol.* 42, 741–754. With permission from Cold Spring Harbor Laboratory Press.

from liver tissue is used. This is in agreement with the idea that transcribed DNA is still found in a nucleosomal structure rather than as naked DNA, which would be rapidly digested by micrococcal nuclease and would not appear in the ladder of nucleosome-sized DNA fragments.

Active or potentially active chromatin shows enhanced sensitivity to DNaseI digestion

To identify differences between transcribed and non-transcribed DNA, many workers have studied the sensitivity of these different regions to digestion with the pancreatic enzyme deoxyribonuclease I (DNaseI). Although this enzyme would eventually digest all the DNA in a cell, if it is applied to chromatin in small amounts for a short period only a small amount of the DNA will be digested. The proportion of transcribed or non-transcribed genes present in the relatively resistant undigested DNA in a given tissue compared to the proportion in total DNA can be used to detect the presence of differences between active and inactive DNA in their sensitivity to digestion with this enzyme.

The fate of an individual gene in this procedure can be followed simply by cutting the DNA surviving digestion with an appropriate restriction enzyme and carrying out the standard Southern-blotting procedure (see Chapter 1, Methods Box 1.1) with a probe derived from the gene of interest. The presence or absence of the appropriate band derived from the gene in the digested DNA provides a measure of the resistance of the gene to DNaseI digestion (**Figure 3.3**; **Methods Box 3.1**).

This procedure has been used to demonstrate that a very wide range of genes that are active in a particular tissue exhibit a heightened sensitivity to DNaseI, which extends over the whole of the transcribed gene and for some distance upstream and downstream of the transcribed region. For example, when chromatin from chick oviduct tissue is digested, the active ovalbumin gene is rapidly digested and its characteristic band disappears from the Southern blot under digestion conditions which leave the DNA from the inactive globin gene undigested. This difference between the globin and ovalbumin genes is dependent upon their different activity in oviduct tissue rather than any inherent difference in the resistance of the genes themselves to digestion. If the digestion is carried out with chromatin isolated from red blood cell precursors in which the globin gene is active and that encoding

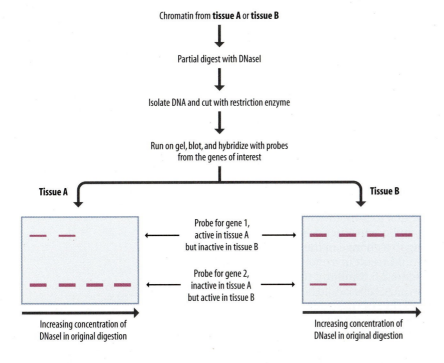

Figure 3.3
Preferential sensitivity of chromatin containing active genes to digestion with DNaseI, as assayed by Southern blotting with specific probes.

Methods Box 3.1
PROBING CHROMATIN STRUCTURE WITH DNaseI (FIGURE 3.3)

- Isolate chromatin (DNA and associated histones and other proteins).
- Partially digest chromatin with DNaseI.
- Purify partially digested DNA by removing protein.
- Digest with restriction enzyme and carry out Southern blotting with probe for gene of interest (see Methods Box 1.1).
- Monitor disappearance of band for DNA of interest in samples where increasing amounts of DNaseI have been used to digest chromatin.

ovalbumin is inactive, the reverse result to that seen in the oviduct is obtained, with the globin DNA exhibiting preferential sensitivity to digestion and the ovalbumin DNA being resistant to such digestion (see Figure 3.3).

It is clear that actively transcribed genes, though packaged into nucleosomes, are in a more open chromatin structure than that found for non-transcribed genes and are hence more accessible to digestion with DNaseI. This altered chromatin structure is not confined to the very active genes such as globin and ovalbumin but appears to be a general characteristic of all transcribed genes, whatever their rate of transcription. If chromatin is digested using conditions that degrade less than 10% of the total DNA, over 90% of transcriptionally active DNA is digested and the DNA of genes encoding rare mRNAs is as sensitive as that encoding abundant mRNAs. The altered chromatin structure of active genes does not therefore appear to be dependent upon the act of transcription itself, since the genes encoding many rare mRNAs will be transcribed only very occasionally.

In agreement with this idea, the altered DNaseI sensitivity of previously active genes persists even after transcription has ceased. For example, the ovalbumin gene remains preferentially sensitive to the enzyme when chromatin is isolated from the oviduct after withdrawal of estrogen when, as we have previously seen, transcription of the gene ceases (see Section 1.4). A similar preferential sensitivity to DNaseI is observed for the genes encoding the fetal forms of globin which are not transcribed in adult sheep cells and for the adult globin genes in mature (14-day) chicken erythrocytes, following the switching off of transcription.

As well as being detectable after transcription has ceased permanently, such increased sensitivity can be detected in genes about to become active prior to the onset of transcription. As discussed previously (see Section 1.4), the transcription of the globin gene in Friend erythroleukemia cells only occurs following treatment of the cells with dimethyl sulfoxide. Increased sensitivity to DNaseI digestion is observed, however, in both the treated cells which transcribe the gene at high levels and in the unstimulated cells which do not.

The altered more open chromatin structure detected by increased DNaseI sensitivity does not therefore reflect the act of transcription itself. Rather, it appears to reflect the ability to be transcribed in a particular tissue or cell type. In cells that have become committed to a particular lineage expressing particular genes, such commitment will be reflected in an altered chromatin structure which will arise prior to the onset of transcription and will persist after transcription has ceased.

The ability of imaginal disc cells in *Drosophila* to maintain their commitment to give rise to a particular cell type in the absence of overt differentiation (see Section 2.1) is likely, therefore, to be due to the genes required in that cell type having already assumed an open chromatin structure. Moreover, a breakdown in this process resulting in a change in the chromatin structure of a particular **regulatory gene** would result in a change in the commitment of the disc cells as illustrated in Figure 2.6. Similarly, the altered chromatin structure of the genes required in cartilage cells would be retained in cells cultured in media not supporting expression of these genes, allowing the restoration of the differentiated cartilage phenotype when the cells are transferred to an appropriate medium (see Section 2.1).

As described above, active or potentially active DNA is still organized in the beads-on-a-string nucleosome structure. However, its enhanced DNaseI sensitivity reflects the fact that it does not form the much more tightly packed 30 nm-fiber structure (see Section 2.4), which is formed by the bulk of DNA that is not about to be or being transcribed. Commitment to a particular pattern of gene expression involves a move from the tightly packed 30 nm-fiber structure to the more open beads-on-a-string structure (**Figure 3.4**). As the DNA in this structure is more accessible, it is therefore more readily digested by DNaseI.

Although the greater accessibility to DNaseI of active or potentially active DNA simply represents an experimental tool for dissecting its chromatin structure, the alteration of chromatin to this more open structure in committed cells is likely to be a necessary prerequisite for gene expression and is therefore of major biological significance. It will allow the *trans*-acting factors which actually activate transcription of the gene access to their appropriate DNA target sequences within it. The different structure of potentially active genes can explain why a particular steroid hormone, such as estrogen, can induce transcription of the ovalbumin gene in the oviduct but activates the vitellogenin gene in the liver (see Section 2.1). The ovalbumin gene in the oviduct will be in an open configuration, allowing induction to occur, whereas the more closed configuration of the vitellogenin gene will not allow access to the hormone–receptor complex and induction will not occur. The reverse situation would apply in liver tissue, allowing induction of the vitellogenin gene but not the ovalbumin gene.

The changes in chromatin structure detected by DNaseI digestion play an important role in establishing the commitment to express the specific genes characteristic of a particular lineage. These changes in chromatin structure are likely to be produced by a variety of epigenetic changes which have been observed to modify DNA or its associated proteins in areas which are about to be or are being transcribed. These biochemical changes are discussed in the next sections.

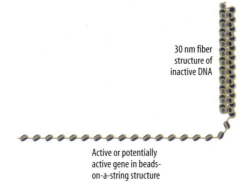

Figure 3.4
Active or potentially active DNA is in the beads-on-a-string chromatin structure whereas transcriptionally inactive DNA is in the more tightly packed 30 nm-fiber structure.

3.2 ALTERATIONS IN DNA METHYLATION IN ACTIVE OR POTENTIALLY ACTIVE GENES

Decreased DNA methylation is associated with active or potentially active genes

Although DNA consists of the four bases – adenine (A), guanine (G), cytosine (C), and thymine (T) – it has been known for many years that these bases can exist in modified forms bearing additional methyl groups. The most common of these in eukaryotic DNA is 5-methyl cytosine (**Figure 3.5**). Between 2 and 7% of the cytosine in mammalian DNA is modified in this way.

Approximately 90% of this methylated C occurs in the dinucleotide CG, where the methylated C is followed on its 3′ side by a G residue. Conveniently, this sequence forms part of the recognition sequence (CCGG) for two restriction enzymes, *Msp*I and *Hpa*II, which differ in their ability to cut at this sequence when the central C is methylated. *Msp*I will cut whether or not the C is methylated and *Hpa*II will only do so if the C is unmethylated. This characteristic allows the use of these enzymes to probe the methylation state of the fraction of CG dinucleotides that is within cleavage sites for these enzymes.

If DNA is digested with either *Hpa*II or *Msp*I, both enzymes will give the same pattern of bands only if all the C residues within the recognition sites are unmethylated. In contrast, if any sites are methylated, larger bands will be obtained in the *Hpa*II digest, reflecting the failure of the enzyme to cut at particular sites (**Figure 3.6**). If this procedure is used in conjunction with Southern-blot hybridization, using a probe derived from a particular gene, the methylation pattern of the *Hpa*II/*Msp*I sites within the gene can be determined.

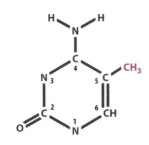

Figure 3.5
Structure of 5-methyl cytosine.

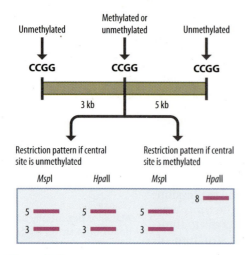

Figure 3.6
Detection of differences in DNA methylation between different tissues, using the restriction enzymes *Msp*I and *Hpa*II.

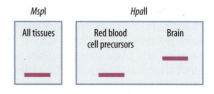

Figure 3.7
Tissue-specific methylation of *Msp*I/*Hpa*II sites in the chicken globin gene results in the methylation-sensitive enzyme, *Hpa*II, producing a band in DNA from red blood cell precursors that is identical to that produced by the methylation-insensitive *Msp*I, but producing a larger band in brain DNA. From Weintraub H, Larsen A & Groudine M (1981) *Cell* 24, 333–344. With permission from Elsevier.

When this is done it is found that although some CG sites are always unmethylated and others are always methylated, a number of sites exhibit a tissue-specific methylation pattern, being methylated in some tissues but not in others. Such sites within a particular gene are unmethylated in tissues where the gene is active or potentially active and methylated in other tissues. For example, a particular site within the chicken globin gene is methylated in a wide variety of tissues and is therefore not digested with *Hpa*I, but is unmethylated and therefore susceptible to digestion in DNA prepared from red blood cell precursors (**Figure 3.7**). Similarly, the tyrosine amino-transferase gene which is expressed only in the liver, is relatively under-methylated in this tissue when compared to other tissues where it is not expressed.

Results obtained by using methylation-sensitive restriction enzymes have been extended by other methods which can analyze all the C residues in the genome, rather than only those within a specific sequence which is a target for a restriction enzyme. Many such methods are based on chemical modification of the DNA using sodium bisulfite. This treatment has no effect on methylated C residues but converts unmethylated C residues to uracil (U). Hence, changes in methylation of particular C residues can be detected by using DNA sequence analysis or by hybridization analysis taking advantage of the fact that methyl-C will bind to G but U will bind to A (**Figure 3.8**).

These methods can potentially be used to analyze a much larger number of C residues in the genome than can be analyzed by restriction enzyme analysis. This can be achieved by large-scale DNA sequence analysis of the bisulfite-treated DNA or by using gene-chip **microarray** technology (see Section 1.2) to hybridize it to DNA chips containing oligonucleotides homologous to many different DNA sequences. Methylation of C residues can also be analyzed across the genome using DNA chips, in conjunction with the chromatin **immunoprecipitation** assay which will be described in Chapter 4 (Section 4.3).

The results of these experiments reinforce those described above; namely that demethylation is characteristic of genes which are active or potentially active. Moreover, high levels of DNA methylation are associated with the inactivation of specific genes both during development and, as discussed in Section 3.6, DNA methylation is also involved in the processes of X-chromosome inactivation and **genomic imprinting**.

Interestingly, the promoter regions of many genes have a 10–20 times higher frequency of CG dinucleotides compared to the rest of the genome. These so-called **CG islands** are found particularly in **housekeeping genes** which are expressed in all tissues. Moreover, they are frequently

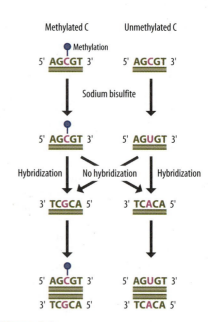

Figure 3.8
Treatment with sodium bisulfite converts unmethylated cytosine (C) residues to uracil (U) but has no effect on methylated C residues. The presence or absence of methylation on particular C residues can therefore be analysed by sodium bisulfite treatment followed by either DNA sequence analysis or hybridization to a potentially complementary DNA sequence, when C will hybridize to G but U will hybridize to A.

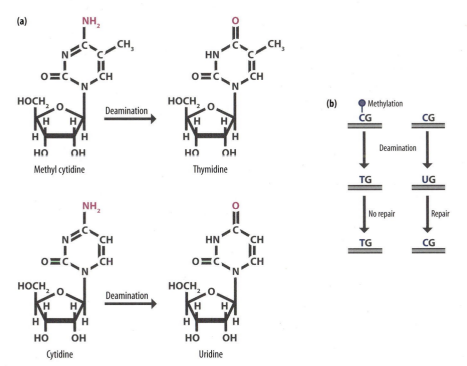

Figure 3.9
Naturally occurring deamination processes convert methyl-C residues to T residues, whereas deamination of an unmethylated C residue produces a U residue (a), which is not normally present in the DNA. Methyl-C residues converted to T residues are not efficiently repaired, whereas conversion of C residues to U residues is recognized efficiently by DNA repair enzymes, which restore the C residues (b).

unmethylated as would be expected from the link between undermethylation and gene activity.

These CG islands appear to arise during evolution due to the vulnerability of C residues to naturally occurring deamination processes. When this occurs for an unmethylated C, it produces a U residue. As U residues do not normally occur in the DNA, this change is recognized by DNA repair enzymes which restore the C residue. However, deamination of methyl-C residues produces T residues, which are indistinguishable from normal T residues (**Figure 3.9a**). Hence, some of these escape detection by DNA repair enzymes. Therefore, over evolutionary time a high proportion of methyl-CG dinucleotides may have been converted into TG dinucleotides (**Figure 3.9b**), resulting in the CG dinucleotide being under-represented in the genome. However, this would not occur in regions where the CG dinucleotide is normally unmethylated, resulting in the clusters of CG residues which occur in the CG islands found in promoter regions (**Figure 3.10**).

In the case of tissue-specific genes which exhibit changes in methylation pattern, under-methylation, like DNaseI sensitivity, is observed prior to the onset of transcription and persists after its cessation. For example, the under-methylation of the chicken globin gene persists in erythrocytes after globin gene transcription has ceased, paralleling the continued sensitivity of the globin gene to DNaseI digestion.

Most interestingly, the region in which unmethylated C residues are found in different genes correlates with that exhibiting heightened DNaseI

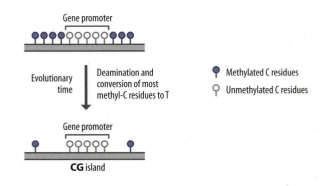

Figure 3.10
Over evolutionary time, many C residues at positions that are methylated (bar with solid circle) are lost from the DNA by conversion of C to T. In contrast, C residues at positions that are unmethylated (bar with open circle) are retained. Clusters of CG dinucleotides (CG islands) therefore occur at regions in the genome, such as gene promoters, where the C residues are predominantly unmethylated.

sensitivity and is also depleted of histone H1 (see Section 2.4 for discussion of histone H1). As with DNaseI sensitivity, therefore, under-methylation is a consequence of commitment to a particular pattern of gene expression and is associated with the change in chromatin structure observed in active or potentially active genes.

DNA methylation plays a key role in regulating chromatin structure

Although under-methylation is associated with the more open chromatin structure of active DNA, this association does not prove that it has a key role in creating this structure. Two lines of evidence suggest, however, that this is indeed the case and that alterations in DNA methylation can affect chromatin structure and gene expression.

The first line of evidence involves introducing DNA containing 5-methyl cytosine into cells. A number of such experiments have shown that such DNA is not expressed, whereas the same DNA which has been demethylated is expressed. These experiments have been carried out using both eukaryotic viruses and cellular genes such as those encoding β- and γ-globin, cloned into plasmid vectors. In these experiments the methylated DNA adopts a DNaseI-insensitive structure typical of inactive genes whereas unmethylated DNA adopts the DNaseI-sensitive structure typical of active genes (**Figure 3.11**). This provides direct evidence for the role of methylation differences in regulating the generation of different forms of chromatin structure.

The second line of evidence involves the effect of artificially induced demethylation of DNA. If methylation differences play a crucial role in the regulation of differentiation, it should be possible to change gene expression by demethylating DNA. This has been achieved in a number of cases by treating cells with the cytidine analog **5-azacytidine**. This analog is incorporated into DNA but cannot be methylated, having a nitrogen atom instead of the carbon atom at position 5 of the **pyrimidine** ring which is normally the target for methylation (see Figure 3.5).

In the most dramatic of these cases, treatment of an undifferentiated fibroblast **cell line** (known as the 10T½ cell line) with this compound results in the activation of key regulatory loci and the cells differentiate into multinucleate, twitching, striated muscle cells (see Section 10.1 for further discussion of this system).

In other cases, although not actually producing altered gene expression, demethylation may facilitate it. If HeLa cells are treated with 5-azacytidine no dramatic effects are observed. However, if such cells are fused with muscle cells, muscle-specific genes are switched on in the treated HeLa cells, a phenomenon which is not observed when untreated HeLa cells are fused with mouse muscle cells. Thus, treatment of the HeLa cells has altered their muscle-specific genes in such a way as to allow them to respond to *trans*-acting factors present in the mouse muscle cells. This type of regulatory process is exactly what would be predicted if methylation has a role in the alteration of chromatin structure and thereby in facilitating interactions with *trans*-acting regulatory factors.

DNA methylation patterns can be propagated stably through cell divisions

The evidence discussed above suggests that DNA methylation plays a central role in the regulation of gene expression, at least in mammals. This conclusion is reinforced by the finding that several different **DNA methyltransferase** enzymes which methylate DNA are essential for normal embryonic development. For example, in mammals there are two major methyltransferase enzymes which can carry out *de novo* methylation of C residues, thereby converting an unmethylated site into a methylated one, and both of these are essential for normal development. Inactivation of the gene encoding DNA methyltransferase 3a (Dnmt3a) results in death a few weeks after birth, whereas mice lacking Dnmt3b survive for only a few days

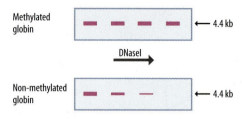

Figure 3.11
Unmethylated DNA introduced into cells adopts an open DNaseI-sensitive configuration, whereas the same DNA when methylated and then introduced into cells adopts a more tightly packed DNaseI-insensitive form. From Keshet I, Lieman-Hurwitz J & Cedar H (1986) *Cell* 44, 535–543. With permission from Elsevier.

(a) *De novo* methylation

(b) Maintenance methylation

Figure 3.12
The DNA methyltransferase enzymes Dnmt3a or Dnmt3b catalyze the *de novo* methylation of unmethylated sites (a) whereas Dnmt1 catalyzes the maintenance methylation of a half-methylated site (b).

after birth. Hence, in mammals DNA methylation is essential for normal development.

The idea that methylation differences might be essential for changing chromatin structure is particularly attractive because of the ease with which such differences can be propagated, allowing cellular commitment to be stable over many generations (see Section 2.1). In double-stranded DNA, the CG dinucleotide will exist as a symmetrical structure:

5'-CG-3'
3'-GC-5'

It has been observed that when one C in this structure is methylated, the C on the opposite strand is also methylated. This effect is achieved by the DNA methyltransferase enzyme Dnmt1. Unlike Dnmt3a and Dnmt3b, which can carry out *de novo* methylation of unmethylated sites (**Figure 3.12a**), Dnmt1 recognizes sites where only one C is methylated (hemimethylated sites) and rapidly methylates the second C residue (**Figure 3.12b**).

The pattern of DNA methylation will therefore be maintained following DNA replication, with the hemimethylated sites produced by replication being rapidly re-methylated (**Figure 3.13**). Similarly, because the **maintenance methylase** is only active on hemimethylated sites, unmethylated sites present in a particular tissue will be propagated through subsequent cell divisions. Once established, a particular pattern of methylation will be maintained accounting for the stability of the committed state.

Such a mechanism also allows readily for the specific loss of methylation sites that must occur during the process of commitment to a particular lineage. Such losses could occur via a specific demethylation event catalyzed by a demethylation enzyme (**Figure 3.14a**). Alternatively, it could occur simply by inhibiting the action of the maintenance methylase at a particular site following cell division (**Figure 3.14b**) and it appears that both these mechanisms are used.

The mechanism involving inhibition of a maintenance methylase would eventually result in the generation of cells at which the site was fully methylated and cells in which it was unmethylated (see Figure 3.14b). This is exactly the pattern frequently observed in embryonic development where a **stem cell** divides to yield one daughter that differentiates and another that maintains the stem cell lineage (**Figure 3.15**). As with other methylation patterns, the new demethylated site in the committed cell would be propagated through subsequent cell divisions.

DNA methylation processes therefore provide a means of explaining the stability of the committed state, while allowing for its modification in suitable circumstances. Unlike DNA deletion events, methylation patterns are not irreversible and could be altered when a cell undergoes transdifferentiation or following nuclear transplantation (see Section 1.3). As we have discussed, however, such changes in the differentiated state normally require dedifferentiation and cell division, exactly as would be necessary for

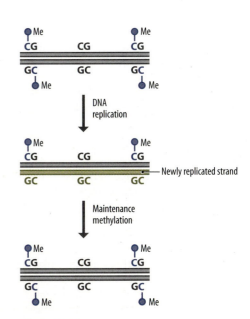

Figure 3.13
Model for the propagation of methylation patterns through cell division.

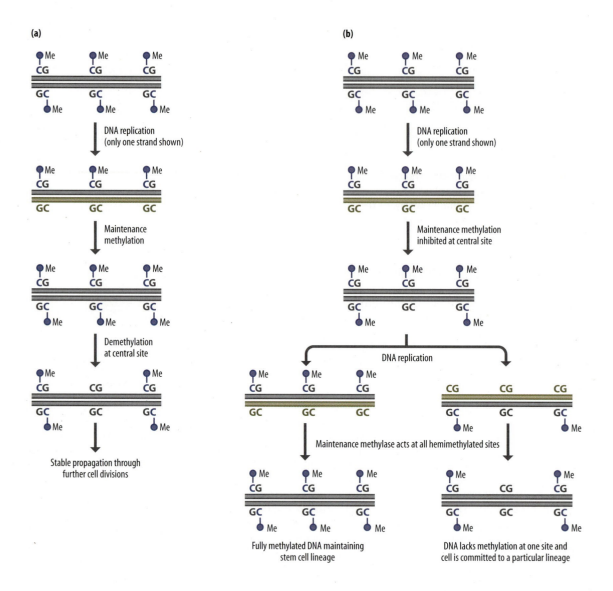

a process dependent on DNA replication and subsequent inhibition of maintenance methylation at particular sites. Hence, models of this type can explain the stable but not irreversible nature of the differentiated state and the frequent involvement of cell division in its reprogramming.

DNA methylation recruits inhibitory proteins that produce a tightly packed chromatin structure

The evidence suggesting an important role for DNA methylation in modulating the difference in chromatin structure between inactive and active genes raises the question of how this is achieved. It is clearly possible for the methylation differences between inactive or potentially active genes to be recognized by proteins since, as discussed above, the proteinaceous enzyme *Hpa*II digests only unmethylated DNA. Under-methylation could promote the binding of proteins which produce a more open chromatin structure (**Figure 3.16a**). Alternatively, an inhibitory protein could bind to methylated DNA and promote a closed chromatin structure (**Figure 3.16b**).

Although both these mechanisms may be used, current evidence supports the second mechanism. A number of different proteins able to bind specifically to methyl-CG have been characterized and shown to play critical roles in the regulation of gene expression. For example, a specific protein, **MeCP2**, has been shown to bind directly to methylated CG but not to unmethylated CG and its binding results in the production of a tightly packed closed chromatin structure and transcriptional repression.

Figure 3.14
Alteration of the methylation pattern to produce an unmethylated site can occur either by specific demethylation (a) or by inhibition of the maintenance methylase and subsequent DNA replication (b).

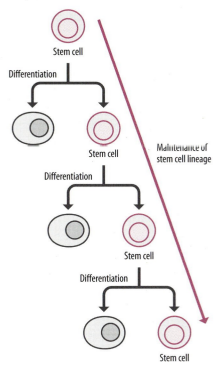

Figure 3.15
Schematic model of a pattern of differentiation which is frequently found in embryonic development. A stem cell divides to produce one daughter cell which differentiates and another daughter which maintains the stem cell lineage.

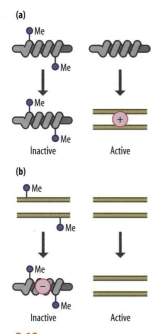

Figure 3.16
The transition from an inactive (wavy line) to an active state (straight line) of the chromatin could take place via an activating protein which binds specifically to unmethylated DNA, thereby activating it (a), or via an inhibitory protein which binds specifically to methylated DNA, thereby repressing it (b).

The importance of MeCP2 and its recognition of methylated C residues is demonstrated by the finding that humans with mutant forms of MeCP2 which are unable to recognize methyl-CG exhibit Rett syndrome. This is a severe developmental disorder leading to mental retardation. Mice lacking another methyl-CG-binding protein, MBD1, also show defects in the nervous system, indicating that several methyl-CG-binding proteins are required for proper development and functioning of the nervous system.

Interestingly, binding of MeCP2 to methyl-CG results in the recruitment of a multi-protein complex which includes a histone deacetylase (HDAC) that can induce the removal of acetyl groups from histones (**Figure 3.17**) (see Section 2.3). As deacetylation of histones is known to be associated with transcriptionally inactive DNA (see Section 3.3), this illustrates the close link between modification of DNA and modification of histones in determining the structure of chromatin. This link is further strengthened by the observations that mice with a defective MeCP2 protein demonstrate enhanced acetylation of histone H3.

Both MeCP2 and MBD1 also associate with histone methylase enzymes resulting in enhanced methylation of histone H3, indicating that DNA methylation can regulate more than one histone modification. In agreement with this, mouse cells lacking the maintenance methylase Dnmt1 exhibit alterations in histone acetylation and methylation as well as greatly

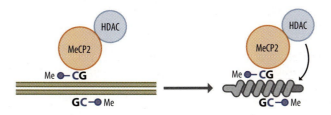

Figure 3.17
Binding of the MeCP2 protein to methyl-CG leads to the recruitment of other proteins including a histone deacetylase (HDAC; see Section 3.3), resulting in the chromatin forming a tightly packed inactive configuration (wavy line).

reduced methylation of the DNA itself. Therefore, DNA methylation may exert its effects on chromatin structure, at least in part, by altering histone modification. The various histone modifications associated with changes in chromatin structure are discussed in the next section.

3.3 MODIFICATION OF HISTONES IN THE CHROMATIN OF ACTIVE OR POTENTIALLY ACTIVE GENES

As discussed in Section 2.3, histones are subject to a number of different modifications; for example, acetylation, methylation, ubiquitination, and phosphorylation. All of these have been implicated in the regulation of chromatin structure and therefore of gene transcription. Each of these modifications will be discussed in turn (Table 3.1).

Acetylation

As described in Section 2.3, in the process of acetylation the free amino group on specific lysine residues at the N-terminus of the histone molecule is modified by one of the hydrogen atoms in the amino group being substituted by an acetyl group (see Table 3.1). This reduces the net positive charge on the histone molecule. Hyperacetylated forms of these histones, containing several such acetyl groups, have been shown to be localized preferentially in active genes exhibiting DNaseI sensitivity while hypoacetylation of the histones is characteristic of transcriptionally inactive regions. Furthermore, treatment of cells with sodium butyrate which inhibits a cellular deacetylase activity and hence increases histone acetylation has been shown to result in DNaseI sensitivity of some regions of chromatin and to activate the expression of some previously silent cellular genes. As with DNA methylation, there is therefore direct evidence that hyperacetylation of histones plays a role in opening the chromatin structure of active or potentially active genes.

Further evidence in favor of a role for histone acetylation in the regulation of gene expression is the finding that several proteins which were characterized as being involved in the activation of transcription were subsequently shown to have histone acetyltransferase (HAT) activity, being capable of adding acetyl groups to histone molecules. For example, it has been shown that both the CBP transcriptional co-activator (which plays a key role in the activation of genes in response to **cyclic AMP** and other stimuli) and the related p300 co-activator have HAT activity (see Section 5.2 for discussion of these co-activators). This directly links this enzymatic activity with the ability to stimulate transcription. Similarly the TAF$_{II}$250

TABLE 3.1 POST-TRANSLATIONAL MODIFICATIONS OF HISTONES AND THEIR EFFECT ON TRANSCRIPTION		
MODIFICATION	MODIFIED SITES	TRANSCRIPTIONAL EFFECT
Acetylation of lysine	H3 positions 2, 4, 9, 14, 18, 56; H4 positions 5, 8, 12, 16, 20; H2A; H2B	Activation
Methylation of lysine	H3 positions 4, 36, 79 H3 positions 9, 27; H4 positions 12, 20	Activation Repression
Methylation of arginine	H3 positions 2, 17, 26; H4 position 3	Activation
Ubiquitylated lysine	H2B position 120 H2A position 119	Activation Repression
Sumoylated lysine	H2B position 5; H2A position 126	Repression
Phosphorylated serine/threonine	H3 positions 3, 10, 11, 28; H4 position 1; H2A; H2B	Activation

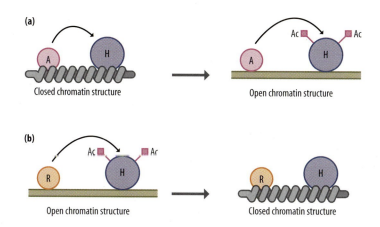

Figure 3.18
(a) An activating molecule (A) can direct the acetylation of histones (H), thereby resulting in a change in chromatin structure from a closed (wavy line) to an open (solid line) configuration. (b) An inhibitory molecule (R) can direct the deacetylation of histones, thereby having the opposite effect on chromatin structure.

subunit of the TFIID transcription factor, which is critical for **basal** transcription of a wide range of genes (see Section 4.1), has been shown to have HAT activity, as has the ATF2 activator transcription factor.

In contrast, HDAC activity has been observed to be involved in the action of the nuclear receptor **co-repressor**, which mediates the inhibitory effect of the thyroid hormone receptor on transcription when thyroid hormone is absent. Thus, this factor has been shown to associate with the Sin3–HDAC protein complex which has the ability to deacetylate histones. The inhibitory effect of the nuclear receptor co-repressor is therefore likely to involve the deacetylation of histones, thereby producing a more tightly packed chromatin structure incompatible with transcription (see Section 5.3 for further discussion of transcriptional repression by the thyroid hormone receptor).

Thus, activating factors can direct the acetylation of histones, thereby opening up the chromatin (**Figure 3.18a**), whereas inhibitory factors can direct histone deacetylation, thereby directing a more tightly packed chromatin structure (**Figure 3.18b**).

These findings therefore link the study of transcription factors (see Chapter 5) with studies on chromatin structure. They suggest that the regulation of histone acetylation/deacetylation and thereby of chromatin structure by specific factors plays a key role in the regulation of gene expression. As noted above (see Section 3.2), the MeCP2 protein which binds specifically to methyl-CG dinucleotides can recruit a HDAC activity, thereby linking histone deacetylation to the repressive effect of DNA methylation.

In addition to activators recruiting acetylases and repressors recruiting deacetylases, there is evidence that the acetylase/deacetylase enzymes can themselves be regulated. This is seen in muscle differentiation where in **myoblasts** the transcriptional activator **MEF2** is associated with HDACs. When differentiation from myoblasts to mature **myotubes** occurs, the HDAC is phosphorylated, which induces it to move to the cytoplasm, thereby freeing MEF2 to activate transcription (**Figure 3.19**) (see Section 10.1 for further discussion of the role of MEF2 in muscle cell-specific gene expression).

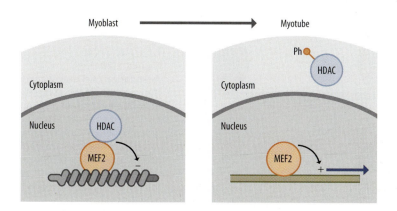

Figure 3.19
In myoblasts, the MEF2 transcriptional activator is associated with an HDAC and transcription of myotube-specific genes is repressed. When differentiation is induced, the HDAC is phosphorylated (Ph) and moves into the cytoplasm, allowing MEF2 to activate the transcription of myotube-specific genes.

As discussed in Section 2.3, the N-terminal domains of the histone proteins which are modified by acetylation project beyond the nucleosome core and could therefore potentially interact with the N-terminal ends of histones in adjacent nucleosomes or with other non-histone proteins. The effects of histone acetylation on chromatin structure may therefore operate in one of two possible ways. First acetylation might act by affecting the association of the histones with each other. A looser association between the histones in adjacent nucleosomes could directly result in improved access to the DNA for factors which can stimulate transcription (**Figure 3.20a**). Alternatively, such a looser association could facilitate displacement of nucleosomes by chromatin-remodeling complexes (see Section 2.2), so indirectly producing improved access for activators (**Figure 3.20b**).

An alternative possibility is that these modifications affect the protein–protein interaction of the histones with other regulatory molecules. This would parallel the role proposed for DNA methylation differences in affecting the binding of positively or negatively acting factors to the DNA (see Figure 3.16; Section 3.2). For example, acetylated histones might be recognized by a positively acting molecule, leading to the destabilization of the 30 nm-fiber structure and transcriptional activation (**Figure 3.21a**). Similarly, acetylation could disrupt an association with an inhibitory molecule involved in maintaining the closed chromatin structure (**Figure 3.21b**).

In agreement with this model of histone-regulatory protein interaction, it has been shown that several activating factors, such as Brg1, contain a region known as the **bromodomain** that binds to histones with much greater affinity when specific lysines in the histones are acetylated. Binding of bromodomain-containing proteins to acetylated histones results in a more open chromatin structure and transcriptional activation (**Figure 3.22**).

Methylation

As well as being modified by acetylation, histones can also be modified by methylation (see Section 2.3). Unlike acetylation, methylation can occur for

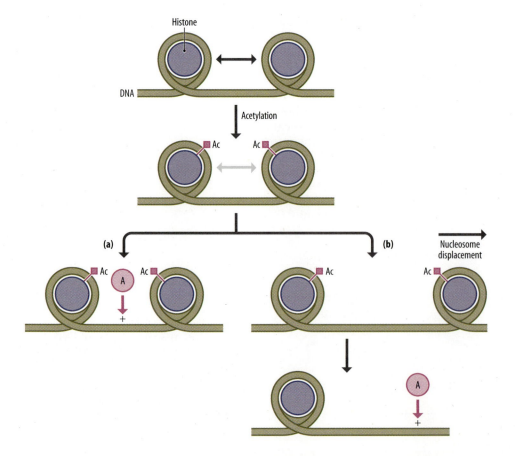

Figure 3.20
Acetylation (Ac) may weaken the strength of the association between histones in adjacent nucleosomes (black arrow versus gray arrow). In turn, this may directly facilitate access of an activator (A) (a) or promote nucleosome displacement so indirectly improving access for an activator (b).

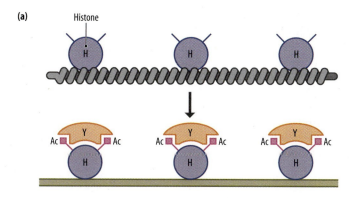

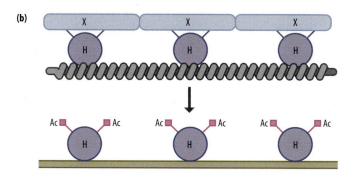

Figure 3.21
Acetylation (Ac) of histones (H) may activate either by promoting the association of the histones with an activating factor (Y), producing a more open chromatin structure (a), or indirectly by disrupting their association with an inhibitory molecule (X) (b).

both lysine and arginine amino acids in the histone molecule and does not affect the molecule's negative charge.

Although the acetylation of histones is associated with a more open chromatin structure, the situation with methylation is more complex. Methylation of specific arginine residues and of some lysine residues is associated with a more open chromatin structure and consequent transcriptional activation. In contrast, however, methylation of other lysine residues produces a more tightly packed chromatin structure and transcriptional repression (see Table 3.1 and **Figure 3.23**).

The methylation of histone H4 on the arginine at position 3 promotes a more open chromatin structure and facilitates transcriptional activation for example by nuclear hormone receptors. A similar effect has also been demonstrated for lysine 4 of histone H3 and ecdysone-dependent gene activation in *Drosophila*. Similarly, lysine 4 methylation of histone H3 was found to be associated with transcriptionally active genes in the human genome and with the transcriptionally active region of the **mating-type locus** in yeast (see Section 10.3 for discussion of the yeast mating-type system).

Interestingly, the adjacent transcriptionally inactive region of the yeast mating-type locus is associated with methylation of histone H3 on the lysine at position 27, indicating that this modification is associated with transcriptional repression (see Table 3.1). Similarly, in the human genome, methylation of histone H3 on lysines 9 and 27 is associated with transcriptionally inactive genes.

It appears that the balance of methylation at different histone sites plays a critical role in determining chromatin structure. For example, the polycomb complex which is involved in transcriptional repression contains a

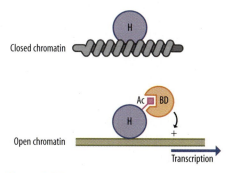

Figure 3.22
Acetylation (Ac) of histones (H) results in binding of bromodomain-containing activator proteins (BD), leading to an open chromatin structure and subsequent transcription.

Figure 3.23
N-terminal region of histone H3 showing the arginine (R) and lysine (K) residues which can be modified by methylation. Methylations shown above the line produce a more open chromatin structure, whereas those shown below the line produce a more closed chromatin structure. Numbers indicate the amino acid position in the histone molecule.

histone methyltransferase enzyme that methylates histone H3 on lysines 9 and 27 (**Figure 3.24**). Conversely, the **trithorax** proteins act to open the chromatin by promoting the methylation of histone H3 on lysine 4 which in turn promotes demethylation at lysines 9 and 27 (**Figure 3.25**) (see Section 4.4 for further discussion of transcriptional regulation by polycomb and trithorax proteins).

As with acetylation, methylation may act by affecting the interaction of the histones with each other or their interaction with other positively or negatively acting proteins. In agreement with this latter possibility, it has been shown that methylation of histone H3 on lysine 9 allows it to be recognized by a protein known as HP1 which cannot recognize the unmethylated histone. HP1 is able to organize the chromatin into a very tightly packed, inactive structure characteristic for example of heterochromatin (see Section 2.5).

As well as being recruited to chromatin where histones are methylated on lysine 9, HP1 can also in turn recruit a histone methyltransferase enzyme. This enzyme can catalyze the lysine 9 methylation of histone H3 in adjacent nucleosomes. Thus, the tightly packed heterochromatin structure produced by HP1 will spread along the DNA producing a large region of tightly packed chromatin (**Figure 3.26**). This will continue until the spreading structure encounters a sequence such as an insulator, which prevents the spread of a particular chromatin structure to an adjacent region (see Section 2.5)

Interestingly, this effect can also explain how the methylation pattern of histones can be propagated through cell division. When DNA with histone H3 methylated on lysine 9 replicates, the nucleosomes will be randomly divided between the daughter chromosomes (**Figure 3.27**). As each daughter chromosome will have only half the correct number of nucleosomes, new nucleosomes will be assembled which will contain unmethylated histone H3. However, the original nucleosomes containing histone H3 methylated on lysine 9 will recruit HP1 and its associated histone methyltransferase, allowing histones in adjacent nucleosomes to be methylated (see Figure 3.27).

This mechanism may therefore allow the pattern of histone H3 methylation to be stably inherited paralleling the mechanisms which exist to allow the transmission of DNA methylation patterns through cell division (see Section 3.2). However, as with DNA methylation patterns such patterns can also be altered. In the case of histone methylation, this can be achieved either by active demethylation by specific enzymes or by inhibiting the methylation of newly assembled nucleosomes which normally occurs during cell division.

Clearly, a key feature of HP1 is its ability to recognize histone H3 which is methylated on lysine 9. HP1 binds to the methylated histone via a domain known as the **chromodomain**, which is found in many proteins that inhibit transcription. Recognition of histones methylated on lysine 9 by chromodomain-containing proteins is therefore a key event in organizing the tightly packed structure of inactive chromatin (**Figure 3.28**).

The recognition of lysine 9 methylated histones by chromodomain-containing proteins contrasts with the role of acetylation in promoting recognition of the histones by proteins containing a bromodomain, described above. Chromodomain proteins therefore recognize histones methylated

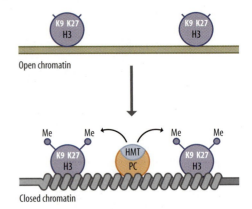

Figure 3.24
The polycomb (PC) complex contains a histone methyltransferase (HMT) activity which methylates histones (H), producing a more tightly packed chromatin structure.

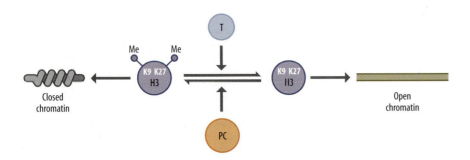

Figure 3.25
Polycomb (PC) proteins methylate histone H3 on lysines (K) 9 and 27, producing a closed chromatin structure, whereas trithorax proteins (T) produce a more open chromatin structure by promoting demethylation at these positions.

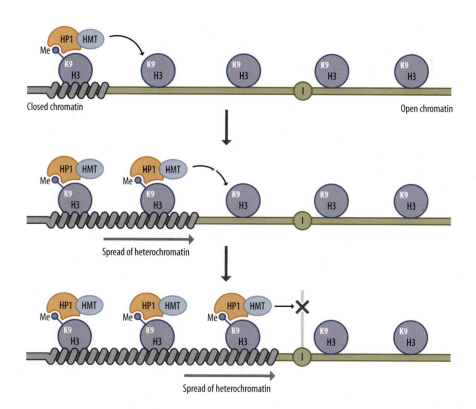

Figure 3.26
Following binding of HP1 to methylated histone H3, it recruits a histone methyltransferase (HMT) enzyme, which in turn methylates histone H3 in adjacent nucleosomes. This promotes the spreading of the tightly packed structure of heterochromatin, which continues until a boundary sequence such as an insulator (I) is encountered.

on specific residues and promote the tight packing of chromatin while the recognition of acetylated histones by bromodomain-containing proteins promotes the more open chromatin structure characteristic of active or potentially active DNA (see Figure 3.28).

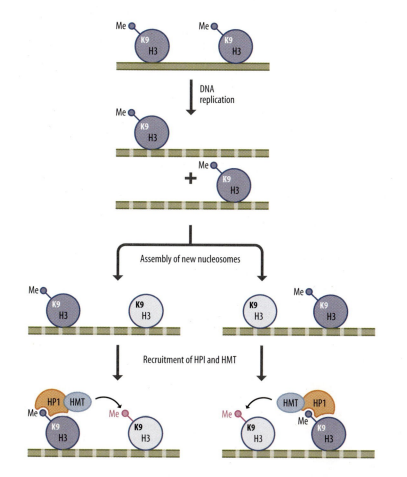

Figure 3.27
Following DNA replication to produce two daughter DNA molecules (dashed lines), pre-existing H3-containing nucleosomes (dark blue) are randomly divided between the two DNA molecules. New nucleosomes (light blue) then assemble on the DNA and these contain unmethylated histone H3. Recruitment of HP1 and histone methyltransferase (HMT) to the original methylated histones produces methylation of the histone H3 in the adjacent newly assembled nucleosomes. This ensures that the pre-existing pattern of histone methylation is inherited when cells replicate.

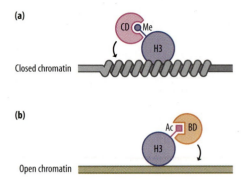

Figure 3.28
(a) Methylation of histone H3 promotes binding of chromodomain-containing proteins (CD) which can induce the closed chromatin structure characteristic of inactive DNA (wavy line). (b) In contrast, acetylation promotes binding of bromodomain-containing proteins (BD) which induce the more open chromatin structure characteristic of active or potentially active DNA.

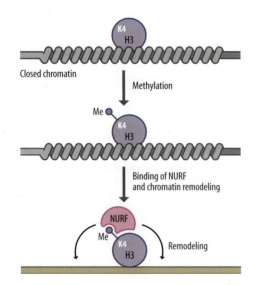

Figure 3.29
Methylation of histone H3 on lysine 4 (K4) allows it to bind the nucleosome-remodeling factor (NURF), which remodels the chromatin to produce a more open chromatin structure.

Interestingly, the nucleosome-remodeling factor (**NURF**), which can affect nucleosome positioning and thereby alter chromatin structure (see Section 2.2), uses a different domain, known as the PHD finger, to recognize methylated histones. Using this domain, NURF recognizes the activating methylation on lysine 4 of histone H3. It then remodels the chromatin to produce a more open chromatin structure, allowing transcription to occur (**Figure 3.29**).

Methylations of histones at different positions can therefore recruit different proteins which bind to the histones via specific protein domains. In turn, these proteins promote either a more tightly packed or a more open chromatin structure.

Ubiquitination and sumoylation

As noted in Section 2.3, addition of the small protein ubiquitin to histones only occurs for histones H2A and H2B. Only a small amount of the H2A in a cell (about 5–10%) exists in this form and such ubiquitination is associated with the repression of gene expression. It has been shown that ubiquitination of histone H2A is catalyzed by a component of the polycomb protein complex which as described above promotes an inactive chromatin structure and transcriptional repression (see also Section 3.6 for further discussion of the role of the polycomb complex in X-chromosome inactivation).

Ubiquitination of H2B has the opposite effect to the ubiquitination of H2A since it stimulates rather than inhibits gene expression (**Figure 3.30**). This is because, as discussed in Section 2.3, ubiquitination of H2B promotes the methylation of histone H3 on lysines at positions 4 and 79. Interestingly, it appears that H2B plays a particularly important role in stimulating the transition of lysines 4 and 79 from a monomethylated state with one methyl group to the trimethylated state where each lysine has three methyl groups (see Section 2.3 for discussion of mono-, di-, and tri-methylation). As methylation on lysines 4 and 79 promotes an open chromatin structure (see

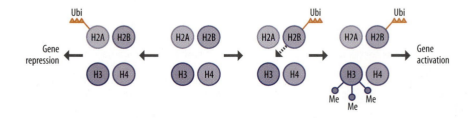

Figure 3.30
Opposite effects of ubiquitination of histones H2A and H2B. Ubiquitination of H2A represses gene expression whereas ubiquitination of H2B promotes methylation of histone H3 and therefore stimulates gene expression.

Figure 3.31
Ubiquitination (Ubi) of H2A and de-ubiquitination of H2B blocks the recruitment of the FACT transcriptional elongation factor and therefore blocks transcriptional elongation by RNA polymerase (a). In contrast, de-ubiquitination of histone H2A and ubiquitination of H2B promotes FACT recruitment and thereby stimulates transcriptional elongation (b).

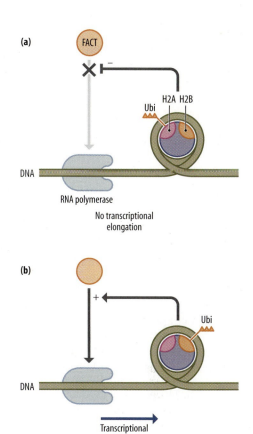

above), this provides an explanation for the stimulatory effect of H2B ubiquitination on gene expression. Moreover, it indicates that modification of one type of histone can modulate the modification of another type of histone molecule.

In addition to their overall opposing effects on chromatin structure and gene expression, ubiquitinated H2A and ubiquitinated H2B also specifically oppose one other in respectively inhibiting or promoting the recruitment of the FACT protein which plays a key role in the elongation of the initial RNA transcript (for further discussion of transcriptional elongation, see Section 4.2). Ubiquitinated H2A blocks recruitment of FACT, thereby inhibiting transcriptional elongation, whereas ubiquitinated H2B promotes such recruitment, thereby stimulating transcriptional elongation (**Figure 3.31**).

As discussed in Section 2.3, lysine residues on histones can also be modified by addition of small ubiquitin-related modifier (SUMO) which, as its name suggests, is a small protein related to ubiquitin. This modification is associated with transcriptional repression due to the recruitment of enzymes which deacetylate histones, thereby resulting in a more closed chromatin structure.

Phosphorylation

Unlike acetylation, methylation, or ubiquitination/sumoylation, phosphorylation targets serine or threonine amino acids in the histone molecule. Such histone phosphorylation is associated with a more open chromatin structure, allowing transcription to occur. For example, it has been shown that when cells are exposed to heat shock (elevated temperature), phosphorylated histone H3 is concentrated at the heat-shock gene loci which are being actively transcribed and is depleted from other loci which are silenced following heat shock.

Similarly, when cells are stimulated to undergo cell division by treatment with **growth factors**, histone H3 is phosphorylated on the serine residue at position 10 in the N-terminus of the protein. These phosphorylated histones are localized to genes which become transcriptionally active following growth factor stimulation, such as the c-*fos* and c-*myc* genes (**Figure 3.32**) (see Sections 11.1 and 11.2 for further discussion of these genes and their role in cell growth/division).

Such growth factor-induced phosphorylation of histone H3 does not occur in human patients with Coffin–Lowry syndrome who lack the phosphorylation enzyme known as Rsk-2, which is responsible for phosphorylating histone H3 following growth factor stimulation. These patients suffer from a number of developmental defects including mental retardation as well as facial and other abnormalities. Moreover, they show a lack of gene activation in response to growth factor stimulation, indicating the importance of Rsk-2-mediated phosphorylation of histone H3 in this process.

In addition to the core histones, phosphorylation also occurs on histone H1. Interestingly, this phosphorylation reduces its ability to interact with

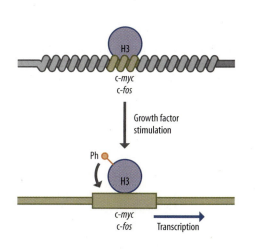

Figure 3.32
Following growth factor stimulation, histone H3 becomes phosphorylated within the nucleosomes bound to the c-*myc* and c-*fos* genes. This is associated with their moving from an inactive closed chromatin structure (wavy line) to a more open structure (solid line), allowing transcription to occur in response to the growth factor.

the HP1 protein mentioned above. As HP1 is important in producing the very tightly packed structure of heterochromatin, phosphorylation of H1 may be a key event in moving from a very tightly packed chromatin structure to a more open one (**Figure 3.33**).

3.4 INTERACTION OF DIFFERENT HISTONE MODIFICATIONS, DNA METHYLATION, AND RNAi

The different histone modifications interact functionally with one another

As discussed in Section 3.3, modifications of the histones can have either stimulatory or inhibitory effects on gene expression by altering chromatin structure. Table 3.1 summarizes the different modifications of the various histones. The modifications at the N-terminus of histone H3 are illustrated in **Figure 3.34**.

Throughout our discussion of these modifications in both Section 2.3 and this chapter (Section 3.3) we have seen examples where different histone modifications interact with one another functionally. At the simplest level, this can involve the fact that some lysine residues (such as residues 9 and 27 in histone H3) can be modified by either acetylation or methylation. As these modifications target the same chemical group, they are mutually exclusive. Moreover, acetylation at these positions promotes an open chromatin structure whereas methylation at the same positions promotes a closed chromatin structure (**Figure 3.35a**).

At a more complex level, modification at one residue can promote or inhibit modifications at adjacent residues. As discussed in Section 2.3, phosphorylation of serine 10 in histone H3 is associated with acetylation of lysines 9 and 14 as well as demethylation of lysine 9. In this region of histone H3, acetylation and phosphorylation are characteristic of open chromatin whereas in closed chromatin the histone H3 is not acetylated or phosphorylated but is methylated on lysine 9 (**Figure 3.35b**). Hence, different modifications on the same histone can functionally interact to produce different patterns of modification, characteristic of active or inactive chromatin.

At a still more complex level, the modification of one histone can affect the modification of another histone. An example of this was discussed in Section 3.3, with ubiquitination of histone H2B promoting methylation of lysines in histone H3. Similarly, methylation of histone H4 on the arginine at position 3 stimulates acetylation of histone H3 and its demethylation at lysines 9 and 27, producing the modification pattern characteristic of open chromatin (**Figure 3.35c**).

These various interactions result in particular patterns of modified histones. In turn, this has led to the idea of a "histone code" in which regulatory proteins are recruited to the DNA by recognizing a pattern of histone modifications rather than a single modification at a single site. This could involve a single protein recognizing multiple different modifications (**Figure 3.36a**). Alternatively, it could involve multiple proteins each of which recognizes an individual modification, with a code reader protein then recognizing the pattern of multiple bound proteins (**Figure 3.36b**). In either case, the end point would be the local modification of chromatin structure to produce either the open beads-on-a-string structure, the more tightly packed structure of the 30 nm fiber, or the even closer packing characteristic of heterochromatin.

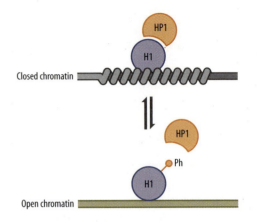

Figure 3.33
Phosphorylation of histone H1 disrupts its association with the HP1 protein. As HP1 promotes a more tightly packed chromatin structure (wavy line), its dissociation produces a more open chromatin structure (solid line).

Figure 3.34
The first 37 amino acids (in the one-letter amino acid code) at the N-terminus of histone H3. The figure shows the lysine (K) and arginine (R) residues modified by methylation together with the lysine (K) residues modified by acetylation and the serine (S) and threonine (T) residues modified by phosphorylation. Note that modifications which produce transcriptional activation are shown above the line and those producing repression are shown below the line.

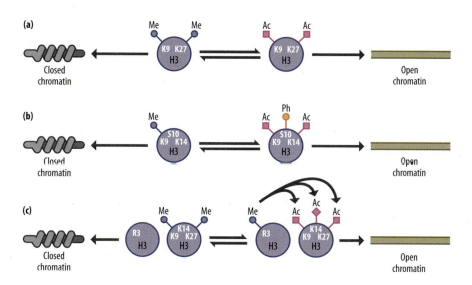

Figure 3.35
Examples of interaction between different modifications of the histones. Panel (a) shows the mutually exclusive acetylation or methylation at lysines (K) 9 and 27. Panel (b) shows interaction between different modifications of the same histone. Histone H3 is either methylated on lysine 9 or acetylated on lysines 9/14 and phosphorylated on serine (S) 10. Panel (c) shows interaction between modifications on different histones with methylation on arginine (R) 3 promoting acetylation of lysines 9, 14, and 27 of histone H3 and inhibiting methylation of histone H3 on lysines 9 and 27.

Histone modifications interact with DNA methylation to regulate chromatin structure

As discussed in Section 3.2, DNA methylation plays an important role in regulating chromatin structure, paralleling the role of histone modifications. Interestingly, as with the different histone modifications, there is clear evidence of an interaction between histone modifications and DNA methylation.

As discussed above (Section 3.3), the HP1 protein binds to histone H3 methylated on lysine 9 and plays a critical role in organizing the very tightly packed chromatin structure characteristic of heterochromatin (see Section 3.3). HP1 has been shown to recruit the DNA methyltransferases, Dnmt1, Dnmt3a, and Dnmt3b, so linking the inhibitory effects of H3 lysine 9 methylation and DNA methylation (**Figure 3.37**). A similar interaction with DNA

Figure 3.36
Multiple different histone (H) modifications could be recognized by a single regulatory protein, which then modifies chromatin structure (a). Alternatively, each modification could be recognized by a different protein, with a "code reader" protein then recognizing the multiple bound proteins (b). Note that in the case illustrated the proteins are recognizing a pattern of methylation, acetylation, and phosphorylation, which will produce a more open chromatin structure.

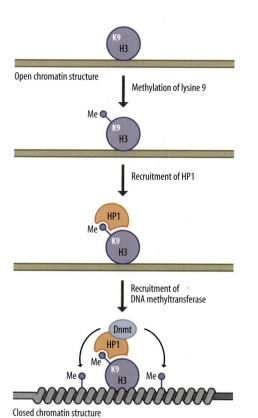

Figure 3.37
Methylation of histone H3 on lysine 9 recruits the HP1 protein which can both direct the chromatin into a more tightly packed structure and also recruit DNA methyltransferase enzymes (Dnmt). These enzymes direct methylation of DNA on C residues, further promoting a tightly packed chromatin structure.

methyltransferases has also been reported for the polycomb complex which can methylate lysines 9 and 27 of histone H3.

Interestingly, an interaction has recently been demonstrated between DNA methylation and one of the variant isoforms of the histones, discussed in Section 2.3. Thus, the histone variant H2A.Z, which is enriched in active genes, is able to protect such genes from DNA methylation while conversely DNA methylation prevents H2A.Z recruitment.

Hence, specific histone modifications not only influence the modification of other sites in the same or different histones but together with variant histone isoforms can also influence methylation of the DNA.

RNAi can induce alterations in chromatin structure

As discussed in Section 1.5 it is now clear that small RNA molecules (RNAi) play a key role in the inhibition of gene expression. These RNAs predominantly act at post-transcriptional levels inducing degradation of their target mRNA or blocking its translation (see Section 7.6). However, the small interfering RNAs (siRNAs) can also repress gene expression at the level of transcription by directing the formation of a tightly packed chromatin structure.

These siRNAs are produced by cleavage of larger double-stranded RNA precursors by the Dicer protein (see Chapter 1, Section 1.5). In cases where the siRNAs repress gene expression post-transcriptionally, they bind a complex known as RNA-induced silencing complex (**RISC**), which contains a number of proteins including a member of the **Argonaute** protein family. The complex of RISC and the siRNA then binds to its target mRNA (**Figure 3.38a**) (see Section 7.6). In contrast, where the siRNA represses gene transcription it binds a different complex which also contains an Argonaute protein and is known as the RNA-induced transcriptional silencing (**RITS**) complex. The siRNA and RITS complex then bind to target genes, resulting in their transcriptional repression (**Figure 3.38b**). The ability of siRNAs to repress gene expression at the transcriptional level was first described in plants, where it appears to be particularly widespread. However, it has also been described in a range of organisms from yeast to mammals.

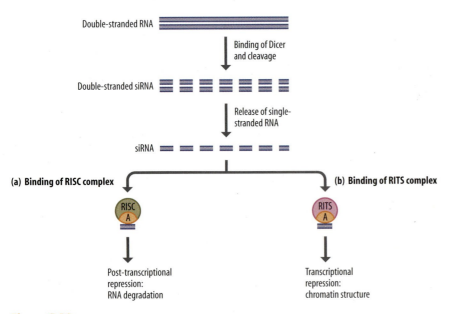

Figure 3.38
Double-stranded RNA is cleaved by the Dicer protein to yield siRNAs. Following binding to the RISC complex, which contains an Argonaute (A) protein, single-stranded siRNAs then bind to their target mRNA and induce its degradation (a). Alternatively, they can bind the RITS complex, which also contains an Argonaute protein. They then bind to target genes and direct the formation of a tightly packed chromatin structure (b).

In particular, siRNAs play a key role in the formation of the very tightly packed chromatin structure of heterochromatin (see Section 2.5). The siRNA molecules can bind to complementary sequences in the target gene and then recruit the HP1 protein. As described above, HP1 plays a key role in the formation of heterochromatin. It has the ability to produce inhibitory methylation of histones (see Section 3.3) and to recruit DNA methyltransferases which methylate the DNA on C residues (see Section 3.2).

The binding of siRNA therefore results in the methylation of histones at inhibitory sites, for example on lysine 9 of histone H3 and the methylation of DNA on C residues, thereby promoting the tightly packed structure of heterochromatin (**Figure 3.39**). As noted above (Section 3.3), methylation of histone H3 on lysine 9 promotes further binding of HP1, which can also catalyze further histone methylation, forming a positive-feedback loop. Hence, a complex interplay of siRNA, histone methylation, and DNA methylation exists which establishes and maintains the tightly packed structure of heterochromatin, as well as propagating it along the chromosome (**Figure 3.40**).

Clearly, the siRNA can achieve its effect on chromatin structure by binding to the DNA of its target gene and then recruiting histone methyltransferases and DNA methyltransferases, as shown in Figures 3.39 and 3.40 (**Figure 3.41a**). However, it is also possible for the small siRNA to bind to the gene by complementary base pairing to an RNA which is being transcribed from the target gene (**Figure 3.41b**). As before, this will result in recruitment of histone methyltransferases and DNA methyltransferases to the DNA leading to a more tightly packed chromatin structure.

The model of siRNA binding to an RNA rather than a DNA target is supported by the finding that many genes targeted by siRNAs also produce antisense transcripts derived from the opposite strand of the DNA to that which produces the protein-coding mRNA. These antisense RNAs can originate in the gene itself or alternatively can initiate in flanking DNA with

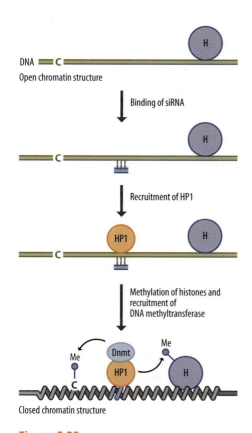

Figure 3.39
Binding of a small interfering RNA (siRNA) to its target gene results in the recruitment of the HP1 protein, which can direct the formation of a tightly packed chromatin structure by methylating histones (H) and recruiting DNA methyltransferases, which then methylate C residues in the DNA.

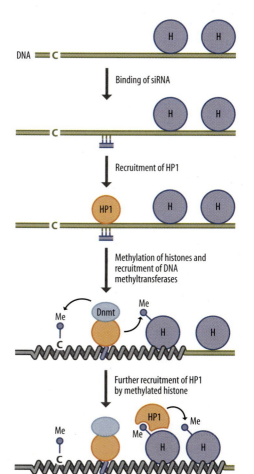

Figure 3.40
Positive-feedback loop in which HP1 recruitment results in methylation of histones which in turn promotes further recruitment of HP1. This ensures the stability of the tightly packed chromatin structure once it is created and its propagation down the chromosome.

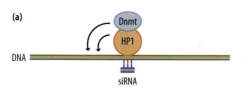

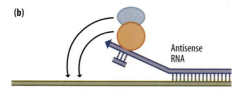

Figure 3.41
Binding of siRNA to its target gene may involve binding to the DNA (a) or to an RNA derived from the gene. In either case, this will lead to the recruitment of chromatin-modifying complexes, such as HP1 and DNA methyltransferases (Dnmt).

antisense transcription continuing into the gene itself. They can then act as a target for siRNA binding (see Figure 3.41b).

Although this mechanism of siRNA binding to an RNA target and recruiting enzymes which alter chromatin structure appears to occur in all eukaryotes, it has been particularly well characterized in plants, such as *Arabidopsis*. Interestingly, in this case the siRNAs are transcribed by a plant-specific RNA polymerase enzyme, known as RNA polymerase IV (also known as IVa), which is not found in animals. Similarly, the larger RNA targets for siRNA have been shown to be transcribed by another plant-specific RNA polymerase, which is known as RNA polymerase V (or IVb). These polymerases have some subunits in common with RNA polymerase II and others which are specific to RNA polymerase IV and/or RNA polymerase V (see Section 4.1 for discussion of RNA polymerases).

3.5 CHANGES IN CHROMATIN STRUCTURE IN THE REGULATORY REGIONS OF ACTIVE OR POTENTIALLY ACTIVE GENES

DNaseI-hypersensitive sites can be identified in active or potentially active genes

So far in this chapter we have seen that regions of chromatin containing active or potentially active genes have a number of distinguishing features, including under-methylation of C residues, histone modifications, and increased sensitivity to digestion with DNaseI. Such changes can extend over the entire region of the gene and some flanking sequences and in the case of DNaseI sensitivity result in an approximately tenfold increase in the rate at which active or potentially active genes are digested.

Following the discovery of increased DNaseI sensitivity, many investigators studied whether within the region of increased sensitivity there might be particular sites which were even more sensitive to cutting with the enzyme and which would therefore be cut even before the bulk of active DNA was digested. The technique used to look for such **DNaseI-hypersensitive sites** is based on that used to look at the overall DNaseI sensitivity of a particular region of DNA (which was described in Methods Box 3.1). Chromatin is digested with DNaseI and a restriction enzyme and then subjected to a Southern-blotting procedure using a probe derived from the gene of interest. As we have seen previously, the overall sensitivity of the gene can be monitored by observing how rapidly the specific restriction enzyme fragment derived from the gene disappears with increasing amounts of the enzyme (see Figure 3.3).

To search for DNaseI-hypersensitive sites, however, much lower concentrations of the enzyme are used and the appearance of discrete digested fragments derived from the gene is monitored (**Figure 3.42**; see **Methods Box 3.2** and compare with Methods Box 3.1). Such specific fragments have at one end the cutting site for the restriction enzyme used and at the other, a site at which DNaseI has cut, producing a defined fragment. Since the position at which the restriction enzyme cuts in the gene is known, the

---

Methods Box 3.2
DETECTING DNaseI-HYPERSENSITIVE SITES (FIGURE 3.42)

- Isolate chromatin (DNA and associated histones and other proteins).
- Digest with very small amounts of DNaseI.
- Purify partially digested DNA by removing protein.
- Digest with restriction enzyme and carry out Southern blotting with probe for gene of interest (see Methods Box 1.1).
- Monitor appearance of specific smaller band due to the presence of a DNaseI-hypersensitive site within the DNA being tested.

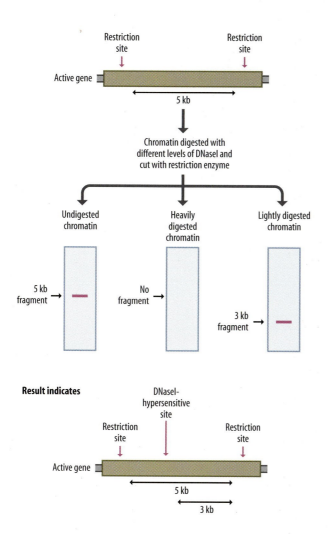

Figure 3.42
Detection of DNaseI-hypersensitive sites in active genes by mild digestion of chromatin to produce a digestion product with a restriction site at one end and a DNaseI hypersensitive site at the other (right-hand panel). More extensive digestion will result in the disappearance of the band (central panel) as in the experiment illustrated in Figure 3.3.

position of the hypersensitive site can be mapped simply by determining the size of the fragment produced.

Using this procedure a very wide variety of genes have been shown to contain such hypersensitive sites, exhibiting a sensitivity to DNaseI digestion tenfold above that of the remainder of an active gene and therefore about 100-fold above that seen in inactive DNA. A representative list of cases in which such sites have been detected is given in **Table 3.2**. As with the increased sensitivity of the gene itself, many hypersensitive sites appear only in tissues where the gene is active. The increased sensitivity of globin DNA in erythrocytes to digestion is paralleled by the presence of hypersensitive sites within the gene in erythrocyte chromatin but not in that of other tissues. Similarly, the ovalbumin gene in hormonally treated chick oviduct also exhibits hypersensitive sites that are not found in other tissues, including erythrocytes (**Figure 3.43**).

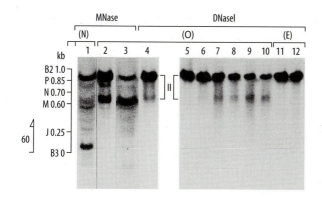

Figure 3.43
Detection of a DNaseI-hypersensitive site in the ovalbumin gene in oviduct tissue (O) but not in erythrocytes (E). Track 4 shows the detection of a lower band caused by cleavage at a hypersensitive site when oviduct chromatin is digested with DNaseI. Note the progressive appearance of this band as increasing amounts of DNaseI are used to digest the oviduct chromatin (tracks 5–10). No cleavage is observed when similar amounts of DNaseI are used to cut erythrocyte chromatin (tracks 11 and 12). The hypersensitive site in oviduct chromatin is also cleaved, however, with micrococcal nuclease (tracks 2 and 3). Track 1 shows the pattern produced by micrococcal nuclease cleavage of naked DNA (N). Courtesy of P Chambon, from Kaye JS, Bellard M, Dretzen G et al. (1984) *EMBO J.* 3, 1137–1144. With permission from Macmillan Publishers Ltd.

TABLE 3.2 EXAMPLES OF GENES CONTAINING DNaseI-HYPERSENSITIVE SITES

	EXAMPLES
Tissue-specific genes	
Immune system	Immunoglobulin, complement C4
Red blood cells	α-, β- and ε-globin
Liver	α-Fetoprotein, serum albumin
Nervous system	Acetylcholine receptor
Pancreas	Preproinsulin, elastase
Connective tissue	Collagen
Pituitary gland	Prolactin
Salivary gland	*Drosophila* glue proteins
Silk gland	Silk moth fibroin
Inducible genes	
Steroid hormones	Ovalbumin, vitellogenin, tyrosine aminotransferase
Stress	Heat-shock proteins
Viral infection	β-Interferon
Amino acid starvation	Yeast HIS 3 gene
Carbon source	Yeast GAL genes, yeast ADH II gene
Others	Histones, ribosomal RNA, 5S RNA, transfer RNA, cellular oncogenes, *c-myc* and *c-ras*, glucose-6-phosphate dehydrogenase, dihydrofolate reductase, cysteine protease, etc.

Similar to the pattern of DNA hypomethylation and the sensitivity of the entire gene to digestion, DNaseI-hypersensitive sites appear to be related to the potential for gene expression rather than always being associated with the act of transcription itself. For example, the hypersensitive sites near the *Drosophila* heat-shock genes are present in the chromatin of embryonic cells prior to any heat-induced transcription of these genes and one of the sites in the mouse α-fetoprotein gene persists in the chromatin of adult liver after the transcription of the gene (which is confined to the fetal liver) has ceased.

DNaseI-hypersensitive sites frequently correspond to regulatory DNA sequences

As with DNA methylation and generally increased sensitivity to DNaseI, the appearance of hypersensitive sites therefore appears to be involved in gene regulation. This idea is reinforced by the location of the hypersensitive sites which can be precisely mapped as described above. Many sites are located at the 5′ end of the genes in positions corresponding to DNA sequences that are known to be important in regulating transcription. For example, a site present at the 5′ end of the steroid-inducible tyrosine amino-transferase gene is localized within the DNA sequence that is responsible for the steroid inducibility of the gene. Even in cases where hypersensitive sites are located far from the site of transcriptional initiation, they appear to correspond to other **regulatory sequences** such as enhancers which can act over large distances (see Section 4.4 for a discussion of enhancers).

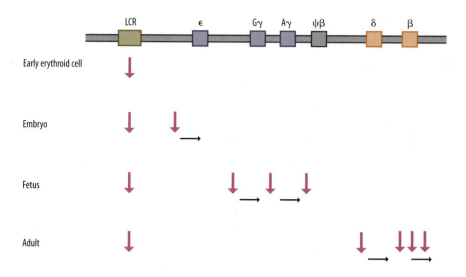

Figure 3.44
The appearance of multiple DNaseI-hypersensitive sites within the LCR (shown as a single arrow for simplicity) in early erythroid cells, precedes the appearance of other hypersensitive sites adjacent to each individual gene which occurs later in development, as these genes are expressed sequentially.

Interestingly, hypersensitive sites show a highly significant spatial and temporal pattern in the cluster of β-globin genes, which was described in Section 2.5. Very early in erythroid development, prior to the expression of the β-globin genes, multiple DNaseI-hypersensitive sites appear in the locus-control region (LCR) (**Figure 3.44**). This is consistent with the role of the LCR in directing the opening up of a large region of DNA allowing the subsequent activation of individual genes in the cluster. Indeed, the ε-globin, γ-globin, and β-globin genes are subsequently expressed successively in embryonic, fetal, and adult erythroid cells. In each case, expression is preceded by the appearance of DNaseI-hypersensitive sites in the regulatory region upstream of the individual gene (see Figure 3.44).

As noted in Section 2.5 the looping pattern of the β-globin cluster changes during erythroid development so that at each stage, the gene(s) being expressed at that stage are closely associated with the LCR to form an active chromatin hub. This involves an association of the LCR region containing multiple hypersensitive sites, with the regulatory region of the active gene(s) that contains the gene-specific hypersensitive site. This further supports the key role of hypersensitive sites in producing a chromatin structure that allows transcription of the appropriate gene(s) to occur.

In the case of the *Drosophila* gene encoding the glue protein Sgs4, the fortuitous existence of a mutant strain of fly has indicated the functional importance of hypersensitive sites. In normal flies this gene contains two hypersensitive sites, 405 and 480 bases upstream of the start of transcription. In the mutant, both sites are removed by a small DNA deletion of 100 bp. Despite the fact that this gene still has the start site of transcription and 350 bases of upstream sequences, no transcription occurs (**Figure 3.45**), indicating the regulatory importance of the region containing the hypersensitive sites.

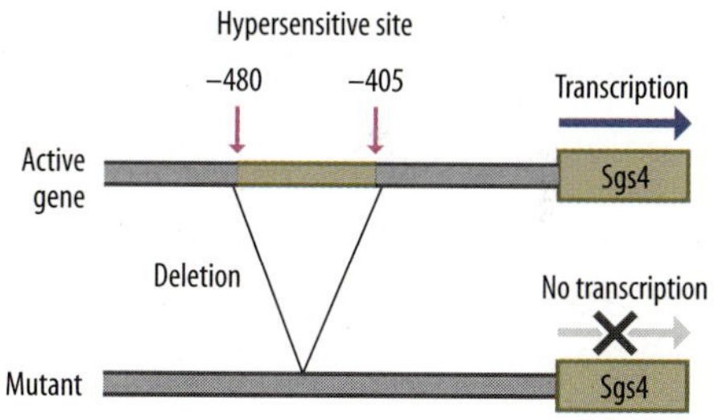

Figure 3.45
Deletion of a region containing the two hypersensitive sites upstream of the *Drosophila sgs4* gene abolishes transcription.

DNaseI-hypersensitive sites represent areas which are either nucleosome-free or have an altered nucleosomal structure

It is clear therefore that hypersensitive sites represent another marker for active or potentially active chromatin and are a feature likely to be of particular importance in gene regulation, being associated with many DNA sequences that regulate gene expression. It is therefore necessary to consider the nature and significance of these sites.

In some cases, DNaseI-hypersensitive sites are likely to be formed where DNA is entirely free of nucleosomes and is therefore highly sensitive to DNaseI digestion This is seen, for example, in the case of the DNaseI-hypersensitive sites within the enhancer element that regulates transcription of the eukaryotic virus **SV40**. When this virus enters cells its DNA, which is circular and only 5000 bases in size, becomes associated with histones in

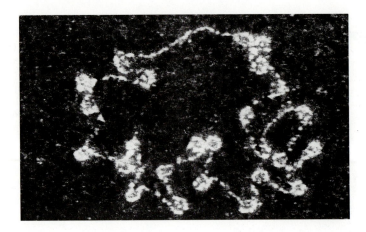

Figure 3.46
Electron micrograph of the SV40 mini-chromosome consisting of DNA and associated histones. Note the region of the enhancer and hypersensitive sites, which appears as a thin filament of DNA free of associated proteins. Courtesy of M Yaniv, from Saragosti S, Moyne G & Yaniv M (1980) *Cell* 20, 65–73. With permission from Elsevier.

a typical nucleosomal structure which can be visualized in the electron microscope as a mini-chromosome. When this is done, however, the region containing the hypersensitive sites remains nucleosome-free and is seen as naked DNA (**Figure 3.46**). A similar lack of nucleosomes in the region of hypersensitive sites is also found in the chicken β-globin gene, the 5′ hypersensitive site of this gene being excisable as a 115 bp **restriction fragment** lacking any associated nucleosome.

Although such cases indicate that hypersensitive sites can be produced by the complete loss of nucleosomes in a particular region, other cases exist where such sites are produced by an alteration in the structure of a nucleosome rather than its complete displacement (**Figure 3.47**). It is clear, however, that whether caused by nucleosome displacement or structural alterations, the changes occurring in hypersensitive sites facilitate the entry of transcription factors or the RNA polymerase itself and allow the initiation of transcription (see Figure 3.47). In view of the critical role played by the changes in nucleosomes which occur at hypersensitive sites, it is necessary therefore to consider the manner in which such alterations in nucleosome positioning or structure are produced.

Chromatin remodeling can be produced by proteins capable of displacing nucleosomes or altering their structure

As discussed in Section 2.2, protein factors have been identified which can displace nucleosomes or alter their structure, thereby facilitating the subsequent binding of transcription factors which stimulate transcription. In

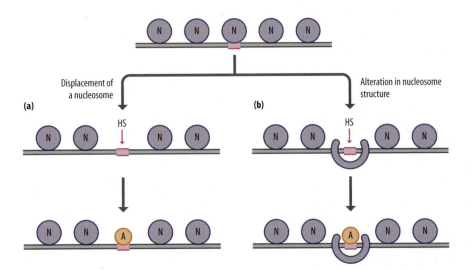

Figure 3.47
A hypersensitive site (HS) can be created either by displacement of a nucleosome (N) (a) or by alteration of its structure (b). In either case such an alteration allows a transcriptional activator (A) to bind to its binding site (pink bar) and activate transcription.

the case of the heat-shock genes, transcription of which is induced by elevated temperature, hypersensitive sites are produced by the binding of the GAGA protein factor to its upstream DNA-binding sites in the gene promoter which results in the displacement of a nucleosome (**Figure 3.48**). This binding of the **GAGA factor** occurs in cells prior to heat treatment and hence hypersensitive sites are present prior to heat treatment. Following heat treatment, a transcription factor known as the heat-shock factor (HSF) binds to this region of DNA and transcription is stimulated. In this case, HSF is only capable of binding following heat shock and stimulation of transcription therefore only occurs following such treatment (**Figure 3.49a**) (see Sections 4.3 and 8.1 for further discussion of HSF and its mechanism of action).

In other cases, however, where the necessary transcription factors are present in all tissues, transcription may follow immediately after the nucleosome-free region is generated, allowing these factors access. In the case of glucocorticoid-responsive genes, the critical regulatory event is the binding

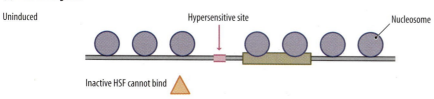

Figure 3.48
The binding of the GAGA factor to its binding site in the heat-shock genes displaces a nucleosome (N), thereby exposing the HSE binding site for the heat-shock factor (HSF). HSF then binds and activates transcription.

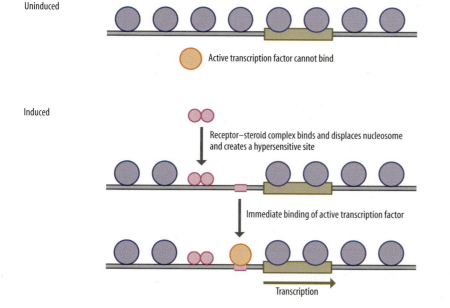

Figure 3.49
Two mechanisms for transcriptional activation. (a) The heat-shock transcription factor (HSF) is activated by heat and binds to a pre-existing nucleosome-free region. (b) The receptor–steroid complex displaces a nucleosome, creating a hypersensitive site, and allowing an active transcription factor to bind. The pink bar indicates the position of a hypersensitive site. Note that although the figure shows the displacement of a nucleosome, it is also possible that generation of the hypersensitive site may involve the alteration of nucleosome structure, as illustrated in Figure 3.47.

of the glucocorticoid receptor–steroid complex to a particular DNA sequence (the **GRE** or glucocorticoid-response element), which displaces a nucleosome or alters its structure and generates a DNaseI-hypersensitive site. Ubiquitous transcription factors present in all tissues, such as NFI and the **TATA box**-binding factor (**TBP**) (see Section 4.1), immediately bind to this region and transcription begins (**Figure 3.49b**) (for further details of the glucocorticoid receptor and its mode of action see Sections 5.1 and 8.1).

Although these two situations appear different in terms of the time at which the hypersensitive site appears relative to the onset of transcription, they illustrate the basic role of hypersensitive sites, namely the displacement of nucleosomes or alteration of their structure and the generation of a site of access for regulatory proteins.

The SWI–SNF and NURF chromatin-remodeling complexes are recruited to the DNA by a variety of different mechanisms

Interestingly, the glucocorticoid receptor–steroid complex does not directly alter nucleosomal structure. Rather, it acts by recruiting the multi-protein complex known as the SWI–SNF complex which, as described in Section 2.2, is able to hydrolyze ATP and use this energy to alter nucleosomal structure, so facilitating the subsequent binding of activator molecules.

The role of the SWI–SNF complex is not confined to steroid-responsive genes and it appears that it can be recruited to a wide variety of different genes by regulatory proteins which bind to these genes. The SWI–SNF complex then acts to alter the chromatin structure of these genes, thereby facilitating their subsequent activation by other transcription factors (**Figure 3.50**). In agreement with this idea, the brahma mutation in *Drosophila* inactivates the SWI2 component of the complex, which produces the ATP-hydrolyzing activity. This results in a failure to activate the homeotic genes which play a key role in determining body pattern and produces a mutant fly with an abnormal body pattern (see Sections 5.1 and 9.2 for discussion of homeotic genes and their regulation).

The SWI–SNF complex is therefore involved in altering the chromatin structure and thereby facilitating the transcriptional activation of genes as diverse as steroid-responsive genes and homeotic genes. Moreover, once nucleosome disruption has been produced by SWI–SNF it can dissociate from the gene, since the alteration in nucleosome structure and DNaseI-hypersensitive site produced by SWI–SNF persists even after it has dissociated.

The activity of the SWI–SNF complex has been shown to be modified by interaction with the linker histone H1. In the absence of histone H1, SWI–SNF promotes the displacement of nucleosomes to the end of the DNA molecule, whereas in the presence of histone H1 SWI–SNF promotes controlled nucleosome displacement. Hence, in addition to its role in the 30 nm-fiber structure of chromatin (see Section 2.4), histone H1 also regulates the displacement of nucleosomes to the correct position on the DNA (**Figure 3.51**).

As with SWI–SNF, the GAGA factor described above also plays a general role in chromatin remodeling and does not act solely on the heat-shock genes. Inactivation of the gene encoding the GAGA factor in *Drosophila* results in a mutant fly with an altered body pattern known as trithorax, which is similar to the brahma mutation in that a wide range of homeotic genes are not activated. Moreover, the GAGA/trithorax factor is also associated with another multi-protein complex, known as NURF, which resembles the SWI–SNF factor in its ability to hydrolyze ATP and alter nucleosome structure (see Sections 3.3 and 4.4 for further discussion of trithorax proteins).

The alteration of chromatin structure at regulatory regions is brought about by multi-protein complexes such as SWI–SNF or NURF which are active on a wide range of genes. These complexes are likely to be recruited to specific gene promoters by interaction with DNA-binding proteins such as the glucocorticoid receptor or GAGA which have already bound to the

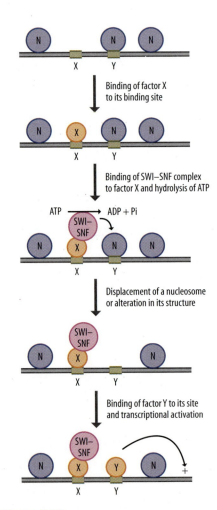

Figure 3.50

A regulatory factor (X) binds to its DNA-binding site (X) and recruits the SWI–SNF complex. This complex hydrolyzes ATP to ADP and inorganic phosphate (Pi) and uses the energy generated to alter the structure of a nucleosome which was masking the binding site for the transcriptional activator (Y). This allows Y to bind to its site and activate transcription.

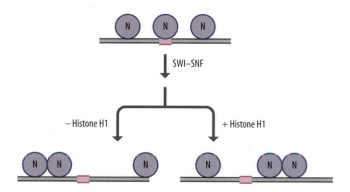

Figure 3.51
In the absence of histone H1, the SWI–SNF complex displaces nucleosomes (N) to the end of the DNA fragment. In contrast, in the presence of histone H1 it promotes the controlled nucleosome displacement required for specific gene activation.

DNA in a gene-specific manner. In turn this binding facilitates the subsequent binding of other activating molecules by altering nucleosome positioning or structure in an ATP-dependent manner.

The SWI–SNF complex has also been shown to be associated with the RNA polymerase II holoenzyme which, as discussed in Section 4.1, is a complex of RNA polymerase II with **basal transcription factors** such as TFIIB, TFIIF, and TFIIH. Moreover, it has been shown that the recruitment of the RNA polymerase II complex containing SWI–SNF factors can result in the opening up of chromatin, indicating that the SWI–SNF complex can also function when brought to the DNA in this way. In agreement with this, the TATA box which binds TFIID and hence recruits the RNA polymerase II complex has been shown to be of critical importance for recruiting the complexes which remodel the chromatin containing the globin gene promoter.

As well as being recruited by specific transcription factors or together with the RNA polymerase II holoenzyme, it has also been shown that SWI–SNF can be recruited to DNA by the SATB1 protein. As SATB1 is involved in the looping of the chromatin (see Section 2.5), this provides a link between such looping and the chromatin-remodeling/gene-regulation functions carried out by SWI–SNF.

Multiple mechanisms may therefore act to recruit the SWI–SNF and NURF complexes to a particular gene. Whatever the manner in which they are recruited, however, it is clear that these ATP-dependent chromatin-remodeling complexes play a key role in opening up the regulatory regions of specific genes so that *trans*-acting transcription factors can bind and regulate their expression. Moreover, such remodeling processes relate closely to the histone modifications discussed in Section 3.3. For example, it has been shown that acetylation of histones facilitates recruitment of SWI–SNF to the interferon-β promoter and such histone modification also prevents SWI–SNF dissociating once it has bound. Similarly, as discussed in Section 3.3, NURF can be recruited to the DNA by recognizing histones methylated at positions associated with an opening of the chromatin structure. Histone modification and chromatin-remodeling complexes therefore appear to act in concert to open up the chromatin for transcription.

Although we have discussed the opening of chromatin structures by these complexes and its role in transcriptional activation, chromatin-remodeling complexes can also produce a more closed chromatin structure leading to transcriptional repression. Hence, these complexes appear to play a key role in the regulation of gene expression via the alteration of chromatin structure.

3.6 OTHER SITUATIONS IN WHICH CHROMATIN STRUCTURE IS REGULATED

In this chapter we have discussed the role of changes in the chromatin structure of individual genes in mediating commitment to a particular differentiated state and thereby allowing the tissue-specific activation of gene expression. However, differences in chromatin structure are also involved in regulating gene expression in two other well-characterized processes, namely X chromosome inactivation and genomic imprinting. Both of these processes involve differences in expression between the two copies of a specific gene which are present on different **homologous chromosomes** in a diploid cell.

In female mammals one of the two X chromosomes is inactivated

The fact that females of mammalian species have two X chromosomes whereas males have one X and one Y chromosome creates a problem of how to compensate for the difference in dosage of genes on the X chromosome, which occurs because females have two copies of each gene whereas males have only one. In mammals, this problem is solved by the process of X-chromosome inactivation.

During the process of embryonic development, one of the two X chromosomes in each female cell undergoes an inactivation process so that the expression of virtually all the genes on this chromosome is inactivated while those on the other chromosome remain active. This results in each female cell having only one active copy of X-chromosome genes, paralleling the situation in male cells which have only one X chromosome to start with (**Figure 3.52**).

This process occurs randomly with individual differentiating cells in the **inner cell mass** of the early female embryo inactivating either the X chromosome inherited from the father (the paternal X chromosome) or that inherited from the mother (the maternal X chromosome) (see Section 9.1 for discussion of inner cell mass cells and their differentiation to form the cells which give rise to the different cell types of the embryo). However, once one or other of the X chromosomes has been inactivated, the inactivation is propagated stably through cell division to all the progeny of that cell (see Figure 3.52).

The active and inactive X chromosomes have a different chromatin structure

The stable propagation of X chromosome inactivation is dependent upon the fact that the inactive and active X chromosomes have a different chromatin structure. Within the inactive X chromosome, the DNA is tightly packed into the highly condensed structure of heterochromatin (see Section 2.5). This high-density structure results in the inactive X chromosome being visible as a distinct element within the cell known as a **Barr body**.

Such a condensed structure has been shown to exhibit decreased sensitivity to digestion with DNaseI, altered histone modification, and enhanced methylation on C residues compared to the equivalent regions on the active X chromosome. For example, 60 of the 61 CG dinucleotides in the CG island (see Section 3.2) located around the promoter of the PGKI gene are methylated on the C residue when the inactive X chromosome is studied, whereas all these sites are unmethylated on the active X chromosome. Moreover,

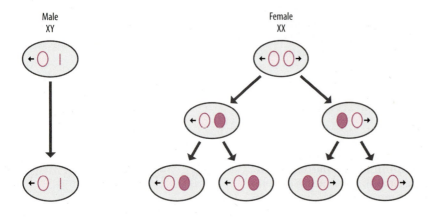

Figure 3.52
X-chromosome inactivation results in one or other of the two X chromosomes becoming inactivated (dark color) in each cell while the other remains active (light color).

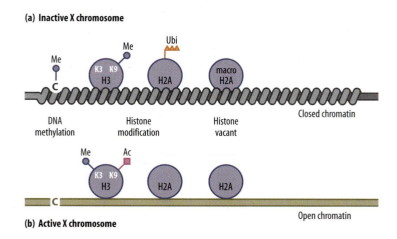

(a) Inactive X chromosome

DNA methylation   Histone modification   Histone vacant   Closed chromatin

(b) Active X chromosome   Open chromatin

Figure 3.53
The inactive X chromosome shows enhanced methylation of C residues in DNA, a different pattern of histone modifications and histone variants (a) compared to the active X chromosome (b).

treatment with 5-azacytidine which leads to demethylation (see Section 3.2) can reactivate previously inactivated X chromosome genes.

As well as changes in DNA methylation, the inactive X chromosome also exhibits the changes in histones which are characteristic of tightly packed heterochromatin (see Sections 3.3 and 3.4). These include methylation of histone H3 on lysines 9 and 27, as well as ubiquitination of histone H2A. In contrast, the inactive X chromosome shows reduced histone acetylation and methylation at residues associated with activation such as lysine 4 of histone H3. Interestingly, the inactive X chromosome is also enriched in the variant histone, macro H2A, which is associated with transcriptional repression (see Section 2.3 for a discussion of histone variants). Hence, compared to the active X chromosome, the inactive X chromosome shows the modifications of DNA and histones associated with a tightly packed chromatin structure (**Figure 3.53**).

Interestingly, in *Drosophila* compensation for the reduced number of X chromosomes in male cells is achieved in the embryo by doubling the transcriptional activity of the genes on the single male X chromosome rather than by inactivating one of the female X chromosomes. As in the case of **X inactivation**, however, this effect appears to involve alterations in chromatin structure since the male X chromosome has a higher level of acetylated histones than either of the female chromosomes (see Section 3.3). Moreover, a HAT enzyme specifically associates with the male X chromosome.

The XIST regulatory RNA is specifically transcribed on the inactive X chromosome

Chromatin structure therefore plays a critical role in differentially regulating the activity of genes on the X chromosomes in females and males and in particular in the maintenance of X-chromosome inactivation through cell division. The onset of X-chromosome inactivation in the embryo requires a particular region of the chromosome known as the **X-inactivation center**. If this region is deleted then X inactivation does not occur. A gene known as **XIST** has been mapped to the X-inactivation center. It has been shown that the inactivation of one copy of the *XIST* gene in mutant mice results in a failure of X-chromosome inactivation on the chromosome which lacks the active *XIST* gene, with the other X chromosome being preferentially inactivated (**Figure 3.54a**). Hence, *XIST* is essential for inactivation of the X chromosome from which it is expressed.

The *XIST* gene is transcribed only from the inactive X chromosome and not on the active chromosome, the opposite pattern to all other genes on

Figure 3.54
Inactivation of one copy of the *XIST* gene in mouse mutants results in the chromosome containing the intact *XIST* gene being preferentially inactivated (a). Mutant mice that transcribe the *XIST* gene on both X chromosomes show inactivation of both X chromosomes (b).

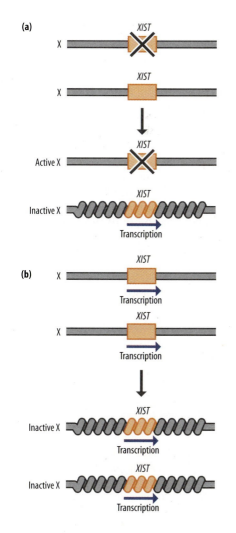

(a)

X   *XIST*

X   *XIST*

Active X   *XIST*

Inactive X   *XIST*   Transcription

(b)

X   *XIST*   Transcription

X   *XIST*   Transcription

Inactive X   *XIST*   Transcription

Inactive X   *XIST*   Transcription

the X chromosome. Its critical role in X inactivation is demonstrated by the finding that mutant mice which show expression of *XIST* from both chromosomes also show inactivation of both X chromosomes. This is exactly as would be expected if the expression of the *XIST* gene does indeed inactivate genes on the X chromosome containing it (**Figure 3.54b**). Moreover, if an active *XIST* gene is placed on a non-X chromosome, the transcription of the genes on this chromosome is inactivated. Hence, *XIST* gene transcription is sufficient to inactivate genes to which it is linked regardless of the nature of these genes.

It has been shown that the transcription of *XIST* early in embryonic development on one of the two chromosomes results in that X chromosome being put into a tightly packed heterochromatin structure (**Figure 3.55**). The *XIST* gene produces a large, 17 kb RNA transcript which is not translated into protein. Rather, multiple copies of the XIST transcript bind all along the length of the inactive X chromosome. As is the case for the much smaller siRNAs (see Section 3.4), the binding of the XIST RNA results in the recruitment to the chromosome of protein complexes which can produce the modifications characteristic of tightly packed heterochromatin and which are incompatible with transcription. Specifically, binding of XIST results in the recruitment of polycomb complexes (see Section 3.3), which can methylate histone H3 on lysines 9 and 27 and also ubiquitinate histone H2A (**Figure 3.56**). In this way, transcription of XIST on one of the two X chromosomes results in a change in chromatin structure which is propagated along the rest of that chromosome, switching off all other genes.

In contrast to the transcription of XIST on the inactive X chromosome, the X-inactivation center on the active X chromosome is transcribed to produce a non-protein-coding transcript, known as TSIX. This transcript is produced from the opposite strand of the DNA to that which produces XIST and the two transcripts overlap (**Figure 3.57**).

Such transcription of TSIX by the active X chromosome is necessary for X-chromosome inactivation, since inactivation of TSIX interferes with this process. Evidently, the overlap between the XIST and TSIX RNA transcripts creates the possibility of their forming a double-stranded RNA, which in turn could result in the production of small interfering RNAs (see Sections 1.5 and 3.4).

Interestingly, inactivation of the Dicer protein, which is required for siRNA production, results in abnormal expression of XIST on the normally active X chromosome. Hence, siRNAs produced from a XIST–TSIX double-stranded RNA may function to inhibit XIST transcription on the normally active X chromosome, perhaps by organizing an inactive chromatin structure at the XIST promoter.

The initiation and maintenance of X chromosome inactivation therefore requires the ability to produce and stably propagate an altered chromatin structure which we have previously seen to be critical in tissue-specific gene regulation. Moreover, like the regulation of the chromatin structure of individual genes it involves processes such as methylation of C residues in the DNA, modification of histones, and the organization of a tightly packed chromatin structure by inhibitory RNAs.

Genomic imprinting involves the specific inactivation of either the maternally or paternally inherited copy of specific genes

Genomic imprinting resembles X-chromosome inactivation in that one of the two copies of specific genes is inactivated while the other remains active. This process differs, however, in that about 80 genes scattered on different

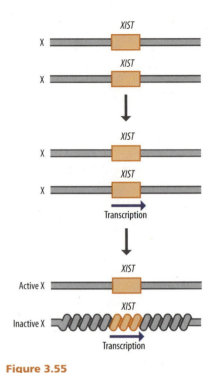

Figure 3.55
In normal mice, transcription of the *XIST* gene on one of the two X chromosomes is associated with its inactivation via a change in chromatin structure.

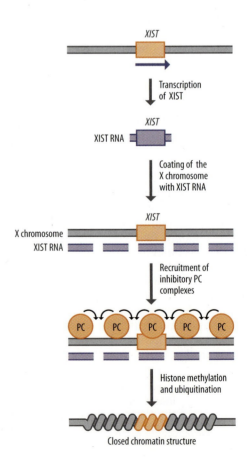

Figure 3.56
Transcription of XIST produces a 17 kb RNA which coats the X chromosome from which it was transcribed. This recruits polycomb (PC) protein complexes which can methylate and ubiquitinate histones, so producing a tightly packed inactive chromatin structure.

Figure 3.57
The XIST and TSIX RNAs are transcribed from opposite strands of the X chromosome and overlap one another (a). XIST is transcribed from the inactive X chromosome and TSIX only from the active X chromosome (b).

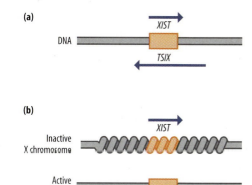

chromosomes have been shown to be imprinted in mammals. Moreover, unlike X inactivation the same copy is always inactivated in all cells and in all organisms whether male or female. It is always the maternally inherited copy of some imprinted genes, such as the genes encoding the insulin-like growth factor 2 (IGF2) protein, the SmN **splicing** protein, and the U2AF35-related protein which are inactivated, with the paternally inherited gene remaining active. Conversely, the paternally inherited copies of other imprinted genes, such as the IGF2 receptor gene and the *H19* gene, are inactivated with the maternally inherited gene remaining active (**Figure 3.58**).

As with X inactivation therefore this process results in all cells having only one functional copy of each of these genes, although in the case of genomic imprinting all cells express the same copy. This process cannot be reversed during embryonic development, even when genetic crosses are used to produce embryos with two imprinted copies of the gene. Embryos which inherit two maternal and no paternal copies of the *Igf2* gene or two maternal but no paternal copies of the *SmN* gene die due to the lack of an active gene and therefore of a functional protein even though two copies of the gene capable of encoding this protein are present in each cell of the embryo.

Despite the lethal effects which can result when imprinting goes wrong, the normal function of this process is unclear. It has been suggested for example that imprinting may represent a means of preventing the parthenogenetic development of the unfertilized egg to produce a haploid embryo with no paternal contribution. Alternatively, it may have evolved because of the conflict between the maternal and paternal genomes in terms of the transfer of nutrients from mother to offspring. Although it is in the paternal interest to promote the growth of the individual fetus, the maternal interest is to restrict the growth of any individual fetus so that other fetuses fathered by different males, either concurrently (in multi-fetal litters) or subsequently, can develop fully. Other theories range from the need for the cell to

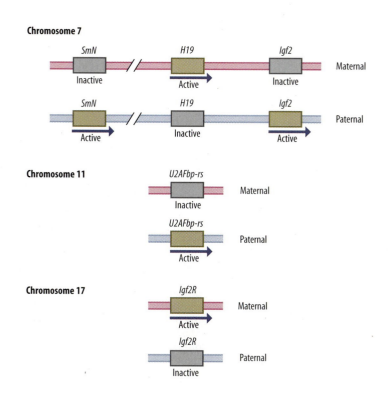

Figure 3.58
Imprinting results in the inactivation of the maternally inherited *SmN, U2AFbp-rs,* and *Igf2* genes and of the paternally inherited *Igf2R* and *H19* genes.

distinguish between the two copies of each chromosome which have been inherited from the mother or the father to a mechanism for producing differences between male and female offspring.

This lack of a clear functional role for genomic imprinting is in contrast to the clear role of X-chromosome inactivation in compensating for the extra X chromosome in females compared to males. Indeed, it has even been suggested that imprinting may not have a function at all but may represent a vestige of an evolutionarily ancient defense system to inactivate foreign DNA, with the imprinted genes having some feature which causes this system to consider them as foreign.

Imprinting involves changes in chromatin structure

Whatever its precise function (if any), it is clear that genomic imprinting resembles X-chromosome inactivation in that specific differences exist in the methylation pattern of CG dinucleotides between the active and inactive copies of the imprinted gene. Moreover, embryos which lack the DNA methyltransferase Dnmt3a (see Section 3.2) fail to carry out genomic imprinting. Hence specific methylation of one of the two copies of the gene is critical for imprinting to occur.

Interestingly, in some cases, the role of C methylation in regulating imprinting can be quite complex. In the case of the *H19* gene, which is silenced on the paternal chromosome, a simple correlation exists with the regulatory region of the gene being methylated on the paternal chromosome but not on the maternal as would be expected (**Figure 3.59**). However, in the case of the adjacent *Igf2* gene, which is silenced on the maternal chromosome, no specific methylation pattern is evident on this chromosome compared to the paternal one. Rather, the activity of the *Igf2* gene is controlled by an imprinting control region (ICR) which is located between the *H19* and *Igf2* genes and which is methylated on the paternal chromosome where the *Igf2* gene is active.

This paradox is resolved by the finding that when the ICR is methylated, a positive regulatory element (known as an enhancer; see Section 4.4) located downstream of the *H19* gene acts at a distance to activate *Igf2* gene expression (see Figure 3.59). In contrast, when the ICR is unmethylated on the maternal chromosome, a protein known as CTCF binds to it. This protein acts as an insulator (see Section 2.5) preventing the positive regulatory element from activating *Igf2* and the gene is therefore silent. In agreement with this idea, mutations in the binding site for CTCF disrupt the correct pattern of expression of the *H19* and *Igf2* genes. Hence, in this chromosomal region, methylation of C residues on the paternal chromosome results in silencing of the *H19* gene and expression of the *Igf2* gene on this chromosome.

As well as DNA methylation, the *Igf2/H19* system also involves several other features of chromatin structure which have been discussed elsewhere. For example, CTCF is able to recruit the CHD8 protein, which is a chromodomain-containing helicase protein (see Section 3.3) able to remodel chromatin structure to a more tightly packed configuration (**Figure 3.60**). This

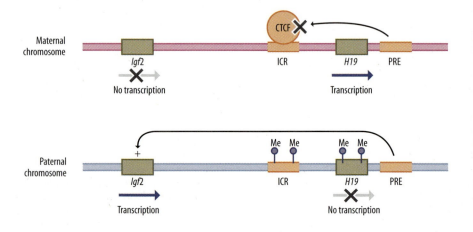

Figure 3.59
Role of C residue methylation (Me) in controlling imprinting of the closely linked *Igf2* and *H19* genes. On the paternal chromosome, the *H19* gene is methylated and is therefore only expressed from the maternal chromosome where it is unmethylated. In contrast, demethylation of the imprinting control region (ICR) on the paternal chromosome prevents binding of an insulator protein (CTCF) and allows a distinct positive regulatory element (PRE) to specifically activate *Igf2* expression only on the paternal chromosome.

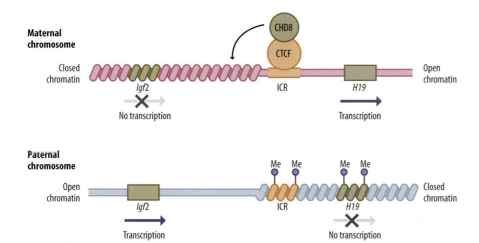

Figure 3.60
The CTCF insulator protein can bind the CHD8 helicase protein which can produce an inactive chromatin structure preventing transcription of the *Igf2* gene on the maternal chromosome. This effect does not occur on the paternal chromosome where DNA methylation of the imprinting control region (ICR) prevents binding of CTCF and therefore of CHD8.

idea that CTCF fulfils its insulator role by altering chromatin structure has been confirmed by studies in the β-globin gene cluster, where binding of CTCF has been shown to promote histone modification and looping of the chromatin (see Section 2.5 for discussion of the β-globin gene cluster). Hence, CTCF can remodel the chromatin, to create a boundary between regions of DNA with different chromatin structures (see Figure 3.60).

The *H19* gene does not produce a protein but rather produces a non-protein-coding RNA which, as described above, has the opposite pattern of expression to that of the protein coding *Igf2* gene (see Figure 3.59). This pattern of protein coding genes being expressed on the opposite chromosome to that expressing a non-protein-coding RNA is seen in several other imprinted gene clusters. For example, the Air non-coding RNA transcript overlaps that of the *Igf2 receptor* gene but is transcribed from the opposite strand of the DNA. The Air RNA is transcribed only from the paternal chromosome and specifically silences the *Igf2 receptor* gene on this chromosome. This results in the *Igf2 receptor* gene being expressed only from the maternal chromosome. In addition, the Air RNA also silences two other imprinted genes on the same chromosome (**Figure 3.61**). Interestingly, these effects have been shown to involve the Air RNA recruiting histone methyltransferases that methylate histone H3 on lysine 9, a modification which, as described above (Section 3.3) is associated with a tightly packed chromatin structure.

Hence in imprinting, as with XIST (see above), transcription of a non-protein-coding RNA on a particular chromosome is involved in silencing protein coding genes on that chromosome by producing a tightly packed chromatin structure incompatible with transcription.

It is clear therefore that imprinting involves similar modifications and changes in chromatin structure to those which are used to regulate cellular commitment and gene transcription in specific cell types as well as X-chromosome inactivation. Moreover, although X-chromosome inactivation in embryonic cells differs from imprinting in that it is random rather

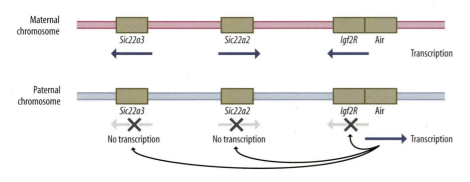

Figure 3.61
The Air noncoding RNA is transcribed only on the paternal chromosome and specifically silences transcription of the *Igf2 receptor, Sic22a2,* and *Sic22a3* genes on that chromosome. Note that Air can silence both the overlapping Igf2R transcript and the more distant *Sic22a2* and *Sic22a3* genes. Arrows indicate direction of transcription.

than specifically inactivating either the paternal or maternal X chromosome, this is not always the case. As discussed in Section 9.1 X-chromosome inactivation in extra-embryonic tissue always targets the paternal X chromosome, although the mechanisms of such inactivation are similar to those in embryonic cells and for example involve XIST.

CONCLUSIONS

A variety of changes take place in the chromatin of genes during the process of commitment to a particular pathway of differentiation. Such changes involve both modification of the DNA itself by under-methylation, to the histones with which it is associated and to the general packaging of the DNA in chromatin. In the last few years, the study of these changes has moved from the simple description of changes such as DNaseI sensitivity to a more mechanistic approach. This has identified three key processes regulating chromatin structure; namely, DNA methylation (see Section 3.2), histone modifications, particularly acetylation and methylation (see Section 3.3), and ATP-dependent remodeling of chromatin structure by complexes such as SWI–SNF and NURF (see Section 3.5). Indeed, these processes are closely linked as described earlier, with DNA methylation stimulating histone changes such as deacetylation and methylation, while in turn such changes in histone modification can regulate SWI–SNF recruitment to the promoter (see Section 3.4).

Although protein complexes able to, for example, methylate DNA or modify histones play a key role in these events, it is clear that regulatory RNA molecules which do not encode proteins are also involved in controlling chromatin structure. These include the siRNAs (see Section 3.4), as well as larger RNAs such as antisense transcripts (see Section 3.4), the XIST RNA (see Section 3.6), and non-coding transcripts involved in imprinting (see Section 3.6).

Although all these RNAs produce an inhibitory effect, other regulatory RNAs can stimulate gene expression (see, for example, Sections 4.4 and 8.1), suggesting that regulatory RNAs may have widespread roles. This may provide an explanation for an unexpected finding of the ENCODE human genome project (see Section 4.3 for discussion of this project). This study demonstrated that a very large proportion of the human genome which cannot encode protein is nonetheless transcribed into RNA. Moreover, it also identified many antisense RNAs which overlap with sense transcripts from protein-coding genes.

The combination of all the various processes discussed in this chapter results in three basic levels of chromatin structure within the cell (**Figure 3.62**). Although the bulk of inactive DNA is organized into a tightly packed 30 nm-fiber structure, active or potentially active genes are organized into a more open beads-on-a-string structure, and short regions within the gene are either nucleosome free or have structurally altered nucleosomes.

The role of these changes in allowing cells to maintain a commitment to a particular differentiated state and to respond differently to inducers of gene expression is well illustrated in the case of the steroid hormones and their effect on gene expression. As described in Section 2.1, different tissues will respond differently to treatment with estrogen and this will occur even though they contain the appropriate receptor for the hormone (see Section 5.1). This is likely to be due to the fact that, in one tissue, certain steroid-responsive genes will be inaccessible within the 30 nm-fiber structure and will therefore be incapable of binding the receptor–hormone complex that is necessary for activation. In other genes, which are in the beads-on-a-string structure and therefore more accessible such binding of the complex to defined sequences in the gene will occur. Even in this case, however, gene activation will not occur as a direct consequence of this interaction, as might be the case in bacteria. Rather, the binding will result in the displacement of a nucleosome from the region of DNA, generating a hypersensitive site and allowing other regulatory proteins to interact with their specific recognition sequences and cause transcription to occur.

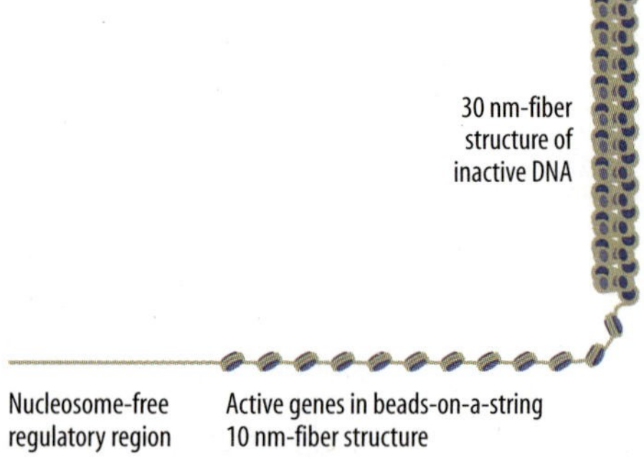

30 nm-fiber structure of inactive DNA

Nucleosome-free regulatory region

Active genes in beads-on-a-string 10 nm-fiber structure

Figure 3.62
Levels of chromatin structure in active and inactive DNA.

Both the action of the receptor–hormone complex and the subsequent binding of other transcription factors to nucleosome-free DNA clearly involve the interaction of regulatory proteins with specific DNA sequences to regulate transcription by RNA polymerase II. The basic process of transcription will be described in Chapter 4 and its regulation by specific transcription factors will be discussed in Chapter 5.

KEY CONCEPTS

- In eukaryotes gene regulation involves long-term changes which allow a cell to become and remain committed to a particular pattern of gene expression.

- These changes occur prior to a gene becoming active and involve an alteration in the chromatin structure of the whole gene from the tightly packed 30 nm-fiber structure to the more open beads-on-a-string structure.

- These changes result in the chromatin of active or potentially active genes exhibiting:
  (a)   Enhanced sensitivity to digestion with DNaseI
  (b)   Under-methylation on specific C residues in the DNA
  (c)   Changes in the post-translational modifications of the histones associated with the DNA.

- At specific regulatory sites within the gene, chromatin-remodeling complexes produce greater changes in the chromatin structure, resulting in the appearance of sites which are hypersensitive to DNaseI digestion.

- As well as its role in global gene regulation, alterations in chromatin structure are also involved in other biological processes such as X-chromosome inactivation and genomic imprinting.

FURTHER READING

General

American Association for Cancer Research Human Epigenome Task Force and the European Union, Network of Excellence Scientific Advisory Board (2008) Moving AHEAD with an international human epigenome project. *Nature* 454, 711–715.

Bernstein BE, Meissner A & Lander ES (2007) The mammalian epigenome. *Cell* 128, 669–681.

Lande-Diner L and Cedar H (2005) Silence of the genes–mechanisms of long-term repression. *Nat. Rev. Genet.* 6, 648–654.

Mohn F & Schubeler D (2009) Genetics and epigenetics: stability and plasticity during cellular differentiation. *Trends Genet.* 25, 129–136.

Reik W (2007) Stability and flexibility of epigenetic gene regulation in mammalian development. *Nature* 447, 425–432.

3.1   Changes in chromatin structure in active or potentially active genes

Stalder J, Groudin M, Dodgson JB, Engel JD & Weintraub H (1980) Hb switching in chickens. *Cell* 19, 973–980.

3.2   Alterations in DNA methylation in active or potentially active genes

Beck S & Rakyan VK (2008) The methylome: approaches for global DNA methylation profiling. *Trends Genet.* 24, 231–237.

Hendrich B & Tweedie S (2003) The methyl-CpG binding domain and the evolving role of DNA methylation in animals. *Trends Genet.* 19, 269–277.

Jones PA & Takai D (2001) The role of DNA methylation in mammalian epigenetics. *Science* 293, 1068–1070.

Klose RJ & Bird AP (2006) Genomic DNA methylation: the mark and its mediators. *Trends Biochem. Sci.* 31, 89–97.

Ooi SK & Bestor TH (2008) The colorful history of active DNA demethylation. *Cell* 133, 1145–1148.

3.3   Modification of histones in the chromatin of active or potentially active genes

Berger SL (2007) The complex language of chromatin regulation during transcription. *Nature* 447, 407–412.

Corpet A & Almouzni G (2009) Making copies of chromatin: the challenge of nucleosomal organization and epigenetic information. *Trends Cell. Biol.* 19, 29–41.

Hartzog GA & Quan TK (2008) Just the FACTs: histone H2B ubiquitylation and nucleosome dynamics. *Mol. Cell* 31, 2–4.

Kouzarides T (2007) Chromatin modifications and their function. *Cell* 128, 693–705.

Martin C & Zhang Y (2005) The diverse functions of histone lysine methylation. *Nat. Rev. Mol. Cell Biol.* 6, 838–849.

Mellor J (2008) On your marks, get set, methylate! *Nat. Cell Biol.* 10, 1249–1250.

Weake VM & Workman JL (2008) Histone ubiquitination: triggering gene activity. *Mol. Cell* 29, 653–663.

3.4   Interaction of different histone modifications, DNA methylation, and RNAi

Cam HP, Chen ES & Grewal SI (2009) Transcriptional scaffolds for heterochromatin assembly. *Cell* 136, 610–614.

Cedar H & Bergman Y (2009) Linking DNA methylation and histone modification: patterns and paradigms. *Nat. Rev. Genet.* 10, 295–304.

Daxinger L, Kanno T & Matzke M (2008) Pol V transcribes to silence. *Cell* 135, 592–594.

Kloc A & Martienssen R (2008) RNAi, heterochromatin and the cell cycle. *Trends Genet.* 24, 511–517.

Kobor MS & Lorincz MC (2009) H2A.Z and DNA methylation: irreconcilable differences. *Trends Biochem. Sci.* 34, 158–161.

Moazed D (2009) Small RNAs in transcriptional gene silencing and genome defence. *Nature* 457, 413–420.

Probst AV, Dunleavy E & Almouzni G (2009) Epigenetic inheritance during the cell cycle. *Nat. Rev. Mol. Cell Biol.* 10, 192–206.

Smith E & Shilatifard A (2007) The A, B, Gs of silencing. *Genes Dev.* 21, 1141–1144.

Taghavi P & van Lohuizen M (2006) Two paths to silence merge. *Nature* 439, 794–795.

Wu JI, Lessard J & Crabtree GR (2009) Understanding the words of chromatin regulation. *Cell* 136, 200–206.

Zamore PD & Haley B (2005) Ribo-gnome: the big world of small RNAs. *Science* 309, 1519–1524.

3.5   Changes in chromatin structure in the regulatory regions of active or potentially active genes

Henikoff S (2008) Nucleosome destabilization in the epigenetic regulation of gene expression. *Nat. Rev. Genet.* 9, 15–26.

Jiang C & Pugh BF (2009) Nucleosome positioning and gene regulation: advances through genomics. *Nat. Rev. Genet.* 10, 161–172.

Kwon CS & Wagner D (2007) Unwinding chromatin for development and growth: a few genes at a time. *Trends Genet.* 23, 403–412.

Lehmann M (2004) Anything else but GAGA: a nonhistone protein complex reshapes chromatin structure. *Trends Genet.* 20, 15–22.

Yaniv M (2009) Small DNA tumour viruses and their contributions to our understanding of transcription control. *Virology* 384, 369–374.

3.6   Other situations in which chromatin structure is regulated

Constancia M, Kelsey G & Reik W (2004) Resourceful imprinting. *Nature* 432, 53–57.

Ercan S & Lieb, J.D (2008) Chromatin proteins do double duty. *Cell* 133, 763–765.

Heard E & Disteche, C.M (2006) Dosage compensation in mammals: fine-tuning the expression of the X chromosome. *Genes Dev.* 20, 1848–1867.

Morison IM, Ramsay JP & Spencer HG (2005) A census of mammalian imprinting. *Trends Genet.* 21, 457–465.

Muers M (2008) Antisense transcripts get involved. *Nat. Rev. Genet.* 9, 898.

Wilkinson LS, Davies W & Isles AR (2007) Genomic imprinting effects on brain development and function. *Nat. Rev. Neurosci.* 8, 832–843.

Wutz A (2007) Xist function: bridging chromatin and stem cells. *Trends Genet.* 23, 457–464.

The Process of Transcription

4

INTRODUCTION

As discussed in Section 1.4, a variety of evidence demonstrates that gene regulation operates primarily at the level of transcription, determining which genes will be transcribed into RNA in specific tissues or in response to specific stimuli. In part, such transcriptional control operates at the level of chromatin structure so that the DNA which is to be transcribed moves to a more open chromatin structure, allowing access to regulatory molecules (see Chapters 2 and 3).

Although such an open chromatin structure is required for the transcription of the gene, the actual process of transcription involves RNA polymerase enzymes that can copy the DNA into RNA together with a variety of transcription factors which can stimulate or inhibit polymerase activity. This process of transcription by RNA polymerases is a major target for the regulation of gene expression. Accordingly, the basic processes of transcription itself will be discussed in this chapter and its regulation by specific transcription factors will be discussed in Chapter 5.

4.1 TRANSCRIPTION BY RNA POLYMERASES

Enzymes which are capable of copying the DNA so that a complementary RNA copy is produced by the polymerization of ribonucleotides are referred to as RNA polymerases. In **prokaryotes** a single RNA polymerase enzyme is responsible for the transcription of DNA into RNA. In eukaryotes, however, this is not the case and several such enzymes exist.

Three RNA polymerase enzymes are found in the nucleus of all eukaryotes and are known as RNA polymerases I, II, and III. In addition, plants contain two further enzymes (RNA polymerases IV and V) which in these organisms are involved in the production of an inhibitory chromatin structure by small interfering RNAs (siRNAs) (see Sections 1.5 and 3.4).

RNA polymerases I, II, and III are large, multi-subunit enzymes and several subunits are held in common between the three enzymes. The three enzymes can be distinguished, however, by their relative sensitivity to the fungal toxin α-amanitin and each of them is active on a distinct set of genes (Table 4.1). All genes capable of encoding a protein, as well as the genes for some **small nuclear RNAs** involved in **RNA splicing** (see Section 6.3), are transcribed by RNA polymerase II. In contrast, the genes encoding the 28, 18, and 5.8S ribosomal RNAs (see Section 6.6) are transcribed by RNA polymerase I and those encoding the **transfer RNAs** (tRNAs) and the 5S ribosomal RNA are transcribed by RNA polymerase III.

In considering the transcriptional regulatory processes that produce tissue-specific variation in specific mRNAs and proteins, our primary concern will therefore be with the regulation of transcription by RNA polymerase II. Transcription by RNA polymerases I and III is also subject to

TABLE 4.1 EUKARYOTIC RNA POLYMERASES

GENES TRANSCRIBED	SENSITIVITY TO α-AMANITIN
I Ribosomal RNA (45S precursor of 28S, 18S, and 5.8S rRNA)	Insensitive
II All protein-coding genes, small nuclear RNAs U1, U2, U3, etc.	Very sensitive (inhibited 1 μg/ml)
III Transfer RNA, 5S ribosomal RNA, small nuclear RNA U6, repeated DNA sequences: Alu, B1, B2, etc.; 7SK, 7SL RNA	Moderately sensitive (inhibited 10 μg/ml)

regulation, however. Moreover, the nature of the components involved in transcription by these polymerases is much simpler than those which are involved in transcription by RNA polymerase II. A prior understanding of the processes involved in basal transcription by RNA polymerases I and III therefore assists a subsequent understanding of the more complex processes involved in basal transcription by RNA polymerase II. Transcription by RNA polymerases I, III, and II will therefore be considered in turn.

Transcription by RNA polymerase I is relatively simple

RNA polymerase I is responsible for the transcription of the tandem arrays of genes encoding ribosomal RNA, such transcription constituting about one half of total cellular transcription. As is the case for all the RNA polymerases, RNA polymerase I has multiple subunits. The structure of the 14-subunit RNA polymerase I from yeast has recently been determined by electron microscopy.

As with all RNA polymerases, RNA polymerase I itself does not recognize the DNA sequences around the start site of transcription. Rather, other protein factors recognize such sequences and then recruit the RNA polymerase by a protein–protein interaction. In the case of RNA polymerase I, the essential DNA sequences which are recognized are located within the 50 bases immediately upstream of the start site of transcription. As in all genes, the sequences adjacent to the transcriptional start site which control the expression of the gene are known as the gene promoter (see Section 4.3 for more detailed discussion of gene promoters). Note that in the case of all the RNA polymerases the site at which transcription begins is denoted +1 with bases within the transcribed region being denoted +100, +200, etc., whereas bases upstream of the start site are denoted as –100, –200, etc., as one proceeds further and further upstream (**Figure 4.1**).

In the case of RNA polymerase I, sequences around –50 are recognized by a protein transcription factor known as upstream binding factor (UBF). Subsequently another regulatory protein known as SL1 (also known as TIF-IB) is recruited via protein–protein interaction with UBF. In turn SL1 recruits the RNA polymerase itself and its associated factors (see Figure 4.1). Hence the initiation of transcription by RNA polymerase I is relatively simple with one essential transcription factor (SL1) being necessary to recruit the RNA polymerase itself. In turn, the binding of this essential factor is facilitated by the prior binding to a specific DNA sequence of another transcription factor, UBF.

Transcription by RNA polymerase III is more complex than for RNA polymerase I

The involvement of a specific transcription factor which acts to recruit the RNA polymerase, and of other factors which recruit the specific factor, is also illustrated by RNA polymerase III. The situation is complicated, however, by the fact that in different genes which are transcribed by RNA polymerase III the essential promoter DNA sequences recognized by the transcription factors which recruit the polymerase can be located either

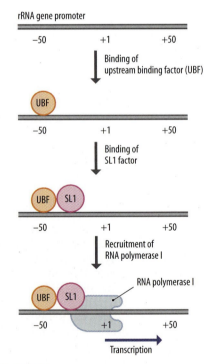

Figure 4.1
Transcription initiation at the ribosomal RNA (rRNA) gene promoter. Binding of upstream binding factor (UBF) is followed by the binding of the SL1 factor, which in turn recruits RNA polymerase I via protein–protein interaction. Note that in this and all subsequent figures, +1 refers to the first base transcribed into RNA, with other + numbers denoting bases within the transcribed region, whereas – signs denote bases upstream of the transcriptional start site.

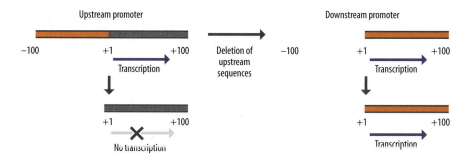

Figure 4.2
Consequences of deleting sequences upstream of the transcriptional start site in genes which have either an upstream promoter or a downstream promoter. Note that when the promoter is located upstream of the transcriptional start site, deletion of upstream sequences results in an absence of transcription. In contrast, when the promoter is located within the transcribed region, deletion of upstream sequences has no effect on transcription.

upstream or downstream of the transcribed region. Genes transcribed by RNA polymerase III can therefore have either an upstream promoter (as is the case for genes transcribed by RNA polymerase I and II) or a downstream promoter located within the transcribed region (**Figure 4.2**).

This type of downstream promoter which is unique to RNA polymerase III was first identified by detailed studies which focused on the genes that encode the 5S RNA of the ribosome. In an attempt to identify the sequences important for the expression of this gene, sequences surrounding it were deleted and the effect on the transcription of the gene in a cell-free system investigated. Somewhat surprisingly, the entire upstream region of the gene could be deleted with no effect on gene expression (**Figure 4.3**). Indeed, deletions within the transcribed region of the 5S gene also had no effect on its expression until a boundary 40 bases within the transcribed region was crossed. By this means, an internal control region essential for the transcription of the 5S RNA gene was defined, located entirely within the transcribed region.

This region of the 5S gene was shown subsequently to bind a transcription factor known as TFIIIA, using a **DNAse I footprinting assay** (**Figure 4.4**) (see Section 4.3 and Methods Box 4.2, for description of this assay of

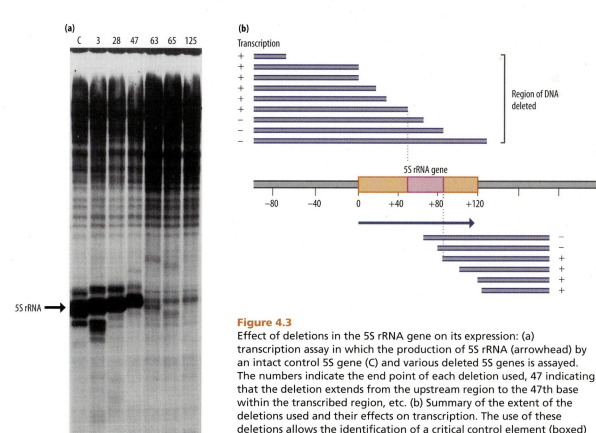

Figure 4.3
Effect of deletions in the 5S rRNA gene on its expression: (a) transcription assay in which the production of 5S rRNA (arrowhead) by an intact control 5S gene (C) and various deleted 5S genes is assayed. The numbers indicate the end point of each deletion used, 47 indicating that the deletion extends from the upstream region to the 47th base within the transcribed region, etc. (b) Summary of the extent of the deletions used and their effects on transcription. The use of these deletions allows the identification of a critical control element (boxed) within the transcribed region of the 5S gene. (a) Courtesy of DD Brown, from Sakonju S, Bogenhagen DF & Brown DD (1980) *Cell* 19, 13–25. With permission from Elsevier.

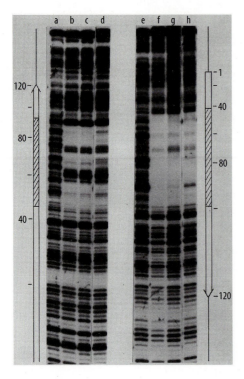

Figure 4.4
Binding of the TFIIIA transcription factor to the internal control region of 5S DNA in a DNaseI footprinting assay in which binding of a protein protects the DNA from digestion by DNaseI and produces a clear region lacking the ladder of bands produced by DNaseI digestion of the other regions (see Methods Box 4.2 for a description of this assay). Tracks a and e show the two DNA strands of the 5S gene in the absence of added TFIIIA whereas tracks b–d and f–h show the same DNA in the presence of TFIIIA. Courtesy of DD Brown, from Sakonju S & Brown DD (1982) *Cell* 31, 395–405. With permission from Elsevier.

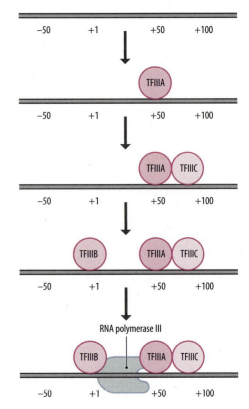

Figure 4.5
Transcription initiation by RNA polymerase III at the 5S rRNA gene promoter. Sequences downstream of the transcription initiation site (+1) are indicated by the + signs; sequences upstream are indicated by the − signs. Following binding of the transcription factor TFIIIA to the internal control sequence, TFIIIC and TFIIIB bind with TFIIIB, then acting to recruit RNA polymerase III and allowing transcription to begin. Note that within the three-dimensional structure TFIIIB interacts directly with TFIIIC, although this is not indicated in the figure.

DNA-protein binding). Subsequently, another transcription factor, TFIIIC, binds to the DNA adjacent to TFIIIA and in turn TFIIIC functions to recruit a further transcription factor, TFIIIB, to form a stable transcription complex (**Figure 4.5**).

This transcriptional complex, which is stable through many cell divisions, promotes the subsequent binding of the RNA polymerase III. The polymerase is recruited via a protein–protein interaction with TFIIIB and binds at the transcriptional start site (see Figure 4.5). The binding of RNA polymerase III is dependent on the presence of the stable transcription complex and not on the precise sequence of the DNA to which it binds since, as discussed above, the region to which the polymerase normally binds can be deleted and replaced by other sequences without drastically reducing transcription.

Although the assembly of transcription complexes on RNA polymerase III genes was first defined on the 5S ribosomal RNA gene, other RNA polymerase III transcription units differ in the details of transcription complex assembly. For example, the genes encoding the tRNAs, which play a key role in translation (see Section 6.6), also have an internal promoter. However, due to differences in the sequence of the promoter, in this case TFIIIA is not required. Rather, TFIIIC binds directly to sequences within the promoter and subsequently recruits TFIIIB. As in the 5S RNA promoters, TFIIIB then recruits the RNA polymerase itself.

Similarly, TFIIIB plays a critical role in transcription of the RNA polymerase III genes which have an upstream promoter. An example of such an

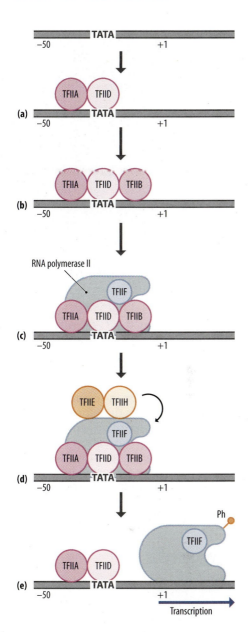

Figure 4.6
Transcription initiation at an RNA polymerase II promoter. Initially, the TFIID factor binds to the TATA box together with another factor TFIIA (a). Subsequently TFIIB is recruited by interaction with TFIID (b) and TFIIB then recruits RNA polymerase II and its associated factor TFIIF (c). TFIIE and TFIIH then bind and TFIIH phosphorylates the C-terminal domain of RNA polymerase II (d). This converts RNA polymerase II into a form which is capable of initiating transcription and RNA polymerase II and TFIIF then move off down the gene, producing the RNA transcript, leaving TFIIA and TFIID bound at the promoter (e).

upstream promoter is found in the gene encoding the small nuclear RNA U6, which unlike the other small nuclear RNAs of the splicesosome (see Section 6.3) is transcribed by RNA polymerase III rather than RNA polymerase II (see Table 4.1).

It is clear therefore that TFIIIB plays an essential role in the transcription of the three different types of RNA polymerase III gene promoter and is the functional equivalent of the RNA polymerase I SL1 factor which acts directly to recruit the RNA polymerase via protein–protein interaction.

Transcription by RNA polymerase II is much more complex than transcription by RNA polymerases I and III

Inspection of the region immediately upstream of the transcriptional start site reveals that a very wide variety of different genes transcribed by RNA polymerase II contain an AT-rich sequence which is found approximately 30 bases upstream of the transcription start site. This TATA box plays a critical role in promoting transcriptional initiation and in positioning the start site of transcription for RNA polymerase II. Its destruction by mutation or deletion effectively abolishes transcription of genes which normally contain it.

Most importantly, this TATA box acts as the initial DNA target site for the progressive assembly of the **basal transcription complex** for RNA polymerase II, which involves considerably more factors than that for RNA polymerase I or III. Initially, the TATA box is bound by the transcription factor TFIID whose binding is facilitated by the presence of another transcription factor TFIIA (**Figure 4.6a**). Interestingly, structural analysis of TFIID has revealed it to have a molecular clamp structure which consists of four globular domains around an accessible groove which can accommodate the DNA to which TFIID binds (**Figure 4.7**).

Subsequently, the TFIID–DNA complex is recognized by another transcription factor TFIIB (**Figure 4.6b**). Structural analysis has shown that TFIIB binds on the opposite side of TFIID to that which is bound by TFIIA (**Figure 4.8** and **Figure 4.9**). This binding of TFIIB is an essential step in the formation of the initiation complex since, as well as binding to TFIID, TFIIB can also recruit RNA polymerase II itself. The binding of TFIIB therefore allows the subsequent recruitment of RNA polymerase II to the initiation complex in association with another factor TFIIF (**Figure 4.6c**). Subsequently, two other factors TFIIE and TFIIH associate with the complex (**Figure 4.6d**).

In particular, the recruitment of TFIIH plays a critical role in allowing the RNA polymerase to initiate transcription. TFIIH is a multi-component complex whose molecular structure has been determined and which appears to play a key role in both transcription and the repair of damaged DNA.

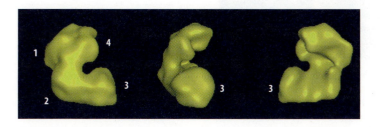

Figure 4.7
Three-dimensional structure of TFIID. Note the globular domains arranged around a groove into which the DNA fits. Courtesy of Patrick Schultz, from Brand M, Leurent C, Mallouh V et al. (1999) *Science* 286, 2151–2153. With permission from The American Association for the Advancement of Science.

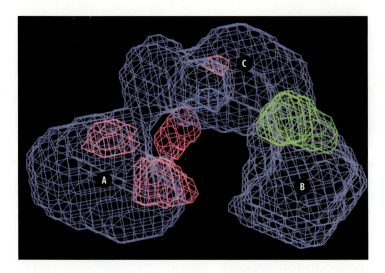

Figure 4.8
Position of TFIIB (green) and TFIIA (red) complexed with TFIID (blue). Courtesy of Eva Nogales, from Andel F 3rd, Ladurner AG, Inouye C et al. (1999) *Science* 286, 2153–2156. With permission from The American Association for the Advancement of Science.

One component of TFIIH has a **kinase** activity. This kinase is capable of phosphorylating the C-terminal domain of the largest subunit of RNA polymerase II, which is known as RPB1. The C-terminal domain of RPB1 contains multiple copies of the sequence Tyr-Ser-Pro-Thr-Ser-Pro-Ser, which is unique to RNA polymerase II and is highly evolutionarily conserved. The kinase activity of TFIIH phosphorylates serine 5 within this repeat and this allows transcription initiation to occur (**Figure 4.10**). Phosphorylation of the C-terminal domain therefore plays a critical role in allowing the polymerase to initiate transcription. Although the dephosphorylated form of RNA polymerase II is recruited to the DNA, its phosphorylation is necessary for transcription to produce the RNA product (**Figure 4.6e**).

Both RNA polymerase II itself and an active transcription complex have been crystallized allowing structural analysis (**Figure 4.11**). This has revealed that as the DNA is transcribed into RNA in the interior of the polymerase molecule, it encounters a wall of protein within the RNA polymerase (labeled

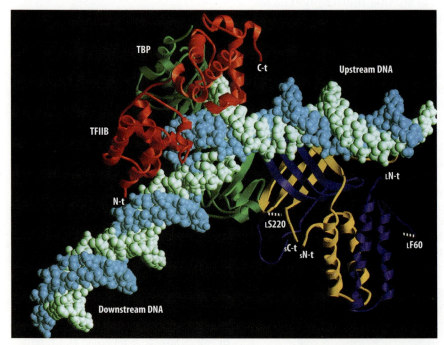

Figure 4.9
Association of the complex of TFIIB (red), the TBP component of TFIID (green), and TFIIA (yellow and purple) with the DNA (light and dark blue). C-t, C-terminus; N-t, N-terminus. Courtesy of JH Geiger, Michigan State University.

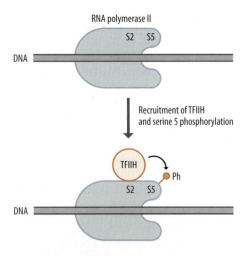

Figure 4.10
Recruitment of TFIIH results in the phosphorylation of RNA polymerase II on serine 5 of its C-terminal domain and allows transcriptional initiation to occur.

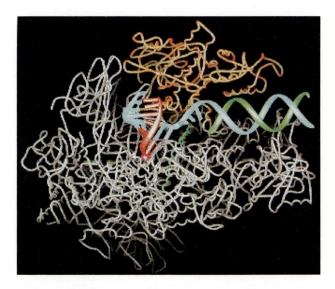

Figure 4.11
The enzyme RNA polymerase II in the act of transcribing DNA. The transcribed strand of the DNA is shown in blue and the non-transcribed strand in green, whereas the RNA transcript is shown in red. Note that the DNA enters from the right, unwinds and makes a right-angled turn as it encounters a wall of protein in the polymerase molecule. Compare with the schematic diagram in Figure 4.12. Courtesy of Aaron Klug, from *Science* (2001) 292 (5523) front cover. With permission from The American Association for the Advancement of Science.

A in **Figure 4.12**). This forces it to make a right-angled turn, exposing the end of the nascent RNA and allowing ribonucleoside triphosphates to be added to it as transcription occurs. Subsequently, the newly formed DNA–RNA hybrid produced as a consequence of transcription encounters another part of RNA polymerase, known as the rudder (labeled B in Figure 4.12). This rudder region forces the separation of the RNA from the DNA. This allows the newly formed part of the RNA molecule to exit and double-stranded DNA to reform (see Figure 4.12).

Interestingly, this complex interaction of the DNA and RNA polymerase is facilitated by TFIIB. Structural studies of the RNA polymerase–TFIIB complex have shown that TFIIB does more than simply recruit the polymerase. By interacting with both DNA-bound TFIID and the polymerase itself in a very precise structure, TFIIB ensures that the DNA and the polymerase molecule are correctly positioned and oriented relative to one another for the DNA to enter the interior of the polymerase, allowing transcription to occur.

As the polymerase moves off down the gene, TFIIF remains associated with it while TFIIA and TFIID remain bound at the promoter, allowing further cycles of recruitment of TFIIB, RNA polymerase II, etc., leading to repeated rounds of transcription (see Figure 4.6e).

Such a role of TFIIA and TFIID in allowing repeated rounds of transcription is of particular interest in view of the finding that some RNA polymerase is found within the cell associated with a large number of proteins including TFIIB, TFIIF, and TFIIH to form a so called **RNA polymerase holoenzyme**. In addition to the stepwise pathway involving progressive recruitment of TFIIB, RNA polymerase, and TFIIH individually, it appears therefore that an alternative pathway exists in which TFIIA and TFIID can recruit a holoenzyme complex containing TFIIB, TFIIE, TFIIF, TFIIH, and RNA polymerase itself (**Figure 4.13**). In addition, the holoenzyme also contains a number of other protein components which appear to be involved either in opening up the chromatin structure so as to allow transcription to occur (see Section 3.5) or in allowing the polymerase complex to be stimulated by transcriptional activators (see Section 5.2).

Transcription by the three different polymerases has a number of common features

Although the three different polymerases each have a different function and different associated transcription factors, the three polymerases themselves are all multi-subunit proteins with several subunits being shared between the different polymerases. Each of the polymerases has two large subunits, known as RPB1 and RPB2, which are related to one another and to the β′ and β subunits of the single bacterial RNA polymerase enzyme (**Figure 4.14**). They also all contain the same ω-like subunit.

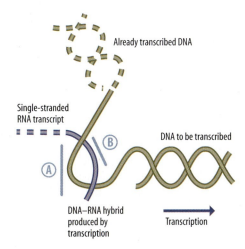

Figure 4.12
Movement of the DNA being transcribed through the RNA polymerase molecule. A indicates the wall within the polymerase protein which forces the DNA–RNA hybrid to make a right-angled turn, thereby allowing transcription to occur by addition of ribonucleoside triphosphates to the end of the RNA chain. B indicates the rudder region of the polymerase which forces the DNA–RNA hybrid to melt, releasing the newly formed RNA and allowing double-stranded DNA to reform. DNA which is about to be transcribed is shown by the solid lines and DNA which has already been transcribed by the dashed lines. The arrow indicates the direction of transcription.

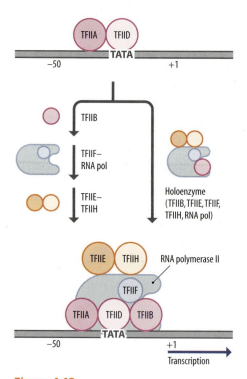

Figure 4.13
Following binding of TFIIA and TFIID to the promoter, the formation of the basal transcription complex for RNA polymerase (RNA pol) II may take place via the sequential recruitment of TFIIB, TFIIF–RNA polymerase II, and TFIIE–TFIIH as indicated in Figure 4.6 or by the recruitment of a holoenzyme containing all of these factors.

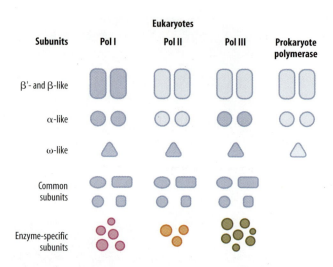

Figure 4.14
Relationship of the various subunits in the three eukaryotic RNA polymerases and the single bacterial (*Escherichia coli*) RNA polymerase. Darker shading indicates that the subunits are identical to one another whereas lighter shading indicates that they are homologous but not identical.

In addition, RNA polymerases I and III share the same two α-like subunits, whereas RNA polymerase II has two α subunits which are related to but distinct from those of the other two polymerases. Again, these ω- and α-like subunits are related to the corresponding subunits of the bacterial enzyme. In addition, however, all three eukaryotic RNA polymerases also contain four common subunits which are not found in the bacterial enzyme and between three and seven subunits which are unique to one or other of the eukaryotic polymerases (see Figure 4.14).

In addition, each of the three polymerases shows a similar pattern of recruitment to the DNA, with one specific transcription factor binding to a target sequence in the gene promoter followed by binding of one or more other proteins which then recruit the polymerase itself. As discussed above, in the case of RNA polymerase II promoters which contain a TATA box, the original binding to DNA is achieved by the TFIID factor which binds to the TATA box. Interestingly, however, although TFIID was originally identified as a single factor it is now clear that it is composed of multiple protein components. One of these proteins known as TATA-binding protein (TBP) is responsible for binding to the TATA box while the other components of the complex, known as TBP-associated factors (**TAFs**), do not bind directly to the TATA box but appear to allow TFIID to respond to stimulation by transcriptional activators (see Section 5.2).

Interestingly, TBP has been shown to have a saddle-like structure, with the concave underside binding the DNA and the convex outer surface being available for interaction with other factors. For example, TFIIA binds to the N-terminal side of the convex outer surface of TBP, whereas TFIIB binds to the C-terminal side of TBP (**Figure 4.15**).

Structural studies of TBP bound to DNA have shown that binding of TBP bends the DNA so that it follows the curve of the saddle (see Figure 4.15). The structure of TFIID (consisting of TBP and the TAFs) bound to DNA resembles that of the core nucleosome of eight histone molecules which

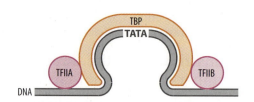

Figure 4.15
Saddle structure of TBP bound to DNA. Binding of TBP bends the DNA so that it follows the concave under surface of the saddle. The convex upper surface of TBP is available to interact with other proteins, such as TFIIA and TFIIB.

forms the normal structure of chromatin (see Section 2.2). This suggests that TFIID bends the DNA at the promoter in a similar way to the bending of DNA around the nucleosomes in the remaining DNA. In this regard, it is of interest that as discussed in Chapter 3 (Section 3.3) the TAF$_{II}$250 subunit of TFIID has histone acetyltransferase activity, allowing it to modulate chromatin structure by acetylating histones.

Although many of the genes transcribed by RNA polymerase II contain a TATA box, a subset of RNA polymerase II genes do not contain the TATA box but instead have an initiator sequence (Inr) located around the transcriptional start site. Paradoxically, however, TBP plays a key role also in the transcription of this class of RNA polymerase II genes. In this case, TBP does not bind to the DNA but is recruited by another DNA-binding protein which binds to the **initiator element** overlapping the transcriptional start site of these genes. TBP then binds to this initiator-binding protein, allowing the recruitment of TFIIB and the RNA polymerase itself, as occurs for promoters containing a TATA box. Hence TBP plays a central role in the assembly of the transcription complex for RNA polymerase II, joining the complex by binding to DNA in the case of TATA box-containing promoters and being recruited via protein–protein interactions in the case of promoters which lack a TATA box (**Figure 4.16**).

These findings indicate therefore that TBP may be a basic transcription factor, which is essential for transcription by RNA polymerase II, paralleling the role of SL1 for RNA polymerase I and of TFIIIB for RNA polymerase III. This idea is supported by the amazing finding that TBP is actually also a component of both SL1 and TFIIIB. Thus SL1 is not a single protein but is actually a complex of four factors one of which is TBP. Hence the recruitment of SL1 to an RNA polymerase I promoter by UBF (see above) actually results in the delivery of TBP to the DNA exactly as in the case of RNA polymerase II promoters which lack a TATA box.

Similarly, TFIIIB is a complex of TBP and two other proteins, Bdpl and Brf1. In the 5S and tRNA genes, TBP is delivered to the promoter as part of the TFIIIB complex following prior binding of either TFIIIA or TFIIIC. Moreover, some RNA polymerase III genes, such as that of the U6 RNA gene contain an upstream promoter with a TATA box and hence in this case TBP can bind directly to the promoter (**Figure 4.17**). Remarkably, following its direct or indirect recruitment to the DNA, TBP can then make a variety of

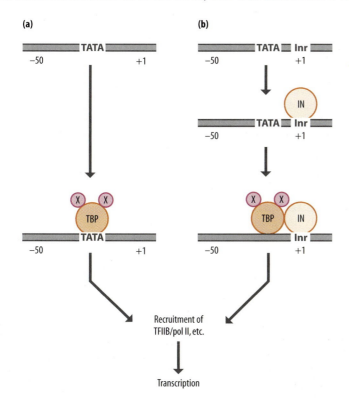

(a)   **(b)**

Recruitment of TFIIB/pol II, etc.

Transcription

Figure 4.16
Transcription of promoters by RNA polymerase II involves the recruitment of TBP (and its associated factors (X) forming the TFIID complex) to the promoter. This may occur by direct binding of TBP to the TATA box where this is present (a) or by protein–protein interaction with a factor (IN) bound to the initiator element (INR) in promoters lacking a TATA box (b).

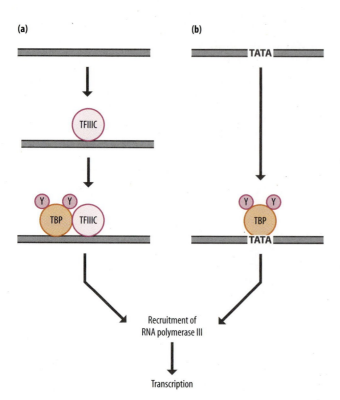

(a)   (b)

TATA

TFIIIC

Y Y
TBP TFIIIC

Y Y
TBP

TATA

Recruitment of
RNA polymerase III

Transcription

Figure 4.17
Transcription of promoters by RNA
polymerase III involves the recruitment of
TBP (and its associated factors (Y) forming
the TFIIIB complex) to the promoter.
This may be achieved by protein–protein
interactions with either TFIIIA and TFIIIC
or TFIIIC alone in the case of promoters
lacking a TATA box (a) or by direct binding
to the TATA box where this is present (b).

further contacts within the transcriptional complexes for RNA polymerases I, II, and III in order to enhance assembly and/or activity of the complexes and these have now been defined by structural analysis (**Figure 4.18**).

The multiple protein–protein and protein–DNA interactions, which we have discussed above, therefore appear to serve merely to recruit TBP to the DNA which, in the case of all three polymerases then leads directly or indirectly to the recruitment of the RNA polymerase itself. It has therefore been suggested that TBP represents an evolutionarily ancient transcription factor whose existence precedes the evolution of three independent eukaryotic RNA polymerases and which therefore plays a universal and essential role in eukaryotic transcription. In agreement with this idea, a TBP homolog has been identified in the Archaebacteria. As these organisms constitute a separate kingdom distinct from prokaryotes and eukaryotes, it is clear that TBP is an evolutionarily ancient protein which predates the divergence of the archaebacterial and eukaryotic kingdoms.

Initially, it was believed that each organism would have only one TBP protein. It has now become clear, however, that other TBP-like proteins exist in multicellular organisms. Thus, all multicellular organisms contain, in addition to TBP itself, a protein known as TBP-like factor (TLF; also known as TRF2), whereas additional TBP-like proteins are found specifically in insects (TRF1) and vertebrates (TRF3).

Interestingly, these TBP-related factors (TRFs) substitute for TBP in the basal transcriptional complex which assembles on certain specific RNA polymerase II-transcribed genes. For example, in the frog **Xenopus**, inactivation of TBP prevents the transcription of only a relatively few genes in the

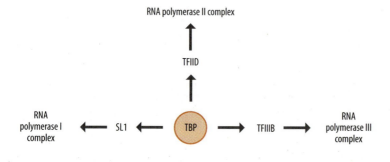

RNA polymerase II complex

TFIID

RNA
polymerase I ← SL1 ← TBP → TFIIIB → RNA
complex   polymerase III
                                           complex

Figure 4.18
As a component of SL1, TFIID, and TFIIIB,
the TBP protein plays a key role in the
transcription-initiation complex for all three
RNA polymerases.

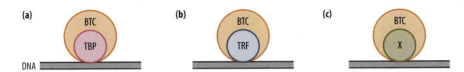

Figure 4.19
The basal transcriptional complex (BTC) which assembles on different promoters can contain TBP (a), a TBP-related factor (TRF) (b), or lack either TBP or a TRF, which are presumably replaced by an unrelated protein (X) (c).

early embryo, with the majority of embryo-specific genes being dependent on TLF/TRF2 or TRF3 for their transcription.

Similarly, TRF3 appears to substitute for TBP in terminally differentiated muscle myotubes. This results in the preferential transcription of muscle-specific genes which can be transcribed in a TRF3-dependent manner, whereas TBP-dependent non-muscle-specific genes are not transcribed (see Section 10.2 for further discussion of muscle-specific gene transcription).

Hence, the basal transcriptional complex which assembles on some genes will contain a TRF rather than TBP and this is required for transcription of these genes (**Figure 4.19a** and **b**). Indeed, in some cases it has been demonstrated that a basal transcriptional complex can assemble without either TBP or a TRF, with another factor presumably fulfilling their role (**Figure 4.19c**). These effects evidently offer a means of differentially regulating the expression of genes with differing requirements for TBP or TRFs in a particular cell type.

It is clear therefore that TBP is an evolutionarily ancient transcription factor, which is involved in transcription by all three RNA polymerases (see Figure 4.18). In the case of some genes, however, transcription can occur in a TBP-independent manner involving either a TRF or a complex lacking TBP or TRFs.

Transcription takes place in defined regions of the nucleus

It has been known for many years that transcription of the ribosomal RNA genes by RNA polymerase I takes place in a defined region of the nucleus. This region is known as the **nucleolus** and can be readily visualized by light or electron microscopy (**Figure 4.20**). It contains several hundred copies of the ribosomal RNA genes which are tandemly repeated along the chromosome. The nucleolus also contains proteins involved in transcription of the ribosomal RNA genes by RNA polymerase I (see above) as well as the processing of the initial RNA transcript to form the mature 28, 18, and 58S ribosomal RNAs.

More recently, however, it has become clear that protein-coding genes, which are being transcribed by RNA polymerase II are also located in specific parts of the nucleus. Moreover, regions of DNA which are not transcribed are also located in specific nuclear regions, distinct from the location of

Figure 4.20
(a) Two-dimensional schematic diagram of the nucleus, showing the nucleolus and the heterochromatin which is located adjacent to the nucleolus and the nuclear envelope but not adjacent to the nuclear pores which link the nucleus and cytoplasm. (b) Three-dimensional reconstruction of a mouse liver nucleus using electron microscope analysis. Heterochromatin appears as darkly stained areas. (b) Courtesy of Christel Genoud, Patrick Schwarb, & Susan Gasser, The Friedrich Miescher Institute.

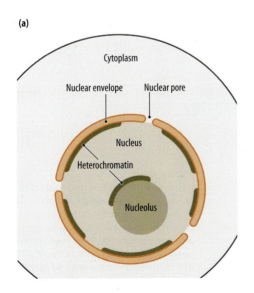

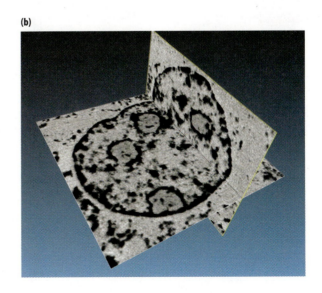

transcribed genes. For example, the tightly packed non-transcribed DNA which is known as heterochromatin (see Section 2.5) is located in specific regions adjacent to the nucleolus and at the **nuclear envelope**, which separates the nucleus from the cytoplasm (see Figure 4.20).

Interestingly, transcription of protein-coding genes by RNA polymerase II has also been shown to occur at specific regions of the cell and some of these are located at the nuclear envelope. However, such transcription close to the nuclear envelope takes place adjacent to the nuclear pores which link the nucleus to the cytoplasm, whereas the non-transcribed heterochromatin is localized to regions of the nuclear envelope between the pores (see Figure 4.20).

The location of different regions of the DNA to different regions of the nucleus is a dynamic process in which a gene can change its position in the nucleus when it is about to be transcribed. Thus, as described in Chapter 2 (Section 2.5) specific regions of the DNA can be thrown into loops containing genes which are about to be transcribed. Such looping can result in the DNA relocating within the nucleus, for example, to a nuclear pore region, where it is actively transcribed (**Figure 4.21**).

Interestingly, transcribed genes on different chromosomes can be relocalized to the same transcriptionally active region (see Figure 4.21). This has led to the idea of transcription factories which contain all the factors required for transcription by RNA polymerase II and in which inter-chromosomal interactions occur between transcriptionally active genes on different chromosomes (see Figure 4.21).

As expected from this, a specific stimulus can result in the inter-chromosomal association between two genes on different chromosomes, both of which are activated by that stimulus. For example, treatment with the steroid hormone estrogen activates both the TFF1 gene on chromosome 21 and the GREB1 gene on chromosome 2. Prior to estrogen treatment, these genes do not co-localize with one another, as expected from their location on different chromosomes. However, following estrogen treatment, it can be shown that these genes are closely associated with one another (**Figure 4.22a**).

Conversely, as described in Chapter 2 (Section 2.5), the gene encoding interferon-γ and the locus containing the genes encoding the **cytokines** interleukin (IL)-4, -5, and -13 are located on different chromosomes and are expressed in a mutually exclusive manner. In early T-cell development when both loci are inactive, they are associated together but they move apart when one locus or the other becomes transcriptionally active (**Figure 4.22b**). Hence, associations between genes on different chromosomes can involve genes which show the same pattern of activation (see Figure 4.22a) or those which are repressed under the same conditions (see Figure 4.22b).

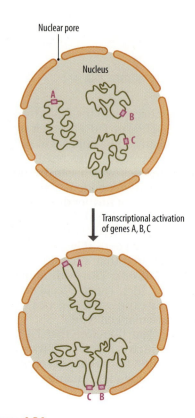

Figure 4.21
When genes become active, looping of chromatin occurs so that the active gene relocates to specific regions of the nucleus which are often located adjacent to nuclear pores. This can result in association of genes on different chromosomes (B and C).

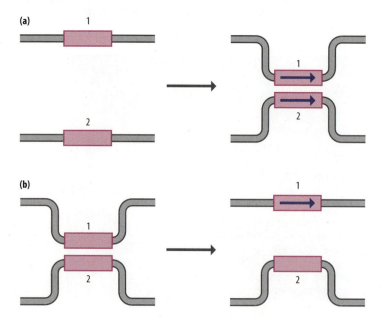

Figure 4.22
Association between genes on different chromosomes (1 and 2) can occur when both genes become active (a). In other cases it can occur when both genes are inactive and the association is lost when one gene becomes active (b). The horizontal arrows in the boxes indicate that transcription is occurring (see also Figure 2.46).

Figure 4.23
Linkage of nucleotides in the RNA chain via a phosphate residue (P) which joins the carbon at the 3' position in the ribose sugar of the upstream nucleotide and the carbon at the 5' position in the ribose sugar of the downstream nucleotide. Note that the first base of the RNA chain is usually an A or G (corresponding to T or C in the DNA), whereas any base (N) can be inserted in the remaining positions in different RNAs depending on the sequence of the DNA being transcribed.

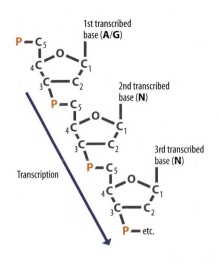

It is clear therefore that transcription occurs at particular regions within the nucleus and that specific genes move to these regions and form or break associations with other genes when they become transcriptionally active.

4.2 TRANSCRIPTIONAL ELONGATION AND TERMINATION

Transcriptional elongation requires further phosphorylation of RNA polymerase II

As discussed in Section 4.1, phosphorylation of serine 5 within the C-terminal repeat region of RNA polymerase II results in the polymerase beginning transcription of the gene. The first base in the DNA which is copied into RNA (the +1 base) is usually a C or T residue, resulting in the RNA beginning with the complementary base; that is, either a G or an A. Subsequently, transcription proceeds with the base complementary to the base in the DNA being inserted into the RNA, i.e. C in the DNA directs insertion of G in the RNA and vice versa, T in the DNA directs insertion of A, and A in the DNA directs insertion of U (which replaces T in the RNA). The bases in the RNA chain are joined to one another by a **phosphodiester bond** between the carbon at the 3' position on the **ribose** sugar of the upstream nucleotide and the carbon at the 5' position of the downstream nucleotide (**Figure 4.23**). Hence, the first base in the chain has a free 5' end and the RNA chain grows in the 5' to 3' direction.

Initially it was thought that once transcription was initiated, the polymerase would transcribe the entire gene without further regulatory processes being required. It is now clear, however, that following transcriptional initiation, transcription proceeds for only approximately 20–30 bases and the polymerase then pauses and does not continue transcribing. Release of this block and continued transcriptional elongation requires phosphorylation of the RNA polymerase II C-terminal repeat region on serine 2 of the Tyr-Ser-Pro-Thr-Ser-Pro-Ser repeated sequence (**Figure 4.24**).

This phosphorylation of serine 2 is closely linked to the modification of the free 5' end of the nascent RNA transcript by addition of a modified G nucleotide in a process known as capping (see Section 6.1). This capping process occurs when the RNA transcript is 20–30 bases long and promotes the binding of the **pTEF-b** kinase protein. The pTEF-b kinase then phosphorylates the C-terminal repeat of RNA polymerase II on serine 2, allowing transcriptional elongation to occur (see Figure 4.24). The linkage between transcription and post-transcriptional processes such as capping is discussed further in Chapter 6 (Section 6.4).

Recent studies have suggested that in many genes the polymerase may pause for an extended period after transcribing 20–30 bases and only then continue transcribing the gene. For example in *Drosophila* embryos approximately 1000 different genes were found to have **stalled polymerases** which

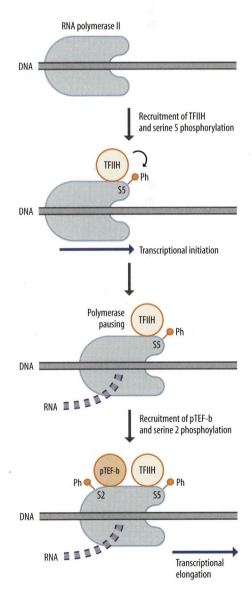

Figure 4.24
Recruitment of TFIIH results in the phosphorylation of RNA polymerase II on serine 5 of its C-terminal domain and allows transcriptional initiation to occur. However, the polymerase pauses and ceases transcribing after producing a short RNA transcript. Subsequently, recruitment of the pTEF-b kinase results in the phosphorylation of polymerase II on serine 2 of the C-terminal domain, allowing transcriptional elongation to produce the complete RNA product.

had initiated transcription but had not extended the transcript beyond 20–30 nucleotides.

Such studies in both *Drosophila* and humans identified three classes of genes in terms of RNA polymerase II distribution. Thus, transcriptionally inactive genes lacked polymerase whilst actively transcribed genes had polymerase molecules distributed over their entire length with slight peaks near the 5′ and 3′ ends of the gene. Most importantly, however, a third class of potentially active genes had a peak of stalled polymerases near the 5′ end of the gene (**Figure 4.25**). Evidently, such pausing and restarting of RNA polymerases offers a significant opportunity for gene regulation at the level of transcriptional elongation and this will be discussed further in Chapter 5 (Section 5.4).

Following the phosphorylation of serine 2 on the stalled polymerase by pTEF-b, transcriptional elongation will occur with the polymerase moving down the gene producing the RNA transcript. As noted in Chapter 3 (Section 3.5), loss of nucleosomes occurs at sites of transcriptional initiation facilitating access for regulatory factors and the RNA polymerase itself. However, this is not the case for the remainder of the gene, so the elongating RNA polymerase must transcribe through nucleosome-packaged DNA.

This appears to be achieved by the elongating polymerase recruiting histone acetyltransferases, which acetylate the histone molecules on nucleosomes in front of the elongating polymerase. This not only opens the chromatin structure (see Section 3.3) but actually displaces the nucleosomes from the DNA. The free histones then associate with other proteins, such as the FACT protein (see Section 3.3). These proteins catalyze reassembly of the nucleosome behind the elongating polymerase, which is followed by deacetylation of the histones (**Figure 4.26**).

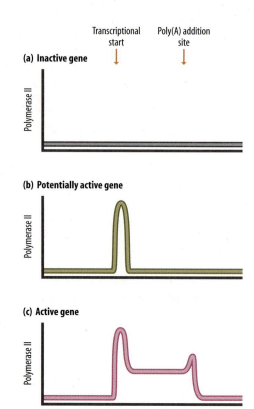

Figure 4.25
Distribution of RNA polymerase II molecules across an inactive gene (a), a potentially active gene (b), and an active gene (c). Note the peak of stalled molecules, which have initiated transcription, on the potentially active gene. The arrows indicate the positions of the transcriptional start site and the site of polyA addition at the 3′ end of the mature transcript (see Section 6.2 for discussion of polyadenylation).

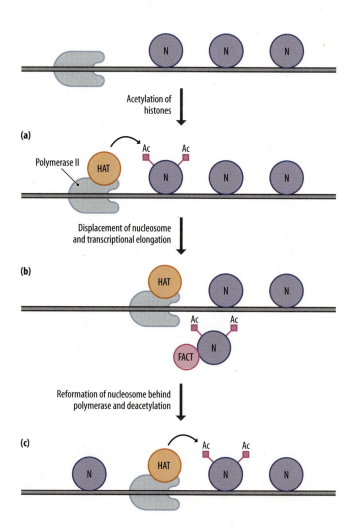

Figure 4.26
The elongating RNA polymerase complex can recruit histone acetyltransferase (HAT) enzymes, which can acetylate (Ac) histone molecules in nucleosomes (N) located in front of the polymerase (a). This results in the displacement of such nucleosomes and their association with the FACT protein, clearing the way for the elongating polymerase (b). Subsequently, the nucleosome reforms behind the polymerase and the histones in it are deacetylated, whereas histones in the next nucleosome in front of the polymerase are acetylated, so that the cycle continues (c).

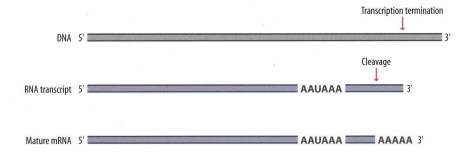

Figure 4.27
The initial RNA transcript is cleaved downstream of the polyadenylation signal (AAUAAA) and a polyA tail added to the free 3′ end. Hence, the 3′ end of the mature mRNA is significantly upstream of the transcriptional termination site.

Termination of transcription occurs downstream of the polyadenylation signal

Clearly the process of transcriptional elongation must eventually end, producing a transcript with a 3′ as well as a 5′ end. In the mature mRNA, the 3′ end is defined by the addition post-transcriptionally of a run of adenosine nucleotides to produce a **poly(A) tail** in a process known as **polyadenylation**. As discussed in Chapter 6 (Section 6.2), this process involves the recognition of a **polyadenylation signal** including the essentially invariant sequence AAUAAA within the RNA followed by internal cleavage of the RNA downstream of this sequence and polyadenylation of the free 3′ end. Hence, the 3′ end of the mature transcript is upstream of the point at which the initial transcript terminates (**Figure 4.27**).

Despite this, however, the polyadenylation signal is also involved in the process of transcriptional termination since mutation of this sequence interferes with normal termination. Two models have been put forward to explain this. The first model known as the allosteric model, postulates that the polymerase undergoes a structural/allosteric change as it transcribes through the polyadenylation signal. This either promotes its association with termination factors and/or inhibits its association with factors that prevent termination, so promoting termination (**Figure 4.28**).

The second model, known as the torpedo model, is based on the idea that cleavage of the nascent transcript occurs while transcription is still going on. This will produce not only an RNA with a free 3′ end for polyadenylation but also a downstream RNA with a free 5′ end which continues to be extended by the transcribing polymerase. It is postulated that an **exonuclease** enzyme binds to this free 5′ end and moves along the RNA, degrading it. When the exonuclease catches up with the polymerase it effectively 'torpedoes' it, resulting in the termination of transcription (**Figure 4.29**).

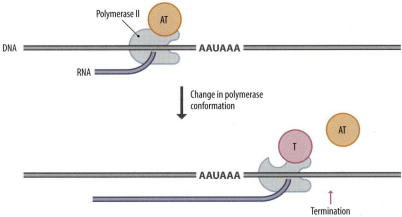

Figure 4.28
Allosteric model of transcriptional termination. Transcription through the polyadenylation signal (AAUAAA) produces a conformational change in the RNA polymerase promoting the dissociation of anti-termination factors (AT) and/or the association of termination factors (T), so resulting in transcriptional termination.

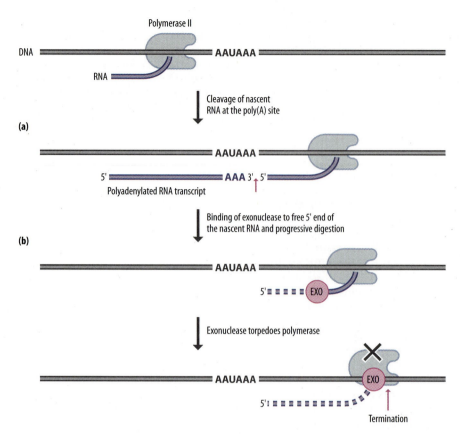

(a)

(b)

Figure 4.29
Torpedo model of transcriptional termination. Following transcription through the polyadenylation signal, the nascent RNA is cleaved to yield an upstream transcript, which will be polyadenylated and processed further to produce the mature mRNA (a). The polymerase continues transcribing, however, extending a downstream transcript with a free 5′ end. A 5′ to 3′ exonuclease (EXO) binds to this transcript at the 5′ end and starts to degrade it (b). Eventually, this exonuclease catches up with the polymerase, and 'torpedoes' it, producing transcriptional termination.

A variety of evidence exists for and against each of these models. Thus, both humans and yeast have been shown to express a 5′ to 3′ exonuclease enzyme, which is essential for transcription termination as required by the torpedo model. In contrast, however, in some systems it is possible for termination to occur without cleavage at the polyadenylation site, which is not compatible with the torpedo model. It is possible therefore that different termination mechanisms may operate in different situations or that some combination of the two models may operate.

Whatever the precise mechanism of transcriptional termination, however, the studies discussed in Section 4.1 and in this section indicate that transcription involves multiple stages; namely initiation, stalling after transcription of a short region, active elongation, and termination (**Figure 4.30**).

4.3 THE GENE PROMOTER

As discussed in Section 4.1, sequences around or just upstream of the transcriptional start site play a critical role in the recruitment of RNA polymerase II to the genes which it transcribes. A number of genes transcribed by RNA polymerase II contain a TATA box approximately 30 bases upstream of the transcriptional start site. Other RNA polymerase II-transcribed genes utilize an initiator (Inr) sequence with the consensus 5′-YCANTYY-3′, with the A residue being the first base which is transcribed (Y is C or T and N is any nucleotide). This sequence is located around the transcriptional start

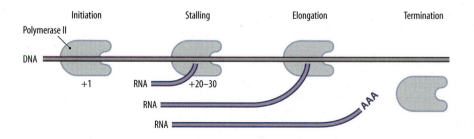

Figure 4.30
Stages in the transcriptional process.

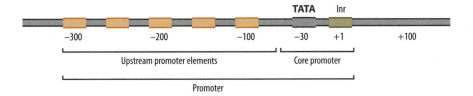

Figure 4.31
The promoter structure of a typical gene transcribed by RNA polymerase II, consisting of the core or basal promoter and multiple upstream promoter elements. Note that in different cases the basal promoter contains either a TATA box or an initiator (Inr) element, both elements, or neither.

site and binds an initiator-binding protein, which then recruits TBP (see Section 4.1 and Figure 4.16).

Although most RNA polymerase II-transcribed genes contain either a TATA box or an Inr sequence, some have both elements while some have neither. Those genes lacking a TATA box or an Inr sequence generally contain a CpG island (see Section 3.2) close to the transcriptional start site and are transcribed at a low level in all cell types with a variable start site of transcription.

The region around or just upstream of the transcriptional start site containing the TATA box and/or the Inr sequence (or the equivalent region in genes where they are absent) is known as the core or basal promoter and serves to recruit the basal transcriptional complex, as described in Section 4.1. The **core promoter** is defined as the minimal region which can direct the initiation of transcription. However, even when a TATA box and/or an Inr sequence are present, this produces only a relatively low rate of transcription. This rate of transcription is increased by the presence of **upstream promoter elements**, which as their name indicates are located upstream of the core promoter. Together, the core promoter and the upstream promoter elements constitute the promoter, which drives transcription of the gene (**Figure 4.31**).

The 70 kDa heat-shock protein gene contains a typical promoter for RNA polymerase II

To further illustrate the nature of RNA polymerase II promoters, we will focus upon the gene encoding the 70 kDa heat-shock protein (hsp70). Exposure of a very wide variety of cells from different organisms to elevated temperature results in the increased synthesis of a few heat-shock proteins, of which hsp70 is the most abundant. Such increased synthesis is mediated in part by increased transcription of the corresponding gene, which can be visualized as a puff within the polytene chromosome of *Drosophila* (see Section 1.4) As described below, examination of the promoter sequences located upstream of the start site for transcription in this gene allows us to identify potential upstream promoter sequences involved in its induction by temperature elevation, as well as those involved in the general mechanism of transcription. The sequences present in this region of the *hsp70* gene which are also found in other genes are listed in **Table 4.2** and their arrangement is illustrated in **Figure 4.32**.

TABLE 4.2 SEQUENCES PRESENT IN THE UPSTREAM REGION OF THE *hsp70* GENE WHICH ARE ALSO FOUND IN OTHER GENES		
NAME	**CONSENSUS**	**OTHER GENES CONTAINING SEQUENCES**
TATA box	TATA(A/T)A(A/T)	Very many genes
CCAAT box	TGTGGCTNNNAGCCAA	α- and β-globin, herpes simplex virus thymidine kinase, cellular oncogenes *c-ras*, *c-myc*, albumin, etc.
Sp1 box	GGGCGG	Metallothionein IIA, type II procollagen, dihydrofolate reductase, etc.
CRE	(T/G)(T/A)CGTCA	Somatostatin, fibronectin, α-gonadotrophin, *c-fos*, etc.
AP2 box	CCCCAGGC	Collagenase, class 1 antigen H-2K^b, metallothionein IIA
Heat-shock element	CTNGAATNTTCTAGA	Heat-inducible genes *hsp83*, *hsp27*, etc.

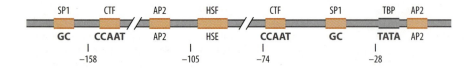

Figure 4.32
Transcriptional control elements in the human *hsp70* gene promoter. The protein binding to a particular site is indicated above the line and the corresponding DNA element below the line. These elements are described more fully in Table 4.2.

A study of this type reveals a number of sequence motifs including the TATA box shared by the *hsp70* gene and other non-heat-inducible genes as well as one which is unique to heat-inducible genes. These will be considered in turn.

The hsp70 gene promoter contains several DNA sequence motifs which are found in a variety of other gene promoters

As discussed in Section 4.1, recruitment of the basal transcriptional complex containing TBP, RNA polymerase II, and other associated factors via initial binding of TBP to the TATA box can produce only a low rate of transcription. This rate is enhanced by the binding of other transcription factors to sites upstream of the TATA box which enhances either the stability or the activity of the basal complex. The binding sites for several of these factors are present in a wide variety of different genes with different patterns of activity. These sites act as targets for the binding of specific factors which are active in all cell types and their binding therefore results in increased transcription in all tissues. The presence or absence of these sequences and their number will determine the transcription rate of a particular gene in the absence of any specific inducing stimulus.

An example of such a sequence is the Spl box, two copies of which are present in the *hsp70* gene promoter (see Figure 4.32). This GC-rich DNA sequence binds a transcription factor known as Spl which is present in all cell types. Similarly, the **CCAAT box** is located upstream of the start site of transcription of a wide variety of genes (including the *hsp70* gene) that are regulated in different ways and is believed to play an important role in allowing transcription of the genes containing it by binding constitutively expressed transcription factors.

The heat-shock element is found only in heat-inducible genes

In contrast to these very widespread sequence motifs, another sequence element in the *hsp70* gene promoter is shared only with other genes whose transcription is increased in response to elevated temperature. This sequence is found 62 bases upstream of the start site for transcription of the *Drosophila hsp70* gene and at a similar position in other heat-inducible genes. This heat-shock element (HSE) is therefore believed to play a critical role in mediating the observed heat-inducibility of transcription of these genes.

To confirm that this is the case, it is necessary to transfer this sequence from the *hsp70* gene to another gene which is not normally heat-inducible and show that the recipient gene now becomes inducible. This was achieved initially by transferring the HSE onto the non-heat-inducible thymidine kinase (*tk*) gene taken from the eukaryotic virus, herpes simplex. When the hybrid gene was introduced into cells and the temperature subsequently raised, increased thymidine kinase production was detected showing that the HSE had rendered the *tk* gene inducible by elevated temperature (**Figure 4.33**).

This experiment therefore proves that the common sequence element found in the heat-inducible genes is responsible directly for their heat-inducibility. The manner in which these experiments were carried out also permits a further conclusion with regard to the way in which this sequence acts. Thus, the HSE used by Pelham was taken from the *Drosophila hsp70* gene and in this cold-blooded organism would be activated normally by the thermally stressful temperature of 37°C. The cells into which the hybrid

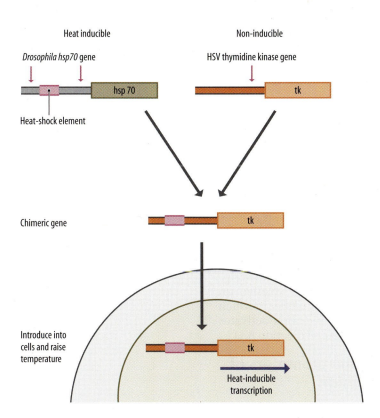

Figure 4.33
Demonstration that the heat-shock element mediates heat inducibility. Transfer of this sequence to a gene (thymidine kinase, tk) which is not normally inducible renders this gene heat-inducible.

gene was introduced, however, were mammalian cells which grow normally at 37°C and only express the heat-shock genes at the higher temperature of 42°C.

In these experiments, the hybrid gene was induced only at 42°C, the heat-shock temperature at which heat-shock genes are induced in the mammalian cell into which it was introduced. It was not induced at 37°C, the temperature at which heat-shock genes are induced in *Drosophila* from which the DNA sequence came. This means that the HSE does not possess some form of inherent temperature sensor or thermostat which is set to go off at a particular temperature, since in this case the *Drosophila* sequence would activate transcription at 37°C even in mammalian cells. Rather, it must act by being recognized by a cellular protein which is activated in response to elevated temperature and, by binding to the heat-shock element, produces increased transcription. Evidently, although the elements of this response are conserved sufficiently to allow the mammalian protein to recognize the *Drosophila* sequence, the mammalian protein will only be activated at the mammalian heat-shock temperature and hence induction will only occur at 42°C.

Hence these experiments not only provide evidence for the importance of the HSE in causing heat-inducible transcription but also indicate that it acts by binding a protein. This indirect evidence that the HSE acts by binding a protein can be confirmed directly by using a variety of techniques which allow the proteins binding to a specific DNA sequence to be analyzed (see below for a description of some of these methods). When an analysis of this type is carried out on the upstream regions of the heat-shock genes, it can be shown that in non-heat-shocked cells TBP is bound to the TATA box and the GAGA factor is bound to upstream sequences (**Figure 4.34a**). The binding of the GAGA factor is believed to displace a nucleosome and create the DNaseI-hypersensitive sites observed in these genes in non-heat-shocked cells (see Section 3.5). By contrast, in heat-shocked cells where high-level transcription of the gene is occurring, an additional protein which is bound to the HSE is detectable on the upstream region (**Figure 4.34b**).

Hence the induction of the heat-shock genes is indeed accompanied by the binding of a protein known as the **heat-shock factor** (HSF) to the HSE as suggested by the experiments described above. The binding of this factor to

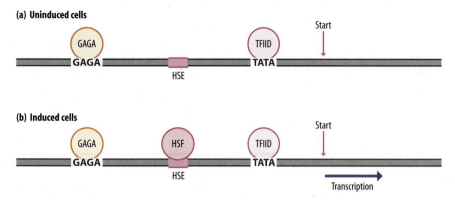

(a) Uninduced cells

(b) Induced cells

Figure 4.34
Proteins binding to the promoter of the *hsp70* gene before (a) and after (b) heat shock. HSE, heat-shock element; HSF, heat-shock factor.

a gene whose chromatin structure has already been altered to render it potentially activatable, results in stimulation of transcription exactly as discussed in Chapter 3 and illustrated in Figure 3.49. In agreement with this, the purified heat-shock factor can bind to the HSE and stimulate the transcription of the *hsp70* gene in a cell-free nuclear extract while having no effect on the transcription of the non-heat-inducible actin gene. The manner in which HSF is activated in response to elevated temperature and then activates transcription of genes with an HSE is described in Chapter 8 (Section 8.1).

Other response elements are found in the promoters of genes with different patterns of expression

An analysis of the hsp70 promoter therefore identifies a number of sequences such as the TATA box and the Spl and CCAAT boxes which are shared with a number of genes with different patterns of expression and which therefore play a role in the general process of transcription. It also identifies the HSE as an element which is shared only with other heat-inducible genes and which plays a key role in producing their inducibility.

A number of similar elements which are found in the promoters of genes activated by other signals have now been identified. These sequences were originally identified by comparison of several different genes which are activated by the same stimulus. More recently, genome-wide projects, such as the ENCODE project in the human genome, have begun to fully map the distribution of such regulatory sequences in the entire genome. In many cases, the identification of such sequences has been followed up by functional analysis showing that they are capable of transferring the specific response to another marker gene. A selection of such sequences is listed in **Table 4.3**.

As indicated in Table 4.3, these sequences act by binding specific proteins which are synthesized or activated in response to the inducing signal. Such transcription factors are discussed further in Chapter 5. It is noteworthy, however, that many of the sequences in Table 4.3 exhibit dyad symmetry, a similar sequence being found in the 5′ to 3′ direction on each strand. The estrogen response element for example, has the sequence:

5′ AGGTCANNNTGACCT 3′
3′ TCCAGTNNNACTGGA 5′

Here, the two halves of the 10-base palindrome is separated by three random bases. Such symmetry in the binding sites for these transcription factors indicates that they bind to the site in a dimeric form consisting of two protein molecules.

Various sequences that confer response to several different signals have therefore been identified. One gene can possess more than one such element, allowing multiple patterns of regulation. Thus comparison of the sequences listed in Table 4.3 with those contained in the *hsp70* gene listed in Table 4.2 reveals that in addition to the HSE this gene also contains the cyclic AMP-response element (CRE), which mediates the induction of a number of genes, such as that encoding somatostatin, in response to treatment with cyclic AMP.

TABLE 4.3 SEQUENCES THAT CONFER RESPONSE TO A PARTICULAR STIMULUS

CONSENSUS SEQUENCES	RESPONSE TO	PROTEIN FACTOR	GENE CONTAINING SEQUENCES
CTNGAATNTTCTAGA	Heat	Heat-shock factor	*hsp70, hsp83, hsp27*, etc.
(T/G)(T/A)CGTCA	Cyclic AMP	CREB/ATF	Somatostatin, fibronectin, α-gonadotrophin, *c-fos, hsp70*
TGAGTCAG	Phorbol esters	AP1	Metallothionein IIA, α-antitrypsin, collagenase
CC(A/T)$_6$GG	Growth factor in serum	Serum response factor	*c-fos, Xenopus* γ-actin
RGRACANNNTGTYCY	Glucocorticoid	Glucocorticoid receptor	Metallothionein IIA, tryptophan oxygenase, uteroglobin, lysozyme
RGGTCANNNTGACCY	Estrogen	Estrogen receptor	Ovalbumin, conalbumin, vitellogenin
RGGTCATGACCY	Thyroid hormone	Thyroid hormone receptor	Growth hormone, myosin heavy chain
TGCGCCCGCC	Heavy metals	Mep-1	Metallothionein genes
AGTTTCNNTTTCNY	Interferon-α	Stat-1, Stat-2	Oligo A synthetase, guanylate-binding protein
TTNCNNNAA	Interferon-γ	Stat-1	Guanylate-binding protein, Fcγ receptor

N indicates that any base can be present at that position; R indicates a purine, i.e. A or G; Y indicates a pyrimidine, i.e. C or T.

Similarly, while genes may share particular elements, flexibility is provided by the presence of other elements in one gene and not another allowing the induction of a particular gene in response to a given stimulus which has no effect on another gene. For example, although the *hsp70* gene and the metallothionein IIA gene share a binding site for the transcription factor AP2, only the metallothionein gene has a binding site for the glucocorticoid receptor which confers responsivity to glucocorticoid hormone induction. Hence only this gene is inducible by hormone treatment. The overall pattern of sequences present will control the basal expression of an individual gene and whether or not it responds to particular stimuli.

In some cases, the sequence elements that confer response to a particular stimulus can be shown to be related to one another. For example, the sequence mediating response to glucocorticoid treatment is similar to that which mediates response to another steroid, namely estrogen. Similarly, both the estrogen- and thyroid hormone-responsive elements contain identical sequences showing dyad symmetry forming a palindromic repeat of the sequence GGTCA. In the estrogen-responsive element, however, the two halves of this dyad symmetry are separated by three bases which vary between different genes, whereas in the thyroid hormone-responsive element the two halves are contiguous (Table 4.4a).

In addition, as well as being arranged as palindromic repeats, the GGTCA core sequence can also be arranged as two direct repeats (Table 4.4b). In this arrangement, the spacing between the two repeats can again regulate which hormone produces a response. Thus, a spacing of four bases forms an alternative thyroid hormone-response element whereas a spacing of one base confers responsivity to 9-*cis*-retinoic acid, two or five bases to all-*trans*-retinoic acid, and a spacing of four bases to vitamin D (see Table 4.4b).

Such similarities in these different hormone response elements are paralleled by a similarity in the individual cytoplasmic receptor proteins which form a complex with each of these hormones and then bind to the corresponding DNA sequence. All of these receptors can be shown to be members of a large family of related DNA-binding proteins whose hormone and

TABLE 4.4 RELATIONSHIP OF VARIOUS HORMONE RESPONSE ELEMENTS	
(a) Palindromic repeats	
Glucocorticoid	RGRACANNNTGTYCY
Estrogen	RGGTCANNNTGACCY
Thyroid	RGGTCA– – –TGACCY
(b) Direct repeats	
9-*cis*-Retinoic acid	$AGGTCAN_1AGGTCA$
All-*trans*-retinoic acid	$AGGTCAN_2AGGTCA$, $AGGTCAN_5AGGTCA$
Vitamin D3	$AGGTCAN_3AGGTCA$
Thyroid hormone	$AGGTCAN_4AGGTCA$
N indicates that any base can be present at that position; R indicates a purine, i.e. A or G; Y indicates a pyrimidine, i.e. C or T; a dash indicates that no base is present, the gap having been introduced to align the sequence with the other sequences.	

DNA-binding specificities differ from one another. Hence a particular DNA sequence confers a response to a particular hormone because it binds the appropriate receptor which also binds that hormone. The exchange of particular regions of these receptor proteins with those of other family members has provided considerable information on the manner in which sequence-specific binding to DNA occurs and this is discussed in Chapter 5 (Section 5.1).

Although the sequence elements shown in Table 4.3 are all involved in the response to particular inducers of gene expression, it is clear that other short sequence elements or combinations of elements are involved in controlling the tissue-specific patterns of expression exhibited by eukaryotic genes. For example, the **octamer motif** (ATGCAAAT), which is found in both the immunoglobulin heavy- and light-chain promoters can confer B-cell-specific expression when linked to a non-regulated promoter. Similarly, short DNA sequences that bind liver-specific transcription factors have been identified in the region of the rat albumin promoter, known to be involved in mediating the liver-specific expression of this gene (see Chapter 10 for further discussion of the mechanisms controling cell type-specific gene transcription).

The proteins binding to short DNA sequence elements can be characterized by a variety of techniques

In discussing short sequence elements which either enhance transcription in all situations or which confer a particular pattern of gene regulation we have assumed that they act by binding specific regulatory proteins. As discussed above, initial studies provided indirect evidence that this is the case for the HSE. However, it is necessary to prove directly that this is so for the HSE and other short DNA sequences. A variety of methods have therefore been devised for demonstrating that regulatory protein(s) bind to a particular DNA sequence and for characterizing these factors. Three of the most important of these will be described in turn.

(a) The DNA mobility-shift assay

When a particular DNA sequence is first identified as a potential transcriptional regulatory element, the next step is usually to carry out a **DNA mobility-shift assay** in which the DNA sequence and a cell extract are mixed.

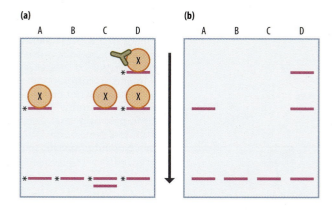

Figure 4.35
DNA mobility-shift assay. Panel (a) shows schematically the factors binding to a radioactively labeled oligonucleotide (*) while panel (b) shows what will be seen when the gel is visualized by autoradiography. In track A, a protein (X) in the extract used has bound to the DNA, resulting in a retarded radioactive complex of lower mobility. In track B, the cell extract used does not have the DNA-binding protein so no retarded complex forms. In track C, a large excess of unlabeled competitor DNA has been added. It binds protein X but no retarded band is seen on autoradiography since the competitor DNA is not radioactive. In track D, an antibody to protein X has been added so that a super-shifted complex of even lower mobility is formed.

Protein binding to the DNA is detected by the DNA moving more slowly in the gel. Hence, protein binding to DNA results in the appearance of a retarded band, giving the technique its alternative names of band-shift or gel-retardation assay (**Figure 4.35**, **track A**) (**Methods Box 4.1**).

Once such a band has been detected, one can for example, carry out the assay using extracts of different cell types or of cells treated in different ways to see if the presence or absence of the DNA-binding activity correlates with the pattern of gene activity conferred by the DNA sequence (**Figure 4.35**, **tracks A** and **B**). Similarly, it is possible to characterize the DNA-binding specificity of the binding factor by adding a large excess of unlabeled DNA sequences as well as the radioactively labeled sequence. If the unlabeled "competitor" sequence can also bind the factor, it will do so and since it is in excess it will prevent the labeled sequence from binding the factor. However, since the competitor DNA is unlabeled, no radioactive band will be detected (**Figure 4.35**, **track C**).

This method can therefore be used to determine whether a novel DNA-binding activity has a similar binding specificity to a known DNA-binding protein and is therefore likely to be identical or closely related to it. This relationship between the binding activity and known proteins can also be probed if an antibody to the known protein is available. Thus, the antibody can be added to the assay to see whether it binds to the DNA-binding factor forming a complex of even lower mobility, which is known as a "super-shifted" complex (**Figure 4.35**, **track D**).

(b) DNaseI footprinting assay

The DNA mobility-shift assay is therefore of great value in characterizing the proteins binding to a specific DNA sequence. However, it does not provide any information as to where the protein binds within the DNA sequence or its position relative to other proteins binding to adjacent DNA sequences in the regulatory region of a particular gene. This is achieved by the DNaseI footprinting assay in which binding of a protein protects its binding site in the DNA from digestion (**Figure 4.36**, **Methods Box 4.2**).

Although technically more difficult to carry out, this method thus offers advantages over the DNA mobility-shift assay. Firstly, it delineates the bases which are actually bound by the binding protein. Secondly, it is possible to

---

Methods Box 4.1
DNA MOBILITY SHIFT ASSAY (FIGURE 4.35)

- Radioactively label DNA fragment or oligonucleotide containing the DNA sequence of interest.
- Mix with whole cell or nuclear extract and incubate.
- Run the mixture on a non-denaturing gel and observe the position of the radioactive bands by autoradiography.
- Detect protein binding to the DNA by the appearance of a retarded band in the autoradiograph (Figure 4.35).

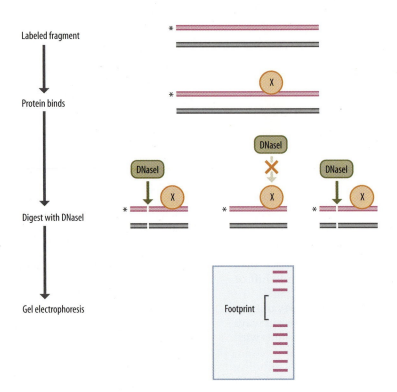

Labeled fragment

Protein binds

Digest with DNaseI

Gel electrophoresis

Footprint

Figure 4.36
DNaseI footprinting assay. A protein (X) that binds to the test sequence, which is radioactively labeled at one end, protects it from digestion by DNaseI. Hence, the DNA fragments corresponding to digestion in this region will not be formed and will appear as a "footprint" in the ladder of bands formed by DNaseI digestion at other, unprotected, parts of the DNA. Hence, protein binding to specific sequences can be detected and localized as a footprint.

visualize multiple footprints on a single DNA molecule thereby elucidating the pattern of proteins bound to the promoter or regulatory region of a particular gene. An example of the use of this assay to analyze the binding of the TFIIIA protein to the internal promoter of the 5S ribosomal RNA gene was described in Section 4.1 (see Figure 4.4).

As with the DNA mobility-shift assay, it is also possible to determine whether a particular binding activity/footprint is observed with extracts prepared from cells exposed to different treatments or from different cell types and relate this to the pattern of expression of the gene being studied. Similarly, the DNA sequence specificity of any binding activity can be studied by adding an excess of an unlabeled oligonucleotide containing a sequence related to that of a particular footprint. If the DNA-binding activity producing the footprint can bind to the unlabeled sequence, it will do so and hence the footprint will disappear since that region of DNA will no longer be protected from digestion. DNaseI footprinting is therefore a valuable technique for examining the interaction of proteins with particular DNA sequences.

(c) Chromatin immunoprecipitation

Both DNA mobility-shift assays and DNaseI footprinting are carried out using purified DNA fragments and cell extracts. They therefore demonstrate what proteins can bind to a DNA sequence, rather than determine which

Methods Box 4.2
DNaseI FOOTPRINTING ASSAY (FIGURE 4.36)

- Radioactively label DNA fragment or oligonucleotide containing the DNA sequence of interest *at one end only.*
- Mix with whole cell or nuclear extract and incubate.
- Digest the mixture with DNaseI to produce a series of DNA fragments each differing by only a single base.
- Run the DNA fragments on a *denaturing* polyacrylamide gel capable of resolving single-base differences and detect radioactive bands by autoradiography.
- Visualize area where protein was bound as a gap in the ladder of DNA fragments due to the protein protecting this region of DNA from digestion (Figure 4.36).

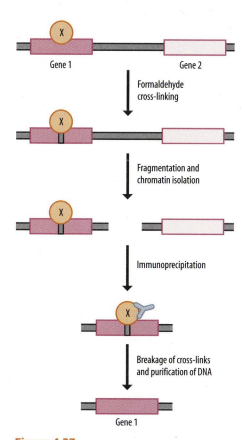

Figure 4.37
Chromatin immunoprecipitation assay in which a DNA fragment (gene 1) is purified on the basis that it binds a particular transcription factor (X) in the chromatin structure of the intact cell. The blue shape shows the antibody used to immunoprecipitate the DNA-bound transcription factor.

> **Methods Box 4.3**
> **CHROMATIN IMMUNOPRECIPITATION (FIGURE 4.37)**
> - Fix living cells with formaldehyde to stably cross-link transcription factors to their DNA-binding sites.
> - Fragment chromatin into small fragments and purify.
> - Immunoprecipitate the transcription factor of interest and its target DNA, using an antibody to the transcription factor.
> - Break the DNA–protein cross-links and isolate DNA.
> - Characterize the isolated DNA.

proteins actually do so in the intact cell. The technique of **chromatin immunoprecipitation** (**ChIP**) overcomes this problem by using an antibody to a particular transcription factor to immunoprecipitate and purify the DNA fragments to which it is bound within the normal chromatin structure (**Methods Box 4.3**, **Figure 4.37**).

Using this method, it is possible to identify whether a particular gene of interest is present in the immunoprecipitate under particular conditions thereby testing whether the transcription factor is bound to the gene in intact cells under those conditions (**Figure 4.38a**). Hence, the effect of specific treatments such as heat shock, steroids, etc., on the binding of a particular factor to a specific DNA-binding site can be investigated under different conditions within the natural chromatin structure of the gene.

In addition, however, the ChIP technique can be used in conjunction with DNA arrays or gene chips (see Section 1.2) in which all the DNA in a particular genome is arrayed on a glass slide. By labeling the immunoprecipitated DNA and hybridizing it to such a filter, all the target genes in the cell for a particular factor can be characterized. This is referred to as ChIP-chip analysis (**Figure 4.38b**).

This genome-wide location analysis, which combines ChIP assays and DNA arrays, will become of increasing importance as the genomes of more and more organisms are sequenced. Thus in yeast, a number of putative regulatory DNA sequences have been identified on the basis of their sequence and their conservation between different yeast strains. By using genome-wide location analysis, such sequences can be tested for their ability to bind specific transcription factors under different conditions. Hence, all the binding sites for a particular factor in the yeast genome can be defined and the effect on such binding of specific treatments can be assessed.

In this way, it is possible to begin to characterize the complex regulatory networks which are characteristic of eukaryotes in which transcription

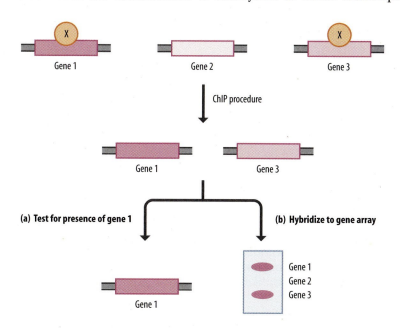

Figure 4.38
The DNA isolated in the ChIP procedure (Figure 4.37) can be tested for the presence of a specific gene (gene 1, panel a) or hybridized to a gene array which will detect all the genes which bind the factor under test (genes 1 and 3, panel b).

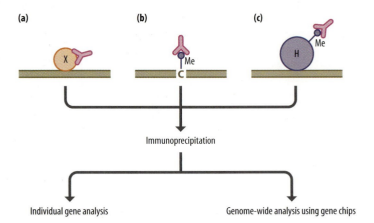

(a)   (b)   (c)

Me

X   C   H

Me

Immunoprecipitation

Individual gene analysis   Genome-wide analysis using gene chips

Figure 4.39
As well as being used to map the binding of a transcription factor to an individual gene or across the genome (a), ChIP analysis can also be used with an antibody specific to methylated (Me) C residues (b) or one specific to a specific histone (H) modification (c) to map these features at the level of a specific gene or across the entire genome. The pink shapes show the antibody in each case.

factors evidently bind at large numbers of different sites across the genome. Indeed, a recent study using genome-wide location analysis showed that the yeast Ste12 transcription factor binds to distinct binding sites and therefore regulates distinct sets of genes during mating compared to filamentous growth. A similar approach aimed at identifying all the regulatory elements in the human genome and defining the factors which bind to them is now being piloted by the ENCODE Project Consortium, although such an analysis is obviously much more complex in humans compared to the less complex yeast.

The use of ChIP in gene control studies is not confined, however, to mapping the distribution of transcription factors (**Figure 4.39a**). Indeed it can be used in any situation where an antibody with a particular specificity is available. For example, by using an antibody specific for methylated C residues, ChIP can be used to identify the distribution of this modification in an individual gene or throughout the genome, if used in conjunction with DNA arrays (**Figure 4.39b**). Hence, ChIP offers an important means of mapping this modification which, as described in Chapter 3 (Section 3.2), plays a key role in controlling chromatin structure and therefore in gene control.

Similarly, in view of the importance of histone modifications in the regulation of chromatin structure (see Chapter 3, Section 3.3) ChIP analysis can also be used to map the distribution of such modifications in cases where a specific antibody is available (**Figure 4.39c**). For example, an antibody specific for histone H3 which is methylated on lysine 9 and one specific for methylation of this histone on lysine 27 have been used to map the location of these modifications and to further strengthen their association with a tightly packed chromatin structure.

Promoter regulatory elements act by binding factors which either affect chromatin structure and/or influence transcription directly

It is clear, therefore, that short DNA sequence elements located near the start site of transcription play an important role in controlling gene expression in eukaryotes. As indicated in Table 4.3 and discussed above, such sequences mediate transcriptional activation by binding a specific protein. This binding may give rise to gene activity in one of two ways (**Figure 4.40**).

First, as discussed in Chapter 3 (Section 3.5) and illustrated by the glucocorticoid receptor, binding of a specific protein may result in displacement of a nucleosome and generation of a DNaseI-hypersensitive site, allowing easy access to the gene for other transcription factors (**Figure 4.40a**). The direct activation of transcription by such factors constitutes the second mechanism of gene induction, and is illustrated both by the binding of other non-regulated factors to glucocorticoid-regulated genes following binding of the receptor and by the binding of the heat-shock factor to its **consensus sequence** in the heat-inducible genes (**Figure 4.40b**). These factors are likely to act by interacting with proteins necessary for transcription, such as TBP or RNA polymerase itself. This interaction facilitates the formation of a

(a) Alteration of chromatin structure

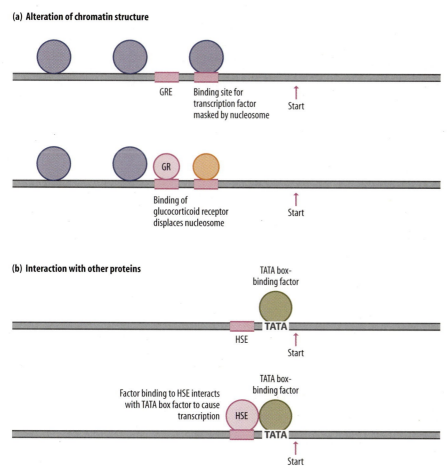

(b) Interaction with other proteins

Figure 4.40
Roles of short sequence elements in gene activation. These elements can either bind a factor that displaces a nucleosome and unmasks a binding site for another factor (a) or can bind a factor that directly activates transcription (b).

stable transcription complex, which may enhance the binding of RNA polymerase to the DNA or alter its structure in a manner which increases its activity (for further discussion see Section 5.2).

It should be noted that these two mechanisms of action are not exclusive. Thus the glucocorticoid receptor whose binding to a specific DNA sequence displaces a nucleosome also contains an **activation domain** capable of interacting with other bound transcription factors (see Section 5.2). Hence, following binding of the receptor, transcription is increased by interactions between the receptor and other bound factors. Similarly, the heat-shock factor can induce the acetylation of histone H4 which results in a more open chromatin structure (see Section 3.3), indicating that it can modulate chromatin structure, as well as stimulating the basal transcriptional complex. Hence, the binding of transcription factors to short DNA sequence elements can activate transcription by two distinct mechanisms, with at least some factors utilizing both of these mechanisms.

Short DNA sequence elements located in the promoter region close to the start site of transcription therefore play a critical role in producing gene transcription. However, important elements involved in the regulation of eukaryotic gene expression are also found at greater distances from the site of initiation of transcription and these elements will be discussed in the next section.

4.4 ENHANCERS AND SILENCERS

Enhancers are regulatory sequences that act at a distance to increase gene expression

The first indication that sequences located at a distance from the start site of transcription might influence gene expression in eukaryotes came with the demonstration that sequences over 100 bases upstream of the

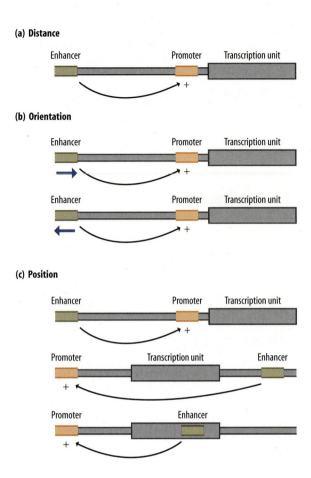

(a) Distance

(b) Orientation

(c) Position

Figure 4.41
Characteristics of an enhancer element which can activate a promoter at a distance (a), in either orientation relative to the promoter (b), and when positioned upstream, downstream, or within a transcription unit (c).

transcriptional start site of the histone H2A gene were essential for its high-level transcription. Moreover, although this sequence was unable to act as a promoter and direct transcription, it could increase initiation from an adjacent promoter element up to 100-fold when located in either orientation relative to the start site of transcription.

Subsequently, a vast range of similar elements have been described in both cellular genes and those of eukaryotic viruses. They have been called enhancers because although they lack promoter activity and are unable to direct transcription themselves, they can dramatically enhance the activity of promoters. Hence, if an enhancer element is linked to a promoter the activity of the promoter can be increased several hundred-fold.

Variation in the position and orientation at which an enhancer element was placed relative to the promoter has led to three conclusions with regard to the action of enhancers. These are: (a) an enhancer element can activate a promoter when placed up to several thousand bases from the promoter; (b) an enhancer can activate a promoter when placed in either orientation relative to the promoter; and (c) an enhancer can activate a promoter when placed upstream or downstream of the transcribed region, or within an intervening sequence which is removed from the RNA by splicing (see Section 6.3). These characteristics constitute the definition of an enhancer and are summarized in **Figure 4.41**. Hence, the typical eukaryotic gene will contain enhancer elements, as well as core and upstream promoter elements (**Figure 4.42**).

Figure 4.42
In addition to promoter elements (see Figure 4.31), eukaryotic genes frequently contain enhancer elements which can be located 50 kb or more upstream or downstream of the transcriptional start site.

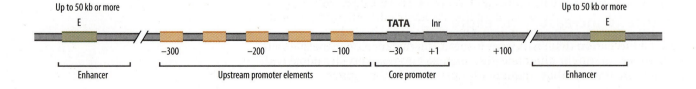

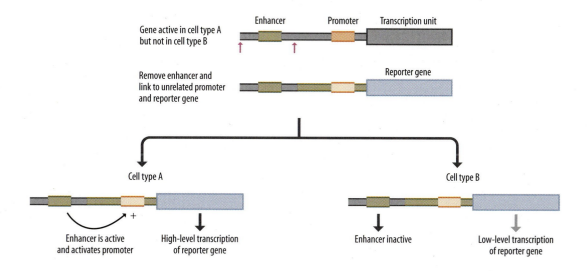

Many enhancers have cell-type- or tissue-specific activity

Although the first enhancers to be discussed were active in all cell types, it was subsequently shown that many genes expressed in specific tissues also contained enhancers. Such enhancers frequently exhibited a tissue-specific activity, being able to enhance the activity of other promoters only in the tissue in which the gene from which they were derived is normally active and not in other tissues (**Figure 4.43**). The tissue-specific activity of such enhancers acting on their normal promoters is likely therefore to play a critical role in mediating the observed pattern of gene regulation.

For example, as previously discussed (see Section 1.3), the genes encoding the heavy and light chains of the antibody molecule contain an enhancer located within the large intervening region separating the regions encoding the joining and constant regions of these molecules. When this element is linked to another promoter, such as that of the β-globin gene, it increases its activity dramatically when the hybrid gene is introduced into B cells. In contrast, however, no effect of the enhancer on promoter activity is observed in other cell types, such as fibroblasts, indicating that the activity of the enhancer is tissue specific.

Similarly, as discussed in Chapter 11 (Section 11.1), specific cancers involve chromosomal rearrangements that link the immunoglobulin enhancer with the gene encoding the c-**Myc** protein, which regulates cellular growth. This results in enhanced expression of the c-Myc gene producing B-cell cancers. This example illustrates the power of enhancer elements to alter gene expression even when removed from their normal location by a chromosomal translocation.

Similar tissue-specific enhancers have also been detected in genes expressed specifically in the liver (α-fetoprotein, albumin, α-1-antitrypsin), the endocrine and exocrine cells of the pancreas (insulin, elastase, amylase), the pituitary gland (prolactin, growth hormone), and many other tissues. The tissue-specific activity of these enhancer elements is likely to play a crucial role in the observed tissue-specific pattern of expression of the corresponding gene.

In the case of the insulin gene, early experiments involving linkage of different upstream regions of the gene to a marker gene and subsequent introduction into different cell types, identified a region approximately 250 bases upstream of the transcriptional start site as being of crucial importance in producing high-level expression in pancreatic endocrine cells. This position corresponds exactly to the position of the tissue-specific enhancer, indicating the importance of this element in gene regulation. Similarly, mutation of conserved sequences within the tissue-specific enhancers of genes expressed in the exocrine cells of the pancreas, such as elastase and chymotrypsin, abolishes the tissue-specific pattern of expression of these genes.

Figure 4.43
A tissue-specific enhancer can activate the promoter of its own or another reporter gene (encoding a protein whose production can readily be assayed) only in one particular tissue and not in others.

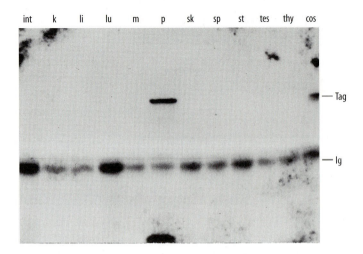

The importance of the enhancer element in the insulin gene in producing tissue-specific gene expression was further demonstrated by experiments in which this enhancer (together with its adjacent promoter) was linked to the gene encoding the large T antigen of the eukaryotic virus SV40, whose production can be measured readily using a specific antibody. The resulting construct was introduced into a fertilized mouse egg and the expression of large T antigen analyzed in all tissues of the transgenic mouse that developed following the return of the egg to the oviduct. Expression of large T antigen was detectable only in the pancreas and not in any other tissue (**Figure 4.44**). Moreover, it was observed specifically in the β cells of the pancreatic islets which produce insulin (**Figure 4.45**). The enhancer of the insulin gene is therefore capable of conferring the specific pattern of insulin gene expression on an unrelated gene *in vivo*.

Hence, like promoter elements, enhancer elements constitute another type of DNA sequence which is involved in the transcription of some genes in all cell types and in the activation of specific genes in a particular tissue or in response to a particular stimulus. As with promoter elements, the binding of constitutively expressed or cell-type-specific proteins to enhancer elements has been demonstrated for many different enhancers including those in the immunoglobulin and insulin genes. This has been achieved using the techniques described in Section 4.3 which can be used to

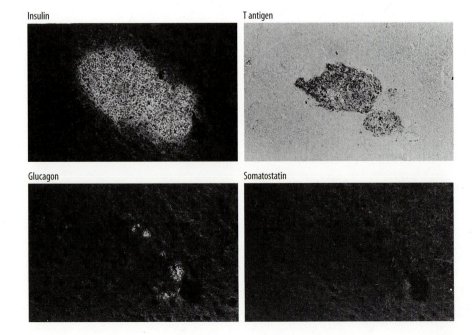

Figure 4.44
Assay for expression of a hybrid gene in which the SV40 T-antigen protein-coding sequence is linked to the insulin gene enhancer and promoter. The gene was introduced into a fertilized egg and a transgenic mouse containing the gene in every cell of its body isolated. Expression of the T antigen (Tag) is assayed by immunoprecipitation of protein from each tissue with an antibody specific for T antigen. Note that expression of the T antigen is detectable only in the pancreas (p) and not in other tissues such as intestine (int), kidney (k), liver (li), lung (lu), marrow (m), skin (sk), spleen (sp), stomach (st), testis (tes), and thymus (thy), indicating the tissue-specific activity of the insulin gene enhancer. The track labeled "cos" contains protein isolated from a control cell line expressing T antigen. The Ig band in all tracks is derived from the immunoglobulin antibody used to precipitate the T antigen. Courtesy of D Hanahan, from Hanahan D (1985) *Nature* 315, 115–122. With permission from Macmillan Publishers Ltd.

Figure 4.45
Immunofluorescence assay of pancreas preparations from the transgenic mice described in the legend to Figure 4.44 with antibodies to the indicated proteins. Note that the distribution of T antigen parallels that of insulin and not that of the other pancreatic proteins. Courtesy of D Hanahan, from Hanahan D (1985) *Nature* 315, 115–122. With permission from Macmillan Publishers Ltd.

determine the protein(s) binding to any specific DNA sequence regardless of whether it is located adjacent to the promoter or within an enhancer.

Therefore, in many cases the tissue-specific expression of a gene will be determined both by the enhancer element and sequences adjacent to the promoter. In the liver-specific pre-albumin gene for example, gene activity is controlled both by the promoter itself which is active only in liver cells and by an upstream enhancer element which activates any promoter approximately tenfold in liver cells and not at all in other cell types. Similarly, in the immunoglobulin genes when the enhancer and the promoter itself are separated both exhibit B-cell-specific activity in isolation but the maximal expression of the gene is observed only when the two elements are brought together.

The importance of enhancer elements in the regulation of gene expression therefore necessitates consideration of the mechanism by which they act.

Proteins bound at enhancers can interact with promoter-bound factors and/or alter chromatin structure

In considering the nature of enhancers, we have drawn a distinction between these elements which act at a distance and the sequences discussed in Section 4.3, which are located immediately adjacent to the start site of transcription. In fact, however, closer inspection of the sequences within enhancers indicates that they are often composed of the same sequences found adjacent to promoters.

For example, the immunoglobulin heavy chain enhancers contain the octamer motif (ATGCAAAT) which as discussed previously is also found in the immunoglobulin promoters. Use of DNA mobility-shift and DNaseI footprinting assays (see Section 4.3) has shown that these promoter and enhancer octamer elements bind the identical B-cell-specific transcription factor (as well as a related protein found in all cell types) and play an important role in the B-cell-specific expression of the immunoglobulin heavy chain gene. Interestingly, within the enhancer, the octamer motif is found within a modular structure containing binding sites for several different transcription factors which act together to activate gene expression (**Figure 4.46**).

The close relationship of enhancer and promoter elements is further illustrated by the *Xenopus hsp70* gene, in which multiple copies of the HSE are located at positions far upstream of the start site and function as a heat-inducible enhancer element when transferred to another gene. Similarly, the heat-shock element of the *Drosophila hsp70* gene which we have used as an example of a promoter motif has been shown to function as an enhancer when multiple copies are placed at a position well upstream of the transcriptional start site.

Enhancers therefore appear to consist of sequence motifs which are also present in similarly regulated promoters and may be present within the enhancer associated with other control elements or in multiple copies. In many cases, an enhancer will therefore consist of an array of different sequence motifs which function together to produce strong activation of transcription. The array of different regulatory proteins which assemble on such an enhancer has been called an **enhanceosome**.

It seems likely, therefore, that enhancers may activate gene expression by either or both of the mechanisms described previously for promoter elements, namely a change in chromatin structure leading to nucleosome displacement or by direct interaction with the proteins of the transcriptional apparatus. Indeed, the enhanceosome of the interferon-β promoter has been shown to function by both these mechanisms. Thus, an enhanceosome complex of regulatory proteins initially assembles on the enhancer. This contains a total of eight different proteins including the DNA-binding

Figure 4.46
Protein-binding sites in the immunoglobulin heavy-chain gene enhancer. O indicates the octamer motif.

(a)

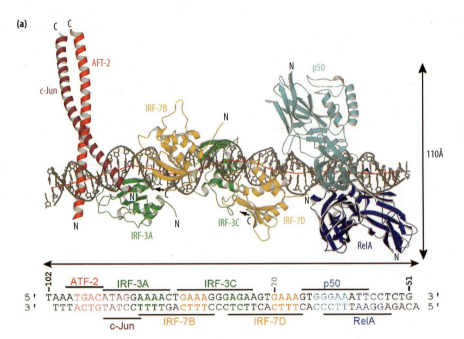

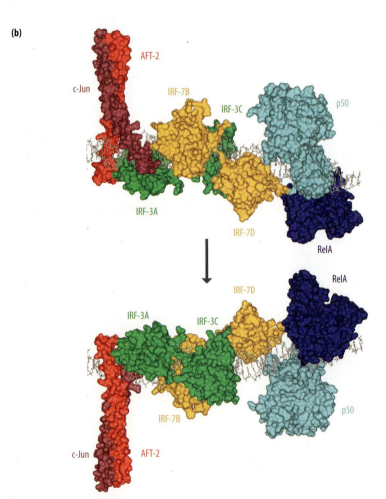

Figure 4.47
Overall structure of the interferon-β enhanceosome. Panel (a) shows a side view of the complex bound to the DNA together with the DNA sequence of the binding site. Panel (b) shows two molecular surface representations (related by 180° rotation and taken from opposite sides of the complex) showing that the eight proteins in the complex form a composite surface for DNA recognition. Courtesy of Steve Harrison, from Panne D, Maniatis T & Harrison SC (2007) *Cell* 129, 1111–1123. With permission from Elsevier.

protein HMG1(Y) and several different transcriptional activators such as **NFκB** and c-Jun. The structure of this complex has recently been defined showing that the eight proteins interact to form a continuous surface which binds to the DNA (**Figure 4.47**).

Subsequently, the enhanceosome complex first recruits a histone acetylase complex stimulating the subsequent recruitment of the chromatin-remodeling complex SWI–SNF. As described in Sections 3.3 and 3.5, this will open up the chromatin and then allows the binding of the complex of RNA polymerase II and associated proteins which will direct enhanced transcription (**Figure 4.48**).

Interestingly, acetylation of histones during the activation of the interferon-β enhancer takes place in two stages, each of which has a different function. Acetylation of histone H4 on lysine 8 allows the recruitment of the SWI–SNF complex, while subsequent acetylation of histone H3 on lysines 9 and 14 results in the recruitment of TFIID.

The enhancer of the interferon-β gene therefore utilizes both DNA-binding transcription factors and the subsequent recruitment of factors modifying chromatin structure to produce enhanced binding of RNA polymerase II and its associated factors. Hence, both promoter and enhancer sequences can act at the level of chromatin structure. Interestingly, however, a recent study showed that histone H3 was trimethylated on lysine 4 at transcriptionally active promoters but was monomethylated at this position at the enhancers of active genes (see Sections 2.3 and 3.3 for discussion of this modification). Hence, both promoters and enhancers exhibit histone modifications characteristic of active genes but the modifications may not always be identical at the different elements.

In the case of chromatin structure changes, it is readily apparent that such changes caused by a protein binding to the enhancer could be propagated over large distances in both directions, causing the observed distance-, position-, and orientation-independence of the enhancer. In agreement with this possibility, DNaseI-hypersensitive sites have been mapped within a number of enhancer elements, including the immunoglobulin enhancer. Moreover, the nucleosome free gap in the DNA of the eukaryotic virus SV40 (see Section 3.5) is located at the position of the enhancer.

At first sight, models involving the binding of protein factors to DNA sequences in the enhancer followed by direct interaction with proteins of the transcriptional apparatus are more difficult to reconcile with the action at a distance characteristic of enhancers. Nonetheless, the binding to enhancers of very many proteins crucial for transcriptional activation suggests that enhancers can indeed function in this manner. Models to explain this postulate that the enhancer serves as a site of entry for a regulatory factor. The factor would then make contact with the promoter-bound transcriptional apparatus either by sliding along the DNA or via a continuous scaffold of other proteins or by the looping out of the intervening DNA (**Figure 4.49**).

Of these possibilities, both the sliding model and the continuous scaffold model are difficult to reconcile with the observed large distances over which enhancers act. Similarly, these models cannot explain the observation that the immunoglobulin enhancer activates equally two promoters placed 1.7 and 7.7 kb away on the same DNA molecule, since they would postulate that sliding or scaffolded molecules would stop at the first promoter (**Figure 4.50**). Moreover, it has been shown that an enhancer can act on a promoter when the two are located on two separate DNA molecules linked only by a protein bridge, which would disrupt sliding or scaffolded molecules.

Such observations are explicable, however, via a model in which proteins bound at the promoter and enhancer proteins have an affinity for one another and make contact via looping out of the intervening DNA. Moreover, such a model can explain the critical importance of DNA structure on the action of enhancers. Thus it has been shown that removal of precise multiples of 10 bases (one helical turn) from the region between the SV40 enhancer and its promoter has no effect on its activity but deletion of DNA corresponding to half a helical turn disrupts enhancer function severely.

Interestingly, some proteins which bind to enhancers actually bend the DNA so that interactions can occur between regulatory proteins bound at distant sites on the DNA (**Figure 4.51**). This is seen in the case of the T-cell receptor α-chain gene enhancer where the LEF-1 factor binds to a site at the center of the enhancer and bends the DNA so that other regulatory transcription factors can interact with one another. Indeed, such effects

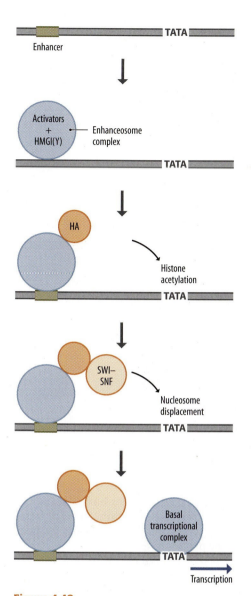

Figure 4.48
Activation of the interferon-β promoter involves binding of the enhanceosome complex to the enhancer. This facilitates the recruitment of a histone acetylase (HA), which in turn recruits the SWI–SNF complex, leading to nucleosome displacement and binding of the basal transcriptional complex.

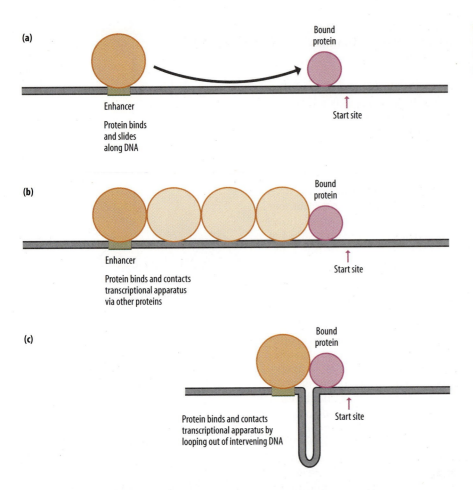

(a)

Bound
protein

Enhancer

Start site

Protein binds
and slides
along DNA

(b)

Bound
protein

Enhancer

Start site

Protein binds and contacts
transcriptional apparatus
via other proteins

(c)

Bound
protein

Protein binds and contacts
transcriptional apparatus by
looping out of intervening DNA

Start site

Figure 4.49
Possible models for the action of enhancers located at a distance from the activated promoter.

are not confined to enhancers since binding of TBP to the TATA box in gene promoters (see Section 4.1) also bends the DNA, suggesting that this may be a widespread mechanism for producing interactions between regulatory proteins.

Hence, regulatory proteins can bind at the enhancer and influence the activity of the basal transcriptional complex via looping or bending of the DNA. Interestingly, it has been shown that in some cases DNA-binding activator proteins bound to the enhancer can actually recruit the basal transcriptional complex (consisting of RNA polymerase II, TFIID, TFIID, etc.) to the enhancer itself. This has been suggested to poise the gene ready for transcription. Subsequently, in response to an activation signal, the polymerase complex is transferred to the promoter via DNA looping and transcription begins (**Figure 4.52**). This effect has recently been demonstrated for the CD80 gene which lacks both an initiator element and a TATA box (see Section 4.1). Recruitment of the basal transcriptional complex via an enhancer may therefore occur, particularly in genes which lack the promoter elements that normally recruit the complex.

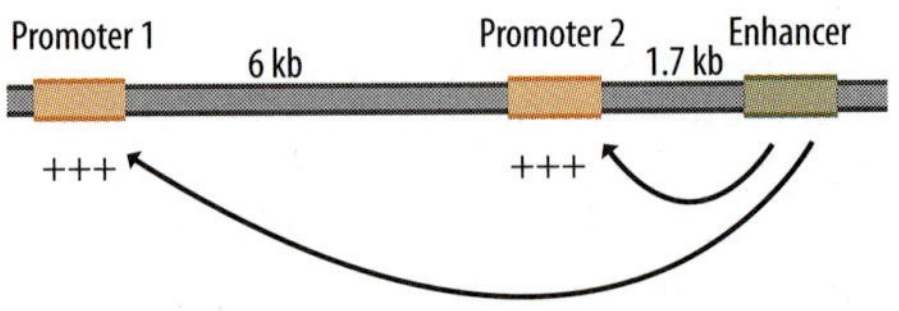

Silencers can act at a distance to inhibit gene expression

Thus far we have assumed that enhancers act in an entirely positive manner. In a tissue containing an active enhancer-binding protein, the enhancer will activate a promoter whereas in other tissues where the protein is absent or inactive, the enhancer will have no effect. Such a mechanism does indeed appear to operate for the majority of enhancers which when linked to

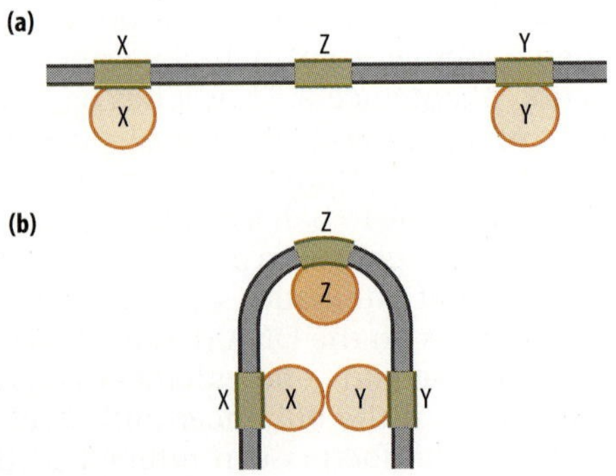

Promoter 1   6 kb   Promoter 2   1.7 kb   Enhancer

+++   +++

Figure 4.50
An enhancer can activate an adjacent promoter and a more distant one equally well.

(a)

X   Z   Y

X   Y

(b)

Z

Z

X X Y Y

Figure 4.51
Binding of a factor (Z) which bends the DNA can produce interactions between other regulatory proteins (X and Y) which bind at distant sites on the DNA.

Figure 4.52
Binding of an activator protein (A) to the enhancer can in some circumstances result in recruitment to the enhancer of the basal transcriptional complex (BTC) consisting of RNA polymerase II and associated proteins. This is believed to produce a situation in which the gene is poised for transcription. In response to a specific signal, the basal transcriptional complex is transferred to the gene promoter by DNA looping and transcription begins.

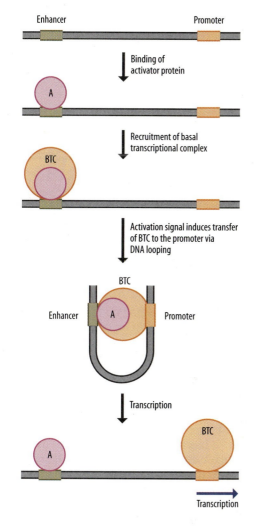

promoters, activate gene expression in one or a few cell types and have no effect in other cell types. In contrast, however, some sequences appear to act in an entirely negative manner, inhibiting the expression of genes which contain them. Following the initial identification of such an element in the cellular **oncogene** c-*myc*, (see Section 11.1), similar elements—referred to as **silencers**—have been defined in genes encoding proteins as diverse as **collagen** type II, growth hormone, and glutathione transferase P. Like enhancers, silencers can act on distant promoters when present in either orientation but have an inhibitory rather than a stimulatory effect on the level of gene expression.

As with enhancer elements it is likely that these silencers act either at the level of chromatin structure by recruiting factors which direct the tight packing of adjacent DNA or by binding a protein which then directly inhibits transcription by interacting with RNA polymerase and its associated factors. Examples of silencers which appear to act in each of these ways have been observed (**Figure 4.53**). For example, the silencer element located approximately 2 kb from the promoters of the repressed mating type loci in yeast plays a crucial role in organizing this region into the tightly packed structure characteristic of non-transcribed DNA (see Section 10.3 for further discussion of the yeast mating-type system).

Interestingly, the silencer element appears to represent a site for attachment of the DNA to the nuclear matrix. Hence, it may act by promoting the further condensation of the 30 nm-chromatin fiber to form a loop of DNA attached to the nuclear matrix which is the most condensed structure of chromatin (see Section 2.5 and Figure 2.40).

This could be achieved by recruiting protein factors which direct tight packing of the adjacent DNA, in the same way as the SWI–SNF complex and the GAGA factor direct the unpacking of the adjacent DNA (see Section 3.5). Indeed, as discussed in Chapter 3 (Section 3.3) specific factors such as polycomb have been identified which have exactly this effect, inducing the ubiquitination of histone H2A and the methylation of histone H3 on lysines 9 and 27, thereby promoting the tight packing of chromatin (**Figure 4.54**). Moreover, inactivation of polycomb by mutation results in aberrant activation of specific genes, exactly as would be expected for a protein

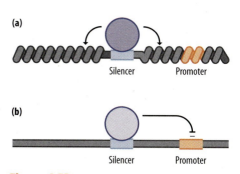

Figure 4.53
A silencer element can inhibit activity of the promoter either (a) by binding a protein which organizes the DNA into a tightly packed chromatin structure or (b) by binding an inhibitory transcription factor which represses promoter activity.

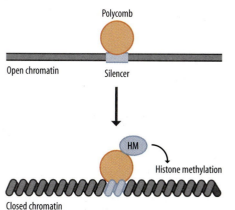

Figure 4.54
Binding of polycomb to a silencer sequence can recruit a histone methyltransferase (HM), which produces a tightly packed chromatin structure by methylating histones.

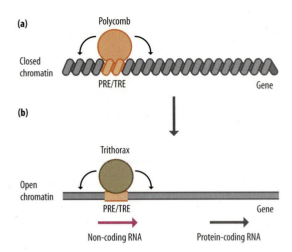

Figure 4.55
Polycomb and trithorax proteins can bind to the same response elements in the DNA (PRE/TRE) with polycomb directing a closed chromatin structure (a) and trithorax directing an open chromatin structure, allowing the transcription of a nearby protein-coding gene (b). The transition from polycomb to trithorax binding requires the transcription of the PRE/TRE to produce a noncoding RNA.

which promotes an inactive chromatin structure (see Sections 9.1 and 9.2 for further discussion of the role of polycomb proteins in development). Hence, chromatin structure is controlled by the antagonistic effects of proteins such as polycomb which direct tight packing of the chromatin and others such as GAGA/trithorax and brahma (see Section 3.3) which direct its opening.

Interestingly, silencer elements which bind polycomb proteins can also bind trithorax proteins. Hence, the same sequence can direct the production of a closed or open chromatin structure depending on the protein complexes it binds under different conditions or in different cell types. These DNA sequences are therefore known as polycomb-response elements/trithorax-response elements.

It has been shown that the transition from polycomb binding to trithorax binding at these sequences can involve the transcription of non-coding RNAs from the response element itself (**Figure 4.55**). Moreover, blocking the synthesis of these non-coding RNAs prevents the transition to an open chromatin structure and transcriptional activation. These non-coding RNAs therefore play a key role in gene activation in contrast to the inhibitory effect on transcription of other non-coding RNAs such as the small interfering RNAs (see Section 3.4) and XIST (see Section 3.6).

As well as acting at the level of chromatin structure, it is also possible for silencers to act by binding negatively acting transcription factors which interact with the basal transcriptional complex of RNA polymerase II and associated factors (see Figure 4.53). For example, the silencer in the gene encoding lysozyme appears to act at least in part by binding the thyroid hormone receptor which in the absence of thyroid hormone has a directly inhibitory effect on gene activity (see Section 5.3).

Hence, just as enhancer elements can act at a distance to enhance gene expression by opening up the chromatin or by directly stimulating transcription, silencers can inhibit gene expression either by promoting a more tightly packed chromatin structure or by directly repressing transcription.

CONCLUSIONS

In this chapter, we have discussed the process of transcription initiated by the three nuclear RNA polymerases which are found in all eukaryotes and each of which has a key role in the transcription of specific genes. Transcription by the three RNA polymerases has a number of common features including the presence of common subunits in the three polymerases, the role of prior DNA binding by specific factors in recruiting the polymerase and the key role of TBP in this process.

Nonetheless, the process of transcription by RNA polymerase II is more complex, paralleling its role in the transcription of protein-coding genes which exhibit highly complex patterns of regulated transcription. For example, the basal transcriptional complex for RNA polymerase II involves far

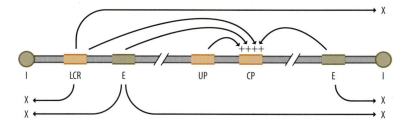

Figure 4.56
Sequences regulating a typical RNA polymerase II gene. The activity of core promoter elements (CP) is influenced by upstream promoter elements (UP), enhancers (E), and locus-control regions (LCR). The inappropriate spread of enhancer and LCR activity to adjacent DNA regions is limited by insulator sequences (I). Note that only positive actions are shown but upstream promoter elements can act negatively, as can silencer elements which resemble enhancers but have the opposite effect. Compare with Figures 4.31 and 4.42.

more factors than the complexes producing transcription by the other RNA polymerases. Moreover, transcriptional elongation by RNA polymerase II is a regulated process, with the **polymerase pausing** 20–30 bases after transcriptional initiation and requiring further signals to proceed with transcriptional elongation.

Perhaps the greatest complexity in RNA polymerase II-mediated transcription is the array of DNA sequences, which control both constitutive and regulated transcription (promoters, enhancers, silencers, etc.). Indeed, it has been argued that the complex pattern of sequences regulating a single gene is one of the defining features of a eukaryote and is necessary to produce the complex patterns of gene expression required in the adult organism and in development.

The promoter, enhancer, and silencer sequences that we have discussed in this chapter can act both by altering chromatin structure and by binding regulatory proteins which interact with the basal transcriptional complex to modify its activity. As such, they complement the actions of the locus-control region (LCR) and insulator sequences which were discussed in Section 2.5 and which act at the level of chromatin structure. These sequences act in the case of LCRs to control the chromatin structure of a DNA region and in the case of insulators to prevent inappropriate spreading of chromatin structure to an adjacent region. Indeed, insulators act in this way to limit the spread of gene activation or repression not only by LCRs but also by enhancers and silencers which as described in Section 4.4 can activate or repress gene expression over long distances.

Hence, a RNA polymerase II gene can be influenced by promoter, enhancer, silencer, LCR, and insulator sequences (**Figure 4.56**). The combined action of such sequences will control the chromatin structure and provide binding sites for transcription factors which can interact directly or indirectly with the basal transcriptional complex to enhance or reduce its transcription. The nature of these transcription factors and the manner in which they control transcription are discussed in the next chapter.

KEY CONCEPTS

- Three RNA polymerases are found in all eukaryotes with each polymerase transcribing different classes of genes.

- RNA polymerase I transcribes the gene for the 45S ribosomal RNA precursor (which is processed to yield 28, 18, and 5.8S ribosomal RNAs), RNA polymerase II transcribes protein-coding genes, and RNA polymerase III transcribes tRNA and 5S ribosomal RNA genes.

- For all three polymerases, assembly of the basal transcriptional complex at the gene promoter requires initial binding of a protein complex to specific DNA sequences with subsequent recruitment of the RNA polymerase itself via protein–protein interaction.

- The TBP transcription factor is a key component of the initial DNA-binding complex for all three RNA polymerases.

- RNA polymerase II pauses after transcribing 20–30 bases and requires further modification in order for elongation to continue along the full length of the gene.

- Transcriptional termination for RNA polymerase II takes place downstream of the polyadenylation signal, which directs the **endonuclease** cleavage of the initial RNA transcript to produce the 3′ end of the mature RNA.

- In genes transcribed by RNA polymerase II, a variety of sequences located at the gene promoter influence the level of gene transcription.

- Other elements located much further upstream or downstream of the gene can also influence transcription either positively (enhancers) or negatively (silencers).

- Both promoter and enhancer/silencer sequences can act either by altering chromatin structure or by binding regulatory proteins which interact with the basal transcriptional complex.

- The precise combination of sequences in the promoter and enhancer/ silencer of individual genes plays a key role in determining whether they are transcribed in all cell types or only in certain cells and/or in specific situations, as well as their rate of transcription in different situations.

FURTHER READING

4.1 Transcription by RNA polymerases

Akhtar A & Gasser SM (2007) The nuclear envelope and transcriptional control. *Nat. Rev. Genet.* 8, 507–517.

Dieci G, Fiorino G, Castelnuovo M et al. (2007) The expanding RNA polymerase III transcriptome. *Trends Genet.* 23, 614–622.

Haag JR & Pikaard CS (2007) RNA polymerase I: a multifunctional molecular machine. *Cell* 131, 1224–1225.

Hahn S (2004) Structure and mechanism of the RNA polymerase II transcription machinery. *Nat. Struct. Mol. Biol.* 11, 394–403.

Jones KA (2007) Transcription strategies in terminally differentiated cells: shaken to the core. *Genes Dev.* 21, 2113–2117.

Kornberg RD (2007) The molecular basis of eukaryotic transcription. *Proc. Natl Acad. Sci. USA* 104, 12955–12961.

Kumaran RI, Thakar R & Spector DL (2008) Chromatin dynamics and gene positioning. *Cell* 132, 929–934.

Reina JH & Hernandez N (2007) On a roll for new TRF targets. *Genes Dev.* 21, 2855–2860.

Schneider R & Grosschedl R (2007) Dynamics and interplay of nuclear architecture, genome organization, and gene expression. *Genes Dev.* 21, 3027–3043.

4.2 Transcriptional elongation and termination

Egloff S & Murphy S (2008) Cracking the RNA polymerase II CTD code. *Trends Genet.* 24, 280–288.

Price DH (2008) Poised polymerases: on your mark...get set...go! *Mol. Cell* 30, 7–10.

Rosonina E, Kaneko S & Manley JL (2006) Terminating the transcript: breaking up is hard to do. *Genes Dev.* 20, 1050–1056.

Workman JL (2006) Nucleosome displacement in transcription. *Genes Dev.* 20, 2009–2017.

4.3 The gene promoter

ENCODE Project Consortium (2007) Identification and analysis of functional elements in 1% of the human genome by the ENCODE pilot project. *Nature* 447, 799–816.

Latchman DS (1999) Transcription Factors: a Practical Approach, 2nd ed. Oxford University Press.

Sandelin A, Carninci P, Lenhard B et al. (2007) Mammalian RNA polymerase II core promoters: insights from genome-wide studies. *Nat. Rev. Genet.* 8, 424–436.

Schones DE & Zhao K (2008) Genome-wide approaches to studying chromatin modifications. *Nat. Rev. Genet.* 9, 179–191.

4.4 Enhancers and silencers

Panne D, Maniatis T & Harrison SC (2007) An atomic model of the interferon-beta enhanceosome. *Cell* 129, 1111–1123.

Pennisi E (2004) Searching for the genome's second code. *Science* 306, 632–634.

Szutorisz H, Dillon N & Tora L (2005) The role of enhancers as centres for general transcription factor recruitment. *Trends Biochem. Sci.* 30, 593–599.

Yaniv M (2009) Small DNA tumour viruses and their contributions to our understanding of transcription control. *Virology* 384, 369–374.

Transcription Factors and Transcriptional Control

5

INTRODUCTION

As discussed in Chapter 4, the expression of specific genes in particular cell types or tissues is regulated by DNA sequence motifs present within promoter or enhancer elements. These elements control the alteration in chromatin structure of the gene that occurs in a particular lineage, or the subsequent induction of gene transcription. It was assumed for many years that such sequences would act by binding a regulatory protein which was only synthesized in a particular tissue or was present in an active form only in that tissue. In turn, the binding of this protein would result in the observed effect on gene expression. Indeed, as described in Chapter 4 (Section 4.3) cell extracts can be used in DNA mobility-shift or DNaseI footprinting assays to show that they contain protein(s) able to bind to a specific sequence.

The isolation and characterization of such factors proved difficult, however, principally because they were present in very small amounts. Hence, even if they could be purified, the amounts obtained were too small to provide much information as to the properties of the protein.

This obstacle was overcome by the cloning of the genes encoding a number of different transcription factors. Two general approaches were used to achieve this. In one approach (**Figure 5.1**) the transcription factor Spl was purified by virtue of its ability to bind to its specific DNA-binding site. The partial amino acid sequence of the protein was then obtained from the small amount of material isolated and was used in conjunction with the **genetic code** to predict a set of DNA oligonucleotides, one of which would encode this region of the protein. The oligonucleotides were then hybridized to a complementary **DNA library** prepared from Spl-containing HeLa cell mRNA. A cDNA clone derived from the Spl mRNA must contain the sequence capable of encoding the protein and hence will hybridize to the probe. In this experiment one single clone derived from the Spl mRNA was isolated by screening a library of 1 million recombinants prepared from the whole population of HeLa cell mRNAs.

An alternative, more direct approach to the cloning of transcription factors was used to clone the gene encoding the NFκB transcription factor, which is involved in regulating the expression of the immunoglobulin genes in B cells (**Figure 5.2**) (see Section 8.2 for discussion of NFκB and its regulation). As in the previous method, a cDNA library was constructed containing copies of all the mRNAs in a specific cell type. However, the library was constructed in such a way that the sequences within it would be translated into their corresponding proteins. This was achieved by inserting the cDNA into the coding region of the bacteriophage β-galactosidase gene, resulting in the translation of the eukaryotic insert as part of the bacteriophage protein.

Most interestingly, these **fusion proteins** were capable of binding DNA with the same specificity as the original transcription factor encoded by the cloned mRNA. The library could therefore be screened directly with the

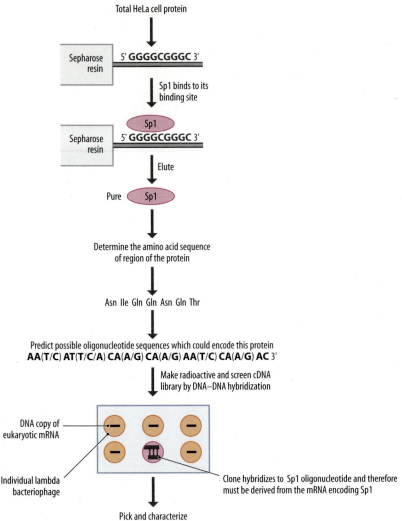

radiolabeled DNA-binding site for a particular transcription factor. A clone containing the mRNA for this factor and hence expressing it as a fusion protein was able to bind the labeled DNA and could therefore be identified and isolated.

Unlike the previous method, this procedure involves DNA–protein rather than DNA–DNA binding and can be used without prior purification of the transcription factor provided its binding site is known. Since most factors are identified on the basis of their binding to a particular site, this is not a significant problem and the use of these two methods has resulted in the isolation of the genes encoding a wide variety of transcription factors.

In turn, this has resulted in an explosion of information on these factors. Thus, once the gene for a factor has been cloned, Southern blotting (see Section 1.3) can be carried out to study the structure of the gene, and Northern blotting or other techniques for RNA analysis (see Section 1.2) can be used to search for RNA transcripts derived from it in different cell types and related genes expressed in other tissues or other species can be identified.

More importantly, considerable information can be obtained from the cloned gene about the corresponding protein and its activity. Not only can the DNA sequence of the gene be used to predict the amino acid sequence of the corresponding protein, but the existence of functional domains within the protein with particular activities can also be defined. As described above, if the gene encoding a transcription factor is expressed in bacteria, it continues to bind DNA in a sequence-specific manner. Hence if the gene is broken up into small pieces and each of these is expressed in bacteria (**Figure 5.3**), the abilities of each portion to bind to DNA, to other proteins or to a

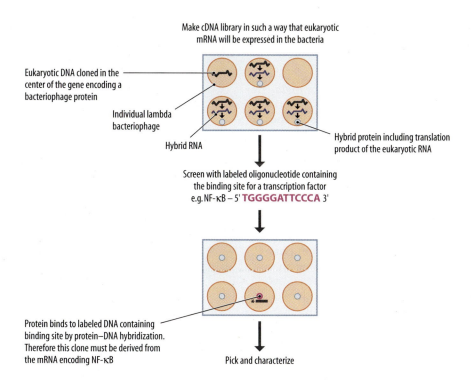

Make cDNA library in such a way that eukaryotic mRNA will be expressed in the bacteria

Eukaryotic DNA cloned in the center of the gene encoding a bacteriophage protein

Individual lambda bacteriophage

Hybrid RNA

Hybrid protein including translation product of the eukaryotic RNA

Screen with labeled oligonucleotide containing the binding site for a transcription factor e.g. NF-κB – 5' **TGGGGATTCCCA** 3'

Protein binds to labeled DNA containing binding site by protein–DNA hybridization. Therefore this clone must be derived from the mRNA encoding NF-κB

Pick and characterize

Figure 5.2
Isolation of cDNA clones for the NFκB transcription factor by screening an expression library with a DNA probe containing the binding site for the factor.

potential regulatory molecule can be assessed using, for example DNA mobility-shift assays, exactly as with cellular extracts (see Section 4.3). This mapping of the **DNA-binding domain** can also be achieved by transcribing and translating pieces of the DNA into protein fragments in the test tube and testing their activity in the same way.

Each of the domains of the transcription factor identified in this way can be altered by mutagenesis of the corresponding DNA encoding that region of the protein and subsequent expression of the mutant protein as before. The testing of the effect of these mutations on the activity mediated by the particular domain of the protein will thus allow the identification of the amino acids that are critical for each of the observed properties of the protein.

In this way large amounts of information have accumulated on individual transcription factors. Rather than attempt to consider each factor individually, we will focus on the properties necessary for such a factor and illustrate our discussion by referring to the manner in which these are achieved in individual cases.

In this chapter, we will first consider in Section 5.1 the manner in which such factors bind to DNA in a sequence-specific manner. Subsequently, in Sections 5.2 and 5.3 we will consider the manner in which the DNA-bound factor influences transcription either positively or negatively by interacting with other transcription factors or with the RNA polymerase itself to regulate transcriptional initiation. Such positive or negative effects can also occur at the level of transcriptional elongation and this will therefore be considered in Section 5.4. The primary focus of this chapter will be on transcription factors which regulate RNA polymerase II transcription since this polymerase transcribes protein-coding genes. However, Section 5.5 will consider transcriptional regulation of genes transcribed by RNA polymerases I and III.

5.1 DNA BINDING BY TRANSCRIPTION FACTORS

Extensive studies of eukaryotic transcription factors have identified several structural elements which either bind directly to DNA or which facilitate DNA binding by adjacent regions of the protein. These motifs will be discussed in turn, using transcription factors that contain them to illustrate their properties.

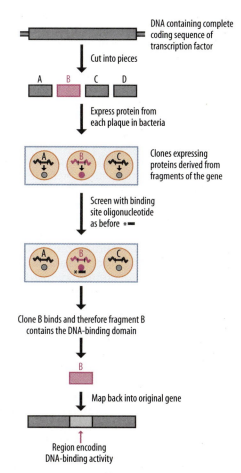

DNA containing complete coding sequence of transcription factor

Cut into pieces

A B C D

Express protein from each plaque in bacteria

Clones expressing proteins derived from fragments of the gene

Screen with binding site oligonucleotide as before

Clone B binds and therefore fragment B contains the DNA-binding domain

B

Map back into original gene

Region encoding DNA-binding activity

Figure 5.3
Mapping of the DNA-binding region of a transcription factor by testing the ability of different regions to bind to the appropriate DNA sequence when expressed in bacteria.

The helix-turn-helix motif is found in a number of transcription factors which regulate gene expression during embryonic development

The small size and rapid generation time of the fruit fly *Drosophila mela-nogaster* has led to it being one of the best-characterized organisms geneti-cally, and a number of mutations which affect various properties of the organism have been described. These include mutations which affect the development of the fly, resulting for example in the production of additional legs in the position of the antennae (**Figure 5.4**). Genes of this type are likely to play a crucial role in the development of the fly and, in particular, in determining the body plan, and are known as homeotic genes (see Section 9.2 for further discussion of the role of homeotic genes in *Drosophila* development).

The critical role for the products of these genes identified genetically sug-gested that they would encode regulatory proteins which would act at par-ticular times in development to activate or repress the activity of other genes encoding proteins required for the production of particular structures. This idea was confirmed when the genes encoding these proteins were cloned. Thus, these proteins were shown to be able to bind to DNA in a sequence-specific manner and to be able to induce increased transcription of genes which contained this binding site. Thus in the case of the homeotic gene *fushi tarazu* (*ftz*), mutation of which produces a fly with only half the normal number of segments, the protein has been shown to bind specifically to the sequence TCAATTAAATGA. When the gene encoding this protein is intro-duced into *Drosophila* cells with a marker gene containing this sequence, transcription of the marker gene is increased. This up-regulation is entirely dependent on binding of the Ftz protein to this sequence in the promoter of the marker gene, since a 1-bp change in this sequence which abolishes bind-ing also abolishes the induction of transcription (**Figure 5.5**).

The product of another homeotic gene, the engrailed protein, binds to the identical sequence to that bound by Ftz. Its binding does not produce increased transcription of the marker gene, however, and indeed it prevents the activation by Ftz. Hence, the expression of Ftz alone in a cell would acti-vate particular genes whereas Ftz expression in a cell also expressing the *engrailed* product would have no effect (**Figure 5.6**). In this way interacting homeotic gene products expressed in particular cells could control the developmental fate of the cells by regulating the expression of specific tar-get genes.

Interestingly, there is evidence that the homeotic genes may be neces-sary not only for the actual production of a specific cell type but also for the long-term process of commitment to a particular cellular phenotype which was discussed in Chapter 2 (Section 2.1). Thus, in the case of the imaginal discs of *Drosophila*, commitment to the production of a particular adult structure was maintained through many cell generations in the absence of differentiation. If during this time, however, a mutation is introduced into one of the homeotic genes in the disc cells inactivating it, when eventually the cells are allowed to differentiate they will produce the wrong structure. For example, if the homeotic gene *ultrabithorax* (*ubx*) is inactivated in a disc cell which normally gives rise to the haltere (balancer), these cells will produce wing tissue when allowed to differentiate. The continual expres-sion of homeotic genes within the cells is essential for their commitment to a particular pathway of differentiation.

A possible molecular mechanism for this is provided by the demonstra-tion that the Ubx protein binds to its own promoter and up-regulates its own transcription. Hence once production of this protein has been induced presumably during the commitment process, it will continue indefinitely and thus maintain this commitment (**Figure 5.7**).

In addition to this mechanism, other processes maintain the chromatin structure of the homeotic genes in an active state so that transcription con-tinues once the gene has initially been activated. This is achieved by mem-bers of the trithorax group of proteins, which include the GAGA and brahma

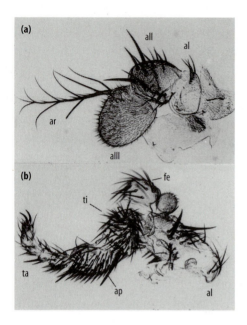

Figure 5.4
Effect of a homeotic mutation, which produces a middle leg (b) in the region that would contain the antenna of a normal fly (a). al, all, and alll: first, second, and third antennal segments; ar, arista; ta, tarsus; ti, tibia; fe, femur; ap, apical bristle. Courtesy of WJ Gehring, from Gehring WJ, *Science* (1987) 236, 1245–1252. With permission from The American Association for the Advancement of Science.

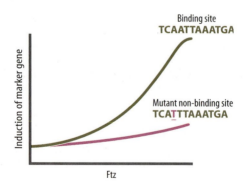

Figure 5.5
Effect of expression of the Ftz protein on the expression of a gene containing its binding site, or a mutated binding site containing a single-base-pair change which abolishes binding of Ftz.

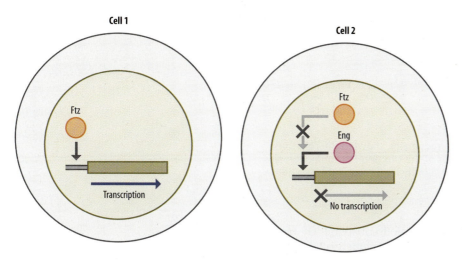

Figure 5.6
Blockage of gene induction by Ftz in cells expressing the engrailed (Eng) protein which binds to the same sequence as Ftz but does not activate transcription.

factors discussed in Chapter 3 (Section 3.5). These factors bind to active homeotic genes and maintain them in an open chromatin structure, allowing transcription to continue.

Conversely, other proteins of the polycomb group bind to homeotic genes and produce an inactive chromatin structure so preventing their inappropriate activation (see Sections 3.3 and 4.4). As might be expected, mutations of the genes encoding trithorax or polycomb group proteins result in gross abnormalities in the fly due, respectively, to a failure to activate the appropriate homeotic genes or their inappropriate activation (**Figure 5.8**) (see Sections 9.1 and 9.2 for further discussion of the role of polycomb and trithorax proteins in development).

Interestingly, as noted in Chapter 2 (Section 2.1), the changes in commitment which occur in imaginal discs when the process of commitment breaks down after the discs are cultured for long periods are precisely those which occur in homeotic mutations. Hence, a change in the chromatin structure and expression of a specific homeotic gene in the imaginal disc will result in a change in the pattern of commitment similar to that which occurs when this gene is mutated.

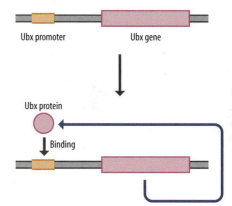

Figure 5.7
The Ubx protein activates its own promoter, producing a positive-feedback loop maintaining high-level production of Ubx.

The helix-turn-helix domain found in homeodomain proteins is a DNA-binding domain

The clear evidence that homeotic gene products regulate both their own genes and other genes by binding specifically to DNA, has led to extensive investigation of their structure in order to identify the region that mediates this sequence-specific DNA binding. When the genes encoding these proteins were first cloned, it was found that they each contained a short related DNA sequence of about 180 bp capable of encoding 60 amino acids (**Figure 5.9**), which was flanked on either side by sequences that differed dramatically between the different genes. This sequence was named the **homeodomain** (also known as the homeobox). The presence of the homeodomain in all these genes suggested that it plays a critical role in mediating their

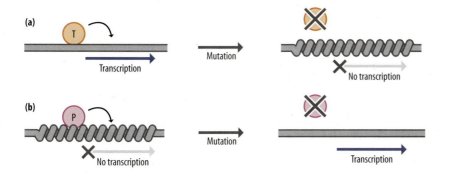

Figure 5.8
(a) Members of the trithorax group of proteins (T) maintain the active chromatin structure (solid line) of transcriptionally active homeotic genes. Inactivation of these proteins by mutation results in an inactive chromatin structure (wavy line), leading to a failure of transcription. (b) Members of the polycomb family (P) maintain the inactive chromatin structure (wavy line) of transcriptionally inactive homeotic genes. Their inactivation by mutation results in an active chromatin structure (solid line), leading to the inappropriate transcription of these genes.

	1	2	3	4	5	6	7	8	9	10	11	12	13	14	15	16	17	18	19	20	21	22	23	24	25	26	27	28	29	30
Antp	Arg	Lys	Arg	Gly	Arg	Gln	Thr	Tyr	Thr	Arg	Tyr	Gln	Thr	Leu	Glu	Leu	Glu	Lys	Glu	Phe	His	Phe	Asn	Arg	Tyr	Leu	Thr	Arg	Arg	Arg
Ubx		Arg																		Thr			His							
Ftz	Ser			Thr																						Ile				

			Helix						**Turn**		**Recognition helix**																			
Antp	Arg	Ile	Glu	Ile	Ala	His	Ala	Leu	Cys	Leu	Thr	Glu	Arg	Gln	Ile	Lys	Ile	Trp	Phe	Gln	Asn	Arg	Arg	Met	Lys	Trp	Lys	Lys	Glu	Asn
Ubx				Met		Tyr								Glu													Leu		Ile	
Ftz			Asp			Asn			Ser		Ser																Ser		Asp	Arg

Figure 5.9
Amino acid sequences of several *Drosophila* homeodomains, showing the conserved helical motifs. Differences between the sequences of the Ubx and Ftz homeodomains from that of Antp are indicated; a blank denotes identity in the sequence. The helix-turn-helix region is indicated.

regulatory function. This suggestion was confirmed subsequently by the use of the homeodomain as a probe to isolate other previously uncharacterized *Drosophila* regulatory genes.

The role of the homeodomain in DNA binding has been confirmed directly by the synthesis of the homeodomain region of the Antennapedia protein without the remainder of the protein, either by expression in bacteria or by chemical synthesis, and showing that it can bind to DNA in the identical sequence-specific manner to that exhibited by the intact protein.

The localization of this DNA binding to a short region of the protein only 60 amino acids in length allows a detailed structural prediction of the corresponding protein. This reveals that the homeodomain contains a so-called **helix-turn-helix** motif which is highly conserved between the different homeodomain-containing proteins. In this motif, a short region which can form an α-helical structure is followed by a β-turn and then another α-helical region. The position of these elements in the homeodomain is shown in Figure 5.9 and a diagram of the helix-turn-helix motif is given in **Figure 5.10**.

The prediction that this structure exists in the DNA-binding homeodomain has been directly confirmed by X-ray crystallographic analysis. Moreover, by carrying out this analysis on the homeodomain bound to its DNA-binding site, it has been shown that the helix-turn-helix motif does indeed contact DNA, with the second helix lying partly within the major groove where it can make specific contacts with the bases of the DNA (**Figure 5.11**). This second helix (labeled the **recognition helix** in the homeodomain sequence in Figure 5.9) can thus mediate sequence-specific binding while the other helix helps to correctly position the recognition helix.

The presence of this structure therefore indicates how the homeodomain proteins can bind specifically to particular DNA sequences, which is the first step in transcriptional activation of their target genes. The role of the helix-turn-helix motif in the recognition of specific sequences in the DNA has been demonstrated directly. For example, a mutation which changes a lysine at position 9 of the recognition helix in the **Bicoid** protein to the glutamine found in the equivalent position of the Antennapedia protein results in the protein binding to DNA with the sequence specificity of an Antennapedia rather than a Bicoid protein. In contrast, exchange of the other amino acids in the recognition helix which differ in the two proteins does not have this effect.

It appears that the lysine at position 9 in Bicoid forms hydrogen bonds with the O6 and N7 positions of a guanine base in the Bicoid-binding site. In contrast, the glutamine at position 9 in Antennapedia forms hydrogen

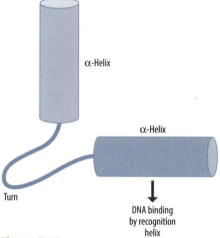

Figure 5.10
The helix-turn-helix motif.

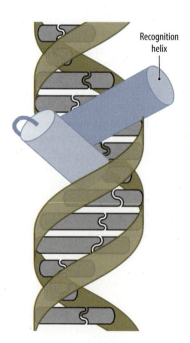

Figure 5.11
Binding of the helix-turn-helix motif to DNA, with the recognition helix in the major groove of the DNA. From Schleif (1988) *Science* 241, 1182–1187. With permission from The American Association for the Advancement of Science.

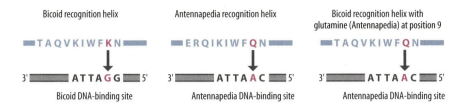

Bicoid recognition helix	Antennapedia recognition helix	Bicoid recognition helix with glutamine (Antennapedia) at position 9
▬TAQVKIWF**K**N▬	▬ERQIKIWF**Q**N▬	▬TAQVKIWF**Q**N▬
3'▬▬ATTA**GG**▬5'	3'▬▬ATTA**AC**▬5'	3'▬▬ATTA**AC**▬5'
Bicoid DNA-binding site	Antennapedia DNA-binding site	Antennapedia DNA-binding site

Figure 5.12
Changing the lysine (K) at position 9 in the Bicoid recognition helix to the glutamine (Q) found in Antennapedia allows recognition of an A residue in the Antennapedia DNA-binding site rather than the G residue found at the corresponding position in the Bicoid DNA-binding site. This results in the hybrid protein binding to the Antennapedia DNA-binding site even though all the other amino acids in the recognition helix are derived from the Bicoid protein.

bonds with the adenine base at the equivalent position in its site. This explains how exchange of these amino acids changes binding-site specificity (**Figure 5.12**).

Hence, not only does the helix-turn-helix motif mediate DNA binding but differences in the precise sequence of this motif in different homeodomains control the precise DNA sequence to which these proteins bind. Clearly, further structural and genetic studies of how this is achieved will throw considerable light on the way in which these proteins function.

The obvious importance of the homeodomain in *Drosophila* prompted a search for proteins containing this element in other organisms. Indeed, homeodomain-containing proteins which are expressed in specific cell types in the early embryo and play a key regulatory role have now been identified in a wide range of organisms, including mammals (see Section 9.3) and yeast (see Section 10.3). Hence, these proteins play a vital role in the processes regulating cellular differentiation and development in a variety of different organisms.

In the POU domain transcription factors, the homeodomain forms part of a larger DNA-binding motif

As well as the homeodomain proteins, another class of regulatory proteins has been identified which contains the homeodomain as one part of a much larger (150–160 amino acids) conserved region known as the POU (pronounced *pow*) domain. Unlike the homeodomain proteins, these regulatory proteins were not identified by mutational analysis or by homology to other regulatory proteins but were characterized as transcription factors having a particular pattern of activity.

Thus the mammalian Oct1 and Oct2 proteins both bind to the octamer sequence (consensus ATGCAAATNA) in the promoters of genes such as the histone H2B gene and those encoding the immunoglobulins and mediate the transcriptional activation of these genes. When the genes encoding these proteins were cloned they were found to possess a 150–160-amino acid sequence that was also found in the mammalian Pit-1 protein which regulates gene expression in the pituitary by binding to a sequence related to but distinct from the octamer. This sequence was also found in the protein encoded by the nematode gene *unc*-86 which is involved in sensory neuron development.

This POU (Pit-Oct-Unc) domain contains both a homeodomain-like sequence and a second conserved domain, the POU-specific domain (**Figure 5.13**). Although there are some differences between different POU proteins, in general the isolated homeodomain of the POU proteins alone is sufficient for sequence-specific DNA binding, but, unlike the classical homeodomain, the binding is of relatively low affinity in the absence of the POU-specific domain. Hence both parts of the **POU domain** are required for high-affinity sequence-specific DNA binding, indicating that the POU homeodomain and the POU-specific domain form two parts of a DNA-binding element which are held together by a flexible linker sequence.

Like the POU homeodomain, the POU-specific domain can also form a helix-turn-helix motif. The recognition helix from the POU-specific domain and that from the POU homeodomain bind to adjacent regions within the major groove of the DNA. This further illustrates the importance of the helix-turn-helix motif in allowing DNA binding by specific transcription factors.

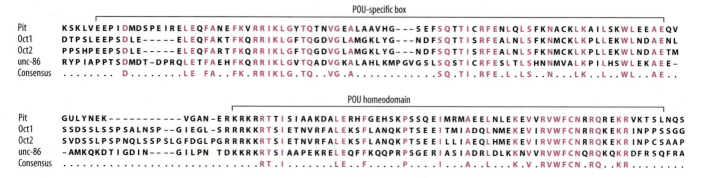

```
                                          POU-specific box
Pit        K S K L V E E P I DMDS P E I R E L E Q F A N E F K V R R I K L G Y T Q T N V G E A L A A V H G - - - S E F S Q T T I C R F E N L Q L S F K N A C K L K A I L S K W L E E A E Q V
Oct1       D T P S L E E P S D L E - - - - - E L E Q F A K T F K Q R R I K L G F T Q G D V G L A M G K L Y G - - - N D F S Q T T I S R F E A L N L S F K N M C K L K P L L E K W L N D A E N L
Oct2       P P S H P E E P S D L E - - - - - E L E Q F A R T F K Q R R I K L G F T Q G D V G L A M G K L Y G - - - N D F S Q T T I S R F E A L N L S F K N M C K L K P L L E K W L N D A E T M
unc-86     R Y P I A P P T S DMDT - D P R Q L E T F A E H F K Q R R I K L G V T Q A D V G K A L A H L KMP G V G S L S Q S T I C R F E S L T L S H N N M V A L K P I L H S W L E K A E E -
Consensus  . . . . . . . . . D . . . . . . . . . L E   F A . . F K . R R I K L G . T Q . . V G . A . . . . . . . . . . . S Q . T I . R F E . L . L S . . N . . . L K . . L . . W L . . A E . .
```

```
                                             POU homeodomain
Pit        G U L Y N E K - - - - - - - - - - - V G A N - E R K R K R R T T I S I A A K D A L E R H F G E H S K P S S Q E I M R M A E E L N L E K E V V R V W F C N R R Q R E K R V K T S L N Q S
Oct1       S S D S S L S S P S A L N S P - - G I E G L - S R R R K K R T S I E T N V R F A L E K S F L A N Q K P T S E E I T M I A D Q L N M E K E V I R V W F C N R R Q K E K R I N P P S S G G
Oct2       S V D S S L P S P N Q L S S P S L G F D G L P G R R R K K R T S I E T N V R F A L E K S F L A N Q K P T S E E I L L I A E Q L H M E K E V I R V W F C N R R Q K E K R I N P C S A A P
unc-86     - A M K Q K D T I G D I N - - - - G I L P N   T D K K R K R T S I A A P E K R E L E Q F F K Q Q P R P S G E R I A S I A D R L D L K K N V V R V W F C N Q R Q K Q K R D F R S Q F R A
Consensus  . . . . . . . . . . . . . . . . . . . . . . . . R T . I . . . . . . L E . . F . . . . . P . . . . I . . A . . L . . . K . V . R V W F C N . R Q . . K R . . . . . . . .
```

Figure 5.13
Amino acid sequences of the POU proteins. The homeodomain and the POU-specific domain are indicated. The final line shows a consensus sequence obtained from the four proteins.

Interestingly, the POU factors illustrate a novel aspect of gene regulation, namely that the sequence of the DNA-binding site to which a factor binds can influence its effect on gene expression. For example, when the Oct1 POU factor binds to its target sequence (ATGCAAAT) in cellular genes, it activates transcription only weakly. However, when it binds to its different target sequence (TAATGART; R=**purine**) in the herpes simplex virus (HSV) immediate-early genes, it binds in a distinct configuration. This allows it to bind the HSV protein VP16, which is a strong activator of transcription and hence strong activation occurs (**Figure 5.14**).

This type of effect is not confined to the recruitment of a viral protein by a cellular factor. Thus, when the Pit-1 protein binds to the prolactin gene promoter it activates transcription. However, the growth hormone promoter has a distinct sequence which binds Pit-1 but produces a different tertiary structure of the Pit-1 protein. This allows it to bind the nuclear receptor co-repressor NCo-R and hence transcription is repressed (**Figure 5.15**). Hence, the sequence of the DNA-binding site can affect the configuration of the bound factor and therefore its ability to recruit other regulatory molecules and to activate or repress transcription.

It is clear, therefore, that the POU proteins represent a family of proteins related to the homeodomain proteins which are likely to play a critical role in development. For example, inactivation of the Pit-1 gene leads to a failure of pituitary gland development, resulting in dwarfism in mouse and humans, whereas the *unc*-86 mutation results in a failure to form specific neurons in the nematode.

Interestingly, the two highly conserved peptide sequences at either end of the POU domain (see Figure 5.13) have been used to isolate novel members of the POU family. Thus, highly degenerate oligonucleotides were prepared which contained all the possible DNA sequences able to encode these conserved sequences (**Figure 5.16**). These were then used in a polymerase chain reaction (PCR) to amplify cDNA prepared from the mRNA of different tissues (see Section 1.2, for another use of the PCR technique, namely to quantitate mRNA levels). The degenerate oligonucleotides amplified cDNAs derived from the mRNAs of novel POU proteins, which like the original POU proteins contained the conserved sequences characteristic of such factors. Many of the novel factors isolated in this way have now also been shown to play critical roles in development.

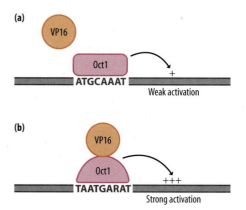

Figure 5.14
The Oct1 factor binds to its distinct binding sites in cellular (a) or herpes simplex virus (b) genes in different structural configurations, only one of which allows binding of the viral activator protein VP16, which strongly activates transcription.

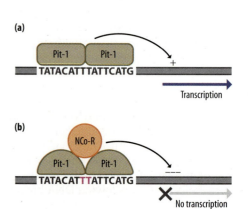

Figure 5.15
The binding sites for two molecules of the Pit-1 factor in the prolactin and growth hormone promoters differ in their sequence and Pit-1 therefore binds in different structural configurations. This results in activation of the prolactin promoter (a), while on the growth hormone promoter, Pit-1 binds the NCo-R co-repressor and hence represses transcription (b).

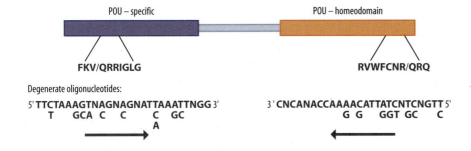

POU – specific

POU – homeodomain

FKV/QRRIGLG

RVWFCNR/QRQ

Degenerate oligonucleotides:

5' TTCTAAAGTNAGNAGNATTAAATTNGG 3'
    T   GCA C C   C GC
                A

3' CNCANACCAAAACATTATCNTCNGTT 5'
          G G   GGT GC   C

Figure 5.16
Isolation of novel members of the POU family on the basis of two conserved amino acid sequences, one at each end of the POU domain. Degenerate oligonucleotides containing all the sequences able to encode the conserved amino acids are used in a PCR with cDNA prepared from mRNA of an appropriate tissue. Novel POU factor mRNAs expressed in that tissue will be amplified on the basis that they contain the conserved sequences and can then be characterized.

Such sequence homology methods of isolating novel transcription factors are being used increasingly to clone novel factors, considerably supplementing the methods described earlier in this chapter. Hence, once several members of a transcription factor family have been identified by conventional means, further members of the family can be identified and characterized in this way. Indeed, as more and more complete genomes are analyzed by DNA sequencing, they can be scanned electronically to identify novel DNA sequences related to those of known transcription factors providing an *in silico* approach to the identification of novel transcription factors.

The two-cysteine–two-histidine (Cys$_2$His$_2$) zinc finger is found in multiple copies in many transcription factors

As discussed in Chapter 4 (Section 4.1), the transcription factor TFIIIA binds to the internal control region of the 5S ribosomal RNA gene and plays a critical role in its transcription by RNA polymerase III. This transcription factor was among the first to be purified. The pure protein was shown to have a periodic repeated structure and to contain between seven and 11 atoms of zinc associated with each molecule of the pure protein.

The basis for this repeated structure was revealed when the gene encoding this protein was cloned and used to predict the corresponding amino acid sequence. This protein sequence contained nine repeats of a 30-amino acid sequence of the form Tyr/Phe-X-Cys-X-Cys-X$_{2-4}$-Cys-X$_3$-Phe-X$_5$-Leu-X$_2$-His-X$_{3-4}$-His-X$_5$, where X is a variable amino acid. This repeating structure therefore contains two invariant pairs of cysteine and histidine residues which were predicted to bind a single zinc atom, accounting for the multiple zinc atoms bound by the purified protein.

This 30-amino acid repeating unit is referred to as a **zinc finger**, on the basis that a loop of 12 amino acids (containing the conserved leucine and phenylalanine residues as well as several basic residues) projects from the surface of the protein. This finger is anchored at its base by the conserved cysteine and histidine residues which directly co-ordinate an atom of zinc (**Figure 5.17**). The binding of zinc by the cysteine and histidine residues has been confirmed directly by X-ray crystallographic analysis of the TFIIIA protein. Such structural studies have also revealed that the finger region consists of two **antiparallel** β-sheets and an α-helix packed against one of the β-sheets, with the α-helix contacting the major groove of the DNA, as occurs for the recognition helix of the homeodomain proteins (**Figure 5.18**).

It is clear, therefore, that like the helix-turn-helix motif, the zinc finger represents a protein structure capable of mediating the DNA binding of transcription factors. Moreover, as described in Section 5.5, it also allows TFIIIA to bind to the 5S ribosomal RNA itself as well as to the gene encoding this RNA. Although originally identified in the RNA polymerase III transcription factor TFIIIA, this motif has now been identified in a number of RNA polymerase II transcription factors and shown to play a critical role in their ability to bind to DNA and thereby influence transcription (**Table 5.1**).

For example, three contiguous copies of the 30-amino acid zinc finger motif are found in the transcription factor Spl, the cloning of which was discussed earlier. The sequence-specific binding pattern of the intact Spl protein can be reproduced by expressing in *Escherichia coli* a truncated protein containing only the zinc finger region, confirming the importance

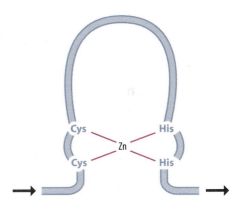

Figure 5.17
Schematic structure of the two-cysteine–two-histidine (Cys$_2$His$_2$) zinc finger.

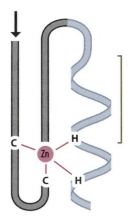

Figure 5.18
Detailed structure of the zinc finger in which two antiparallel β-sheets (gray-green) are packed against an adjacent α-helix (blue). The bracket indicates the region which contacts the major groove of the DNA. From Evans RM & Hollenberg SM (1988) *Cell* 52, 1–3. With permission from Elsevier.

TABLE 5.1
TRANSCRIPTIONAL REGULATORY PROTEINS CONTAINING CYS$_2$ HIS$_2$ ZINC FINGERS

ORGANISM	GENE	NUMBER OF FINGERS
Drosophila	kruppel	4
	hunchback	6
	snail	4
	glass	5
Yeast	ADR1	2
	Swi5	3
Xenopus	TF111A	9
	Xfin	37
Mammal	NGF-1a (Egr1)	3
	MK1	7
	MK2	9
	Evi 1	10
	Sp1	3

of this region in DNA binding. Similarly, the *Drosophila* **Kruppel** protein, which is vital for proper thoracic and abdominal development (see Section 9.2), contains four zinc finger motifs. A single mutation which results in the replacement of the conserved cysteine in one of these fingers by a serine which cannot bind zinc leads to the complete abolition of the function of the protein. This results in a mutant fly whose appearance is indistinguishable from that produced by complete deletion of the gene.

Within the zinc finger DNA-binding domain it is clear that the amino acids at the N-terminus of the α-helix immediately preceding the first histidine residue play a key role in DNA binding. Thus, the *Drosophila* Krox 20 factor binds the DNA sequence 5'-GCGGGGGCG-3'. The protein has three zinc fingers, with the central finger contacting the GGG bases at the centre of the binding site and the outer fingers each contacting one of the flanking GCG triplets.

The two outer fingers have a glutamine residue at position 18 and an arginine at position 21, whereas the central finger has a histidine and threonine at these positions. Changing the histidine and threonine in the central finger to their equivalents in the outer fingers produces a protein which cannot bind to its normal DNA-binding sequence. Instead, it binds to the sequence 5'-GCGGCGGCG-3', indicating that changing the amino acids in the central finger has changed its binding specificity to that of the outer fingers (**Figure 5.19**).

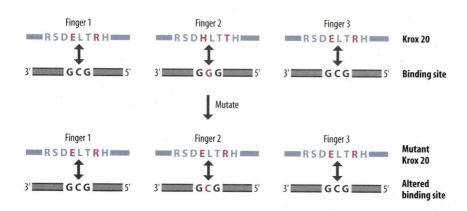

Figure 5.19
The three zinc fingers in the Krox 20 protein each contact three bases in the DNA-binding site. Mutating two amino acids in the central finger to their equivalents in the outer fingers changes its binding specificity from GGG to the GCG sequence normally bound by the outer fingers.

Further studies on these lines have made it possible to predict the DNA-binding specificity of zinc fingers with particular amino acids at specific positions and to design new fingers with novel DNA-binding specificities. As well as being of importance in understanding how zinc-finger factors bind to DNA, these studies also lead to potential therapeutic mechanisms to control gene expression in human diseases using designer fingers with specific DNA-binding specificities. Such applications are discussed further in Chapter 12 (Section 12.5).

The zinc finger clearly represents a DNA-binding element which is present in variable numbers in many regulatory proteins. Indeed, the linkage between the presence of this motif and the ability to regulate gene expression is now so strong that, as with the homeodomain, it has been used as a probe to isolate the genes encoding new regulatory proteins. The DNA sequence encoding the Kruppel zinc finger for example, has been used in this way to isolate Xfin, a 37-finger protein expressed in the early *Xenopus* embryo (**Figure 5.20**).

Hence, zinc-finger proteins are likely to be involved in controlling development in vertebrates as well as in *Drosophila*, where numerous proteins involved in regulating development such as Kruppel, Hunchback, and Snail contain zinc fingers. The interactions of these proteins with the homeodomain proteins, which contain the alternative DNA-binding helix-turn-helix motif, are of central importance in the development of *Drosophila* and other organisms (see Section 9.2).

The nuclear receptors contain two copies of a multi-cysteine zinc finger distinct from the Cys$_2$His$_2$ zinc finger

Throughout this work we have noted that the effect of steroid hormones on mammalian gene expression is one of the best-characterized examples of gene regulation. Thus the steroid-regulated genes were among the first to be shown to be regulated at the level of gene transcription (see Section 1.4) by means of the binding of a specific receptor to a specific DNA sequence (see Section 4.3), resulting in a change in chromatin structure and the generation of a DNaseI-hypersensitive site (see Section 3.5). When the genes encoding the DNA-binding receptors for the various steroid hormones, such as glucocorticoid and estrogen were cloned, they were found to constitute a family of proteins encoded by distinct but related genes. In turn, these proteins were related to other receptors which mediated the response of the cell to hormones such as thyroid hormone, retinoic acid, or vitamin D, leading to the idea of an evolutionarily related family of genes encoding nuclear

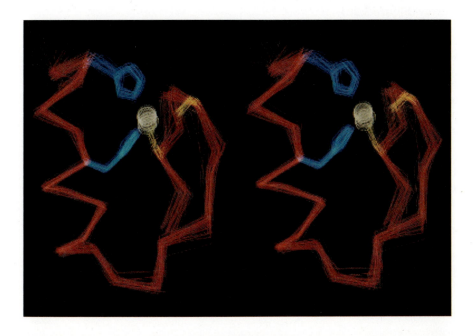

Figure 5.20
Structure of the Cys$_2$His$_2$ zinc finger of the Xfin factor. The cysteine residues are shown in yellow, the histidine side chains in blue, and the zinc atom in white. Courtesy of Peter Wright, from Lee MS, Gippert GP, Soman KV et al. (1989) *Science* 245, 635–637. With permission from The American Association for the Advancement of Science.

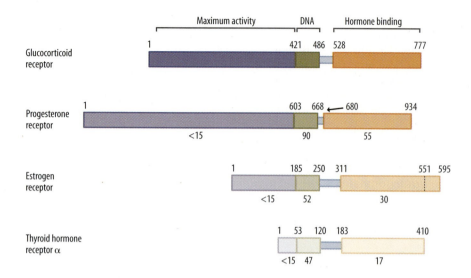

Figure 5.21
Domain structure of individual members of the nuclear receptor superfamily. The proteins are aligned on the DNA-binding domain, which shows the most conservation between different receptors. The N-terminal activation domain and the C-terminal hormone-binding domain are also indicated. The percentage homologies in each domain of the receptors to that of the glucocorticoid receptor are indicated underneath.

receptors known as the steroid-thyroid hormone receptor gene superfamily or the nuclear receptor gene superfamily (**Figure 5.21**).

When the detailed structures of the members of this family were compared (see Figure 5.21), it was found that each had a multi-domain structure, which included a central highly conserved domain. On the basis of experiments in which truncated versions of the receptors were introduced into cells and their activities measured, it was shown that this conserved domain mediated the DNA-binding ability of the receptor while the C-terminal region was involved in the binding of the appropriate hormone and the N-terminal region was involved in producing maximal induction of transcription of target genes.

Sequence analysis of the DNA-binding domain in a variety of receptors showed that it conformed to a consensus sequence of the form $Cys-X_2-Cys-X_{13}-Cys-X_2-Cys-X_{15-17}-Cys-X_5-Cys-X_9-Cys-X_2-Cys-X_4-Cys$. Like the cysteine–histidine finger described above, the DNA binding of this element is dependent upon the presence of zinc or a related heavy metal such as cadmium. Moreover, this element can be drawn as two conventional zinc fingers in which four cysteines replace the two-cysteine–two-histidine structure of the conventional finger in binding zinc and which are separated by a linker region containing the 15–17 variable amino acids (**Figure 5.22**). Such a structure is supported by spectrographic analysis of this region of the receptor, which clearly demonstrates the presence of two zinc atoms each co-ordinated by four cysteines in a tetrahedral array (**Figure 5.23**).

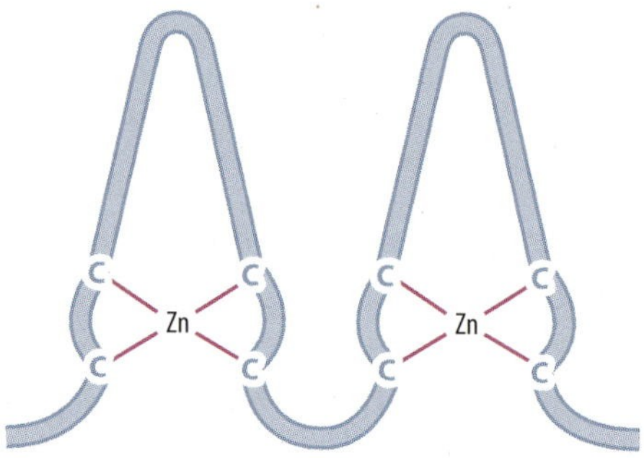

However, such structural analysis also indicates that the two fingers in the steroid receptors interact with one another to form a single structural element (see Figure 5.23). This contrasts with the situation in the cysteine–histidine fingers where each finger forms a separate structural unit. Moreover, the multi-cysteine finger cannot be converted to a cysteine–histidine finger by substituting two of its cysteines with histidines. Thus while the multi-cysteine domain is clearly similar to the cysteine–histidine domain in its co-ordination of zinc, it is distinct in its lack of histidines and conserved phenylalanine and leucine residues as well as in its structure. The two elements are therefore unlikely to be evolutionarily related.

Whatever the precise relationship of the multi-cysteine domain to the cysteine–histidine finger, it is clear that, like this type of finger, the multi-cysteine domain in the nuclear receptors is involved in mediating DNA

Figure 5.22
Schematic structure of the four-cysteine zinc finger.

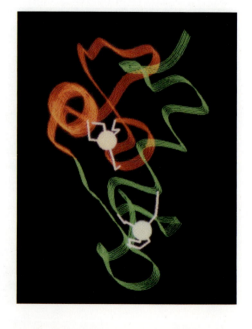

Figure 5.23
Model of the two Cys_4 zinc fingers of the glucocorticoid receptor. The two fingers are shown in red and green respectively with the two zinc atoms in white. Courtesy of Robert Kaptein, from Härd T, Kellenbach E, Boelens R et al. (1990) *Science* 249, 157–160. With permission from The American Association for the Advancement of Science.

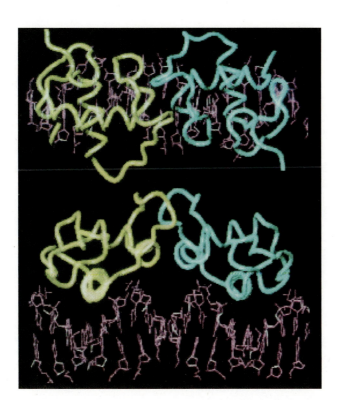

Figure 5.24
Two views of the estrogen receptor binding to DNA. The two zinc fingers are shown in green and blue with the DNA in purple. Courtesy of Daniela Rhodes, from Schwabe JW, Chapman L, Finch JT & Rhodes D (1993) *Cell* 75, 567–578. With permission from Elsevier.

binding (**Figure 5.24**). Similar single domains containing multiple cysteines separated by non-conserved residues have also been identified in other DNA-binding proteins, such as the yeast transcription factors **GAL4**, PPRI, LAC9, etc., which all contain a cluster of six invariant cysteines, and in the adenovirus transactivator, E1A, which has a cluster of four cysteines within the region that mediates *trans*-activation (**Table 5.2**).

The existence of a short DNA-binding region in a number of different nuclear receptors which bind distinct but related sequences (see Section 4.3) has allowed a dissection of the elements in this structure that are important in sequence-specific DNA binding. Thus, as illustrated in Table 4.4, the sequences that confer responsiveness to glucocorticoid or estrogen treatment are distinct but related to one another. If the cysteine-rich region of the estrogen receptor is replaced by that of the glucocorticoid receptor, a chimeric receptor is obtained which has the DNA-binding specificity of the glucocorticoid receptor but, because all the other regions of the protein are derived from the estrogen receptor, continues to bind estrogen. Hence this hybrid receptor induces the expression of glucocorticoid-responsive genes (which carry its DNA-binding site) in response to treatment with estrogen (to which it binds) (**Figure 5.25**).

Further so-called finger-swap experiments using smaller parts of this region have shown that this change in specificity can also be achieved by the exchange of the N-terminal four-cysteine finger together with the region

TABLE 5.2
TRANSCRIPTIONAL REGULATORY PROTEINS WITH MULTIPLE CYSTEINE FINGERS

FINGER TYPE	FACTOR	SPECIES
Cys_4Cys_5	Nuclear receptors	Mammals
Cys_4	E1A	Adenovirus
Cys_6	GAL4, PPRI, LAC9	Yeast

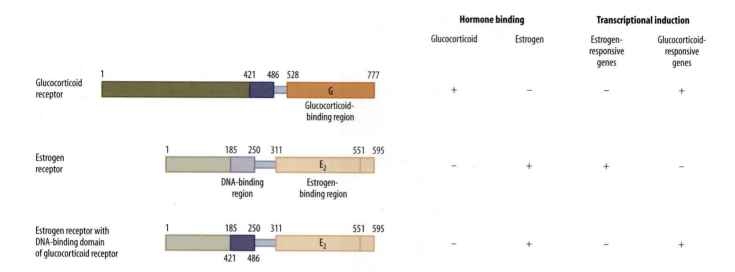

immediately following it, which are therefore critical for determining the sequence-specific binding to the DNA.

These findings have been further refined by exchanging individual amino acids in this region of the glucocorticoid receptor for their equivalents in the estrogen receptor. As shown in **Figure 5.26**, the alteration of the two amino acids between the third and fourth cysteines of the N-terminal finger to their estrogen receptor equivalents results in a glucocorticoid receptor that switches on estrogen-responsive genes. Hence the change of only two critical amino acids within a protein of 777 amino acids can completely change the DNA-binding specificity of the receptor.

The specificity of the hybrid receptor for estrogen-responsive genes can be further enhanced by changing another amino acid located in the linker region between the two fingers (see Figure 5.26), indicating that this region also plays a role in controlling the specificity of binding to DNA.

In contrast to the effect of mutations in the first finger and adjacent region, further alteration of five amino acids in the second finger is sufficient to change the binding specificity of the receptor, such that it now recognizes the thyroid hormone receptor-binding sites (see Figure 5.26). Thyroid hormone-binding sites do not differ from those of the estrogen receptor in sequence but only in the spacing between the two halves of the palindromic DNA-recognition sequence (see Table 4.4a). Hence, these findings indicate that the second finger is critical for mediating protein–protein interactions between the two copies of the receptor that bind to the two halves of the **palindromic sequence** (see Section 4.3) and thus for controlling the optimal spacing of these halves for binding of the particular receptor.

By studying the multiple related nuclear receptors and their relationship with the related DNA sequences to which they bind, it has therefore been possible to determine the critical role of both the first zinc finger and its adjacent helix in controlling the sequence to which these receptors bind

Figure 5.25
Effect of exchanging the DNA-binding domain (blue) of the estrogen receptor with that of the glucocorticoid receptor on the binding of hormone and gene induction by the hybrid receptor.

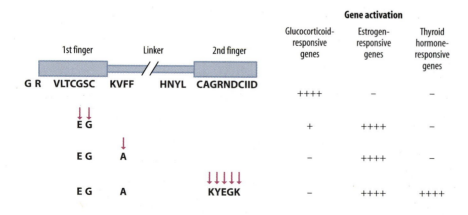

Figure 5.26
Effect of amino acid substitutions in the zinc-finger region of the glucocorticoid receptor on its ability to bind to and activate genes which are normally responsive to different steroid hormones.

and of the second zinc finger in determining the spacing of adjacent sequences, which is optimal for the binding of a homodimer of each receptor. Interestingly, structural studies of the two zinc fingers have shown that the key amino acids for DNA binding at the N-terminal of the first finger are located in an α-helical region, further supporting the key role of such α-helices in sequence-specific DNA binding. Conversely, the key amino acids in the second finger which determine the optimal spacing of the two halves of the DNA-recognition sequence are located on the surface of the molecule. Hence, they can control the interaction of two receptor molecules to form a homodimer able to bind to the two halves of the recognition sequence.

As well as determining the nature of the receptor homodimers which bind to the palindromic repeats illustrated in Table 4.4a, the DNA-binding domain also plays a critical role in determining the receptors which bind to the direct repeat sequences illustrated in Table 4.4b. Thus, when these direct repeats are separated by only one base, they can bind a homodimer of the retinoid-X-receptor (RXR) and hence confer response to 9-*cis*-retinoic acid which binds to this receptor (**Figure 5.27**). In contrast, the RXR homodimer cannot bind to the response elements when the direct repeats are separated by two, three, four or five base pairs. On these response elements the DNA-binding domain of RXR interacts with the DNA-binding domain of another member of the nuclear receptor family to form heterodimers which can bind to each of these sites.

In these heterodimer combinations, the effect of RXR is suppressed and the response of the heterodimer is determined by the other component. For example, a spacing of two or five bases binds a heterodimer of RXR and the retinoic acid receptor (RAR) and results in a response to all-*trans*-retinoic acid which binds to RAR. A spacing of 4 bp binds a heterodimer of RXR and the thyroid hormone receptor and therefore responds to thyroid hormone whereas a spacing of 3 bp binds a heterodimer of RXR and the vitamin D receptor, leading to a response to vitamin D (see Figure 5.27).

Hence, the different DNA-binding domains of the various nuclear receptors produce different patterns of homodimer and heterodimer binding to different binding sites, thereby allowing the diverse members of the nuclear receptor family to produce a wide variety of different responses.

The leucine zipper is a dimerization domain which allows DNA binding by the adjacent basic domain

In earlier sections of this chapter, we have examined how the presence of unusual structural motifs such as the zinc finger, in several different regulatory proteins led to the identification of the crucial role of these elements in DNA binding. A similar approach has led to the identification

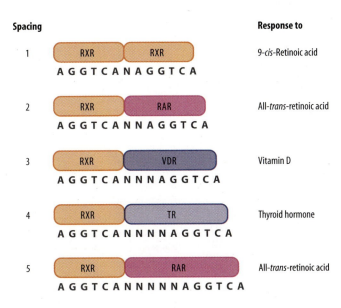

Spacing

1   RXR   RXR   9-*cis*-Retinoic acid
A G G T C A N A G G T C A

2   RXR   RAR   All-*trans*-retinoic acid
A G G T C A N N A G G T C A

3   RXR   VDR   Vitamin D
A G G T C A N N N A G G T C A

4   RXR   TR   Thyroid hormone
A G G T C A N N N N A G G T C A

5   RXR   RAR   All-*trans*-retinoic acid
A G G T C A N N N N N A G G T C A

Response to

Figure 5.27
Patterns of nuclear receptor heterodimers which form on various directly repeated response elements with different spacings (N) between the two halves of the repeat. Note that the response of the element is determined by the nature of the receptor which associates with RXR. RAR, retinoic acid receptor; VDR, vitamin D receptor; TR, thyroid hormone receptor.

C/EBP	**L**	TSDNDR	**L**	RKRVEQ	**L**	SRELDT	**L**	RGIFRQ	**L**
Jun B	**L**	EDKVKT	**L**	KAENAG	**L**	SSAAG	**LL**	REQVAQ	**L**
Jun	**L**	EEKVKT	**L**	KAQNSE	**L**	ASTANM	**L**	REQVAQ	**L**
GCN4	**L**	EDKVEE	**LL**	SKNYH	**L**	EHEVAR	**L**	KKLVGER
Fos	**LL**	QAETDQ	**L**	EDEKSA	**L**	QTEIAN	**L**	LKEKEK	**L**
Fra 1	**LL**	QAETDK	**L**	EDEKSG	**L**	QREIIE	**L**	QKQKER	**L**
c-Myc	VQAEEQK	**L**	ISEEDL	**L**	RKRREQ	**L**	KHKLEQ	**L**
n-Myc	**L**	QAEEHQ	**LLL**	LEKEK	**L**	QARQQQ	**L**	LKKIEHA
l-Myc	**L**	VGAEKKMATEKRQ	**L**	RCRQQQ	**L**	QKRIAY	**L**

Figure 5.28
Alignment of the leucine-rich region in several cellular transcription factors. Note the conserved leucine residues (L) which occur every seven amino acids.

of another such motif, the **leucine zipper**. Thus, in studies of the gene encoding the transcription factor C/EBPα, which is involved in stimulating the expression of several liver-specific genes, it was noted that it contained a region of 35 amino acids in which every seventh amino acid was a leucine. Similar runs of leucine residues were also noted in the yeast transcriptional regulatory protein **GCN4** as well as in the **proto-oncogene** proteins Myc, **Fos**, and **Jun**, which were originally identified on the basis of their ability to transform cultured cells to a cancerous phenotype (see Sections 11.1 and 11.2) and are believed to act by regulating the transcription of other cellular genes (**Figure 5.28**).

It was proposed that the leucine-rich region would form an α-helix in which the leucines would occur every two turns on the same side of the helix. The hydrophobic side chains of the leucine residues in one molecule would then interact with the corresponding side chains in another molecule. This would promote interdigitation of the helices in the two molecules, so allowing the formation of a dimeric molecule (**Figure 5.29**). In agreement with this idea, replacement of individual leucine residues in C/EBPα with valine or isoleucine residues abolishes the ability of the protein to form a dimer. In turn, these mutations also prevent the binding of the protein to its specific recognition sequence.

Unlike the helix-turn-helix motif, however, the leucine zipper does not bind directly to DNA. Rather, by facilitating the dimerization of the protein it provides the correct protein structure for DNA binding by the adjacent region of the protein which is rich in basic amino acids that can interact directly with the acidic DNA (**Figure 5.30**). In agreement with this idea, mutations in the **basic DNA-binding domain** abolish the ability of the protein to bind to the DNA without abolishing its ability to dimerize. Similar juxtapositions of a basic DNA-binding domain and the leucine zipper are also found in the Fos and Jun oncogene proteins and in the yeast transcription factor GCN4, where a single 60-amino acid region contains all the information needed for both dimerization and sequence-specific DNA binding. Hence the leucine zipper has a role similar to that of the second zinc finger in the nuclear receptors (see above) which modulates the activity of the DNA-binding region rather than being involved directly in binding.

Following dimerization mediated via the leucine zipper, the transcription factor will form a symmetric dimer with the bifurcating basic regions contacting the DNA (see Figure 5.30). These basic regions form α-helical structures which track in opposite directions along the dyad symmetric structure of the DNA-recognition site. They form a clamp or scissors grip around the DNA similar to the grip of a wrestler on his opponent, resulting in very tight DNA–protein binding.

In some transcription factors, the basic DNA-binding domain is found associated with a helix-loop-helix dimerization domain

Although originally identified in leucine-zipper-containing proteins, the basic DNA-binding domain has also been identified by homology comparisons in a number of other transcriptional regulatory proteins which lack the leucine zipper. In this case, however, the basic domain is associated with an adjacent region that can form a **helix-loop-helix** structure. This motif is distinct from the helix-turn-helix motif in the homeodomain factors (see

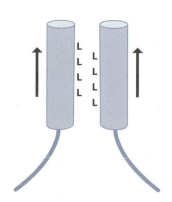

Figure 5.29
Model of the leucine zipper and its role in the dimerization of two molecules of a transcription factor. The arrows indicate that the two α-helices are in a parallel (rather than antiparallel) orientation relative to one another.

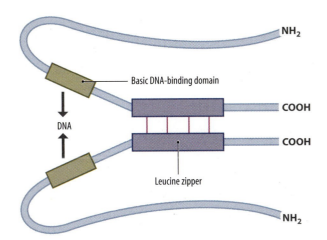

Figure 5.30
Model for the structure of the leucine zipper and the adjacent DNA-binding domain following dimerization of the transcription factor C/EBP.

above) and consists of two **amphipathic** helices (containing all the charged amino acids on one side of the helix) separated by an intervening non-helical loop. Although originally thought to be the DNA-binding domain of these proteins, this helix-loop-helix motif is now known to play a similar role to the leucine zipper in mediating protein dimerization and facilitating DNA binding by the adjacent basic DNA-binding motif.

The helix-loop-helix motif with its adjacent basic DNA-binding region is present in a number of different transcription factors expressed in different tissues. For example, it is present in a number of factors which are critical for the correct development of the nervous system (see Section 10.2). Similarly, this motif is found in several of the factors which control muscle development such as the **MyoD** transcription factor whose artificial expression in an undifferentiated fibroblast cell line can induce it to differentiate to skeletal muscle cells by activating the expression of muscle-specific genes. Thus, the *MyoD* gene is likely to be the critical regulatory locus which is activated by treatment of these cells with 5-azacytidine, allowing this agent to induce these cells to differentiate into muscle cells (see Sections 3.2 and 10.1).

Dimerization between factors provides an additional level of regulation

The leucine-zipper and helix-loop-helix structures therefore act to mediate the dimerization of the transcription factors which contain them, so forming a dimeric molecule which is able to bind to DNA via the adjacent basic DNA-binding domain. This ability provides an additional aspect to gene regulation by such proteins. Thus, in addition to the formation of a dimer by two identical factors, it is possible to envisage the formation of a heterodimer between two different factors which might have different properties in terms of sequence-specific binding and gene activation compared to homodimers of one or other of the two factors.

Such homo- and heterodimerization, resulting in binding to different response elements, occurs in the case of the nuclear receptors as discussed above. In the case of basic domain-containing factors, an example of this type is seen in the case of the related **oncoproteins** Fos and Jun (see Section 11.2). Thus Jun can bind as a homodimer to the activator protein 1 (**AP1**)-recognition sequence, TGAGTCAG, which mediates transcriptional induction by phorbol esters (see Table 4.3). In contrast, Fos cannot bind to DNA alone but can form a heterodimer with the Jun protein. This heterodimer binds to the AP1-recognition site with a 30-fold greater affinity than the Jun homodimer and is considerably more effective in enhancing transcription of genes containing the binding site. Both hetero- and homodimer formation and DNA binding are dependent on the leucine-zipper motif which is found in both proteins. Hence dimerization by the leucine-zipper motif allows two different complexes with different binding affinities and different activity to form on the identical DNA-binding site (**Figure 5.31**).

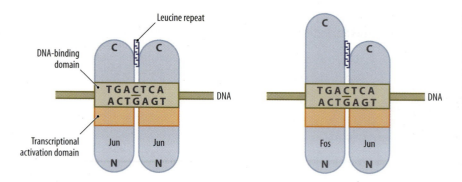

Figure 5.31
Model for DNA binding by the Jun homodimer and the Fos–Jun heterodimer. From Turner R & Tjian R (1989) *Science* 243, 1689–1694. With permission from The American Association for the Advancement of Science.

The failure of the Fos protein to form homodimers and its inability to bind to DNA in the absence of Jun has been shown to be due to differences in its leucine-zipper region from that of Jun. Thus if the leucine-zipper region of Fos is replaced by that of Jun, the resulting protein can dimerize. This dimerization allows the chimeric protein to bind to DNA through the basic region of Fos which is therefore a fully functional DNA-binding domain.

As well as having a positive role, heterodimerization can also have a negative role. Thus, the ability of the MyoD factor to stimulate gene expression is inhibited by heterodimerization with the **Id** factor, which has a helix-turn-helix motif but no basic domain. Since DNA binding requires the co-operation of two basic domains within the heterodimer, the MyoD–Id heterodimer cannot bind to DNA. Hence MyoD cannot activate the expression of muscle-specific genes and thereby promote the production of skeletal muscle cells in the presence of the Id factor (**Figure 5.32**) (for further discussion of gene regulation in skeletal muscle by MyoD and other factors see Section 10.1).

It is clear, therefore, that the ability of the leucine-zipper and the helix-loop-helix motif to facilitate the formation of different dimeric complexes between different transcription factors is likely to play a crucial role in the regulation of gene expression by producing complexes with different binding affinities and different activities. Indeed, a study of 49 human leucine-zipper-containing proteins demonstrated that they show very clear specificities in terms of which proteins pair with one another, even though they have very similar leucine-zipper motifs. Hence, such heterodimerization is very specific, further supporting the idea that it plays a critical role in regulating gene expression.

Other domains can also mediate DNA binding

As the genes encoding more and more transcription factors are isolated and characterized, it has become clear that while the DNA-binding domains of many factors fall into the four classes discussed above, not all do so. Hence additional types of DNA-binding motifs exist. Interestingly, however, as the structures of these DNA-binding domains are progressively understood,

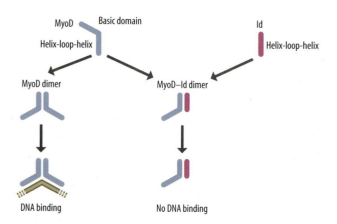

Figure 5.32
DNA binding by the MyoD protein is inhibited by the Id factor which contains the helix-loop-helix dimerization motif but not the basic DNA-binding domain.

relationships between the different classes of DNA-binding domain have emerged. Thus structural analysis of the Ets DNA-binding domain, which is present for example in the *ets*-1 proto-oncogene (see Section 11.2) and the mouse PU-1 gene, has shown it to be identical to the winged-helix-turn-helix or forkhead DNA-binding motif identified in the *Drosophila* forkhead factor and the mammalian liver transcription factor HNF-3. Hence all these factors share a common winged-helix-turn-helix motif.

In turn, this motif as its name suggests, contains a helix-turn-helix structure which is also present in the homeodomain proteins discussed above. These two DNA-binding motifs are thus related to one another. However, the winged-helix-turn-helix motif also contains a β-sheet structure with two loops which appear as wings protruding from the factor, thereby giving it its name. In the majority of winged-helix-containing proteins, it is the helix-turn-helix motif which is responsible for DNA binding. However, in the winged-helix-containing protein hRFX1 it is the β-sheet wing structure which binds to the DNA rather than the helix-turn-helix motif, indicating that members of the winged-helix family can use one of two distinct structures to bind to DNA.

In another example of the relationship of different groups of factors, the existence of the POU DNA-binding domain containing a POU-specific domain in addition to a POU homeodomain (see above) is paralleled in the Pax family of vertebrate transcription factors, many of which contain both a homeodomain and a so-called **paired domain**, both of which contribute to high-affinity DNA binding. Unlike the POU-specific domain, however, the paired domain can be found as an isolated DNA-binding domain without the homeodomain, both in some members of the PAX family and in several *Drosophila* factors including the paired gene in which it was originally identified (**Figure 5.33**).

Interestingly, both the *Drosophila* Pax factor eyeless and the mammalian factors Pax3 and Pax6 (see Section 12.1) play a key role in the development of the eye in these organisms, despite the very different structures of the eye in insects and mammals.

It is clear therefore that a number of different structures exist which can mediate sequence-specific DNA binding and many of these are related to one another. The features of the most common DNA-binding motifs and their associated **dimerization domains** are summarized in **Table 5.3**. Each of these DNA-binding motifs is common to a number of different transcription factors, with differences in the precise amino acid sequence of the motif in each factor controlling the precise DNA sequence to which it binds and hence the target genes for the factor.

Interestingly, the four major DNA-binding motifs we have discussed (the helix-turn-helix, the basic domain, and the two types of zinc finger), although unrelated to one another, all utilize an α-helical region for DNA binding with amino acids in this region having a key role in determining binding specificity. This leads to the question as to why the different families actually exist rather than there being one universal DNA-binding domain.

This may simply reflect the fact that these domains evolved independently in different factors and have been retained since they efficiently

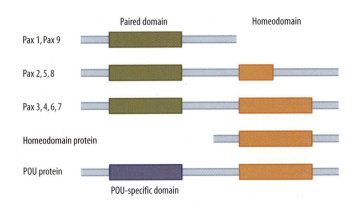

Figure 5.33
In different paired domain proteins, the paired domain can be present without a homeodomain or with either a full-length or truncated homeodomain. The structures of a classic homeodomain and a POU domain are shown for comparison.

TABLE 5.3
TRANSCRIPTION FACTOR DNA-BINDING AND DIMERIZATION DOMAINS

DOMAIN	ROLE	FACTORS CONTAINING DOMAIN	COMMENTS
Homeodomain	DNA binding	Numerous *Drosophila* homeotic genes, related genes in other organisms	DNA binding mediated via helix-turn-helix motif
Cysteine–histidine zinc finger	DNA binding	TFIIIA, Kruppel, Sp1, etc.	Multiple copies of finger motif
Cysteine–cysteine zinc finger	DNA binding	Nuclear receptor family	Single pair of fingers, related motifs in adenovirus E1A and yeast GAL4, etc.
Basic element	DNA binding	C/EBP, *c-fos*, *c-jun*, GCN4	Often found in association with leucine zipper
Leucine zipper	Protein dimerization	C/EBP, *c-fos*, *c-jun*, GCN4, *c-myc*	Mediates dimerization which is essential for DNA binding by adjacent domain
Helix-loop-helix	Protein dimerization	*c-myc*, *Drosophila* daughterless, MyoD, E12, E47	Mediates dimerization which is essential for DNA binding by adjacent domain

fulfilled their function. Alternatively, the different motifs may be important because of differences in the DNA-binding requirements of the factors that contain them. Thus, a DNA-binding domain with multiple Cys_2His_2 zinc fingers may be optimal for factors such as TFIIIA which need to contact an extended regulatory region. In contrast, a basic domain, which can only bind to DNA as a dimer, would be appropriate where a factor is regulated by homodimerization with itself or heterodimerization with another family member.

5.2 ACTIVATION OF TRANSCRIPTION

Although binding to DNA is generally a necessary prerequisite for the activation of transcription, it is clearly not in itself sufficient for this to occur. Following binding, the bound transcription factor must somehow regulate transcription, for example by directly activating the RNA polymerase itself or by facilitating the binding of other transcription factors and the assembly of a stable transcriptional complex. Although some transcription factors can inhibit transcription, the majority of factors defined so far act to activate transcription. In this section, we will discuss in turn the features of activating transcription factors that produce this activation and the manner in which they do so. The manner in which specific factors inhibit transcription will be discussed in Section 5.3.

Activation domains can be identified by "domain-swap" experiments

It is clear from Section 5.1 that transcription factors have a modular structure in which a particular region of the protein mediates DNA binding, while another may mediate binding of a co-factor such as a hormone, and so on. It seems likely,therefore, that a specific region of each individual transcription factor will be involved in its ability to up-regulate transcription following DNA binding.

In the majority of cases, it is clear that such activation regions are distinct from those which produce DNA binding. This domain-type structure is seen clearly in the yeast transcription factor GCN4, which mediates the induction of the genes encoding the enzymes of amino acid biosynthesis in response to amino acid starvation. If a 60-amino acid region of this protein containing the DNA-binding region is introduced into cells, it can bind to the DNA of GCN4-responsive genes but fails to activate transcription. Although DNA binding is necessary for transcriptional activation to occur, it is not therefore sufficient, and gene activation must be dependent upon a region of the protein that is distinct from that mediating DNA binding.

Unlike the DNA-binding region, the region of a transcription factor that mediates gene activation cannot therefore, be identified on the basis of a simple assay of for example the ability to bind to DNA or another protein. Rather, a functional assay of gene activation following binding to DNA is required. Activation regions have therefore been identified on the basis of so-called domain-swap experiments in which the DNA-binding region of one transcription factor is combined with various regions of another factor and the ability to activate transcription of a gene containing the binding site of the first factor is assessed (**Figure 5.34**). Following binding of the hybrid factor to the target gene binding site, gene activation will occur only if the hybrid factor also contains an activation domain provided by the second factor. Hence the activation domain can be identified in this manner.

In the case of the yeast transcription factor GCN4 discussed above, if a 60-amino acid region outside the DNA-binding domain is linked to the DNA-binding region of the bacterial regulatory protein LexA the hybrid factor will activate transcription in yeast from a gene containing the binding site for LexA whereas neither the LexA DNA-binding domain nor the GCN4 region will do so alone. This region of GCN4 therefore contains an activation domain which can increase transcription following DNA binding and is separate from the region of the protein that normally mediates DNA binding (**Figure 5.35a**).

Following its initial use in yeast, similar domain-swapping experiments have also been used to identify the activation domains of mammalian transcription factors. In the glucocorticoid receptor, for example, two independent regions, each able to produce gene activation, have been identified in this way (**Figure 5.35b**).

The success of domain-swap experiments is further proof of the modular nature of transcription factors, allowing the DNA-binding domain of one factor and the activation domain of another to co-operate together to produce gene activation. This is analogous to the exchange of the DNA-binding domain of the glucocorticoid and estrogen receptors, which allows the creation of a hybrid receptor that binds to estrogen-responsive genes through the DNA-binding domain but is responsive to the presence of glucocorticoid through the steroid-binding domain of the protein (see Figure 5.25).

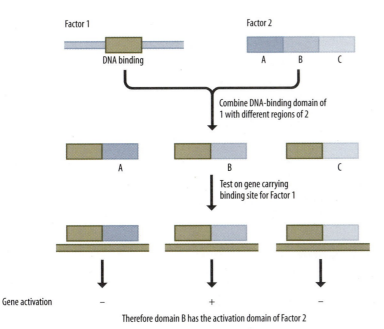

Figure 5.34
Domain-swapping experiment, in which the activation domain of Factor 2 is mapped by combining different regions of Factor 2 with the DNA-binding domain of factor 1 and assaying the hybrid proteins for the ability to activate transcription of a gene containing the DNA-binding site of Factor 1.

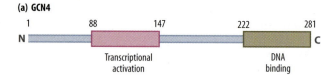

(a) GCN4

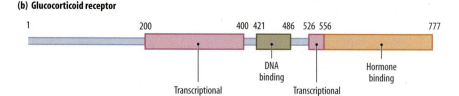

Transcriptional activation

DNA binding

(b) Glucocorticoid receptor

Transcriptional activation

DNA binding

Transcriptional activation

Hormone binding

Figure 5.35
Structure of the yeast GCN4 factor (a) and the mammalian glucocorticoid receptor (b), indicating the distinct regions that mediate DNA binding or transcriptional activation.

An extreme example of this modularity is provided by the herpes simplex virus *trans*-activating protein VP16, discussed in Section 5.1, which transcriptionally activates the viral immediate-early genes during lytic infection of mammalian cells. Although this protein contains a very potent activation region which can strongly induce gene transcription when fused to the DNA-binding domain of GAL4, it contains no DNA-binding domain and cannot bind to DNA itself. Rather, following infection it forms a complex with the cellular octamer-binding protein, Oct1. Oct1 provides the DNA-binding domain which allows binding to the sequence TAATGARAT (R=purine) in the viral promoters and activation is achieved by the activation domain of VP16. Hence, in this instance, DNA-binding and activation motifs actually reside on separate molecules (**Figure 5.36**).

Several different classes of activation domain exist

Using the domain-swapping approaches described above, three major classes of activation domain have been identified (**Table 5.4**). The most common of these classes of activation domain typically contains a very high proportion of acidic amino acids, resulting in a strong net negative charge, and is referred to as the acidic activation domain. For example, in the N-terminal activating region of the glucocorticoid receptor, 17 acidic amino acids are contained in an 82-amino acid region. Similarly, the activating region of the yeast factor GCN4 contains 17 negative charges in an activating region of only 60 amino acids. This has led to the idea that activation regions consist of so-called acid blobs or negative noodles which are necessary for the activation of transcription. However, although the net negative charge in acidic activation regions is likely to be of importance in their ability to stimulate transcription, it appears that other features such as the presence of several conserved hydrophobic residues are also necessary for transcriptional activation.

Indeed, studies of the acidic activation domain of the VP16 transcription factor have suggested a two-stage process of transcriptional activation involving both the negatively charged residues and the conserved

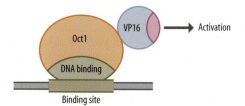

Activation

Figure 5.36
Activation of gene transcription by interaction of the cellular factor Oct1, which contains a DNA-binding domain, and the herpes simplex virus VP16 protein, which contains an activation domain but cannot bind to DNA.

TABLE 5.4 TRANSCRIPTION-FACTOR ACTIVATION DOMAINS	
DOMAIN	**FACTORS CONTAINING DOMAIN**
Acidic region	Yeast GCN4, GAL4, nuclear receptors, etc.
Glutamine-rich region	SP1, Oct1, Oct2, AP2, etc.
Proline-rich region	CTF/NF1, AP2, c-*jun*, Oct2, etc.

hydrophobic residues. Thus, VP16 can stimulate transcription by interacting with the $TAF_{II}31$ component of TFIID (also known as $TAF_{II}40$; see below for the role of the TAF components of TFIID as targets for transcriptional activators). This occurs initially via a long-range electrostatic interaction between the negatively charged acidic residues of the VP16 activation domain and positively charged residues in $TAF_{II}31$. This initial interaction produces a conformational change in the VP16 activation domain, forming an α-helical structure which brings together three hydrophobic residues in the activation domain that then interact with $TAF_{II}31$ (**Figure 5.37**). Such a two-step interaction mechanism mediated, respectively, by acidic and hydrophobic amino acids has subsequently been demonstrated for cellular acidic activators indicating that it is not unique to the viral VP16 transactivator.

Although acidic domains have been identified in a wide range of transcriptional activators from yeast to human (see Table 5.4), it is clear that this type of structure is not the only one which can mediate transcriptional activation. Thus, of the two regions of the human Spl transcription factor that can mediate activation of transcription, neither is particularly rich in negatively charged acidic amino acids. Instead, each of these two domains is particularly rich in glutamine residues, and the intactness of the glutamine-rich region is essential for transcriptional activation. Similar sequences have also been identified in the homeotic proteins Antennapedia and Cut, in Zeste, another *Drosophila* transcriptional regulator, and in the POU proteins Oct1 and Oct2, indicating that this type of activating region is not confined to a single protein.

A further type of activation domain has been identified in the transcription factor CTF/NF1 which binds to the CCAAT box present in many eukaryotic promoters (see Table 4.2). The activation domain of this protein is not rich in acidic or glutamine residues but instead contains numerous proline residues, forming approximately one-quarter of the amino acids in this region. Similar proline-rich regions are found in other transcription factors, such as AP2 and Jun, indicating that, as with glutamine-rich domains, this element is not confined to a single protein.

Hence, as with DNA-binding domains, it is clear that several distinct protein motifs are involved in the activation of transcription (see Table 5.4).

How is transcription activated?

The similarity of activation domains from yeast, *Drosophila*, and mammalian transcription factors discussed above suggests that common mechanisms may mediate transcriptional activation in a wide range of organisms. This is supported by the observations that mammalian transcription factors such as the glucocorticoid receptor can activate a gene carrying their binding site in cells of *Drosophila*, tobacco plants, and mammals. Indeed, yeast and mammalian factors can co-operate together in gene activation. For example, a gene bearing binding sites for GAL4 and the glucocorticoid receptor is synergistically activated by the two factors in mammalian cells, so that the activation observed with the two factors together is greater than the sum of that observed with each factor independently. This co-operation between two factors which come from widely different species, and would therefore never normally interact, suggests that they both function by interacting with some highly conserved component of the basal transcriptional complex (see Section 4.1) which is involved in the basic process of transcription in all different species.

In fact, there is evidence that activators can interact with a variety of different targets within the basal transcriptional complex and these will be discussed in turn. Clearly such interactions could stimulate transcription by increasing the binding of a particular component of the basal transcriptional complex so enhancing its assembly (**Figure 5.38a**). Alternatively,

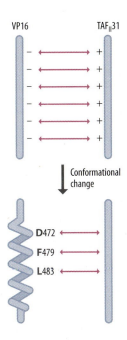

Figure 5.37
An initial electrostatic interaction occurs between negatively charged residues in the acidic activation domain of VP16 and positively charged residues in $TAF_{II}31$. This produces a conformational change in the VP16 activation domain resulting in an α-helical structure. In turn, this brings the hydrophobic residues asparagine (D) at position 472, phenylalanine (F) 479, and leucine (L) 483 close to one another, allowing them to bind to $TAF_{II}31$.

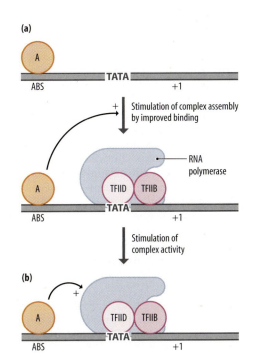

Figure 5.38
An activator (A) bound to its binding site (ABS) can stimulate either complex assembly (a) or the activity of the already assembled complex (b).

stimulation could occur by the activator altering the conformation of an already bound factor so stimulating the activity and/or stability of the complex (**Figure 5.38b**). It is clear that both of these mechanisms are actually used and they could evidently operate whether the complex actually assembles in a fully stepwise fashion in which each component binds sequentially or in a situation where several components bind together in a holoenzyme complex with RNA polymerase itself (see Section 4.1 for further discussion of these alternative models of complex assembly). In discussing the various targets within the complex, however, we shall consider them in the order in which they bind in the stepwise model.

Activators can interact with TFIID

As described in Chapter 4 (Section 4.1), the binding of the TFIID complex (containing TATA-binding protein (TBP) and associated proteins) to the TATA box is the initial stage of complex assembly in both the stepwise and holoenzyme assembly models. TFIID therefore constitutes an obvious target for activating molecules. Indeed, early studies indicated that both the recruitment of TFIID to the promoter and its conformation when bound are affected by activator molecules. Increased binding of TFIID would evidently result in enhanced binding of the other components of the basal transcription complex which bind subsequently. Alternatively, alterations in TFIID configuration might act by improving its ability to recruit these other factors or by directly enhancing its activity within the assembled basal transcription complex (**Figure 5.39**). Such findings are complicated, however, by the fact that, as discussed in Section 4.1, TFIID is a multi-protein complex consisting of the DNA-binding component TBP and a variety of TAFs.

Indeed it is likely that both TBP itself and one or more of the TBP-associated factors (TAFs) can be targets for transcriptional activators (**Figure 5.40**). Thus, acidic activators have been shown to interact directly with TBP. Single-amino acid mutations in such activators which abolish interaction with TBP also abolish the ability to activate transcription, supporting an important role in this effect. It appears that such activators act by enhancing the rate of recruitment of TBP to the promoter, hence increasing the rate of assembly of the basal transcriptional complex. This ability of activators to enhance TBP recruitment has now been demonstrated in intact cells, as well as in cellular extracts, using the chromatin immunoprecipitation (ChIP) assay described in Section 4.3.

Although there is thus evidence that activators can interact directly with TBP, it is clear that in some circumstances activation requires interaction with one or more of the TAFs. Thus in many cases, stimulation of transcription *in vitro* does not occur with purified TBP alone but only when the full TFIID complex is present, indicating that such activation requires interaction with the TAFs. Most interestingly, different types of activation domain appear to contact different TAFs. Thus, for example, glutamine-rich activators such as Sp1 can bind TAF$_{II}$110, while acidic activators such as VP16 bind TAF$_{II}$31 (see above) and multiple activators including proline-rich activators bind TAF$_{II}$55. Hence, different components within TFIID can be targeted by different classes of activator (**Figure 5.41**).

The interaction between TAFs and transcriptional activators is further complicated by the existence of cell type-specific TAFs. For example, B lymphocytes express a form of TAF$_{II}$130 known as TAF$_{II}$105, which is not expressed by other cell types. Similarly, mutation of TAF$_{II}$250 inhibits the expression of specific genes involved in the cell cycle, without affecting the expression of other genes. Hence, different promoters and different cell types differ in their requirement for different TAFs, paralleling the existence of different TBP-like factors, which was discussed in Section 4.1.

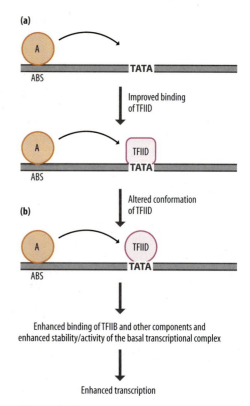

Figure 5.39
An activator (A) bound to its binding site (ABS) can enhance the binding of TFIID to the TATA box, thereby improving the rate of assembly of the basal transcriptional complex by facilitating the subsequent binding of other components which is dependent upon prior binding of TFIID (a). In addition, the activator can also alter the configuration of TFIID so stimulating its activity either by increasing its ability to recruit other components of the complex or by enhancing its ability to stimulate transcription (b).

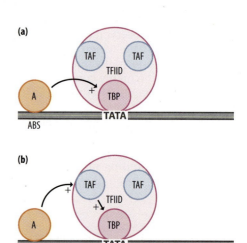

Figure 5.40
An activating molecule (A) can interact with TFIID either by interacting with the TATA-binding protein (TBP) directly (a) or by interacting indirectly with TBP via TBP-associated factors (TAFs) (b).

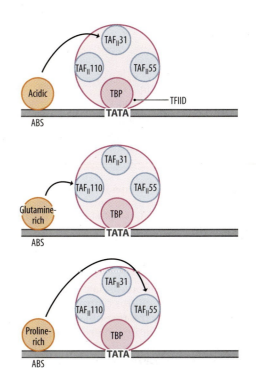

Overall, it is clear that TAFs function as critical intermediaries between activators and the basal transcription complex. They are often referred to therefore as co-activators since they do not bind to DNA directly but mediate the effects of DNA-bound activators (see below).

Activators can interact with TFIIB

Although there is considerable evidence for the interaction of activators and TFIID, it is clear that activators can also act to stimulate the assembly/activity of the basal transcriptional complex after TFIID has bound. As discussed in Chapter 4 (Section 4.1) the binding of TFIID to the promoter is followed by the binding of TFIIB. As with TBP and the TAFs, there is clear evidence for the interaction of TFIIB with activators. Thus TFIIB can be purified from a mixture of proteins on a column containing a bound acidic activator. Moreover, this interaction is of importance for activation of transcription since mutations in acidic activators which abolish the interaction with TFIIB prevent the activator from stimulating transcription.

As with TFIID, such interactions can both stimulate the binding of TFIIB to the promoter and alter its configuration when bound, so improving its ability to recruit the other components of the basal transcriptional complex, such as the RNA polymerase which binds subsequently (**Figure 5.42**). Hence, as with the various components of TFIID, the single TFIIB **polypeptide** is evidently a target for transcriptional activators.

Activators can interact with the mediator and SAGA complexes

As discussed in Chapter 4 (Section 4.1), binding of TFIIB allows the subsequent recruitment of the RNA polymerase itself together with the TFIIF factor. Interestingly, the repeat-containing C-terminal domain of the large subunit of RNA polymerase II, which is involved in transcriptional initiation and elongation (see Sections 4.1 and 4.2), is also implicated in the response to transcriptional activators. Thus, deletion of this region from RNA polymerase II prevents enhanced transcription in response to transcriptional activators while increasing the number of repeated elements in this domain enhances the response to activators.

Despite this, however, it is unlikely that activators contact the RNA polymerase directly. Rather, work in yeast and subsequently in mammalian cells has identified a **mediator** complex containing over 20 polypeptides. The C-terminal domain of RNA polymerase II projects from the rest of the molecule, allowing it to interact with the mediator complex and this interaction is required for the response to transcriptional activators. It is clear therefore that activators can act by interacting with the mediator complex which in turn stimulates RNA polymerase activity (**Figure 5.43**).

Interestingly, electron microscope analysis of the mediator complex bound to RNA polymerase II suggests that the mediator partially envelopes the polymerase. This would allow it to receive signals from activators and transmit them to the polymerase (**Figures 5.44** and **5.45**). As with TFIID, it

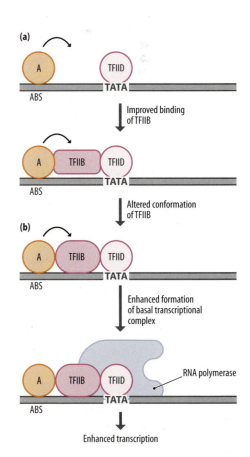

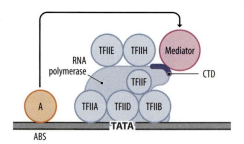

Figure 5.43
Activators appear to interact with the RNA polymerase indirectly via a mediator complex which binds to the C-terminal domain (CTD) of RNA polymerase II.

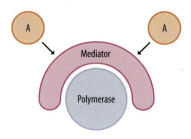

Figure 5.44
Schematic diagram of the mediator complex bound to RNA polymerase II indicates that the mediator partially envelops the polymerase. Hence, it could receive signals from activators (A) and transmit them to the polymerase.

appears that different classes of activators interact with different components of the mediator complex. Thus, mutation of different mediator components affects the response to different transcriptional activators.

The mediator complex has also been shown to stimulate the ability of TFIIH to phosphorylate the C-terminal region of RNA polymerase. It is possible, therefore, that activators may enhance the ability of the mediator to stimulate this effect of TFIIH, which is necessary for RNA polymerase II to initiate transcription (**Figure 5.46**).

The mediator complex has been shown to be associated with the RNA polymerase holoenzyme prior to binding to the DNA and the interaction between activators and the mediator has been shown to stimulate the assembly of the basal transcriptional complex on the DNA. Hence, as discussed in Chapter 4 (Section 4.1) this holoenzyme contains not only RNA polymerase II and basal transcription factors such as TFIIB, TFIIE, TFIIF, and TFIIH, but also contains mediator components which respond to transcriptional activators as well as the SWI–SNF complex which can alter chromatin structure (see Section 3.5).

In addition to the mediator, another multi-protein complex known as SAGA (Spt-Ada-Gcn5-acetyltransferase) also plays a key role in transcriptional activation. This complex contains components such as TAFs which can respond to transcriptional activators (see above), as well as components such as the Gcn5 protein which can acetylate histones (see Section 2.3). Hence, the **SAGA complex** can link transcriptional activators to the basal transcriptional complex and to the production of alterations in chromatin structure, via acetylation of histones (**Figure 5.47**).

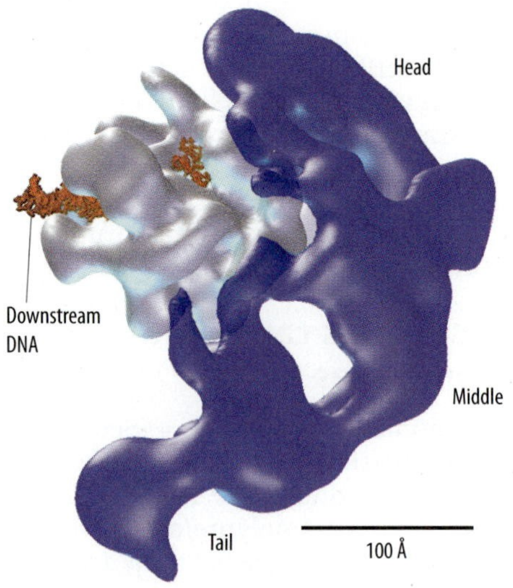

Activators can interact with co-activators

As discussed above, transcriptional activators often contact their ultimate targets indirectly, acting, for example, via TAFs, the mediator, or SAGA complexes. This has led to the concept of co-activators which do not bind to DNA but act to transmit the signal from the DNA-bound transcriptional activator to the basal transcriptional complex.

Indeed, co-activators have been identified in a number of different situations involving specific transcriptional activators. Such co-activators are recruited to the DNA by a protein–protein interaction with the DNA-bound activator and then activate transcription. Perhaps the best known co-activator is CREB-binding protein (CBP). As its name suggests, CBP was first

Figure 5.45
Structural model showing the interaction of the mediator (dark blue) with the polymerase (light blue). DNA is shown in orange. The bar indicates 100 Å. Courtesy of Francisco Asturias, The Scripps Research Institute.

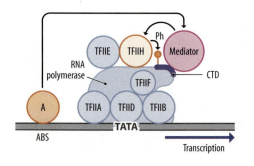

Figure 5.46
As well as affecting transcriptional initiation via the mediator, it is possible that activators act by stimulating the ability of the mediator to interact with TFIIH and stimulate it to phosphorylate (Ph; orange spot) the C-terminal domain (CTD) of RNA polymerase II.

defined as a protein that binds to the **CREB** transcription factor which activates specific genes in response to cyclic AMP (see Section 8.2 for further discussion of the cyclic AMP/CREB system).

Most importantly, CBP only binds to CREB when CREB has been phosphorylated on the serine at position 133. Since such phosphorylation is necessary for CREB to activate transcription, this indicated that recruitment of CBP was likely to play a key role in mediating transcriptional activation by CREB (see Section 8.2). Hence, CBP does not bind to DNA itself but is recruited to the DNA by DNA-bound phosphorylated CREB and then activates transcription.

CBP has now been shown to play a key role in mediating transcriptional activation via a number of DNA-binding transcription factors which are discussed in this book, including the nuclear receptor family (Section 4.1), NFκB (Section 8.2), **p53** (Section 11.3), and MyoD (Sections 5.1 and 10.1). Hence, it is clear that CBP and its close relative p300 play a key role as co-activators for a number of transcriptional activators (**Figure 5.48**). Indeed, the widespread use of CBP by transcription factors activated by different signaling pathways may result in competition between these pathways for limited amounts of CBP. This in turn would account for the phenomenon in which simultaneous stimulation of two different pathways such as those stimulated by glucocorticoid and phorbol esters results in each pathway inhibiting the other so that no transcriptional activation occurs (see Section 8.1 and Figure 8.12).

Although CBP can interact with members of the nuclear receptor family, it is not the only co-activator to do so. In several cases, specific co-activators can bind to the receptors only after hormone treatment. Hence, the binding of specific co-activators to these receptors only after hormone treatment allows the receptors to activate transcription in a hormone-dependent manner (see Section 8.1).

It is likely that co-activators can act to stimulate transcription via at least two distinct mechanisms, following recruitment to the DNA by a DNA-bound activator. Both these mechanisms have been shown to operate in the case of CBP. Thus, CBP has been shown to bind to various components of the basal transcription complex such as TBP, TFIIB, and the RNA polymerase holoenzyme, indicating that it can bridge the gap between CREB and the basal complex, allowing CREB to activate transcription (**Figure 5.49a**).

Moreover, as discussed in Sections 2.3 and 3.3, CBP also has histone acetyltransferase activity, indicating that it can also stimulate transcription

Figure 5.47
Transcriptional activators (A) can stimulate the basal transcriptional complex and alter chromatin structure via the SAGA complex.

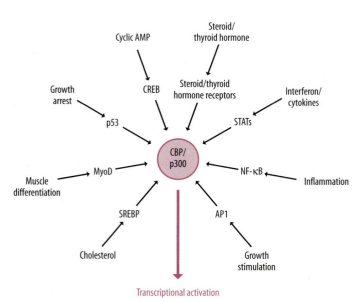

Figure 5.48
A variety of activating transcription factors stimulate transcription via the closely related CBP and p300 co-activators.

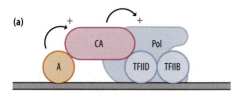

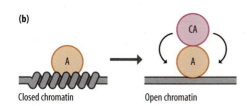

Figure 5.49
A co-activator (CA) may act (a) by linking an activator (A) to the basal transcriptional complex or (b) by promoting the conversion of a closed chromatin structure (wavy line) to a more open structure (solid line). In both cases, the co-activator will be recruited to the DNA by a DNA-bound activator. Pol, polymerase.

by acetylating histones and therefore opening up the chromatin structure (**Figure 5.49b**). Hence, co-activators can act either via linking the activator to the basal complex or by modifying chromatin structure.

Activators can interact with modulators of chromatin structure

As described above, transcriptional activators can interact with components of the basal transcriptional complex either directly or indirectly via molecules such as co-activators or the mediator complex. Such interactions can enhance transcriptional activity by enhancing the assembly of the basal transcriptional complex or its activity. In addition, from the cases discussed above it is clear that in some situations such interactions can also alter chromatin structure. Thus, interaction of activators with the SAGA complex or the CBP co-activator can lead to chromatin opening by stimulating their histone acetyltransferase activity.

A clear example of this is seen in the case of the PHO4 transcription factor in yeast. This factor binds to the PHO5 gene promoter following phosphate starvation and evicts four nucleosomes, allowing subsequent activation of the promoter. Interestingly, removal of the acidic activation domain of PHO4 prevents this nucleosome eviction following binding of the truncated PHO4 protein. However, the ability to evict nucleosomes can be restored by linking the activation domain of VP16 to the truncated PHO4 molecule (**Figure 5.50**).

In this case, the acidic activation domains of PHO4 and VP16 can therefore alter chromatin structure, as well as interacting with the basal transcriptional complex. Indeed, in the case of VP16 it has been demonstrated that it can alter the activity of the SWI–SNF chromatin-remodeling complex (see Section 3.5). Thus, in the absence of VP16, SWI–SNF can cause nucleosomes to slide along the DNA whereas in the presence of VP16 it can drive their actual eviction to produce a nucleosome-free region, as occurs in the PHO5 promoter.

It is clear therefore that transcriptional activators can produce altered chromatin structure both via factors such as CBP or SAGA components

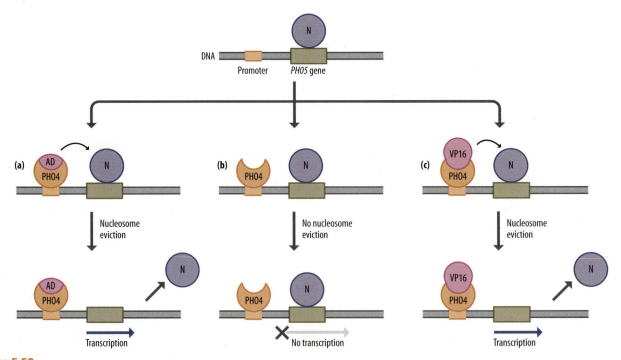

Figure 5.50
Binding of the PHO4 transcription factor to the *PHO5* gene promoter results in nucleosome (N) eviction, mediated by the activation domain (AD) of PHO4 (a). Such nucleosome eviction does not occur with a truncated PHO4 protein lacking the activation domain (b). However, it can be restored by linking the truncated PHO4 protein to the activation domain of VP16 (pink) (c).

which have histone acetyltransferase activity and via interaction with complexes which can remodel chromatin such as SWI–SNF (**Figure 5.51**).

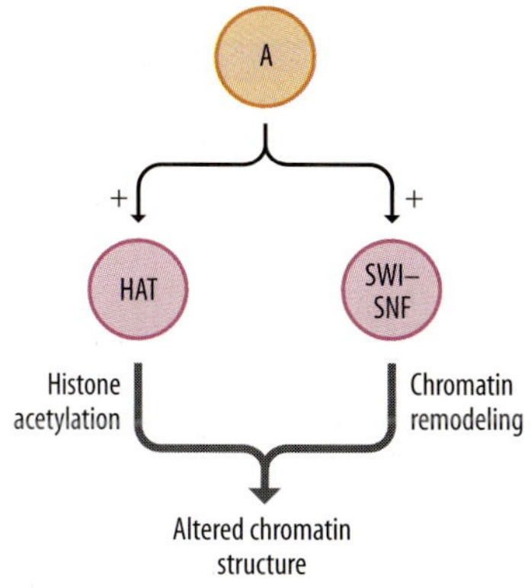

Activators have a multitude of targets

There exists therefore a bewildering array of potential targets for activator molecules, including the RNA polymerase–mediator complex, co-activators, chromatin-remodeling complexes, TFIIB, and the different components of TFIID. Indeed, other components of the basal transcriptional complex such as TFIIA, TFIIE, and TFIIH have also been observed to interact directly with transcriptional activators. All these various possibilities are not mutually exclusive, however. Thus, it is likely that all these components can act as a target either for the same activating factor or different activating factors.

Several possibilities may account for such a wide range of activator targets. For example, it is possible that different organisms differ in the preferred target for transcriptional activators. Indeed, while TAFs associated with TBP appear to be of critical importance for transcriptional activation in multicellular organisms such as *Drosophila*, they may be of much less importance in yeast, Alternatively, it has been suggested that the sole requirement for an activator is to be able to bind to DNA and then bind any component of the basal transcriptional complex and hence recruit it to the DNA. Thus, an activator can function simply by interacting with any component of the complex and thereby enhancing its binding to the promoter.

It is most probable, however, that a multiplicity of targets for activators is required to produce the strong synergistic activation of transcription which is observed when different activating factors are added together compared to the level observed when each factor is added separately. Thus, in a number of cases the binding of two transcription factors to different sites in the promoter can produce very strong activation of transcription, whereas each factor alone produces much weaker activation (**Figure 5.52a**).

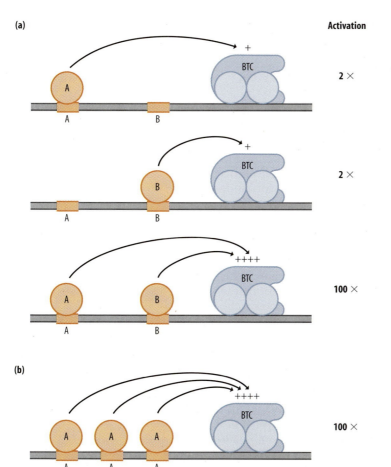

Figure 5.51
Transcriptional activators (A) can interact with factors having histone acetyltransferase (HAT) activity and with chromatin-remodeling complexes, such as SWI–SNF, to produce an altered chromatin structure.

Figure 5.52
Binding of two activators (A and B) can produce much stronger, synergistic, activation of transcription by the basal transcriptional complex (BTC), compared to the binding of either factor alone (a). Strong activation can also be produced by the binding of multiple copies of the same activator (b).

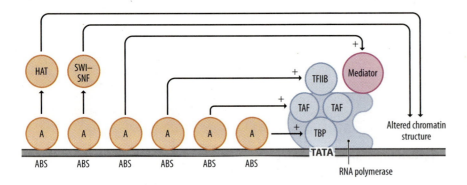

Figure 5.53
Synergistic activation of transcription by activator molecules (A) interacting with different components of the basal transcriptional complex, such as the mediator complex, TFIIB or the different components of TFIID (TBP and non-DNA-binding TAFs), as well as interacting with histone acetyltransferase (HAT) enzymes and complexes such as SWI–SNF to alter chromatin structure. Note that these interactions may be direct or indirect via co-activator molecules.

Similarly, strong enhancement of transcription is observed when multiple copies of a single factor can bind the target DNA (**Figure 5.52b**). In turn, these synergistic effects are likely to play a key role in producing strong transcriptional activation in response to specific stimuli and/or at specific stages of embryonic development.

The complexity of the transcription complex allows different activating factors or different molecules of the same activating factor to contact different targets within the basal transcriptional complex either directly or indirectly, as well as altering chromatin structure, so ensuring the great enhancement of transcriptional activity, which is the ultimate aim of activating molecules (**Figure 5.53**).

5.3 REPRESSION OF TRANSCRIPTION

Although the majority of transcription factors described so far act in a positive manner, a number of cases have now been reported in which a transcription factor exerts an inhibitory effect on transcription and several possible mechanisms by which this can be achieved have been described (**Figure 5.54**).

Repressors can act indirectly by inhibiting the positive effect of activators

The simplest means of achieving repression is for a repressor to prevent an activating molecule from binding to DNA. This can occur by a negative factor promoting a tightly packed chromatin structure which does not allow activating molecules to bind (**Figure 5.54a**) (see Chapter 3 for a discussion of the effect of chromatin structure on gene expression). An example of this effect was discussed in Section 5.1 in which the polycomb repressor binds to *Drosophila* homeotic genes and organizes them into an inactive chromatin structure incapable of binding activators (see Figure 5.8).

Alternatively, the negatively acting factor can bind specifically to the binding site of the activator so preventing it binding (**Figure 5.54b**). This effect is seen in the β-interferon promoter where the binding of several positively acting factors is necessary for gene activation. Another factor acts negatively by binding to this region of DNA and simply preventing the positively acting factors from binding. In response to viral infection, the negative factor is inactivated, allowing the positively acting factors to bind and transcription occurs. A similar example was described in Section 5.1, whereby the DNA binding of the engrailed gene product inhibits gene activation by preventing the binding of the Ftz activator protein (see Figure 5.6).

In a related phenomenon (**Figure 5.54c**) repression is achieved by formation of a complex between the activator and the repressor in solution, preventing the activator binding to the DNA. This is seen in the case of the inhibitory factor Id which dimerizes with the muscle determining factor MyoD via its helix-loop-helix motif and prevents it binding to DNA and activating transcription (see Section 5.1, Figure 5.32, and Section 10.1).

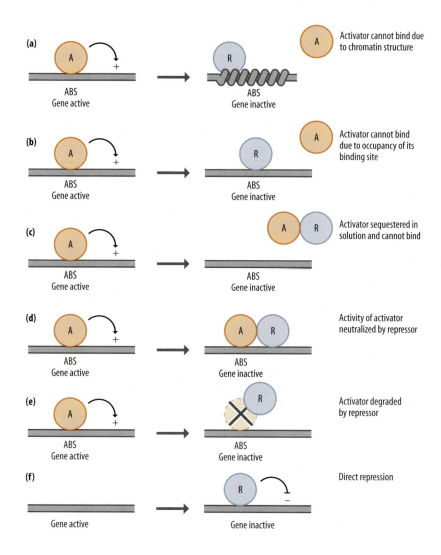

Figure 5.54
Mechanisms by which an inhibitory factor (R) can repress transcription. These involve inhibiting an activator (A) binding to its binding site (ABS) in the DNA either by producing an inactive chromatin structure (a), by competing for the DNA-binding site (b), by sequestering the activator in solution (c), by inhibiting the ability of bound activator to stimulate transcription (d), by degrading the activator (e), or by direct repression (f).

In addition to inhibiting DNA binding by these different means, a negative factor can also act by interfering with the activation of transcription mediated by a bound factor in a phenomenon known as **quenching** (**Figure 5.54d**). This can occur by the inhibitory factor binding to DNA adjacent to the activator (**Figure 5.55a**). Thus, in the case of the promoter driving expression of the c-*myc* gene, a negatively acting factor *myc*-PRF binds to a site adjacent to that occupied by a positively acting factor *myc*-CF1 and prevents it from activating c-*myc* gene expression.

Alternatively this effect can occur by the inhibitor binding to the activator itself and blocking its ability to activate transcription (**Figure 5.55b**). This is seen in the case of the **MDM4** oncogenic protein which binds to the anti-oncogenic p53 protein and masks its activation domain, so preventing it from activating transcription (see Section 11.3).

As well as blocking DNA binding or transcriptional activation, a repressor can also act by producing the degradation of the activator (**Figure 5.54e**). An example of this is the **MDM2** oncogenic protein, which rather than quenching activation by p53 induces the degradation of the p53 protein thereby preventing it from activating transcription (see Section 11.3). This effect is achieved indirectly by MDM2 catalyzing the modification of p53, by the addition of the small protein ubiquitin which targets it for degradation by protease enzymes (**Figure 5.56a**). In contrast, the AEBP1 transcription factor, which regulates adipocyte differentiation, is itself a protease enzyme and therefore can produce transcriptional repression by directly degrading the activator (**Figure 5.56b**).

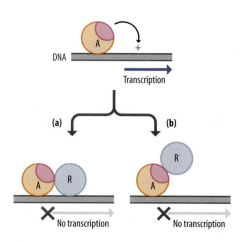

Figure 5.55
An inhibitory factor (R) can block the ability of a DNA-bound activator to stimulate transcription via its activation domain (pink) either by binding to DNA adjacent to the activator (a) or by binding to the activator itself (b).

(a)

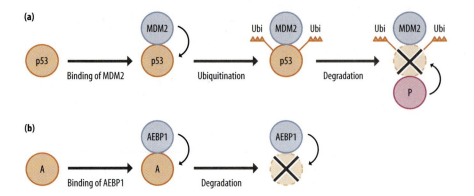

(b)

Figure 5.56
Degradation of an activator can be produced indirectly by a repressor such as MDM2, which ubiquitinates the p53 activator, thereby promoting its degradation by protease enzymes (P) (a). Alternatively, it can be produced directly by a repressor, such as AEBP1, which itself has protease activity and can therefore degrade the activator (A) after binding to it (b).

Repressors can act directly by inhibiting the assembly or activity of the basal transcriptional complex

In all the cases discussed so far the negative factor exerts its inhibitory effect in an essentially passive manner by neutralizing the action of a positively acting factor, by preventing either its DNA binding or its activation of transcription, or by inducing its degradation. It is clear, however, that some factors can have an inherent actively negative action on transcription which does not depend upon the neutralization of a positively acting factor (**Figure 5.54f**).

For example, the *Drosophila* **eve** (even-skipped) protein is able to repress transcription from a promoter lacking DNA-binding sites for any activating proteins, indicating that it has a directly negative effect on transcription. A similar direct inhibitory effect has been defined in the case of the mammalian c-*erbA* gene which encodes the thyroid hormone receptor, a member of the nuclear receptor family (see Sections 4.3 and 5.1). Thus, the binding of this receptor to its specific DNA-binding site in the absence of thyroid hormone results in the direct inhibition of transcription. This effect is mediated by the NCo-R nuclear receptor co-repressor.

As with co-activators, co-repressors do not bind to DNA but are recruited by a protein–protein interaction with a DNA-bound repressor. The NCo-R co-repressor binds to the thyroid hormone receptor in the absence of hormone and inhibits transcription. In the presence of thyroid hormone, however, the receptor undergoes a conformational change which results in the release of the co-repressor and allows co-activators such as CBP to bind (see Section 5.2). This then allows the receptor to activate rather than repress transcription (**Figure 5.57**).

This example indicates the critical role of co-activators and co-repressors in regulating gene expression. Thus, thyroid hormone can modulate gene expression by controlling whether its target factor binds a co-activator or a co-repressor. Similarly, as discussed in Section 5.1, the different configurations a factor adopts on different DNA-binding sites can alter its effect on transcription by altering its ability to bind co-activators or co-repressors.

The thyroid hormone receptor example also indicates that the distinction between activators and repressors is not a precise one. In some cases, the same factor can directly stimulate or directly inhibit transcription depending on the circumstances, with this effect being controlled via the regulated binding of co-repressor and co-activator molecules.

Interestingly, two alternative forms of the thyroid hormone receptor exist which are encoded by alternatively spliced mRNAs. One of these (α-1) contains the hormone-binding domain and can mediate gene activation in response to thyroid hormone, while the other form (α-2) contains another protein sequence instead of part of this domain (**Figure 5.58a**). Therefore the α-2 form cannot respond to the hormone, although since it contains the DNA-binding domain it can bind to the binding site for the receptor in hormone-responsive genes. By doing so, it prevents binding of the α-1 form and

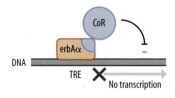

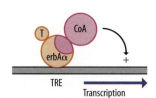

Figure 5.57
The thyroid hormone receptor encoded by the c-*erbA*α gene represses transcription in the absence of thyroid hormone (T) via a specific inhibitory domain which recruits an inhibitory co-repressor (CoR). Following binding of hormone, the protein undergoes a conformational change. This releases the co-repressor and leads to binding of a co-activator (CoA) which allows the receptor to activate transcription.

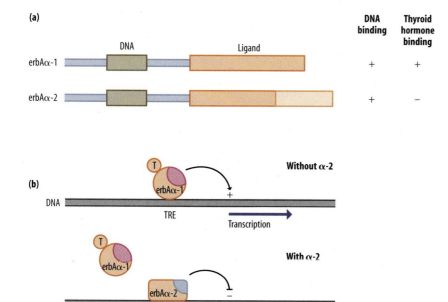

Figure 5.58
(a) Relationship of the erbAα-1 and α-2 proteins. Note that only the α-1 protein has a functional thyroid hormone binding-domain. (b) Inhibition of erbAα-1 binding and of gene activation in the presence of the α-2 protein.

therefore blocks the induction of the gene in response to thyroid hormone (**Figure 5.58b**). These two alternatively spliced forms of the transcription factor which are made in different amounts in different tissues, therefore mediate opposing effects on thyroid hormone-dependent gene expression, further reinforcing the dual activator/repressor role of the thyroid hormone receptor.

In a number of directly acting active transcriptional repressors such as even-skipped and the thyroid hormone repressor, a specific region of the protein has been shown to be able to confer the ability to inhibit gene expression upon the DNA-binding domain of another protein when the two are artificially linked, paralleling the similar behavior of activation domains in such domain-swap experiments (see Section 5.2).

As noted in Chapter 3 (Section 3.3) the NCo-R which binds to the **inhibitory domain** of the thyroid hormone receptor associates with the Sin3–RPD3 protein complex which can alter chromatin structure by deacetylating histones. Hence, some inhibitory domains may function at least in part by recruiting molecules which can alter chromatin structure.

In other cases, however, they are likely to inhibit transcription by reducing the formation and/or stability of the basal transcriptional complex, either by interacting directly with the basal complex or by recruiting a co-repressor which then interacts with the basal complex. Such inhibitory domains therefore act in a manner similar to activation domains but reduce rather than enhance the rate of transcription.

In considering active repressors, we have so far considered factors which produce their effect by binding to DNA and then inhibiting transcription either by directly or indirectly interacting with the basal transcriptional complex to inhibit its activity or by altering chromatin structure (**Figure 5.59a**). Indeed, the binding of the thyroid hormone receptor to the silencer element in the chicken lysozyme gene is responsible for the inhibitory effect of this DNA sequence (see Section 4.4).

In other cases, however, active inhibitory factors can bind directly to the basal transcriptional complex and inhibit its activity (**Figure 5.59b**). One example of this is the Dr1 factor which binds to TBP and prevents TFIIB from binding (**Figure 5.60a**), thereby preventing assembly of the basal transcriptional complex on TATA box-containing promoters (see Section 4.1). In addition, the assembly of the basal complex is also inhibited by binding of the Mot1 factor. Like Dr1, the Mot1 factor also targets TBP but

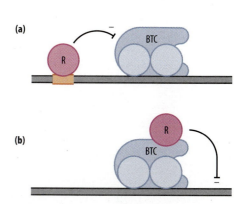

Figure 5.59
An inhibitory transcription factor (R) can repress transcription by binding to a specific DNA sequence and inhibiting the activity of the basal transcriptional complex (a) or by binding directly to the complex and inhibiting its activity (b).

rather than inhibiting the binding of TFIIB, it displaces TBP from the DNA (**Figure 5.60b**).

Interestingly, the TFIIA factor, which was described in Chapter 4 (Section 4.1), appears to act by binding to TFIID and preventing the binding of Dr1 and Mot1, thereby promoting the binding of TFIIB and preventing TBP being displaced from the DNA. In agreement with this idea, the requirement for TFIIA in artificially reconstituted "test tube" transcription systems disappears as TFIID is purified away from these inhibitory molecules, whereas it plays a key role in the cell where these inhibitors are present.

Overall therefore, inhibitory factors are likely to play a critical role in the regulation of transcription both by inhibiting the activity of positively acting factors and in other cases by having a direct inhibitory effect on transcription (see Figure 5.54). Hence, the balance between positively and negatively acting factors, both within the basal transcriptional complex and bound to specific DNA-binding sites, will play a key role in determining the transcription rate of a particular gene in different circumstances.

5.4 REGULATION AT TRANSCRIPTIONAL ELONGATION

Regulation of transcription can occur at the elongation stage, as well as at initiation

In the majority of cases where increased transcription of a particular gene has been demonstrated, it is likely that such increased transcription is mediated by an increased rate of initiation of transcription by RNA polymerase caused by the action of the positively acting transcription factors which were described in Section 5.2. Hence, in a tissue in which a gene is being transcribed actively, a large number of polymerase molecules will be moving along the gene at any particular time, resulting in the production of a large number of transcripts. Such a series of nascent transcripts being produced from a single transcription unit can be visualized in the **lampbrush chromosomes** of amphibian oocytes, the nascent transcripts associated with each RNA polymerase molecule increasing in length the further the polymerase has proceeded along the gene, resulting in the characteristic nested appearance (**Figure 5.61**).

By contrast, in tissues where a gene is transcribed at very low levels, initiation of transcription will be a rare event and only one or a very few polymerase molecules will be transcribing a gene at any particular time. Similarly, the absence of transcription of a particular gene in some tissues will result from a failure of RNA polymerase to initiate transcription in that tissue (**Figure 5.62**).

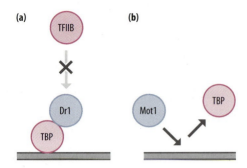

Figure 5.60
The basal transcriptional complex can be inhibited by the binding to TBP of the Dr1 factor which prevents recruitment of TFIIB (a) or by the Mot1 factor which displaces TBP from the DNA (b).

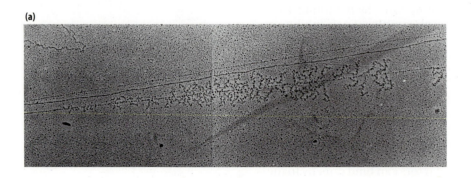

Figure 5.61
Electron micrograph (a) and summary diagram (b) of a lampbrush chromosome in amphibian oocytes, showing the characteristic nested appearance produced by the nascent mRNA chains attached to transcribing RNA polymerase molecules. The bar indicates 1 μm. Courtesy of RS Hill, from Hill RS & Macgregor HC (1980) *J. Cell Sci.* 44, 87–101. With permission from the Company of Biologists Ltd.

Although the majority of cases of transcriptional regulation are likely to occur at the level of initiation, the existence of RNA polymerase pausing after initiating transcription, which was discussed in Chapter 4 (Section 4.2), clearly offers the possibility for control at the level of transcriptional elongation. In such cases, transcriptional control may operate by releasing a block to elongation of the nascent transcript.

This form of regulation is responsible, for example, for the tenfold decline in mRNA levels encoding the cellular oncogene c-*myc* (for discussion of the c-*myc* oncogene see Sections 11.1 and 11.2) which occurs when the human pro-myeloid cell line HL-60 is induced to differentiate into a granulocyte-type cell. If nuclear run-on assays (see Section 1.4) are carried out using nuclei from undifferentiated or differentiated HL-60 cells, the results obtained vary depending on the region of the c-*myc* gene whose transcription is being measured. Thus the c-*myc* gene consists of three **exons**, which appear in the mRNA and which are separated by intervening sequences that are removed from the primary transcript by RNA splicing. If the labeled products of the nuclear run-on procedure are hybridized to the DNA of the second exon, the levels of transcription observed are approximately tenfold higher in nuclei derived from undifferentiated cells than in nuclei from differentiated cells.

Hence the observed differences in c-*myc* RNA levels in these cell types are indeed produced by differences in transcription rates. However, if the same labeled products are hybridized to DNA from the first exon of the c-*myc* gene, virtually no difference in the level of transcription of this region in the differentiated compared to the undifferentiated cells is observed.

Comparison of the rates of transcription of the first and second exons in undifferentiated and differentiated cells indicates that regulation takes place at the level of transcriptional elongation rather than initiation. Although similar numbers of polymerase molecules initiate transcription of the c-*myc* gene in both cell types, the majority terminate in differentiated cells near the end of exon 1, do not transcribe the remainder of the gene, and hence do not produce a functional RNA. In contrast, in undifferentiated cells most polymerase molecules that initiate transcription transcribe the whole gene and produce a functional RNA. Hence the fall in c-*myc* RNA in differentiated cells is regulated by means of a block to elongation of nascent transcripts (**Figure 5.63**).

Interestingly, the rapid inhibition of transcriptional elongation following differentiation is supplemented several days after differentiation by an inhibition of c-*myc* transcription at the level of initiation. Similar effects on transcriptional elongation have been seen in several other cellular oncogenes such as c-*myb*, c-*fos*, and c-*mos*, (see Sections 11.1 and 11.2 for discussion of cellular oncogenes), indicating that this mechanism is not confined to a single oncogene and may be quite widespread.

Interestingly, a protein affecting the rate of transcriptional elongation has been identified in the case of the human immunodeficiency virus (**HIV**-1). Early in infection with this virus, a low level of transcription occurs from the HIV-1 promoter and many of the transcripts terminate very close

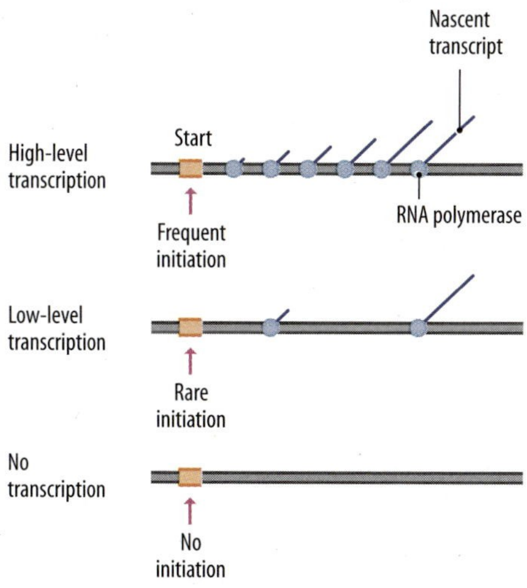

Figure 5.62
Regulation of transcriptional initiation results in differences in the number of RNA polymerase molecules transcribing a gene and therefore in the number of transcripts produced.

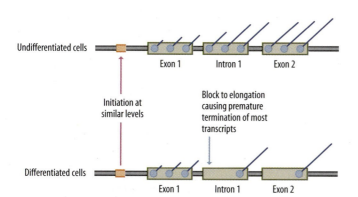

Figure 5.63
Regulation of transcriptional elongation in the c-*myc* gene. A block to elongation at the end of exon 1 results in most transcripts terminating at this point in differentiated HL-60 cells.

to the promoter producing very short RNAs. Subsequently, the viral Tat protein binds to these HIV-1 RNAs at a region known as Tar, which is located at +19 to +42 relative to the start site of transcription at nucleotide +1. This binding produces two effects. Firstly, there is a large increase in transcriptional initiation leading to many more RNA transcripts being initiated. In addition, however, Tat also overcomes the block to elongation leading to the production of a greater proportion of long transcripts which can encode the viral proteins.

Tat therefore produces a large increase in the production of HIV transcripts able to encode viral proteins both by stimulating transcriptional initiation and by overcoming a block to transcriptional elongation close to the promoter (**Figure 5.64**). As well as the viral Tat protein, cellular proteins which can stimulate transcriptional elongation have also been identified. Indeed, such an effect is involved in the increased expression of the gene encoding the heat-inducible hsp70 protein that occurs following exposure of cells to elevated temperature and which results from the binding of the heat-shock factor (HSF) to the promoter of this gene (see Section 4.3). Similarly, other transcriptional activators have also been shown to act by stimulating both transcriptional initiation and elongation and both these effects are dependent upon their transcriptional activation domains, which were described earlier (see Section 5.2).

Factors which regulate transcriptional elongation target the C-terminal domain of RNA polymerase II

A key target for factors which regulate transcriptional elongation is the phosphorylation of the C-terminal domain (**CTD**) of RNA polymerase II. As described in Section 4.2, such phosphorylation of the CTD on serine 2 is a key event required for transcriptional elongation by RNA polymerase II. Interestingly, the HIV Tat protein has been shown to be able to interact with the pTEF-b kinase protein which phosphorylates the CTD on serine 2 allowing transcriptional elongation to occur. Hence, Tat attracts this kinase to the HIV promoter, ensuring phosphorylation of RNA polymerase II and producing transcriptional elongation (**Figure 5.65**).

The involvement of CTD phosphorylation in regulating transcriptional elongation is also seen in the case of the **zebrafish** protein, Foggy. Thus, the Foggy protein interacts with the non-phosphorylated form of RNA polymerase II to block transcriptional elongation. However, when RNA polymerase II is phosphorylated, it is no longer inhibited by Foggy allowing transcriptional elongation to proceed (**Figure 5.66**). When this activity of Foggy is

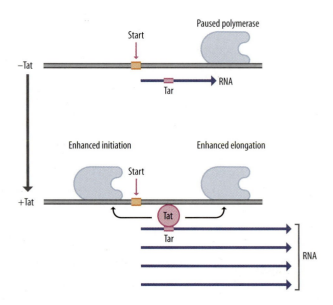

Figure 5.64
The HIV Tat protein acts both by enhancing the rate of transcriptional initiation by RNA polymerase and by enhancing the rate of elongation by overcoming polymerase pausing close to the promoter. Both these effects are achieved by the Tat protein binding to a specific region (Tar) of the nascent RNA. Arrows indicate nascent HIV RNA transcripts.

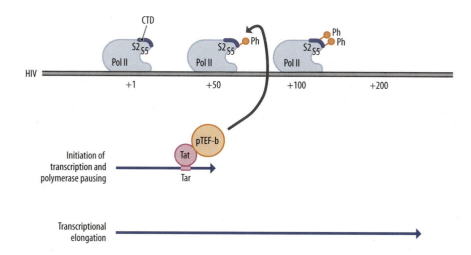

Figure 5.65
Phosphorylation (Ph, orange spot) of serine 5 in the C-terminal domain (CTD; shaded) of RNA polymerase II (Pol II) allows transcriptional initiation on the HIV promoter but the polymerase stalls after transcribing approximately 50 bases. Binding of the viral Tat protein to the Tar sequence in the nascent RNA recruits the cellular kinase, pTEF-b. This phosphorylates the C-terminal domain on serine 2, allowing further transcriptional elongation to occur.

inactivated by mutation, the resulting zebrafish fail to produce the correct number of dopamine synthesizing neurons during development. This indicates that the regulation of transcriptional elongation by Foggy is likely to be critical for the correct expression of genes involved in the production of this cell type. Moreover, it more generally demonstrates that regulation at the level of transcriptional elongation can play a key role in controlling gene expression during development.

The viral Tat protein and the cellular Foggy protein illustrate that regulators of transcriptional elongation can act to either stimulate or inhibit this process. Indeed, in yeast a pair of antagonistic regulators of transcriptional elongation have been identified. Thus, the Fkh2p factor stimulates transcriptional elongation and processing of the RNA by enhancing phosphorylation of the CTD of RNA polymerase II on serine 2 and serine 5. In contrast, Fkh1p inhibits such phosphorylation and thereby inhibits transcriptional elongation and **RNA processing** (**Figure 5.67**).

Hence, transcriptional elongation appears to be regulated by the balance between factors which stimulate elongation of the transcript and those which inhibit it. As many of these factors regulate transcriptional elongation by controlling the phosphorylation of the CTD of RNA polymerase II, they also regulate post-transcriptional processes which are regulated by such phosphorylation, thereby linking RNA transcript production with its proper processing (see Section 6.4 for discussion of the linkage of transcription with post-transcriptional events and the role of the CTD in these processes).

Figure 5.66
The Foggy protein acts by interacting with the RNA polymerase II transcription complex to prevent transcriptional elongation (E) following transcriptional initiation (I). Phosphorylation (Ph; orange spot) of RNA polymerase II on serine (S) 2 of the C-terminal domain (CTD; shaded) prevents this inhibitory effect of Foggy, allowing transcriptional elongation to proceed.

5.5 REGULATION OF TRANSCRIPTION BY RNA POLYMERASES I AND III

In this chapter, we have discussed the very wide range of transcription factors which regulate transcriptional initiation and elongation by RNA polymerase II, paralleling the very wide variety of patterns of expression in genes transcribed by RNA polymerase II. The much simpler patterns of expression of genes transcribed by RNA polymerases I and III (see Section 4.1) are paralleled by their simpler regulation.

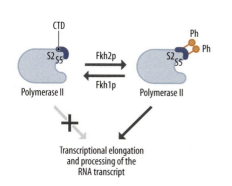

Figure 5.67
The yeast transcription factors Fkh1p and Fkh2p have opposite effects on the phosphorylation (Ph; orange spot) of the C-terminal domain (CTD; shaded) of RNA polymerase II and thereby produce opposite effects on transcriptional elongation and processing of the resulting RNA.

Transcription by RNA polymerases I and III can be regulated by alterations in chromatin structure

Interestingly, as with polymerase II, transcription by polymerases I and III can be regulated by changes in chromatin structure. Thus, in yeast, glucose starvation leads to a reduction in ribosome production. This is achieved by the silencing of the ribosomal DNA genes by their packaging into the tightly packed chromatin structure, known as heterochromatin (see Section 2.5). This is achieved by a multi-protein complex (known as eNoSc, for energy-dependent nucleolar silencing complex), which deacetylates histone H3 and also modifies lysine 9 of this histone by dimethylation, both of which are modifications characteristic of tightly packed chromatin (see Section 3.3) (**Figure 5.68**).

Transcription by RNA polymerases I and III can be regulated by altering the expression or activity of components of the basal transcriptional complex

In a number of situations, transcription by RNA polymerases I and III is regulated by altering either the expression level or the activity of basal transcription factors which are involved in the expression of all the genes transcribed by a particular polymerase (**Figure 5.69**). This is evidently possible because these polymerases predominantly transcribe genes involved in ribosome production. They can therefore be regulated more globally than RNA polymerase II, which transcribes a wide range of genes involved in different functions.

An example of the altered activity of basal transcription factors (**Figure 5.69a**) occurs during mitosis when the transcription of the ribosomal genes is inhibited by regulating the activity of the basal transcription factors UBF and SL1/TIF-1B which are essential for transcription by RNA polymerase I (see Section 4.1). Thus, upon entry into mitosis, both factors are phosphorylated which decreases their activity and hence blocks transcription of the ribosomal genes by RNA polymerase I. After the cell completes mitosis, both factors are dephosphorylated and this activates SL1. However, ribosomal gene transcription does not occur until entry into the **G1 phase** of the next cell cycle since it also requires phosphorylation of UBF on a different region of the protein, which is necessary for its activation (**Figure 5.70**).

As well as being regulated as cells proceed through the cell cycle and mitosis, enhanced activity of RNA polymerase I as well as of RNA polymerase III is required in growing cells compared to resting cells. Thus, as discussed in Chapter 11 (Sections 11.3 and 11.4) **anti-oncogene** proteins restrain cellular growth by inhibiting the activity of specific basal transcription factors for RNA polymerases I and III. In contrast, the c-Myc oncogene protein has the opposite effect and stimulates the activity of these factors,

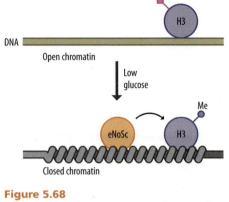

Figure 5.68
In yeast, low glucose concentrations result in binding of the eNoSc (energy-dependent nucleolar silencing complex) which is able to deacetylate and methylate histone H3, producing an inactive chromatin structure.

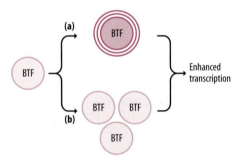

Figure 5.69
Transcription by RNA polymerase I or RNA polymerase III can be enhanced by increasing the activity of a basal transcription factor (BTF) (a) or increasing its expression (b).

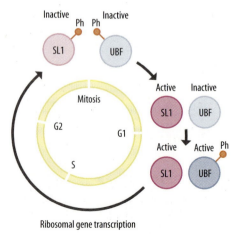

Ribosomal gene transcription

Figure 5.70
Regulation of transcription of the ribosomal RNA genes by RNA polymerase I. During mitosis, transcription is blocked by phosphorylation (Ph; orange spot) of the UBF and SL1 basal transcription factors. Following mitosis both factors are dephosphorylated which activates SL1. Subsequently, UBF is phosphorylated on a different region of the protein which activates it and allows transcription to occur during the cell cycle.

so enhancing RNA polymerase I- and III-dependent transcription and stimulating cellular growth (see Section 11.4).

Interestingly, the fact that the TBP factor is involved in transcription by all three RNA polymerases (see Section 4.1) appears to be used to regulate the balance between transcription by the different RNA polymerases. Thus, as discussed in Section 5.3, the Dr1 inhibitory factor binds to TBP and inhibits its ability to recruit other transcription factors to TATA box-containing RNA polymerase II gene promoters. Interestingly, Dr1 also affects recruitment to RNA polymerase III promoters but has no effect on the activity of TBP within the RNA polymerase I transcriptional complex. Hence Dr1 may regulate the balance between the transcription of the ribosomal genes which are the only genes transcribed by RNA polymerase I and all the other genes in the cell (**Figure 5.71a**).

Interestingly, Dr1 appears also to selectively inhibit transcription of polymerase II genes which have a TATA box, while actually stimulating transcription of genes containing an initiator element (see Sections 4.1 and 4.3, for discussion of the TATA box and initiator elements). Hence, Dr1 may also switch transcription between different classes of RNA polymerase II-transcribed genes (**Figure 5.71b**).

As well as being altered by altering the activity of individual basal transcription factors, the activity of RNA polymerase III is also altered in specific situations by altering the expression level of its basal transcription factors (see Section 4.1) (**Figure 5.69b**). Thus the activation of genes transcribed by RNA polymerase III following treatment of cells with serum is due to enhanced levels of active TFIIIC whereas the down-regulation of polymerase III transcription during the differentiation of **embryonic stem (ES) cells** is due to a decrease in TFIIIB levels (see Section 9.1 for further discussion of gene regulation in ES cells).

Regulation of transcription by RNA polymerase III can involve specific transcription factors binding to RNA as well as to DNA

As with RNA polymerase II, the regulatory processes discussed above involve factors which exert their effects on RNA polymerase I- and III-mediated gene expression by binding at the DNA level. In addition, however, in the case of RNA polymerase III, the presence of transcriptional control elements within the transcribed region (see Section 4.1), and hence in both the DNA and the RNA product, offers another potential mechanism of gene regulation. This, is exploited in a unique regulatory mechanism involving the basal transcription factor TFIIIA which binds to the internal regulatory sequence of the 5S ribosomal RNA genes (see also Section 5.1).

Thus, the frog *Xenopus laevis* contains two types of 5S gene: the oocyte genes which are transcribed only in the developing oocyte before fertilization, and the somatic genes which are transcribed in cells of the embryo and adult. The internal DNA control region of both these types of genes binds TFIIIA, whose binding is necessary for their transcription (see Section 4.1) However, sequence differences between the two types of gene (**Figure 5.72**) result in a higher affinity of the TFIIIA factor for the somatic compared to the oocyte genes. Hence, the oocyte genes are only transcribed in the oocyte where there are abundant levels of TFIIIA and not in other cells where the levels are only sufficient for activity of the somatic genes.

In the developing oocyte, TFIIIA is synthesized at high levels and transcription of the oocyte genes begins. As more and more 5S RNA molecules containing the TFIIIA-binding site accumulate in the maturing oocyte, they bind the transcription factor. This factor is thus sequestered with the 5S RNA into storage particles and is unavailable for transcription of

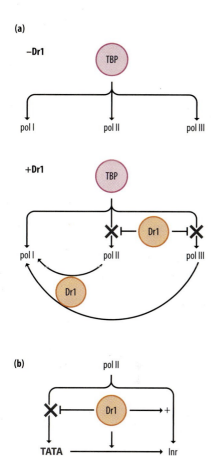

(a)

−Dr1

(b)

Figure 5.71

The Dr1 factor inhibits the activity of TBP within the basal transcription complex of genes transcribed by RNA polymerase II and III but not in that of the ribosomal genes which are transcribed by RNA polymerase I. Hence Dr1 may regulate the balance between the transcription of ribosomal genes by RNA polymerase I and that of all other genes in the cell by RNA polymerases II and III (a). In addition, Dr1 also stimulates genes transcribed by RNA polymerase II which contain an initiator element (Inr), while repressing those which contain a TATA box (b). Hence, it can switch transcription between different classes of genes transcribed by RNA polymerase II.

+46   +98

TCGGAAGCC**AA**GCAGGGTCGGGCCTGGTTAGTACTTGGATGGGAGACCGCCTGG
          G T

Figure 5.72

Sequence of the internal control element of the *Xenopus* somatic 5S genes. The two base changes in the oocyte 5S genes which result in a lower affinity for TFIIIA are indicated below the somatic sequence.

the 5S **rRNA genes**. Hence the level of free transcription factor falls below that necessary for the transcription of the oocyte genes and their transcription ceases, while transcription of the higher-affinity somatic genes is unaffected (**Figure 5.73**).

A simple mechanism involving differential binding of a transcription factor to two related genes is thus used in conjunction with the binding of the same factor to its RNA product, to modulate expression of the 5S genes.

CONCLUSIONS

The findings discussed in this chapter indicate the key roles played by specific transcription factors in the activation and repression of transcription, at the levels of transcriptional initiation and transcriptional elongation. The ability to bind to DNA and/or other proteins and then stimulate or repress transcription, allows transcription factors to play key roles in a wide range of biological processes. Chapters 8, 9, and 10 will discuss, respectively, the gene-control processes that occur in response to specific signals during embryonic development and in different types of differentiated cells, in all of which transcription factors play a key role.

KEY CONCEPTS

- DNA sequences which affect the rate of transcription by RNA polymerase II act by binding regulatory proteins known as transcription factors.

- Transcription factors have a modular structure in which different domains of the protein are responsible for different functions, such as DNA binding or transcriptional activation.

- A number of different DNA-binding domains have been identified and used to classify transcription factors into families with related DNA-binding domains.

- Following DNA binding, transcription factors can either activate or repress transcription.

- Specific activation domains in transcription factors activate transcription by stimulating the assembly/activity of the basal transcription complex either directly or indirectly via co-activators.

- Inhibitory transcription factors can repress transcription either indirectly by blocking the effect of a positively acting factor or directly via the basal transcriptional complex.

- Transcription factors can modulate the rate of transcriptional elongation, as well as of transcriptional initiation.

- Transcription by RNA polymerases I and III is also subject to transcriptional control although this is less complex than for RNA polymerase II.

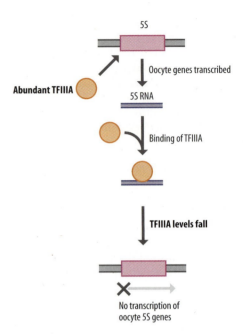

Figure 5.73
Sequestration of the TFIIIA factor by binding to 5S RNA made in the oocyte, results in a fall in the level of the factor, switching off transcription of the oocyte-specific 5S genes. The somatic 5S genes, which have a higher-affinity binding site for TFIIIA, continue to be transcribed.

FURTHER READING

General

Latchman DS (2008) Eukaryotic Transcription Factors, 5th ed. Elsevier/Academic Press.

Kadonaga JT (2004) Regulation of RNA polymerase II transcription by sequence-specific DNA binding factors. *Cell* 116, 247–257.

5.1 DNA binding by transcription factors

Amoutzias GD, Robertson DL, Van de Peer Y & Oliver SG (2008) Choose your partners: dimerization in eukaryotic transcription factors. *Trends Biochem. Sci.* 33, 220–229.

Garvie CW & Wolberger C (2001) Recognition of specific DNA sequences. *Mol. Cell* 8, 937–946.

Klug A (2005) The discovery of zinc fingers and their development for practical applications in gene regulation. *Proc. Jap. Acad. Ser. B Phys. Biol. Sci.* 81, 87–102.

5.2 Activation of transcription

Baker SP & Grant PA (2007) The SAGA continues: expanding the cellular role of a transcriptional co-activator complex. *Oncogene* 26, 5329–5340.

Chadick JZ & Asturias FJ (2005) Structure of eukaryotic Mediator complexes. *Trends Biochem. Sci.* 30, 264–271.

Conaway RC, Sato S, Tomomori-Sato C et al. (2005) The mammalian Mediator complex and its role in transcriptional regulation. *Trends Biochem. Sci.* 30, 250–255.

Malik S & Roeder RG (2005) Dynamic regulation of pol II transcription by the mammalian Mediator complex. *Trends Biochem. Sci.* 30, 256–263.

Spiegelman BM & Heinrich R (2004) Biological control through regulated transcriptional coactivators. *Cell* 119, 157–167.

5.3 Repression of transcription

Auble DT (2009) The dynamic personality of TATA-binding protein. *Trends Biochem. Sci.* 34, 49–52.

Nagy L & Schwabe JW (2004) Mechanism of the nuclear receptor molecular switch. *Trends Biochem. Sci.* 29, 317–324.

Rosenfeld MG, Lunyak VV & Glass CK (2006) Sensors and signals: a coactivator/corepressor/epigenetic code for integrating signal-dependent programs of transcriptional response. *Genes Dev.* 20, 1405–1428.

Schwartz YB & Pirrotta V (2007) Polycomb silencing mechanisms and the management of genomic programmes. *Nat. Rev. Genet.* 8, 9–22.

Toledo F & Wahl GM (2006) Regulating the p53 pathway: in vitro hypotheses, *in vivo veritas. Nat. Rev. Cancer* 6, 909–923.

5.4 Regulation at transcriptional elongation

Core LJ & Lis JT (2008) Transcription regulation through promoter-proximal pausing of RNA polymerase II. *Science* 319, 1791–1792.

Margaritis T & Holstege FC (2008) Poised RNA polymerase II gives pause for thought. *Cell* 133, 581–584.

Price DH (2008) Poised polymerases: on your mark...get set...go! *Mol. Cell* 30, 7–10.

5.5 Regulation of transcription by RNA polymerases I and III

Grummt I & Ladurner AG (2008) A metabolic throttle regulates the epigenetic state of rDNA. *Cell* 133, 577–580.

Post-transcriptional Processes

6

INTRODUCTION

The transcription of DNA into RNA represents the first stage in the process of gene expression (see Chapter 4) and is the major control point regulating which genes are expressed (see Chapter 5). However, the process of transcription is supplemented by a series of post-transcriptional events that are necessary to produce a functional mRNA which is able to be translated into protein (Figure 6.1).

Even while transcription is still proceeding, the nascent RNA is capped at its 5′ end by a modified guanosine residue and is subsequently cleaved near its 3′ end followed by addition of up to 200 adenosine residues in a process known as polyadenylation. Subsequently, intervening sequences or introns (which interrupt the protein-coding sequence in both the DNA and the primary transcript of many genes) are removed by a process of RNA splicing. Although this produces a functional mRNA, the spliced molecule must then be transported from the nucleus, where these processes occur, to the cytoplasm, where it can be translated into protein.

As each of these stages in gene expression is essential for the production of a functional mRNA, they could potentially be a target for control processes which regulate gene expression. Accordingly, post-transcriptional processes are described in this chapter and their role in regulating gene expression is discussed in Chapter 7.

6.1 CAPPING

The capping process modifies the 5′ end of the RNA transcript

Following initiation of the primary RNA transcript, transcription proceeds by the progressive addition of ribonucleotides to the RNA chain so that an RNA molecule is produced with bases complementary to those present in the DNA being transcribed. Even before this process is complete, the 5′ end of the nascent mRNA is modified by the addition of a guanine residue which is not encoded by the DNA. The bond joining this guanine to the 5′ end of the RNA differs from the standard bonds linking the nucleotides in the RNA chain in two ways. Firstly, three phosphate molecules separate the G residue from the first base in the chain whereas only a single phosphate separates each residue in the rest of the chain. Secondly, the phosphate residues are joined via a 5′ to 5′ bond rather than the standard 3′ to 5′ bond which links the nucleotides in the chain (Figure 6.2).

This addition of an extra G residue by a 5′–5′ bond is achieved by the enzyme guanylate transferase. Subsequently, the enzyme guanine methyltransferase adds a methyl group to position 7 of the purine ring of the added G residue to produce 7-methyl guanine (Figures 6.2 and 6.3).

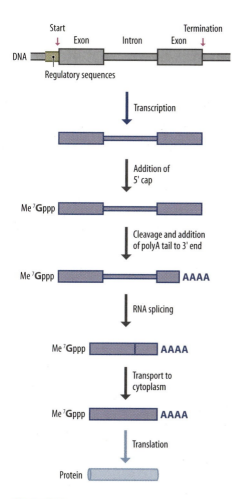

Figure 6.1
Stages in eukaryotic gene expression which could be regulated.

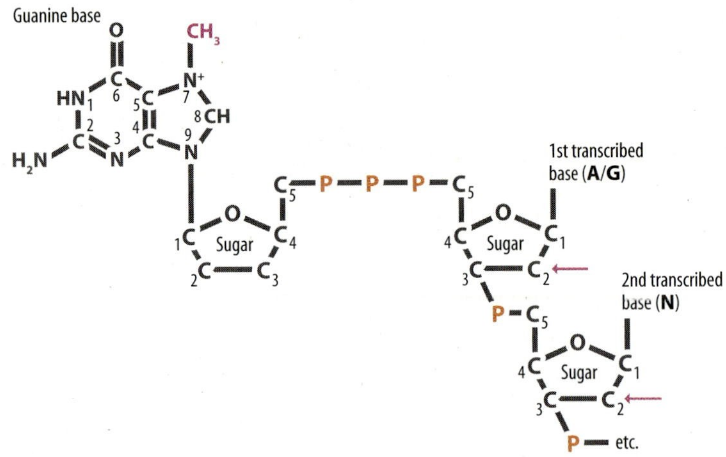

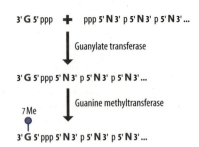

Figure 6.2
Structure of the cap which is present at the 5′ end of eukaryotic mRNAs. Note that the cap structure consists of a guanine residue which is linked via a 5′ to 5′ bond and three phosphate residues to the first transcribed base which is normally an A or a G residue. This linkage is in contrast to the normal 3′ to 5′ link with a single phosphate residue which joins the sugar residues in the nucleotide chain. The G residue is also methylated at the 7 position on the guanine base. Arrows indicate the positions in the ribose sugars, which can also be methylated.

Figure 6.3
Formation of the cap involves a guanylate transferase enzyme forming the 5′ to 5′ bond and a guanine methyltransferase enzyme adding a methyl group (Me) at the 7 position on the guanine base.

This basic cap structure is found in yeast whereas in higher eukaryotes further modifications can occur with additional methyl residues being added to the first and sometimes also the second of the two transcribed nucleotides immediately following the 7-methyl guanine. These methyl residues are added at the 2′ hydroxyl group on the ribose sugar component of the nucleotide (see Figure 6.2).

The cap enhances translation of the mRNA by the ribosome

Capping occurs only in eukaryotes and not in prokaryotes. It is believed to be necessary due to a difference in the process of translation between eukaryotic and prokaryotic mRNAs (see Section 6.6). Prokaryotic mRNAs are in general polycistronic, which means that several different proteins are translated from the same mRNA. The ribosome can initiate translation to produce protein internally within an mRNA molecule, being guided by specific sequences (**Shine–Dalgarno sequences**) within the mRNA which are complementary to the 16S RNA of the small ribosomal subunit. In contrast, eukaryotic mRNAs are generally **monocistronic** with only one protein being derived from each mRNA. These mRNAs therefore lack the guide sequences present in prokaryotic mRNAs and translation is initiated by the small (40S) subunit of the ribosome binding to the 5′ end of the mRNA by recognizing the methylated cap (**Figure 6.4**).

(a) Prokaryotic

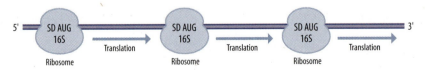

(b) Eukaryotic

Figure 6.4
Translational initiation in prokaryotes (a) and eukaryotes (b). Note that prokaryotic mRNAs are polycistronic with several proteins being translated from a single mRNA. Initiation is achieved by the 16S RNA of the ribosome binding to a complementary sequence known as the Shine–Dalgarno (SD) sequence at various points within the mRNA molecule. Translation is then initiated at an AUG codon adjacent to the Shine–Dalgarno sequence. In contrast, eukaryotic mRNAs are monocistronic with only a single protein being translated from the mRNA. Translation is initiated by the ribosome binding to the cap structure at the 5′ end of the mRNA and moving along the mRNA to initiate translation at an AUG codon which is located in an appropriate surrounding sequence.

Interestingly, in prokaryotes the Shine–Dalgarno sequence which is recognized by the 16S ribosomal RNA is located within 10 bases upstream of the AUG sequence at which translation is initiated (see Section 6.6). In contrast, in eukaryotes the AUG initiation codon may be located several hundred bases away from the 5′ end of the RNA. Following binding at the cap, the 40S subunit of the ribosome migrates along the mRNA until it encounters the appropriate AUG initiation codon. At this point the large (60S) ribosomal subunit binds and translation begins.

However, translation is not initiated at the first AUG which is encountered by the small ribosomal subunit. Rather, the AUG triplet must be set within the appropriate consensus sequence, which is related to GCCA/ GCCAUGG. This so-called scanning hypothesis, in which the ribosome moves along the mRNA looking for this sequence, was originally propounded by Kozak. The rules which determine whether a particular sequence surrounding an AUG codon will allow the ribosome to initiate transcription are known as Kozak's rules.

In eukaryotes, it is therefore necessary for the 5′ end of mRNAs which are to be translated to be recognized by the translational apparatus and distinguished from the much more abundant non-translated RNAs, such as the ribosomal RNAs. This is achieved by specifically modifying the RNAs to be translated by the capping process. As the first step of translation, the cap on the mRNA is recognized by the translational initiation factor eIF4E. As discussed in Section 6.6, this results in the recruitment of other translation initiation factors and of the small ribosomal subunit itself.

As well as its function in allowing recognition by the ribosome, the cap structure also has an important role in protecting the 5′ end of the mRNA from attack by exonuclease enzymes which would otherwise recognize a free 5′ end and would digest the RNA in a 5′ to 3′ direction. Indeed, as discussed in Section 6.7, the removal of the cap structure is one of the steps in mRNA degradation by which the mRNA molecule is ultimately broken down. This protection from degradation is achieved by the binding to the cap of the **cap-binding complex** (CBC). CBC binds to the capped RNA in the nucleus, protects it from degradation and facilitates its transport to the cytoplasm. Following transport to the cytoplasm, CBC is displaced from the cap by eIF4E allowing the mRNA to be translated into protein (**Figure 6.5**). Interestingly, eIF4E and CBC both use similar structural motifs to bind to the cap, with the 7-methyl guanine ring being stacked between two aromatic amino acids in a cap-binding pocket of the protein (**Figure 6.6**).

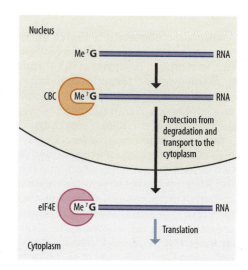

Figure 6.5
A cap-binding complex (CBC) binds to the 5′ cap in the nucleus. Following transport to the cytoplasm it is replaced by the eIF4E translational initiation factor.

(a) eIF4E

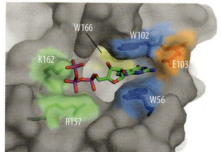

(b) CBC

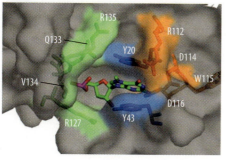

Figure 6.6
Common structure of the cap-binding pockets in the eIF4E (a) and CBC (b) proteins. Aromatic amino acids forming a sandwich around the cap are shown in blue, those binding the functional groups of the guanine residue are shown in orange, those stabilizing the 7-methyl group are shown in yellow, and those binding the triphosphate are shown in green. Courtesy of George Brownlee, from Fechter P & Brownlee GG (2005) *J. Gen. Virol.* 86, 1239–1249. With permission from The Society for General Microbiology.

More recently, a further role for capping has been described involving it acting as a checkpoint regulating elongation of the RNA transcript. As described in Section 4.2, following initiation and production of the first few nucleotides of the RNA, the RNA polymerase pauses and it is at this point that the RNA is capped. Once capping has occurred the pTEF-b kinase is recruited and it phosphorylates RNA polymerase II, so that transcription continues and the full RNA transcript is produced. Hence, a checkpoint mechanism operates in which a full RNA transcript is not produced unless the RNA is correctly capped (**Figure 6.7**). The coupling of transcriptional and post-transcriptional processes is discussed further in Section 6.4.

6.2 POLYADENYLATION

The polyadenylation process modifies the 3′ end of the RNA transcript

As discussed above, the 5′ end of the mRNA differs from that of the primary transcript only by the addition of the single modified G residue forming the cap. In contrast, however, at the 3′ end, the mRNA molecule is much shorter than the original transcript and differs from it in terminating with a run of up to 200 adenosine residues which are added post-transcriptionally in a process known as polyadenylation (**Figure 6.8**). This process involves firstly the cleavage of the RNA transcript and then the subsequent addition of the A residues at the resulting free 3′ end.

The site at which cleavage occurs (the polyadenylation site) is flanked by two conserved sequence elements (**Figure 6.9**). Upstream of the polyA site the RNA contains the essentially invariant sequence AAUAAA and downstream of the polyA site it contains a less well conserved region which is rich in G and U residues. These two sequence elements are recognized by two protein complexes, known as the cleavage- and polyadenylation-specificity factor (**CPSF**) and the cleavage-stimulation factor (**CstF**) (see Figure 6.9). Following the binding of these complexes they are thought to interact with one another and the cutting of the RNA occurs in the region between them. Note that this cutting is an endonucleolytic cleavage where cutting occurs within the RNA chain rather than the exonucleolytic cleavage which would occur following removal of the cap (see Section 6.1 above) where degradation begins at a free end of the RNA molecule. Subsequent to cleavage, the enzyme polyA polymerase then adds the run of A residues to the free 3′ end (see Figure 6.9).

Polyadenylation enhances the stability of the mRNA

The RNA transcript is therefore modified at both ends, firstly by capping at the 5′ end and subsequently by cleavage and polyadenylation at the 3′ end. It is believed that the primary role of the polyadenylation process is to protect the mRNA from degradation by exonucleases which would otherwise attack its free 3′ end and rapidly degrade it. In many cases, an mRNA which is to be degraded is first deadenylated allowing its 3′ end to be attacked (see Section 6.7). In addition, it appears that the presence of the polyA tail can also regulate the efficiency by which an mRNA is translated. Hence like the cap, the polyA tail appears to have a dual function in regulating both translation of the mRNA into protein and its degradation.

Interestingly, not all mRNAs encoding proteins are polyadenylated. In particular, the mRNAs encoding the histone molecules which play a critical role in chromatin structure are not polyadenylated. In this case, however,

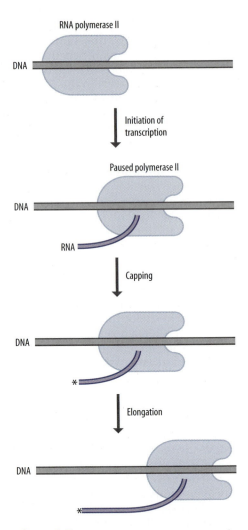

Figure 6.7
Following initiation of transcription and production of the first few bases of the RNA, the polymerase pauses and only continues transcription once the nascent RNA has been capped. Compare with Figure 4.24 and note that capping is essential for recruitment of the pTEF-b kinase which is required for transcriptional elongation.

Figure 6.8
Modifications at the 5′ and 3′ end of RNA polymerase II transcripts. Note that the 5′ end of the RNA is modified by the addition of a single G residue in the capping process whereas the 3′ end is cleaved with removal of a large stretch of RNA and subsequent addition of up to 200 A residues to the free 3′ end in the process of polyadenylation.

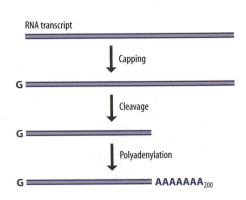

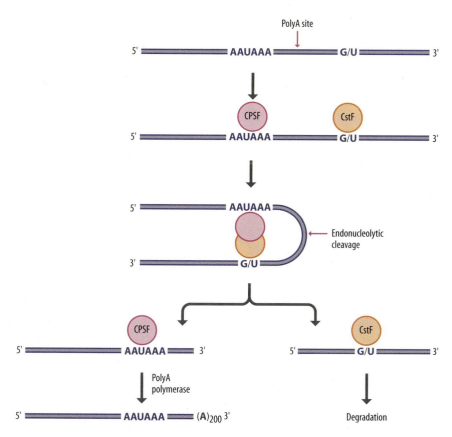

Figure 6.9
The process of polyadenylation involves the binding of protein factors to the AAUAAA signal upstream of the polyA site (CPSF) and to the G/U rich sequence downstream of the polyA site (CstF). CPSF and CstF then interact with one another leading to cleavage in between them at the polyadenylation site. Subsequently the sequence downstream of the polyA site is degraded while the free 3′ end of the upstream region is polyadenylated by the enzyme polyA polymerase.

the 3′ end of the mRNA is protected from degradation since it can fold itself into a double-stranded stem-loop structure which protects the mRNA from degradation. Interestingly, the folding and unfolding of this stem loop plays a critical role in regulating the stability of the histone mRNA. For example, the histone mRNA becomes highly stable during **S phase** of the cell cycle when DNA is being synthesized and more histones are required while being unstable at other points of the cell cycle.

Such an example is, however, the exception and the vast majority of RNA species transcribed by RNA polymerase II are polyadenylated at their 3′ ends following capping at the 5′ end. Indeed, even the mRNAs encoding the minor isoforms of the histones, such as H3.3 and H2A.Z (see Section 2.3), are polyadenylated and consequently do not exhibit decreased stability at phases of the cell cycle outside S phase (**Figure 6.10**).

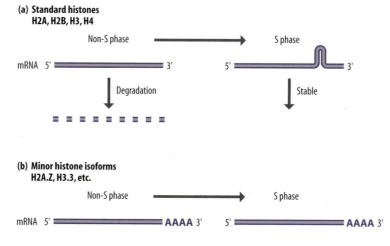

Figure 6.10
The mRNAs encoding the standard forms of the histones are not polyadenylated but are stabilized during S phase of the cell cycle by a stem-loop structure which forms at the 3′ end (a). In contrast, the mRNAs for the minor isoforms of the histones are polyadenylated and are therefore stable throughout the cell cycle (b).

6.3 RNA SPLICING

RNA splicing removes intervening sequences and joins exons together

Probably the most unexpected finding to have been made during the modern era of gene cloning was the discovery that the protein-coding regions of genes transcribed by RNA polymerase II are interrupted by DNA sequences which do not encode parts of the protein. These intervening sequences are known as introns because although they are transcribed into RNA, their RNA product remains *in* the nucleus whereas the coding regions or exons *exit* the nucleus.

When the DNA is transcribed both the exons and the introns which lie between them are transcribed into a single primary RNA transcript. As well as being modified at its 5′ and 3′ ends by capping and polyadenylation, this primary transcript must therefore also have its introns removed before it is transported to the cytoplasm. Otherwise, if it is translated, irrelevant amino acid sequences will appear in the middle of the protein or translation will be prematurely terminated due to the presence of a stop codon within the intervening sequence. The process whereby these introns are removed is known as RNA splicing (**Figure 6.11**).

Initial studies on RNA splicing concentrated on a DNA sequence analysis of the introns and the adjacent regions of the exons. Such studies revealed that virtually all eukaryotic introns for genes transcribed by RNA polymerase II begin with the bases GU and end with the bases AG (**Figure 6.12**). Although other features of the intron and the adjacent regions of the exon have been defined, such as a run of pyrimidine residues (the **polypyrimidine tract**) located adjacent to the AG sequence within the intron, these are much less well conserved than the GU/AG sequences.

Although this finding was of interest, it did not of itself provide any functional insights into the mechanisms by which splicing occurred. These were initially provided by studies using nuclear extracts which could carry out the splicing reaction in the test tube. These experiments showed that when a simple RNA molecule containing two exons interrupted by an intron was added to the splicing reaction, it was possible to produce the spliced product with the two exons joined together and a free intron. Most interestingly, however, the intron emerged in a **lariat** structure in which its 5′ end had folded back on itself and formed a 5′ to 2′ bond with an A residue located 18–40 bases upstream of the AG sequence at the 3′ end of the intron (**Figure 6.13**). This A residue is known as the **branch point**.

This can be explained on the basis that splicing proceeds in two transesterification steps (see Figure 6.13). In the first of these, the ester bond between the 5′ phosphate of the intron and the 3′ oxygen of the upstream exon is exchanged for an ester bond between the 5′ phosphate of the intron and the 2′ oxygen of the branch point A residue. In the second reaction, the ester bond between the 5′ phosphate of the downstream exon and the 3′ oxygen of the intron is exchanged for an ester bond between the 5′ phosphate of the downstream exon and the free 3′ oxygen of the upstream exon. This joins the upstream exon to the downstream exon and releases the intron as a lariat (see Figure 6.13). The reaction has now achieved its purpose by removing the intron from between the two exons and joining them together in a splicing reaction.

Specific RNAs and proteins catalyze the process of RNA splicing

In considering this process, one might ask why the upstream exon does not simply diffuse away while the downstream exon is still linked to the intron and cannot be joined to it. The answer to this question is that this reaction does not take place with the RNA isolated within the nucleus. Rather it takes place in a complex structure known as the spliceosome which consists of

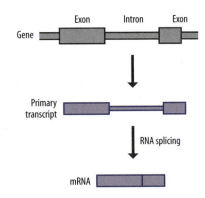

Figure 6.11
Removal of an intervening sequence (intron) from the primary RNA transcript by RNA splicing.

Figure 6.12
Exon/intron junctions in RNA polymerase II transcripts. Note that the first two bases of the intron are normally GU whereas the last two bases are normally AG.

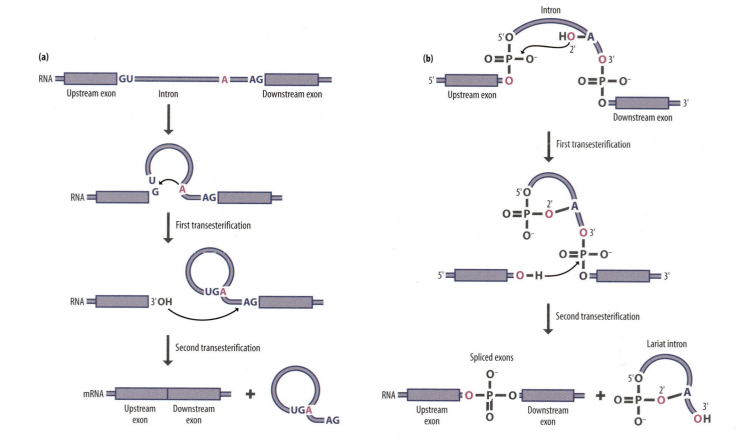

five different RNAs and over a hundred proteins, which together catalyze this process and hold the various intermediates in the splicing process in the correct orientation relative to one another. The three-dimensional structure of the **spliceosome** has been defined. This shows that the spliceosome consists of large and small subunits, with a tunnel or channel in between that can accommodate the RNA that is about to undergo splicing (**Figure 6.14**).

The RNA components of the spliceosome are small RNA molecules ranging from 56 to 217 bases in size and known as the U RNAs because they are rich in uridine residues. Each of these U RNAs has a specific role in the splicing process and each of them is associated with specific proteins forming **small nuclear ribonucleoprotein particles** (snRNP) which consist of a U RNA with its associated proteins. Some of these proteins, such as the Sm proteins, are common to all the different U snRNPs whereas others such as the U1 snRNP A protein are associated with only one specific RNP particle.

Figure 6.13
Mechanism of RNA splicing. Panel (a) shows a schematic diagram while (b) shows the chemical reactions involved. Splicing is initiated by a first transesterification reaction which exchanges the ester bond between the 5′ phosphate of the intron and the 3′ oxygen of the upstream exon for an ester bond between the 5′ phosphate of the intron and the 2′ oxygen of the A residue at the branch point. This is followed by a second transesterification reaction which exchanges the ester bond between the 5′ phosphate of the downstream exon and the 3′ oxygen of the intron for an ester bond between the 5′ phosphate of the downstream exon and the free 3′ oxygen of the upstream exon. This joins the two exons together and releases the intron as a lariat.

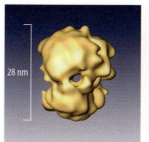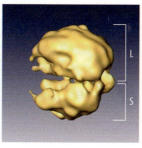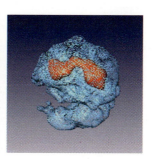

Figure 6.14
Left and middle panels: two different views of the three-dimensional structure of the spliceosome at 20 Å resolution. Note the large (L) and small (S) subunits and the cleft between them which accommodates the RNA to be spliced. Right-hand panel: position of the RNAs involved in catalyzing the splicing process (red) within the spliceosome (blue). Courtesy of Ruth Sperling, The Hebrew University of Jerusalem.

The splicing process is initiated by the binding of the U1 snRNP to the 5′ **splice site** (the junction between the upstream exon and the beginning of the intron) (**Figure 6.15a**). Subsequently the U2 accessory factor (U2AF) binds to the polypyrimidine tract located between the branch point A and the AG at the end of the intron. This then results in the U2 snRNP binding at the adjacent branch point (**Figure 6.15b**). Next, the RNA is bound by the U5 snRNP which binds to the upstream exon and by the single snRNP which contains the U4 and U6 RNAs bound to one another. This binding results in the displacement of the U1 snRNP from the 5′ splice site where it is replaced by the U6 snRNP (**Figure 6.15c**). This results in the release of the U1 snRNP and of the U4 RNA, which is now no longer base-paired to U6.

The U6 and U2 snRNPs then interact with one another bringing the 5′ splice site close to the branch point. This is followed by the separation of the upstream exon from the intron, lariat formation, and the joining of the upstream exon to the downstream exon as outlined in Figure 6.13. The U2 snRNP, U6 snRNP, and U5 snRNP (which has moved from the upstream exon to the intron during this process) are released with the intron (**Figure 6.15d**).

Interestingly, in the final stage of splicing, an RNA-unwinding/helicase enzyme known as Prp22 binds to the downstream exon (see Figure 6.15d). It then moves along the spliced RNA in a 3′ to 5′ direction, releasing it from the proteins/RNAs of the spliceosome.

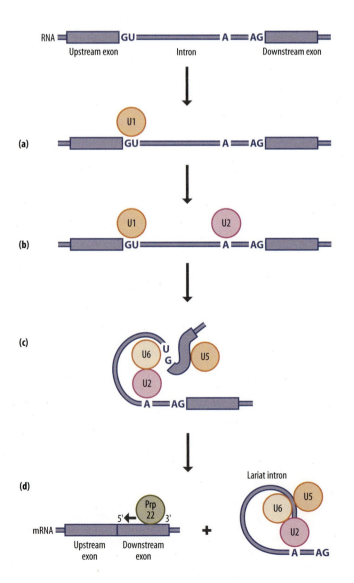

Figure 6.15
Involvement of snRNP particles bearing different U RNAs in RNA splicing. Splicing is initiated by binding of the U1 particle to the 5′ splice site (a) followed by binding of the U2 particle to the branch point (b). This is followed by binding of the U5 particle to the upstream exon and of the U4/U6 particle to the 5′ splice site with release of the U1 particle and U4 RNA. U6 and U2 particles then interact with one another (c) bringing the 5′ splice site adjacent to the branch point. This is followed by separation of the upstream exon from the downstream exon and lariat formation followed by subsequent separation of the intron from the downstream exon with exon joining. The free intron is released in association with the U2, U5, and U6 particles (d), with the U5 particle having moved from the upstream exon to a position within the intron. An RNA helicase (Prp22) binds to the downstream exon and moves along the spliced RNA in a 3′ to 5′ direction releasing it from the spliceosome.

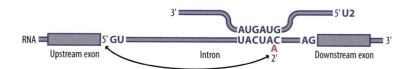

Figure 6.16
The U2 RNA base pairs with the branch point region of the mRNA. The A residue at the branch point is left unpaired and so is displaced from the double-stranded structure, facilitating the formation of a 2′ to 5′ bond with the first base of the intron.

Several of the interactions between the U snRNPs and the RNA which is being spliced are mediated by complementary base pairing between the U RNA in the RNP and the RNA undergoing splicing. For example, the U1 RNA base-pairs to the 5′ splice site in the initial phase of splicing, thereby allowing the U1 RNP to bind to the 5′ splice site. Similarly, the later displacement of the U1 RNP by the U4/U6 RNP is mediated by U6 binding to the 5′ splice site.

Of most interest, however, is the binding of the U2 RNP to the RNA sequence around the branch point. Not only is this mediated by the U2 RNA but the A at the branch point does not participate in the base pairing. It is therefore displaced from the double-stranded structure, facilitating the formation of the bond between its 2′ hydroxyl group and the 5′ end of the intron (**Figure 6.16**).

Hence, the entire process of splicing takes place in the tightly ordered structure of the spliceosome and this explains for example why the upstream exon cannot simply diffuse away after its separation from the intron/downstream exon. Indeed, it has been demonstrated that a specific component of the spliceosome hSLu7 binds tightly to the upstream exon and holds it in close proximity to the AG of the correct 3′ splice site so allowing selection of the correct 3′ splice site to occur.

In addition to the U RNAs and the proteins associated with them in the U snRNPs, the spliceosome also contains other proteins which are not directly associated with the U RNAs. The most intensively studied of these are a series of proteins which are rich in serine and arginine residues and which are therefore known as **SR proteins** (S=serine and R=arginine in the one-letter code for amino acids).

These proteins are believed to serve both in recruiting the snRNPs to the RNA to be spliced and in determining which splice sites are joined to one another. It appears that the SR proteins bind to sequences within the exons, known as exon-splicing enhancers (ESEs) and ensure that the exon is included in the final spliced RNA rather than being spliced out.

As the vast majority of RNAs contain multiple exons and introns (rather than the simple two-exon, one-intron model we have been considering) this role of SR proteins is obviously highly critical to ensure that the correct exons are joined to one another. Thus, the splicing process must ensure that exon 1 is joined to exon 2 which in turn is joined to exon 3 rather than, for example, exon 1 being joined to exon 3 resulting in exon 2 being missing from the mRNA and therefore producing a protein lacking a particular region. Mutations in ESEs which prevent binding of SR proteins result in the affected exon being spliced out in a process known as **exon skipping** (**Figure 6.17**)

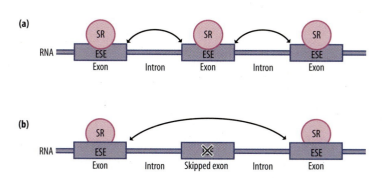

Figure 6.17
SR proteins bind to exon-splicing enhancers (ESE) in each exon, facilitating the correct pattern of splicing (a). The loss of such an ESE by mutation (X) results in skipping of the affected exon (b).

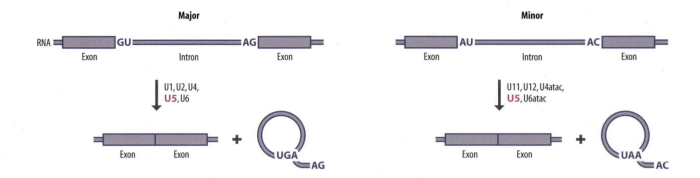

and this occurs in specific human diseases (see Sections 12.3 and 12.5 for discussion of human diseases that involve abnormalities in RNA splicing and their potential therapy).

In view of their role in determining which splice sites will be joined to one another, it is not surprising that the SR proteins also play a key role in the process known as alternative splicing in which different combinations of exons can be joined to one another in different situations (see Section 7.1 for discussion of alternative RNA splicing).

It was initially thought that all introns in genes transcribed by RNA polymerase II began with the GU sequence and ended with the AG sequence. Although this is true for the vast majority of introns, a small number begin with the sequence AU and end with AC (**Figure 6.18**). These introns are spliced by a mechanism which is remarkably similar to the GU–AG pathway described above and involves, for example, exactly the same use of an A residue at the branch point.

However, splicing of AU–AC introns involves only the U5 snRNP of those used in the GU–AG pathway and instead utilizes different snRNPs, containing respectively the U11, U12, U4atac, and U6atac RNAs (see Figure 6.18). Interestingly, it has been suggested that some splicing of AU–AC introns takes place in the cytoplasm indicating that this minor splicing pathway may have a distinct function to the major GU–AG pathway, which is exclusively nuclear.

In discussing both the major and minor pathways of splicing, we have considered cases in which a single RNA transcript contains all the exons which are to be joined together and all the introns which need to be removed. However, it is also possible in certain cases for **trans-splicing** to occur between exons in different mRNA molecules to produce a chimeric mRNA (**Figure 6.19**).

A case of this type has been described in normal human endometrial cells and involves the *JAZF1* gene and the *JJAZ1* gene which on its own

Figure 6.18
A minority of introns begin with AU and end with AC as opposed to the GU and AG found in most introns. Removal of both these types of intron involves lariat formation and the U5 RNA, but the other U RNAs involved are different in the two cases.

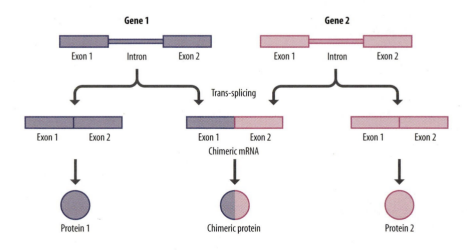

Figure 6.19
As well as the normal splicing of exons in a single RNA molecule it is also possible for exons in different RNA molecules, encoded by different genes, to be joined together to produce a chimeric mRNA. This mRNA can encode a protein distinct from that produced by each gene alone.

encodes a polycomb protein involved in the regulation of chromatin structure (see Sections 3.3 and 4.4 for discussion of polycomb proteins). These two genes are located on different chromosomes and each produces their own specific spliced mRNA in several different cell types. In addition, however, in endometrial cells 5′ exons of the JAZF1 RNA transcript are trans-spliced to 3′ exons of the JJAZ1 RNA transcript to produce a chimeric mRNA.

The chimeric JAZF1/JJAZ1 mRNA produced in this way encodes a distinct protein which protects the cells expressing it from **programmed cell death** (**apoptosis**). Most importantly, this trans-splicing event appears to be regulated since it occurs only in endometrial cells and not in other cells where the gene is expressed. Moreover, it is enhanced by specific hormones and by **hypoxia** (see Section 11.2 for discussion of the involvement of a JAZF1/JJAZ1 fusion protein in endometrial cancer).

6.4 COUPLING OF TRANSCRIPTION AND RNA PROCESSING WITHIN THE NUCLEUS

Transcriptional initiation and elongation are coupled to post-transcriptional processes

Although we have so far discussed transcription, capping, polyadenylation, and splicing as if they were entirely separate processes, it is clear that they are in fact tightly coupled together within the nucleus. The first evidence that this was the case was the finding that deletion of the C-terminal domain (CTD) of RNA polymerase II (see Section 4.1) not only reduces transcription but also interferes with post-transcriptional processes such as capping, splicing, and polyadenylation, indicating that RNA polymerase II is also involved in these events.

As described in Chapter 4 (Section 4.1), phosphorylation of the CTD of RNA polymerase II leads to the polymerase, together with TFIIF, initiating transcription and moving off down the gene, leaving behind the other basal transcription factors, such as TFIIB and TFIID (**Figure 6.20a**). Interestingly, these basal factors are then progressively replaced by factors which are involved in capping, splicing, and polyadenylation and which bind specifically to the phosphorylated form of the RNA polymerase II CTD.

As soon as transcription begins, capping enzymes bind to the CTD of RNA polymerase II, allowing the 5′ end of the RNA to be capped as described in Section 6.1 (**Figure 6.20b**). Subsequently, the SR proteins and other components of the splicing complex (see Section 6.3) bind to the CTD, allowing splicing to occur (**Figure 6.20c**). Finally, components of the polyadenylation complex (see Section 6.2) interact with the transcribing complex and polyadenylate the RNA (**Figure 6.20d**).

Interestingly, the recruitment of these various factors is closely coupled to the phosphorylation of the CTD of RNA polymerase II. As noted in Sections 4.1 and 4.2, this region contains multiple copies of the amino acid sequence Tyr-Ser-Pro-Thr-Ser-Pro-Ser. It has been shown that phosphorylation of serine 5 in this sequence (which stimulates transcriptional initiation) is also essential for recruitment of capping factors. Moreover, subsequent phosphorylation of serine 2 (which stimulates transcriptional elongation) is also necessary for recruitment of factors involved in 3′-end processing and polyadenylation (**Figure 6.21**).

The dual role of CTD phosphorylation in transcriptional elongation and processing of the RNA transcript therefore couples the two events, ensuring that the elongating RNA transcript is correctly processed. Indeed, this linkage provides the explanation for the relationship between capping and transcriptional elongation which was discussed in Section 6.1. Thus, following phosphorylation of the CTD on serine 5 by the kinase activity of TFIIH, capping of the RNA transcript occurs and this stimulates phosphorylation of the CTD on serine 2 by pTEF-b, allowing not only the recruitment of 3′ processing factors but also stimulating transcriptional elongation itself.

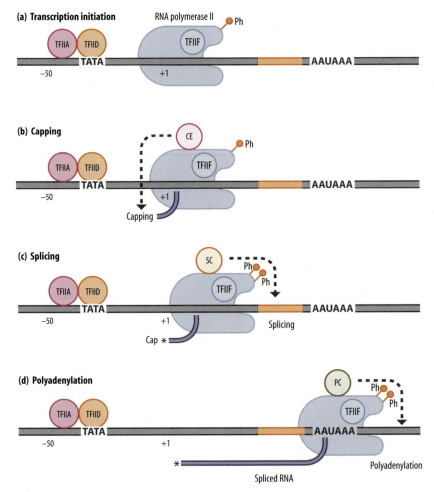

(a) Transcription initiation

(b) Capping

(c) Splicing

(d) Polyadenylation

Figure 6.20
As the RNA polymerase II complex transcribes the gene (a), it becomes progressively associated with the proteins that mediate capping (b; CE, capping enzymes), splicing (c; SC, splicing complex), or polyadenylation (d; PC, polyadenylation complex). This allows the RNA to be capped, spliced, and polyadenylated as it is being transcribed. Note the progressive phosphorylation (Ph; orange circles) of the RNA polymerase in the transcribing complex (see also Figure 4.24), which is essential for the recruitment of the capping, splicing, and polyadenylation factors.

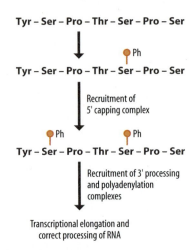

Figure 6.21
The C-terminal domain of RNA polymerase II contains multiple copies of the sequence Tyr-Ser-Pro-Thr-Ser-Pro-Ser. Phosphorylation of this sequence on serine 5 is essential for recruitment of the 5' capping complex, whereas phosphorylation on serine 2 is essential for recruitment of the 3' processing and polyadenylation complexes. Phosphorylation (Ph) is indicated by the orange circles.

Another aspect of this linkage between transcriptional elongation and post-transcriptional processes involves the effect of histone modifications which, as described in Chapter 3 (Section 3.3), regulate chromatin structure. The addition of three methyl residues to the arginine at position four in histone H3 has been shown to be associated with a more open chromatin structure. Interestingly, however, trimethylation of this arginine residue has also been shown to enhance the recruitment of the FACT protein which is involved in transcriptional elongation (see Section 4.2) and of the U2 snRNP (see Section 6.3), so enhancing both transcriptional elongation and splicing of the resulting transcript (**Figure 6.22**).

Post-transcriptional processes can interact with one another

As well as these interactions between factors involved in post-transcriptional processes and the transcriptional complex, there is also evidence that the factors involved in different post-transcriptional processes interact with one another. For example, the CBC (see Section 6.1) can interact with the spliceosome (Section 6.3), linking the processes of capping and splicing. Similarly, the U2 snRNP (Section 6.3) has been shown to interact with the

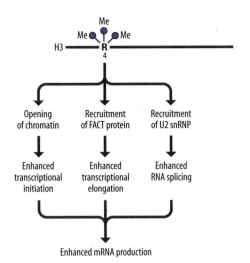

Figure 6.22
Trimethylation (blue circles) of the arginine (R) at position 4 in histone H3 not only results in a more open chromatin structure but also stimulates transcriptional elongation and enhances RNA splicing.

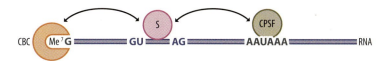

CPSF polyadenylation factor (Section 6.2), thereby coupling splicing with polyadenylation. These interactions will obviously ensure that all the modifications of the RNA transcript which we have discussed are closely linked with one another and with the process of transcription itself (**Figure 6.23**).

The transcription complex, as it moves down the gene, therefore attracts the factors which are needed to post-transcriptionally modify the RNA and this occurs while the RNA is being transcribed, paralleling the translation of the RNA while it is being transcribed, which occurs in prokaryotes. Once the coupled processes of transcription, capping, splicing, and polyadenylation have taken place, the RNA is fully mature and ready to be transported to the cytoplasm.

6.5 RNA TRANSPORT

RNA transport is coupled to other post-transcriptional processes

Although the processes of capping, polyadenylation, and RNA splicing take place in the nucleus, the process of translation, whereby the mature mRNA is converted into protein, takes place in the cytoplasm. The processed RNA must therefore be transported through the nuclear membrane from the nucleus to the cytoplasm before it can be translated into protein.

A number of proteins which associate with the RNA in the nucleus and accompany it into the cytoplasm have been identified and are believed to play a critical role in mediating the transport of the RNA from the nucleus to the cytoplasm which occurs via pores in the nuclear membrane. In the insect *Chironomus tentans* this process can be directly visualized in the electron microscope with a ribonucleoprotein particle consisting of multiple proteins associated with mRNA being observed passing through the nuclear pore into the cytoplasm (**Figure 6.24**).

As described in Chapter 4 (Section 4.1), genes are transcribed in specific regions of the nucleus which are frequently located adjacent to nuclear pores, so facilitating the ultimate transport of the RNA to the cytoplasm once it is fully processed. The actual transport of the fully processed mRNA though the nuclear pore involves it becoming associated with the **RNA exporter complex** (REC) which consists of two proteins: TAP (also known as NXF1) and p15 (also known as NXT1). Most interestingly, the REC complex appears to be recruited to the mRNA via other factors which have already bound to it. For example, REC can associate with the SR proteins, which, as described in Section 6.3, bind to **exon splicing enhancers** and are involved in RNA splicing.

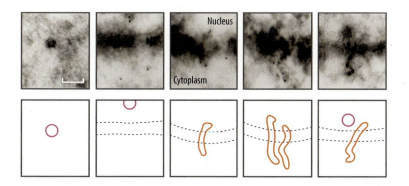

This association of REC and SR proteins is regulated by phosphorylation of the SR proteins. SR proteins initially bind to the RNA in a phosphorylated form which cannot recruit REC. Following splicing of the mRNA, however, SR proteins are dephosphorylated and therefore can bind REC, allowing transport to the cytoplasm to occur. This system effectively couples transport of the mRNA to its splicing, ensuring that intron-containing RNAs are not transported (**Figure 6.25**).

This linkage between splicing and transport is supplemented in higher eukaryotes by a linkage between capping (see Section 6.1) and mRNA transport. A complex of proteins involved in transport, known as TREX, is recruited to the mRNA by binding to the CBC located at the 5′ end of the mRNA. This results in TREX being located at the 5′ end of the mRNA. In turn, this leads to the 5′ end of the RNA passing through the nuclear pore first (**Figure 6.26**), so allowing the rapid initiation of mRNA translation into protein.

Although some proteins involved in mRNA transport are likely to be involved in the transport of all mRNAs, there are also transport proteins which are only required for the transport of certain mRNAs. For example, studies in yeast showed 1000 distinct mRNAs associated with the Ycal export protein while 1150 mRNAs associated with the Mex67 export protein. Only 349 of these mRNAs overlap and can associate with both proteins. Moreover, the proteins encoded by Mex67-associated mRNAs are distinct from those

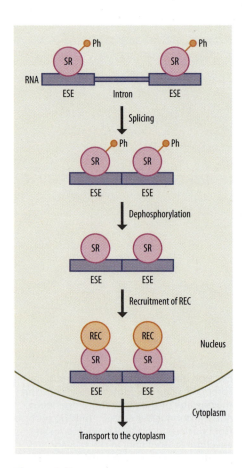

Figure 6.25
Dephosphorylation of SR proteins bound to exon-specific enhancers (ESEs) occurs following successful splicing of the RNA. The REC complex which catalyzes transport of the mRNA to the cytoplasm can only be recruited when the SR proteins have been dephosphorylated, so linking transport of the mRNA to its successful splicing. Phosphorylation is indicated by the orange circles.

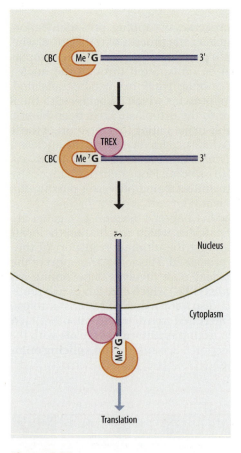

Figure 6.26
The TREX complex is recruited to the 5′ end of the mRNA by the CBC. The location of TREX at the 5′ end of the mRNA ensures that this end is the first to pass into the cytoplasm, allowing rapid initiation of mRNA translation.

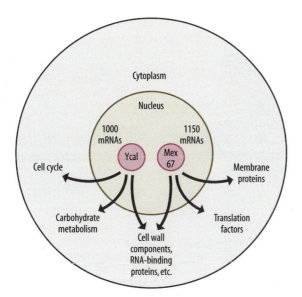

Figure 6.27
In yeast the export proteins Ycal and Mex67 are involved in the transport of distinct sets of mRNAs encoding proteins with distinct biological functions.

whose mRNAs associate specifically with Ycal. For example, Ycal-associated mRNAs encode proteins involved in the cell cycle or carbohydrate metabolism while mRNAs encoding translation factors or membrane proteins associate specifically with Mex67 (**Figure 6.27**). Hence, the mRNAs encoding functionally linked proteins can be exported together, facilitating their translation in parallel and the subsequent functional association of their corresponding proteins.

6.6 TRANSLATION

Translation of the mRNA takes place on cytoplasmic ribosomes

The process of translating the mRNA into protein takes place on defined cytoplasmic organelles known as ribosomes, which in eukaryotes consist of a large subunit of 60S in size and a small subunit of 40S. Each of these subunits contains both specific RNAs and proteins. The large subunit consists of 28, 5.8, and 5S ribosomal RNAs associated with 49 different proteins while the small subunit consists of an 18S ribosomal RNA and 33 associated proteins (**Figure 6.28**).

Although prokaryotic ribosomes are slightly smaller (70S in size), they also have two subunits (50 and 30S), each with their own specific ribosomal RNA (23 and 16S respectively). Moreover, they have a fundamentally similar

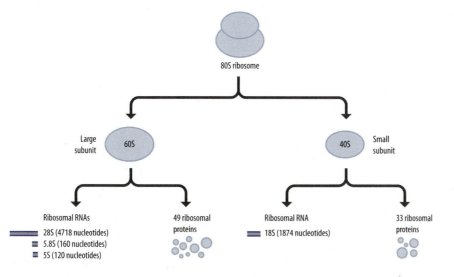

Figure 6.28
The 80S ribosome consists of 60 and 40S subunits, each with specific ribosomal RNAs and proteins.

(a)

(b)

(c)

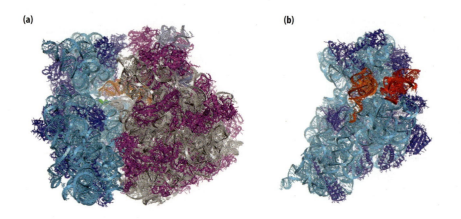

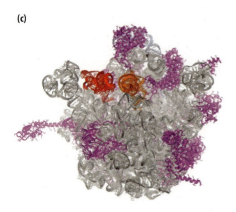

structure which has recently been characterized in atomic detail by X-ray crystallography (**Figure 6.29**).

Translational initiation involves initiation factors binding to the cap

In eukaryotes, translation is initiated by the binding of the **eukaryotic initiation factor** (**eIF**), eIF4E, to the cap structure at the 5′ end of the mRNA, replacing the CBC (see Section 6.1) (**Figure 6.30a**). Subsequently, other eIF

Figure 6.29
All-atom structure of the 70S ribosome from the bacterium *Thermus thermophilus* complexed with mRNA (green) and with tRNAs in the P site (orange) and the E site (red) (see Figure 6.32 for schematic diagram of these sites and their function). (a) Structure of the intact ribosome; (b) 30S small ribosomal subunit; (c) 50S large ribosomal subunit. The 16S rRNA of the small subunit is shown in cyan and its proteins are shown in blue. The 23S rRNA of the large subunit is shown in gray and its proteins are shown in purple. Courtesy of Carl Gorringe & Harry Noller, University of California.

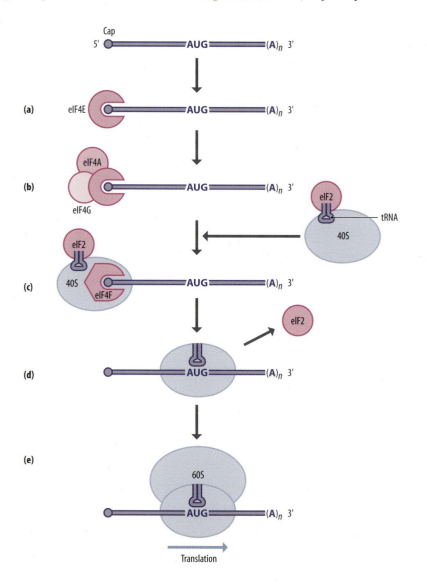

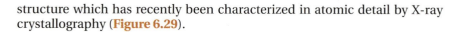

Figure 6.30
Mechanism of translational initiation. Initial binding of eIF4E to the cap structure (a) is followed by the binding of eIF4G and eIF4A (b). The complex of eIF4A, E, and G is known as eIF4F. The eIF4F complex is then recognized by a complex consisting of the 40S ribosomal subunit, the initiator tRNA and eIF2 (c). The 40S subunit then migrates along the RNA until it reaches the initiator AUG codon (d). This is followed by release of eIF2 and binding of the 60S large ribosomal subunit. Note that for clarity eIF4A, E, and G are shown as a single eIF4F complex in (c).

factors, such as eIF4G and eIF4A/B, bind to the 5′ end of the RNA (**Figure 6.30b**). Together, eIF4A, eIF4E, and eIF4G form a complex which is known as eIF4F. The binding of the various components of eIF4F is followed by the binding of a complex consisting of the small 40S ribosomal subunit, a transfer RNA (tRNA) molecule which carries the amino acid methionine, and the initiation factor eIF2 (**Figure 6.30c**).

Subsequently as described in Section 6.1, the 40S subunit migrates down the mRNA until it finds an AUG initiation sequence in the mRNA which is located in the appropriate context for the initiation of translation (**Figure 6.30d**). At this point, eIF2 is released from the complex and the 60S large ribosomal subunit binds to the small subunit. This forms the complete 80S ribosome which is able to initiate translation (**Figure 6.30e**). Several other factors including eIF5, eIF5b, and eIF6 play an essential role in the joining of the large and small ribosomal subunits.

Translational elongation involves base-pairing of triplet codons in the mRNA with tRNA anticodons

A key role in the process of translation is played by the tRNA molecules. These RNAs are unique in that they can bind an amino acid at their 3′ end (**Figure 6.31**). They therefore function to bring the amino acids to the ribosome for incorporation into protein. The tRNAs are RNA molecules from 74 to 95 bases in length which can fold into a highly defined secondary structure (see Figure 6.31). In this structure, one specific loop contains the so-called **anticodon**. Each tRNA molecule which binds a different amino acid has a distinct anticodon of three bases that can pair with the complementary three-base codon within the mRNA. For example, the **initiator tRNA** described above contains the anticodon sequence CAU and therefore binds to the initiator AUG codon. This initiator tRNA carries a methionine amino acid linked to it and the chain of the protein molecule thus begins with a methionine residue (**Figure 6.32a**).

Subsequently, a further tRNA molecule bearing the appropriate anticodon is recruited to bind to the next three-base sequence in the mRNA immediately downstream of the initiating AUG sequence. This binding site for tRNA is known as the A (amino acid) site of the ribosome whereas at this stage the initiator tRNA occupies the adjacent P (peptide) site of the ribosome. As each particular tRNA containing a specific anticodon also binds to a specific amino acid, this tRNA will deliver a particular amino acid to the mRNA-bound ribosome with the nature of the amino acid being specified by the nature of the three-base codon immediately following the AUG (**Figure 6.32b**).

This second tRNA is recruited to the ribosome by a specific eukaryotic **elongation factor** (eEF) eEF1. The structure of the ribosome allows two specific sites (the A and P sites) within it to be occupied by tRNA molecules which are bound to adjacent three base sequences in the mRNA. Once this has occurred, an enzymatic activity associated with the large ribosomal subunit forms a peptide bond between the two amino acids which are bound to the two adjacent tRNA molecules. In this manner, the growth of the peptide chain of the protein molecule begins.

Subsequently, another elongation factor eEF2 catalyzes the move of the ribosome three more bases down the mRNA molecule (**Figure 6.32c**). This moves the tRNA carrying the nascent peptide chain to the P site. At the same time, the initiator tRNA moves into the E (exit site) on the ribosome and it is then released from the mRNA. The ribosome structure illustrated in Figure 6.29 shows tRNAs occupying the P and E sites.

These movements allow a further tRNA molecule to be recruited to the ribosome-bound mRNA by eEF1 and occupy the now vacant A site (**Figure 6.32d**). This tRNA contains an anticodon which can bind to the next three-base codon in the mRNA sequence and has a specific amino acid bound to it. Once again, a peptide bond is formed between the two-amino acid chain and this third amino acid with the resulting three-amino acid peptide being transferred to the new tRNA.

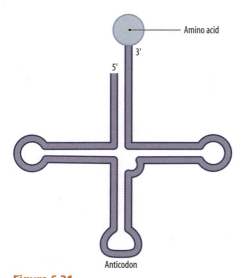

Figure 6.31
Cloverleaf structure of tRNA with stem structures containing paired bases and loops containing unpaired bases. Note the unpaired bases in the anticodon which will bind to the complementary bases in the codon of the mRNA, thereby delivering the appropriate amino acid, which is bound to the 3′ end of the tRNA, to the ribosome.

In this manner, the amino acid chain of the protein is gradually built up (see Figure 6.32). As each tRNA with a particular anticodon binds a specific amino acid, amino acids are added to the chain in accordance with the anticodon of the tRNA which is recruited. In turn this is determined by the

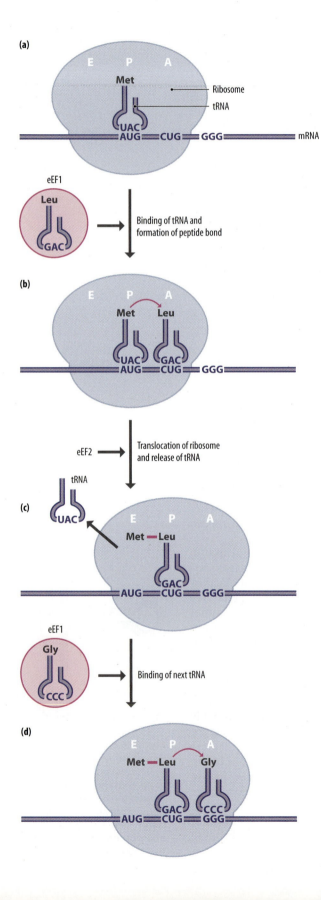

Figure 6.32
Translation elongation. Following the arrival of the ribosome at the initiator AUG codon and binding of the initiator tRNA bearing a methionine amino acid to the AUG codon in the P site of the ribosome (a; see also Figure 6.30), a second tRNA is recruited in association with the elongation factor eEF1 and binds to the next codon in the A site of the ribosome via its corresponding anticodon (b). Note that this codon may have any sequence depending on the next amino acid to be incorporated into the protein, a CUG codon which recruits tRNA with the anticodon sequence CAG and bearing a leucine amino acid is shown as an example. Subsequently, a peptide bond is formed between the methionine and the next amino acid (leucine in this case) with translocation of the methionine to bond with the leucine residue. The elongation factor eEF2 then initiates translocation of the ribosome three bases along the mRNA and the first tRNA moves to the E site of the ribosome and is released (c). The cycle then repeats itself with a new tRNA being recruited to the next three bases (d) with consequent peptide-bond formation producing a three-amino acid peptide.

TABLE 6.1
THE GENETIC CODE

BASE 1	BASE 2			
	A	C	G	U
A	AAA Lys AAC Asn AAG Lys AAU Asn	ACA Thr ACC Thr ACG Thr ACU Thr	AGA Arg AGC Ser AGG Arg AGU Ser	AUA Ile AUC Ile AUG Met AUU Ile
C	CAA Gln CAC His CAG Gln CAU His	CCA Pro CCC Pro CCG Pro CCU Pro	CGA Arg CGC Arg CGC Arg CGU Arg	CUA Leu CUC Leu CUG Leu CUU Leu
G	GAA Glu GAC Asp GAG Glu GAU Asp	GCA Ala GCC Ala GCG Ala GCU Ala	GGA Gly GGC Gly GGG Gly GGU Gly	GUA Val GUC Val GUG Val GUU Val
U	UAA Stop UAC Tyr UAG Stop UAU Tyr	UCA Ser UCC Ser UCG Ser UCU Ser	UGA Stop UGC Cys UGG Trp UGU Cys	UUA Leu UUC Phe UUG Leu UUU Phe

codon present in the mRNA and so the information in the mRNA is gradually translated into protein in accordance with the genetic code (**Table 6.1**).

As can be seen from Table 6.1, many amino acids are encoded by multiple codons. In some cases, this is because the amino acid concerned can bind two or more tRNAs which have different anticodons and therefore bind to different codons in the mRNA.

In addition, however, in some cases the same tRNA can bind to different codons in the mRNA, so allowing these different codons to direct the insertion of the same amino acid. This so-called **wobble effect** normally involves unusual pairing between the third base of the codon in the mRNA and the first base of the anticodon in the tRNA. Because of the unusual structure of the tRNA at this point, a U residue in the wobble position of the codon can form either a normal base pair with an A residue or an unusual base pair with a G residue in the tRNA (**Figure 6.33a** and **b**). Similarly, inosine, which is a modified purine base that is found in tRNA, can pair with A or G residues in the wobble position of the mRNA (**Figure 6.33c** and **d**).

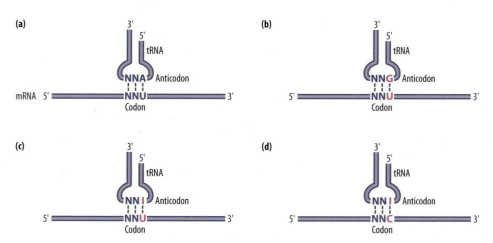

Figure 6.33
Binding of the codon in the mRNA to the anticodon in the tRNA can involve a normal base pair between a U residue at the third position of the codon and an A residue at the first position of the anticodon (a) or an unusual U–G base pair (b). Similarly, an I (inosine) residue at the first position of the anticodon can pair with either a U residue (c) or a C residue (d) at the third position of the codon. These mechanisms allow one tRNA to recognize more than one codon in the mRNA, allowing these different codons to direct incorporation of the same amino acid.

Translational termination occurs at specific stop codons

Ultimately, the moving ribosome encounters one of the three stop codons (UAA, UAG, or UGA) which are not recognized by a specific tRNA and a **release factor** (eRF1) catalyzes the release of the completed polypeptide from the ribosome. Interestingly, although it is a protein, this release factor resembles a tRNA molecule in its shape. It can therefore bind to the A site in the ribosome when this site contains a stop codon. As there is now no amino acid in the A site, a water molecule is added to the end of the peptide chain which is bound at the P site. This results in its release from the ribosome and the termination of translation (**Figure 6.34**).

Interestingly, binding of release factors to the ribosome can also occur even in the absence of a stop codon in the A site. This is observed when the wrong tRNA is bound in error by the mRNA resulting in a codon/anticodon mismatch in the A site. This would evidently result in the insertion of an incorrect amino acid into the protein. Binding of release factors under these conditions therefore provides a proofreading mechanism prematurely terminating the translation of a protein which would contain an incorrect amino acid.

Although we have described the movement of a single ribosome down the mRNA, multiple ribosomes will be moving along the mRNA at the same time forming a structure known as a **polyribosome**. Interestingly, a polyA-tail-binding protein, which is bound to the polyA tail at the 3′ end of the mRNA, can interact with the eIF4G component of the translation-initiation complex located at the 5′ end of the mRNA. This means that the polyribosome-bearing mRNA can assume a circular configuration. In turn, this

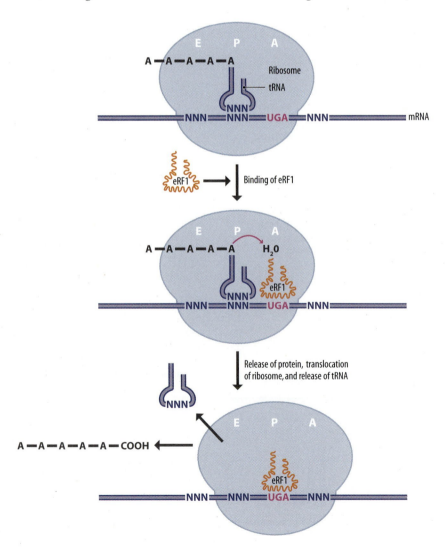

Figure 6.34
The UAA, UAG, or UGA stop codons produce translational termination by directing the binding of the eRF1 release factor rather than a tRNA.

(a)

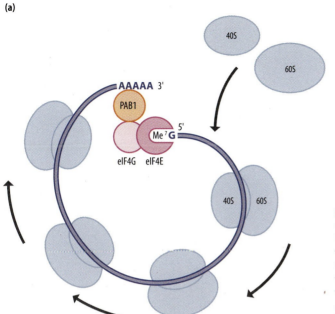

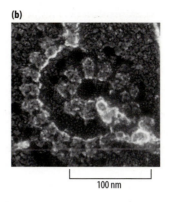

100 nm

Figure 6.35
(a) The PAB1 protein binds to the polyA tail of the mRNA and interacts with the eIF4G factor bound at the 5′ end. This circularizes the mRNA and facilitates reinitiation of protein synthesis by ribosomes which have dissociated at the 3′ end of the mRNA. Note that multiple ribosomes are translating the mRNA at the same time forming a polyribosome. (b) Electron micrograph showing an actual polyribosome from a eukaryotic cell. Courtesy of John Heuser, Washington University School of Medicine.

allows ribosomes which complete translation and terminate at the 3′ end of the mRNA to reinitiate a new round of translation at the 5′ end of the mRNA (Figure 6.35).

The production of the final protein molecule completes the process which began with the transcription of the gene and results in the genetic information encoded in the DNA and subsequently in the RNA being converted into a functional protein molecule.

6.7 RNA DEGRADATION

RNA degradation occurs in both the nucleus and the cytoplasm

As discussed in previous sections, mechanisms exist to tightly couple transcription and the various post-transcriptional processes so that a correctly processed mRNA is transported to the cytoplasm. RNA molecules which fail at any stage in this process, for example by being incorrectly spliced, will be degraded in the nucleus. This process is carried out by a multi-protein complex known as the **exosome** which contains a 3′ to 5′ exonuclease that digests the incorrectly spliced RNA beginning at its 3′ end. Such intranuclear degradation is also the fate of introns, which are removed from the transcript in the normal process of splicing (Figure 6.36).

However, even where an mRNA is successfully transported to the cytoplasm, its ultimate fate is to be broken down and degraded. In the case of functional mRNAs, this occurs after the mRNA has been translated one or more times by the ribosome to produce functional protein. However, a second RNA degradation mechanism also exists in the cytoplasm which is known as **nonsense-mediated RNA decay**. This recognizes and degrades mRNAs that are non-functional because they have a premature stop codon, which would prevent them being translated to produce a functional protein. These two processes will be discussed in turn (see Figure 6.36).

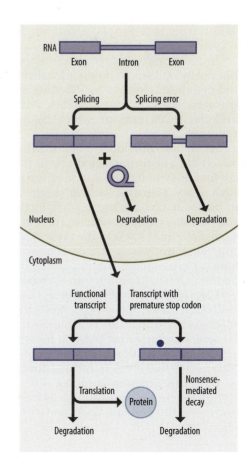

Figure 6.36
Both introns removed from the mRNA in the normal process of splicing and incorrectly spliced RNAs are degraded in the nucleus. Following transport of correctly spliced RNAs to the cytoplasm, functional transcripts are translated into protein and subsequently degraded. Non-functional transcripts with premature stop codons (circle) are rapidly degraded in a process known as nonsense-mediated decay.

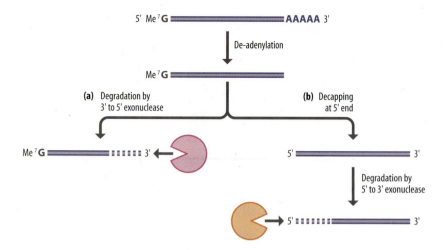

Figure 6.37
De-adenylation of an mRNA can be followed by degradation at its exposed 3′ end by a 3′ to 5′ exonuclease (a) or by subsequent decapping at the 5′ end followed by degradation by a 5′ to 3′ exonuclease (b).

RNA degradation in the cytoplasm involves prior de-adenylation and decapping of the mRNA

As noted in previous sections, functional mRNAs which are transported to the cytoplasm will have a **5′ cap** and 3′ polyA tail. The predominant process for degrading these mRNAs begins with the removal of the polyA tail in a process known as de-adenylation. This process renders the mRNA subject to attack by the 3′ to 5′ exonuclease of the exosome complex (see above) which degrades it from its 3′ end. The exosome complex therefore plays a key role both in the degradation of non-functional RNAs in the nucleus and in the turnover of functional mRNAs in the cytoplasm.

Moreover, the removal of the polyA tail leads to de-adenylation-dependent decapping of the RNA. This removes the 5′ cap of the mRNA and not only prevents its further translation but also renders it susceptible to degradation by 5′ to 3′ exonucleases, which degrade the mRNA from its 5′ end (**Figure 6.37**).

Interestingly, this degradation of the mRNA does not take place throughout the cytoplasm. Rather it is localized to small granular structures within the cytoplasm, known as P-bodies. These P-bodies contain the proteins involved in decapping the mRNA, such as the DCP1 activator of decapping and the DCP2 enzyme which catalyzes the actual decapping process (**Figure 6.38**).

As well as providing a site for the degradation of functional mRNAs which have been translated one or more times and have completed their useful life, P-bodies are also involved in the process of nonsense-mediated RNA decay. This process involves the same decapping and de-adenylation enzymes involved in the degradation of functional mRNAs. It also involves the exosome complex which as described above also catalyzes the degradation of aberrant RNAs in the nucleus as well as the cytoplasmic turnover of functional mRNAs. In this case, however, it targets non-functional mRNAs which have entered the cytoplasm but which contain a premature stop codon that would produce an aberrant truncated protein.

Clearly, the critical question is how the cell recognizes these abnormal mRNAs in the cytoplasm and directs them to the RNA degradation pathway. It seems likely that this is achieved by recognizing that the stop codon is located at an incorrect place in the mRNA. In functional mRNAs, the stop codon for translation is usually located in the last exon of the transcript. This means that it will be located downstream of a protein complex, known as the **exon junction complex** (EJC). This complex binds to each exon during splicing at a position 20–24 bases upstream of the junction between two exons and remains associated with it during transport to the cytoplasm.

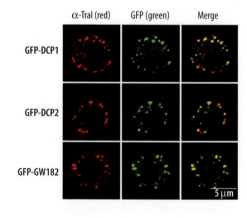

α-Tral (red)   GFP (green)   Merge

GFP-DCP1

GFP-DCP2

GFP-GW182

5 μm

Figure 6.38
Confocal microscopy of *D. melanogaster* cells showing localization of specific proteins to P bodies. Cells were transfected with DNA constructs expressing green fluorescent protein (GFP) linked to the decapping activator DCP1, the decapping enzyme DCP2, or the P body marker GW182. The localization of each of these proteins is visualized as green fluorescence (central panels), while the localization of the endogenous P body protein Tra-1 (as determined by staining with a specific antibody) is visualized as red fluorescence (left-hand panels). Note the similar specific localization of all the proteins which is confirmed with extensive overlap being observed when red and green fluorescence patterns are merged (right-hand panels). Courtesy of Elisa Izaurralde, Max Planck Institute for Developmental Biology.

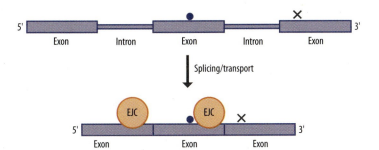

Figure 6.39
During the process of splicing the exon junction complex (EJC) binds to each exon just upstream of the splice junction. Normally the stop codon for translation (X) is located in the last exon which has not undergone a splicing event at its 3′ end. The stop codon is therefore downstream of the last EJC. In contrast, an abnormally located stop codon (blue circle) can be located upstream of an EJC.

A normally located stop codon would be located downstream of the last EJC complex, whereas an abnormally located codon would be found upstream of it (**Figure 6.39**).

During the process of translation, the ribosome normally displaces successive EJCs and then reaches the correctly located stop codon where it terminates translation. On an RNA with an incorrectly located stop codon, the ribosome will terminate translation prematurely, leaving an EJC still bound to the aberrant mRNA. This is thought to result in the recruitment of the surveillance complex (SURF) which in turn recruits the enzymes involved in de-adenylation, decapping, and RNA degradation (**Figure 6.40**).

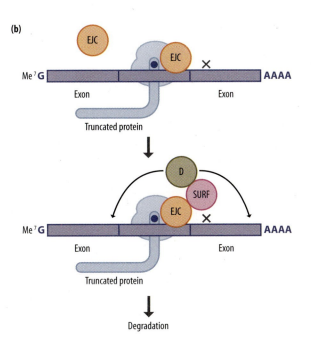

Figure 6.40
During the process of translation of a functional mRNA, the ribosome displaces the EJCs as it moves down the mRNA prior to terminating at the normal stop codon (X) in the last exon (a). In contrast, where there is an abnormally located stop codon (blue circle) the ribosome will terminate translation at this stop codon, before it has displaced all the EJCs. The remaining EJC will then recruit the surveillance complex (SURF) which in turn recruits enzymes that degrade the mRNA (D) (b).

CONCLUSIONS

This chapter has described the processes which modify the RNA transcript at its 5′ and 3′ ends, remove its intervening sequences, transport it to the cytoplasm and eventually translate it into protein. Clearly, since all these events are necessary for protein production, they represent targets for gene regulatory processes. Similarly, the process of RNA degradation will not only ensure that non-functional mRNAs are rapidly degraded but will also control the amount of protein produced from a functional mRNA before it is degraded. The regulation of gene expression at these various post-transcriptional stages is discussed in the next chapter.

KEY CONCEPTS

- The process of transcription of the DNA into RNA is complemented by a number of post-transcriptional events which produce a functional mRNA able to be transported to the cytoplasm and translated into protein.

- Shortly after transcription is initiated, the nascent RNA transcript is modified by addition of a cap to its 5′ end.

- The RNA transcript is cleaved at its 3′ end and a run of A residues is added in a process known as polyadenylation.

- Intervening sequences or introns are removed from the RNA by the process of RNA splicing.

- The processes of transcription, capping, splicing, and polyadenylation are tightly coupled within the nucleus.

- The fully spliced and correctly modified mRNA is recognized by export proteins which transport the mRNA to the cytoplasm.

- The mRNA is translated into protein in the cytoplasm by the ribosome.

- The process of nonsense-mediated decay rapidly degrades non-functional mRNA in the cytoplasm.

- After one or more rounds of translation by the ribosome, functional mRNAs are de-adenylated and decapped which exposes them to degradation by nucleases.

FURTHER READING

6.1   Capping

Fechter P & Brownlee GG (2005) Recognition of mRNA cap structures by viral and cellular proteins. *J. Gen. Virol.* 86, 1239–1249.

Hernandez G (2009) On the origin of the cap-dependent initiation of translation in eukaryotes. *Trends Biochem. Sci.* 34, 166–175.

Kozak M (1986) Point mutations define a sequence flanking the AUG initiator codon that modulates translation by eukaryotic ribosomes. *Cell* 44, 283–292.

6.2   Polyadenylation

Barabino SML & Keller W (1999) Last but not least: regulated Poly(A) tail formation. *Cell* 99, 9–11.

Marzluff WF, Wagner EJ & Duronio RJ (2008) Metabolism and regulation of canonical histone mRNAs: life without a poly(A) tail. *Nat. Rev. Genet.* 9, 843–854.

6.3   RNA splicing

Azubel M, Wolf SG, Sperling J & Sperling R (2004) Three-dimensional structure of the native spliceosome by Cryo-electron microscopy. *Mol. Cell* 15, 833–839.

McManus CJ & Graveley BR (2008) Getting the message out. *Mol. Cell* 31, 4–6.

Query CC (2009) Spliceosome subunit revealed. *Nature* 458, 418–419.

Rowley JD & Blumenthal T (2008) The cart before the horse. *Science* 321, 1302–1304.

Sharp PA (2005) The discovery of split genes and RNA splicing. *Trends Biochem. Sci.* 30, 279–281.

Stark H & Luhrmann R (2006) Cryo-electron microscopy of spliceosomal components. *Annu. Rev. Biophys. Biomol. Struct.* 35, 435–457.

Wahl MC, Will CL & Luhrmann R (2009) The spliceosome: design principles of a dynamic RNP machine. *Cell* 136, 701–718.

6.4   Coupling of transcription and RNA processing within the nucleus

Moore MJ & Proudfoot NJ (2009) Pre-mRNA processing reaches back to transcription and ahead to translation. *Cell* 136, 688–700.

Orphanides G & Reinberg D (2002) A unified theory of gene expression. *Cell* 108, 439–451.

6.5   RNA transport

Kohler A & Hurt E (2007) Exporting RNA from the nucleus to the cytoplasm. *Nat. Rev. Mol. Cell Biol.* 8, 761–773.

6.6   Translation

Dinman JD (2009) The eukaryotic ribosome: current status and challenges. *J. Biol. Chem.* 284, 11761–11765.

Fredrick K & Ibba M (2009) Errors rectified in retrospect. *Nature* 457, 157–158.

Korostelev A & Noller HF (2007) The ribosome in focus: new structures bring new insights. *Trends Biochem. Sci.* 32, 434–441.

6.7   RNA degradation

Eulalio A, Behm-Ansmant I & Izaurralde E (2007) P bodies: at the crossroads of post-transcriptional pathways. *Nat. Rev. Mol. Cell Biol.* 8, 9–22.

Franks TM & Lykke-Andersen J (2008) The control of mRNA decapping and p-body formation. *Mol. Cell* 32, 605–615.

Garneau NL, Wilusz J & Wilusz CJ (2007) The highways and byways of mRNA decay. *Nat. Rev. Mol. Cell Biol.* 8, 113–126.

Goldstrohm AC & Wickens M (2008) Multifunctional deadenylase complexes diversify mRNA control. *Nat. Rev. Mol. Cell Biol.* 9, 337–344.

Houseley J & Tollervey D (2009) The many pathways of RNA degradation. *Cell* 136, 763–776.

Schmid M & Jensen TH (2008) The exosome: a multipurpose RNA-decay machine. *Trends Biochem. Sci.* 33, 501–510.

Post-transcriptional Regulation

7

INTRODUCTION

The evidence discussed in Chapter 1 (Section 1.4) indicates that in mammals at least the primary control of gene expression lies at the level of transcription. However, a number of cases exist where changes in the rate of synthesis of a particular protein occur without a change in the transcription rate of the corresponding gene or where post-transcriptional controls operate as a significant supplement to transcriptional control. This indicates that in these cases regulatory processes are operating at the level of one or more of the post-transcriptional events described in Chapter 6. Indeed, in some lower organisms post-transcriptional regulation may constitute the predominant form of regulation of gene expression.

In the sea urchin for example, the nuclear RNA contains many more different RNA species than are found in the cytoplasmic mRNA. A large proportion of the transcribed genes give rise to RNA products that are not transported to the cytoplasm and do not function as a mRNA. Interestingly, however, this process is regulated differently in different tissues; for instance, an RNA species which is confined to the nucleus (where it will ultimately be degraded) in one tissue being transported to the cytoplasm and functioning as a mRNA in another tissue. Indeed, up to 80% of the cytoplasmic mRNAs found in the embryonic **blastula** are absent from the cytoplasmic RNA of adult tissues such as the intestine but are found in the nuclear RNA of such tissues.

Although such post-transcriptional regulation in mammals does not appear to be as generalized, as in the sea urchin, some cases exist where changes in cytoplasmic mRNA levels occur without alterations in the rate of gene transcription. Post-transcriptional regulation of this type may be more important in controlling variations in the level of mRNA species expressed in all tissues than in the regulation of mRNA species that are expressed in only one or a few tissues. For example, in the experiments of Darnell and colleagues, which demonstrated the importance of transcriptional control in the regulation of liver-specific mRNAs (see Section 1.4), tissue-specific differences in the levels of the mRNAs encoding actin and tubulin (which are expressed in all cell types) were observed in the absence of differences in transcription rates. Clearly, these and other cases where mRNA levels alter in the absence of changes in transcription rates indicate the existence of post-transcriptional control processes and require an understanding of their mechanisms.

In principle, such post-transcriptional regulation could operate at any of the many stages between gene transcription and the translation of the corresponding mRNA in the cytoplasm which were described in Chapter 6. Indeed, the available evidence indicates that in different cases regulation can occur at any one of these levels. Each of these will now be discussed in turn.

7.1 ALTERNATIVE RNA SPLICING

RNA splicing can be regulated

The finding that the protein-coding regions of eukaryotic genes are split by intervening sequences (introns), which must be removed from the initial transcript by RNA splicing of the protein-coding exons (see Section 6.3), led to much speculation that this process might provide a major site of gene regulation. In theory, an RNA species transcribed in several tissues might be correctly spliced to yield functional RNA in one tissue and remain unspliced in another tissue. An unspliced RNA would either be degraded within the nucleus or, if transported to the cytoplasm, would be unable to produce a functional protein due to the interruption of the protein-coding regions (**Figure 7.1**).

Several cases of such **processing/discard decisions** have now been described in *Drosophila* and a similar pathway in which unspliced RNAs are specifically degraded within the nucleus has been characterized in yeast. Hence, in lower organisms there appear to be regulatory pathways in which a particular transcript is spliced to produce a functional mRNA in one situation while remaining unspliced and then being degraded within the nucleus in another situation.

Alternative splicing represents a major regulatory process which supplements transcriptional control

Although processing/discard decisions of the type described above are not widespread in mammals, numerous cases of **alternative RNA splicing** have been described both in mammals and other organisms. In this process, a single gene is transcribed in several different tissues, the transcripts from this gene being processed differentially to yield different functional mRNAs in the different tissues (**Figure 7.2**). In many cases, these RNAs are translated to yield different protein products.

It is noteworthy that this mechanism of gene regulation involves not only regulation of processing but also regulation of transcription. Thus, the primary transcript of an alternatively spliced gene is frequently transcribed in only a restricted range of cell types and not in many other cells. The cells which express the primary transcript then splice it in different ways to produce different mRNAs.

Cases of alternative RNA processing occur in the genes involved in a wide variety of different cellular processes, ranging from genes which regulate embryonic development or sex determination in *Drosophila* to those involved in muscular contraction or neuronal function in mammals. Indeed, recent studies of the whole human genome concluded that over 90% of human genes with multiple exons are alternatively spliced. A representative selection of such cases is given in **Table 7.1**.

For convenience, cases of alternative RNA processing can be divided into three groups: (a) situations where the 5′ end of the alternatively processed transcripts is different; (b) situations where the 3′ end of the alternatively processed transcripts is different; (c) situations where both the 5′ and 3′ ends of the alternatively processed transcripts are identical.

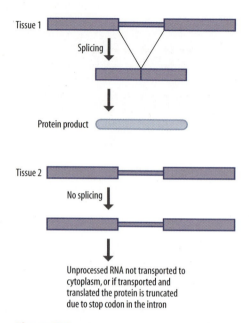

Figure 7.1
The absence of RNA splicing of a transcript in a particular tissue results in a lack of production of the corresponding protein.

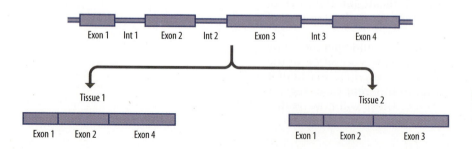

Figure 7.2
Alternative splicing of the same primary transcript in two different ways results in two different mRNA molecules. Int, intron.

TABLE 7.1
SOME EXAMPLES OF ALTERNATIVE SPLICING WHICH ARE REGULATED DEVELOPMENTALLY OR TISSUE-SPECIFICALLY

PROTEIN	SPECIES	NATURE OF TRANSCRIPTS WHICH UNDERGO ALTERNATIVE SPLICING	CELL TYPES CARRYING OUT ALTERNATIVE SPLICING
(a) Immune system			
Immunoglobulin heavy-chain IgD, IgE, IgG, IgM	Mouse	3′ end differs	B cells
Lyt-2	Mouse	Same transcript	T cells
(b) Enzymes			
Alcohol dehydrogenase	*Drosophila*	5′ end differs	Larva and adult
Aldolase A	Rat	5′ end differs	Muscle and liver
α-Amylase	Mouse	5′ end differs	Liver and salivary gland
(2′5′) oligo A synthetase	Human	3′ end differs	B cells and monocytes
(c) Muscle			
Myosin light chain	Rat, mouse, human, chicken	5′ end differs	Cardiac and smooth muscle
Myosin heavy chain	*Drosophila*	3′ end differs	Larval and adult muscle
Tropomyosin	Mouse, rat, human, *Drosophila*	Same transcript	Different muscle cell types
Troponin T	Rat, quail, chicken	Same transcript	Different muscle cell types
(d) Nerve cells			
Calcitonin/CGRP	Rat, human	3′ end differs	Thyroid C cells or neural tissue
Myelin basic protein	Mouse	Same transcript	Different glial cells
Neural cell adhesion molecule	Chicken	Same transcript	Neural development
Preprotachykinin	Bovine	Same transcript	Different neurons
(e) Others			
Fibronectin	Rat, human	Same transcript	Fibroblasts and hepatocytes
Early retinoic acid-induced gene 1	Mouse	Same transcript	Stages of embryonic cell differentiation
Thyroid hormone receptor	Rat	Same transcript	Different tissues

(a)   Situations where the 5′ end of the alternatively processed transcripts is different

In situations where the 5′ end of the transcripts is different, two alternative primary transcripts are produced by transcription from different promoter elements and these are then processed differentially. In several situations, differential splicing is controlled simply by the presence or absence of a particular exon in the primary transcript (**Figure 7.3a**). An example of this is the mouse α-amylase gene. In the salivary gland where transcription of the gene takes place from an upstream promoter, the exon adjacent to this promoter is included in the processed RNA and a downstream exon is omitted. In the liver where the transcripts are initiated 2.8 kb downstream and do not contain the upstream exon, the processed RNA includes the downstream exon (**Figure 7.4**).

(a)

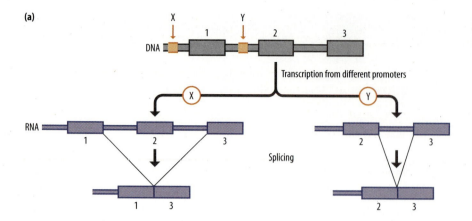

(b)

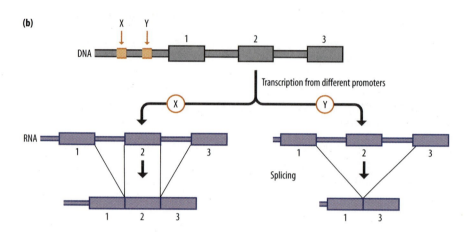

Figure 7.3
The use of different promoters (X and Y) to produce two different primary transcripts can produce an alternative splicing pattern simply because one of the alternatively spliced exons (exon 1) is only present in one of the two primary transcripts (a). However, it can also occur where both primary transcripts contain each of the alternatively spliced exons (b).

However, other cases involving the use of different promoters are more complex, with each of the alternative primary transcripts containing both the alternatively spliced exons (**Figure 7.3b**). A case of this type, the myosin light-chain gene, is illustrated in **Figure 7.5**.

In such cases it is possible that the different primary transcripts fold into different secondary structures which favor the different splicing events (**Figure 7.6**). Another mechanism, however, is suggested by the link between transcription and splicing, which was discussed in Chapter 6 (Section 6.4), where the transcribing polymerase and its associated factors can recruit factors involved in RNA splicing. It is possible therefore that the transcription complexes which assemble on the two different promoters have some differences in the proteins which associate with the RNA polymerase. In turn, this could result in the recruitment of different splicing complexes, leading to the observed difference in splicing which occurs (**Figure 7.7**).

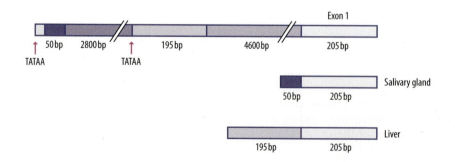

Figure 7.4
Alternative splicing at the 5′ end of α-amylase transcripts in the liver and salivary gland. The two alternative start sites for transcription are indicated (TATAA) together with the 5′ region of the mRNAs produced in each tissue.

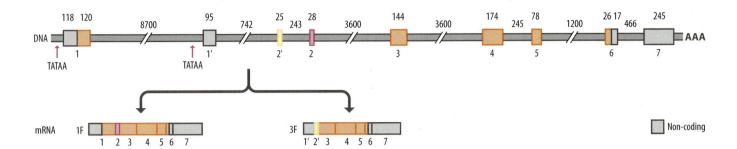

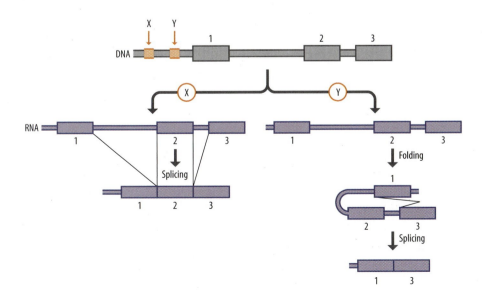

Regardless of whether either or both of these models is correct, it is clear that cases of alternative splicing arising from differences in the site of transcriptional initiation represent further examples of transcriptional regulation. Thus, the variation in RNA splicing is secondary to the primary choice involving the selection of different promoters in the different tissues. This is not so for the other categories of alternative processing event.

(b) Situations where the 3′ end of the alternatively processed transcripts is different

As well as cases where the 5′ end of the alternatively spliced transcripts is different, there are a number of cases where the 5′ end is the same but the 3′ end of the alternatively spliced transcripts is different. As described in Chapter 6 (Section 6.2), after the primary transcript has been produced it is rapidly cleaved and a polyA tail is added. In many genes the process of cleavage and polyadenylation occurs at a different position within the primary transcript in different tissues and the different transcripts are then differentially spliced.

The best-defined example of this process occurs in the genes encoding the immunoglobulin heavy chain of the antibody molecule and plays an important role in the regulation of the antibody response to infection. Early in the **immune response**, the antibody-producing B cell synthesizes membrane-bound immunoglobulin molecules, the interaction of which with antigen triggers proliferation of the B cell and results in the production of more antibody-synthesizing cells. The immunoglobulin produced by these cells is secreted, however, and can interact with antigen in tissue fluids, triggering the activation of other cells in the immune system.

The production of membrane-bound and secreted immunoglobin molecules is controlled by the alternative splicing of different RNA molecules differing in their 3′ ends (**Figure 7.8**). The longer of these two molecules

Figure 7.5
Alternative splicing of the myosin light-chain transcripts in different muscle cell types produces two mRNAs (1F and 3F) differing at their 5′ ends. The two alternative start sites of transcription used to produce each of the RNAs are indicated (TATAA), together with the intron/exon structure of the gene. Note that both primary RNA transcripts contain the alternatively spliced exons 2′ and 2, although only one of these is found in each of the alternatively spliced mRNAs.

Figure 7.6
Different primary RNA transcripts produced from different promoters (X and Y) may fold in different ways, resulting in different splicing patterns.

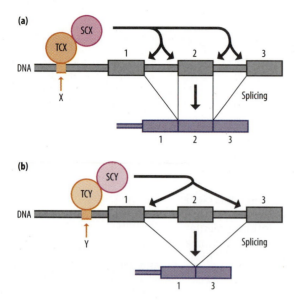

Figure 7.7
Different transcription complexes (TC) may assemble at different promoters (X and Y). In turn these may recruit different splicing complexes (SC), resulting in different splicing patterns of the transcripts from each promoter.

contains two exons encoding the portion of the protein that anchors it in the membrane. When this molecule is spliced both these two exons are included but a region encoding the last 20 amino acids of the secreted form is omitted. In the shorter RNA, the two transmembrane-domain-encoding exons are absent and the region specific to the secreted form is included in the final mRNA.

If the polyadenylation site used in the production of the shorter immunoglobulin RNA is artificially removed, thus preventing its use, the expected decrease in the production of secreted immunoglobulin is paralleled by a corresponding increase in the synthesis of the membrane-bound form of the protein (**Figure 7.9a**). This indicates that the choice of splicing pattern is controlled by which polyadenylation site is used; removal of the upstream site resulting in increased use of the downstream site and increased production of the mRNA encoding the membrane-bound form.

This switch in the polyadenylation site is dependent on an increase in concentration of one of the subunits of the polyadenylation factor CstF (see Section 6.2) which occurs during B-cell development. CstF binds preferentially to the polyA site of the mRNA encoding the membrane form compared to that of the secreted form. Early in B-cell development, when CstF levels are low, it binds to the polyA site for the membrane form and so this

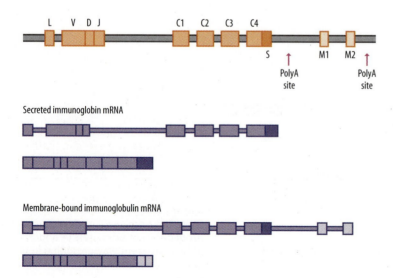

Figure 7.8
Alternative splicing of the immunoglobulin heavy-chain transcript at different stages of B-cell development. The two unspliced RNAs produced by use of the two alternative polyadenylation sites in the gene are shown, together with the spliced mRNAs produced from them.

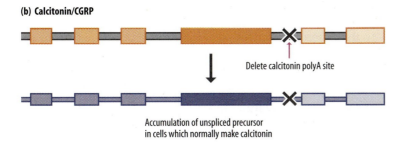

(a) Immunoglobulin heavy chain

Delete secreted polyA site

Membrane-bound form produced in cells
which normally make secreted form

(b) Calcitonin/CGRP

Delete calcitonin polyA site

Accumulation of unspliced precursor
in cells which normally make calcitonin

Figure 7.9
Effect of deleting the more upstream
of the two polyadenylation sites in
the immunoglobulin heavy-chain (a)
and calcitonin/*CGRP* genes (b) on the
production of the alternatively spliced
RNAs derived from each of these genes.

mRNA is produced. As the levels of CstF rise, binding and cleavage occur at
the polyA site for the secreted mRNA (**Figure 7.10**).

   This finding indicates that in at least some cases of this type the primary
regulatory event is that determining the site of cleavage and polyadenylation and that, as with cases of differential promoter usage,
alternative RNA splicing is regulated by differences in the transcript
produced in different tissues.

   Not all cases of alternative RNA splicing where the 3′ end of the alternatively spliced RNAs varies are of this type, however. This conclusion has
emerged from studies of the gene encoding the calcium regulatory protein,

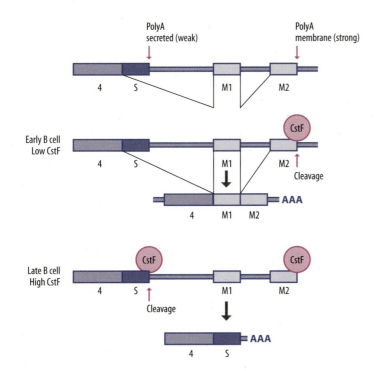

PolyA
secreted (weak)

PolyA
membrane (strong)

4   S   M1   M2

Early B cell
Low CstF

CstF

4   S   M1   M2   Cleavage

4   M1   M2   AAA

Late B cell
High CstF

CstF   CstF

4   S   M1   M2

Cleavage

4   S   AAA

Figure 7.10
Role of the polyadenylation factor CstF
in the regulated polyadenylation of the
immunoglobulin transcript. At low levels
of CstF, it binds preferentially to the
membrane polyA site. Following cleavage
at this site, splicing joins exons 4, M1 and
M2. At higher levels of CstF, it also binds at
the weaker secreted polyA site, leading to
loss of exons M1 and M2 by cleavage of the
mRNA. The rise in the level of CstF, which
occurs in B-cell development therefore
results in a switch from the membrane-
bound to the secreted form.

calcitonin. When the gene encoding the calcitonin protein (which is a small peptide of 32 amino acids) was isolated it was found that it also had the potential to produce an RNA encoding an entirely different peptide of 36 amino acids, which was named calcitonin-gene-related peptide (CGRP). Unlike calcitonin, which is produced in the thyroid gland, CGRP is produced in specific neurons in the brain and peripheral nervous system. These two peptides are produced by alternative splicing of two distinct transcripts differing in their 3′ ends (**Figure 7.11**).

This case differs from that of the immunoglobulin heavy chain, however, in that deletion of the polyadenylation site used in the shorter, calcitonin-encoding, RNA does not result in an increase in CGRP expression in cells normally expressing calcitonin. Instead, large unspliced transcripts utilizing the downstream (CGRP) polyadenylation site accumulate in these cells (**Figure 7.9b**). Although in CGRP-producing cells such transcripts would normally be spliced to yield CGRP mRNA, this does not occur in these experiments in cells normally producing calcitonin, and so these unspliced precursors accumulate. This suggests that in the calcitonin/*CGRP* gene the use of different polyadenylation sites is secondary to the difference in RNA splicing, suggesting the existence of tissue-specific splicing factors whose presence or absence in a specific tissue determines the pattern of calcitonin/CGRP RNA splicing.

(c)   Situations where both the 5′ and 3′ ends of the alternatively processed transcripts are identical

The existence of tissue-specific splicing factors which regulate alternative splicing is also indicated by the existence of a third group of cases where a transcript with identical 5′ and 3′ ends is spliced differently in different tissues and which therefore cannot be explained by differential usage of promoters or polyadenylation sites.

Although the initial reports of such cases were confined to the eukaryotic DNA viruses such as SV40 and adenovirus, many cases involving the cellular genes of higher eukaryotes were described subsequently. In one such case involving the skeletal muscle *troponin T* gene, the same RNA can be spliced in up to 64 different ways in different muscle cell types (**Figure 7.12**). The existence of tissue-specific splicing factors acting on this gene is indicated by the finding that the artificial introduction and expression of this gene in non-muscle cells or myoblasts which do not normally express the gene results in the complete removal of exons 4–8, whereas in muscle cells (myotubes) the correct pattern of alternative splicing seen with the endogenous gene is reproduced faithfully.

It should be noted, however, that the production of 64 possible mRNAs is certainly not the most dramatic example of multiple mRNAs being

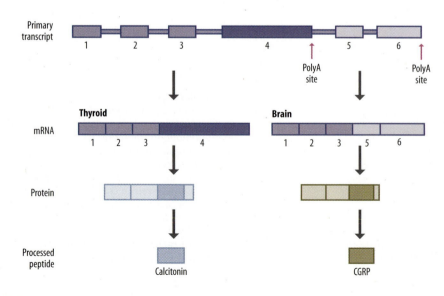

Figure 7.11
Alternative splicing of the calcitonin/*CGRP* gene in brain and thyroid cells. Alternative splicing followed by proteolytic cleavage of the protein produced in each tissue yields calcitonin in the thyroid and CGRP in the brain.

produced by alternative splicing. For example, the vast number of different alternative exons found in the *Dscam* gene of *Drosophila* could result in over 38,000 different mRNAs being produced, which is greater than the number of different genes in this organism (approximately 14,000)! This illustrates the extraordinary power of alternative splicing to produce multiple mRNAs encoding related but distinct proteins from a single gene.

Alternative RNA splicing involves specific splicing factors that promote or inhibit the use of specific splice sites

The idea that certain factors are necessary for each particular pattern of splicing in genes such as calcitonin/*CGRP* or troponin, begs the question of how such factors act. It is likely that these factors recognize *cis*-acting sequences within the RNA transcript itself. Clearly, the interaction of these factors with such sequences could produce alternative splicing either by promoting splicing at the site of the *cis*-acting sequence at the expense of the alternative splice site (**Figure 7.13a**) or by inhibiting splicing at the site of binding and thereby promoting the use of the alternative splice site (**Figure 7.13b**).

Both of these types of mechanism appear to be used in different cases. Indeed, examples of each of these mechanisms can be seen in the hierarchy of alternatively spliced genes which regulates sex determination in *Drosophila*. In this hierarchy, each gene product controls the alternative splicing of the next gene in the pathway, resulting in different protein products in males and females. In turn, these products differentially regulate the splicing of the next gene in the hierarchy, leading ultimately to the production of a male or female fly (**Figure 7.14**).

For example, the *Sxl* gene is differentially spliced in males and females, with the product of the female-specific mRNA not only controlling the splicing of the next gene in the hierarchy, *tra*, but also promoting its own female-specific splicing. Mutations which affect this autocatalytic function of Sxl on its own RNA map in the intron between exons 2 and the male-specific exon 3, indicating that the Sxl protein acts by inhibiting the male-specific splicing of exons 2 and 3. Similarly, Sxl prevents the use of the male-specific exon 2 in the *tra* gene by binding within intron 1 and preventing the binding of the constitutive splicing factor U2AF. As U2AF normally recruits the U2 snRNP particle which is essential for removal of all introns (see Section 6.3), this prevents the male-specific splicing event.

In contrast, the action of the *tra* and *tra-2* gene products on the splicing of the *dsx* transcript appears to be mediated by promoting the use of the female-specific splice event rather than the constitutive male-specific splicing event. In this case, mutations in *dsx* which affect its splicing map in the intron between exon 3 and the female-specific exon 4, indicating that the alternative splicing factor promotes the splicing of exons 3 and 4 (see below).

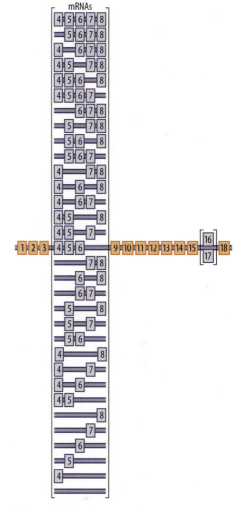

Figure 7.12
Alternative splicing of the four combinatorial exons (4–8) and the two mutually exclusive exons (16 and 17) can result in up to 64 distinct mRNAs from the rat *troponin T* gene. From Breitbart & Nadal–Ginard (1987) *Cell* 49, 793–803. With permission from Elsevier.

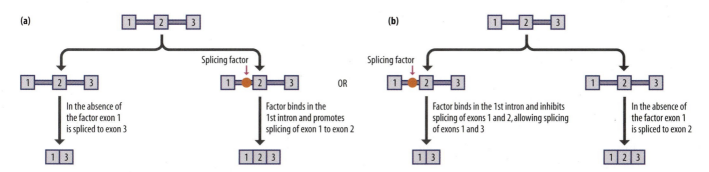

Figure 7.13
Possible models by which an alternative splicing factor can affect splicing by binding to a *cis*-acting sequence. In (a), the factor acts by promoting the use of the weaker of the two potential splicing sites while in (b) it acts by inhibiting use of the stronger of the two sites so that the other, weaker, site is used.

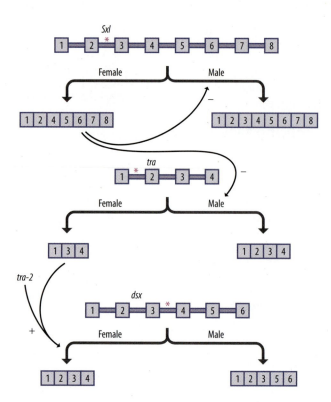

Figure 7.14
Schematic diagram (not to scale) of the
hierarchy of alternatively spliced genes
which controls sex determination in
Drosophila. Female-specific splicing of the
Sxl transcript produces a protein which
inhibits the male-specific splicing of its
own RNA and that of the *tra* gene. In turn,
the protein produced by the *tra* female-
specific transcript, in conjunction with the
tra-2 gene product, promotes the female-
specific splicing of the *dsx* transcript. The
sites of mutations in the *Sxl* and *dsx* genes
which affect their sex-specific splicing are
indicated by an asterisk.

Splicing factors can therefore act either negatively, like Sxl, to inhibit the use of a particular splice site, or positively, like Tra, to promote the use of a particular splice site (see Figure 7.13).

As well as responding to the presence or absence of a specific factor, specific RNA sequences can also regulate the pattern of alternative splicing in response to cellular signaling pathways. For example, when the **calcium/calmodulin-dependent protein kinase** type IV (CaMKIV signaling) is activated, inclusion of a specific exon (STREX) in transcripts from the BK potassium channel gene is repressed and the exon is spliced out. In contrast, in the absence of active CaMKIV, this exon is included in the RNA (**Figure 7.15a**). This effect is mediated by a 54 bp sequence from the 3′ splice site upstream of the regulated STREX exon. Mutation of this sequence abolishes regulation of splicing by CaMKIV whereas transferring it to a constitutionally spliced exon renders it sensitive to CaMKIV regulation (**Figure 7.15b**). This sequence thus acts as a CaMKIV-responsive RNA element (CaRRE) regulating splicing of the STREX exon in response to CaMKIV activation. The regulation of alternative splicing by specific cellular signaling pathways is discussed further in Chapter 8 (Section 8.5).

Factors regulating alternative splicing have been identified by genetic and biochemical methods

The evidence for the existence of alternative splicing factors and the identification of *cis*-acting sequences with which they interact have led to many attempts to identify these factors. In *Drosophila*, which is genetically very well characterized, the main approach to this problem has been a genetic one. As discussed above, genes such as *Sxl* and *tra* were originally identified by the fact that mutations within them disrupted the process of sex determination but are now known to act by controlling alternative splicing. The products of these genes are alternative splicing factors and their study allows a unique insight into the nature of such factors.

The sequencing of the *Sxl* and *tra-2* genes has revealed that the proteins they encode contain one or more copies of a ribonucleoprotein (RNP) consensus sequence that is found in a wide variety of RNA-binding proteins, such

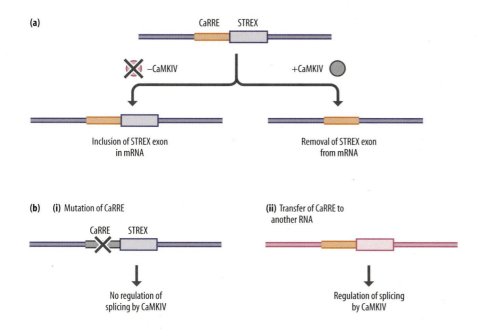

Figure 7.15
Regulation of splicing in BK potassium channel gene transcripts by the kinase enzyme CaMKIV. (a) In the presence of active CaMKIV, the STREX exon is spliced out of the BK gene RNA, whereas in the absence of active CaMKIV this splicing event is inhibited. (b) The response to CaMKIV is mediated by a 54 bp sequence from the 3′ splice site upstream of STREX, which is known as CaMKIV-response RNA element (CaRRE). Inactivation of this sequence prevents regulation of splicing by CaMKIV (i) whereas its inclusion in a completely different RNA (pink) causes splicing of that RNA to be regulated by CaMKIV (ii).

as those of the mammalian spliceosome, and which constitutes an RNA-binding domain. Indeed, structural analysis of *Sxl* bound to the *tra* mRNA has shown that the RNA binds to a V-shaped cleft within the *Sxl* protein (**Figure 7.16**). Sxl and Tra-2 therefore influence alternative splicing by binding directly to *cis*-acting sequences in the alternatively spliced RNAs discussed above.

Hence, specific factors have been identified in *Drosophila* which are expressed only in a particular situation and which regulate specific alternative splicing events. Similar cell-type-specific factors regulating alternative splicing have also been identified in vertebrates. For example, in mammals neuronal cells specifically express splicing factors such as neuronal polypyrimidine tract-binding protein (nPTB; also known as PTB2), Nova 1, and Nova 2, which play a key role in producing neuron-specific alternative splicing events (see Section 10.2 for further discussion of these factors).

Interestingly, however, in vertebrates it appears that other tissue-specific splicing events depend on quantitative variations in the levels of splicing factors which are present in all tissues rather than on splicing factors which are

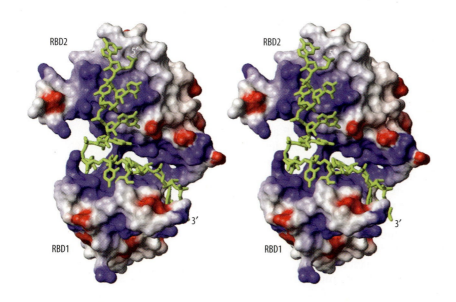

Figure 7.16
Two views of the structure of the Sx1 splicing protein bound to an RNA molecule (green). Courtesy of Yutaka Muto & Shigeyuki Yokoyama, from Handa N, Nureki O, Kurimoto K et al. (1999) *Nature* 398, 579–585. With permission from Macmillan Publishers Ltd.

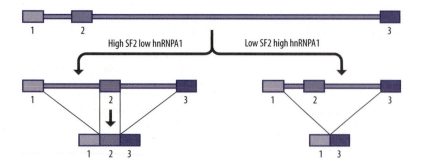

Figure 7.17
The pattern of splicing can be regulated by the balance in levels between the constitutively expressed SF2 and heterogeneous nuclear ribonuclear protein (hnRNP) A1 proteins. A high ratio of SF2 to hnRNPA1 favors use of the proximal exon (2) whereas a low ratio favors the distal exon (3).

specifically exposed in a particular cell type. For example, the SF2 factor is a member of the SR protein family discussed in Chapter 6 (Section 6.3), which is present in all cells and is essential for the basic process of splicing itself. However, its concentration has been shown to influence which of two competing upstream splice sites is joined to a downstream site. High concentrations of SF2 favor the more proximal of the two sites whereas low concentrations favor the more distal site (**Figure 7.17**). Interestingly the constitutively expressed heterogeneous nuclear ribonuclear protein A1 (**hnRNP**A1) protein binds to RNA before it is spliced and has the opposite effect, favoring the use of the more distal site. In this situation, the outcome of a specific alternative splicing event could be different in two different tissues depending on the relative concentration of SF2 and hnRNPA1 in each tissue even though both factors were present in both tissues (see Figure 7.17).

Such a system in which the different outcome of alternative splicing events in each tissue is controlled by quantitative differences in the relative levels of constitutively expressed factors can readily be fitted into the mechanistic models illustrated in Figure 7.13, simply by suggesting that the two factors have opposite effects on the relative strengths of the two competing splice sites. Indeed, it is likely that SR proteins such as SF2 can influence splice-site selection in two ways, both involving the recruitment of the U snRNP particles which are essential for splicing to occur (see Section 6.3). Thus, SR proteins can bind to the 5′ splice site and promote the binding of the U1 snRNP to this site (**Figure 7.18a**). In addition, as noted in Chapter 6 (Section 6.3), they can also bind to exon-splicing enhancers (ESEs) within the exon downstream of the regulated splice site and stimulate the binding of the U2AF protein to the branch point, thereby recruiting the U2 snRNP particle (**Figure 7.18b**).

Both tissue-specific factors and variations in constitutively expressed factors can therefore regulate alternative splicing. Indeed as described above, both Sxl and SR proteins such as SF2 can act by regulating the recruitment of U2AF. It is not surprising therefore that some cases of alternative splicing are regulated by the interaction of SR proteins with specifically expressed factors. For example, in the case of the *dsx* gene, the splicing of exon 3 to the female-specific exon 4 does not occur in the absence of the female-specific proteins tra and tra-2 (**Figure 7.19**). This is because U2AF binds only weakly to the branch point of this intron and the binding site of SR proteins in exon 4 is weak and too far away for SR to enhance U2AF binding (see Figure 7.19). In the presence of tra and tra-2, however, the interaction of SR with the exon 4 sequence is stabilized, allowing SR in turn to promote binding of U2AF and splicing of exon 3 to exon 4 (see Figure 7.19). As well as an RNA-binding domain found in many RNA-binding proteins with different functions (see above), the SR proteins and tra/tra-2 also contain the domain rich in serine (S) and arginine (R) residues which gives the SR proteins their name.

Overall therefore, alternative splicing involves the interaction of a number of factors including both sequences within the RNA itself and constitutively expressed or tissue-specific *trans*-acting protein factors. Each splice site within the RNA will have a particular strength depending on several factors. The first of these will be its ability to bind general splicing factors, such as the U1 snRNP or U2AF with those sites which bind these factors more strongly

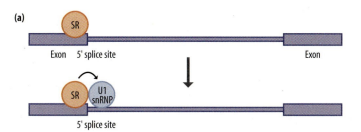

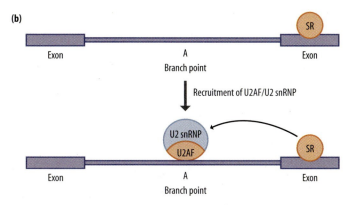

Figure 7.18
SR proteins such as SF2 can stimulate the use of a particular splice site either by binding to the 5′ splice site and promoting binding of the U1 snRNP to this site (a) or by binding to the downstream exon sequence and promoting recruitment of U2AF and hence of the U2 snRNP to the branch point (b).

being stronger splice sites (**Figure 7.20a**). Secondly, the strength of a splice site will depend on the presence or absence of ESEs (**Figure 7.20b**). Thirdly, there will also be an effect of position; i.e. whether the splicing event involves a proximal or distal downstream exon (**Figure 7.20c**).

Which splicing event actually occurs in a particular tissue or cell type will then be determined both by these features of the competing splice sites and the presence or absence of tissue- or cell-type-specific splicing factors and/or the relative levels of different constitutively expressed factors involved in splicing (see Figure 7.20).

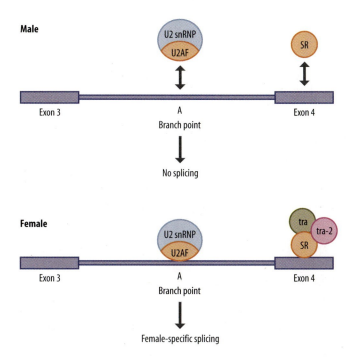

Figure 7.19
In male *Drosophila*, SR protein binds weakly to the female-specific exon 4 sequence and hence cannot promote binding of the U2 snRNP to the branch point. In females, however, tra and tra-2 stabilize the binding of SR and hence allow it to stimulate U2 snRNP binding with consequent splicing of exon 3 and exon 4.

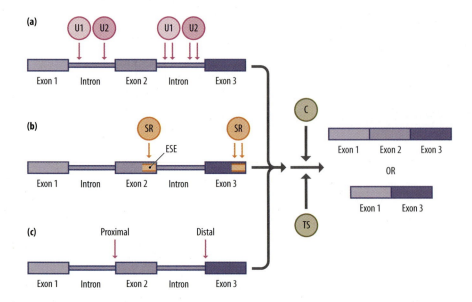

Figure 7.20
A number of features of the RNA itself control the strength of a specific splice site. These include the ability to recruit splicing factors such as the U1 and U2 snRNPs (a), the ability of exon-specific enhancers (ESE) to bind SR proteins (b), or the relative positions (proximal or distal) of the competing splice sites (c). These features will interact with the presence or absence of tissue-specific (TS) splicing factors and the level of constitutive (C) splicing factors to determine the splicing pattern.

The processes of transcription and alternative splicing interact with one another

It is clear therefore that alternative splicing is a process which is influenced both by features of the RNA itself and by factors which interact with it. However, as discussed in Chapter 6 (Section 6.4), the process of splicing is closely linked to transcription itself and this is another factor which needs to be considered. Indeed, as discussed above, the use of alternative promoters for transcription of a particular gene can affect the pattern of alternative splicing by producing RNAs which fold in different ways or by recruiting different splicing complexes (see Figures 7.6 and 7.7).

However, transcriptional events can also affect alternative splicing even when only a single promoter is used to transcribe the gene. For example, given that the splicing complex is recruited to the RNA while transcription is still occurring, the rate of transcriptional elongation can influence which alternative splicing event occurs. In a situation where an upstream 3′ splice site is weaker than a downstream one, a low rate of transcriptional elongation will favor the use of the weaker site since it can be used before the downstream site has even been transcribed. In contrast, with a faster rate of transcriptional elongation, both splice sites will be present in the RNA transcript which is being spliced, resulting in the stronger splice site being used (**Figure 7.21**).

A further link between transcription and alternative splicing is provided by the arginine methyltransferase enzyme CARM1. This enzyme can methylate the arginine at position 17 in histone H3 which, as described in

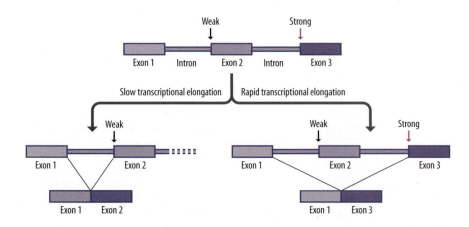

Figure 7.21
A slow rate of transcriptional elongation may result in the use of a weaker upstream splice site because the stronger downstream site has not yet been transcribed when splicing occurs.

Chapter 3 (Section 3.3 and Table 3.1), results in a more open chromatin structure compatible with transcription. Similarly, arginine methylation also modulates the activity of the CBP and p300 transcriptional co-activators (see Section 8.3).

However, as well as regulating transcription, CARM1 can also modulate splicing by methylating splicing factors such as the U1C protein, (which is part of the U1 snRNP) and the SmB protein (which is found in several snRNPs). This methylation of splicing proteins promotes the omission of certain exons which would otherwise be included in the mRNA, thereby promoting specific alternative splicing events (**Figure 7.22**).

In summary, therefore, the outcome of alternative splicing in different cell types or tissues will depend on a number of factors such as sequences in the RNA itself, the presence or absence and concentration of particular splicing factors, and the interaction of the splicing processes with other events such as transcription.

Alternative RNA splicing is a very widely used method of supplementing transcriptional control

The cases discussed above indicate the use of alternative splicing in a wide variety of biological processes. In mammals such splicing has been shown to regulate the immune system's production of antibodies, the production of neuropeptides such as CGRP and the tachykinins, substance P and substance K, as well as the synthesis of the different forms of several of the major sarcomere muscle proteins. Similarly, in *Drosophila* much of the posterior body plan is determined by developmentally regulated differential splicing of the *ultrabithorax* gene, while sex determination is also controlled by differential splicing of a hierarchy of genes in males and females (see above).

Although these cases were defined by identifying a gene of interest in a particular process and then showing that it was alternatively spliced, the generality of alternative splicing has now been demonstrated by genome-wide methods. For example, the microarray systems used to study global gene-expression patterns which were described in Chapter 1 (Section 1.2) can be adapted to investigate alternative splicing events. If mRNA from different tissues is used to probe an array containing oligonucleotides derived from various exons, their presence or absence in the mRNA can be determined by the pattern of hybridization (**Figure 7.23**). The use of methods of this type, using sequences of thousands of human genes, has allowed the

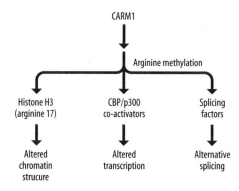

Figure 7.22
The CARM1 arginine methyltransferase enzyme can regulate chromatin structure, transcription, and alternative splicing by methylating specific proteins involved in these processes.

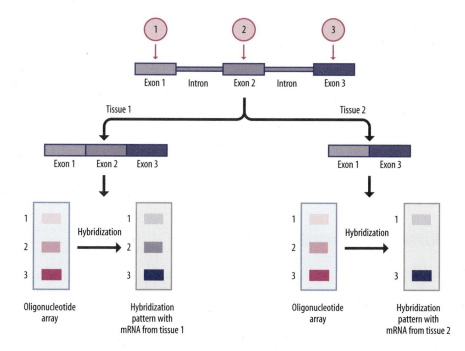

Figure 7.23
Microarray systems can be used to probe whether a particular exon is included or excluded from the final mRNA in different tissues. Oligonucleotides (1, 2, and 3) are selected from each of the exons and hybridized with mRNA prepared from each tissue.

conclusion mentioned above that over 90% of human genes with multiple exons exhibit alternative splicing in different situations.

The widespread use of alternative splicing in mammals does not refute the conclusion discussed in Chapter 1 (Section 1.4) that gene regulation occurs primarily at the level of transcription. Clearly, alternative splicing represents a response to a requirement for the production of related but different forms of a gene product in different tissues. It therefore supplements the regulation of transcription of the gene responsible for producing the different forms. The immunoglobulin heavy-chain gene, which produces both membrane-bound and secreted forms of the protein at different stages of B-cell development is transcribed only in B cells and not in other cell types, while the transcription of the *troponin T* gene which produces multiple different isoforms in different muscle cell types is confined to differentiated muscle cells.

Indeed, a genome-wide analysis of the proteins produced by alternative splicing in species as diverse as humans and *Drosophila* indicated that such alternative splicing inserts or deletes whole functional domains of proteins more often than would be expected by chance. This general analysis reinforces the specific examples such as troponin and the immunoglobulins and indicates the key role of alternative splicing in producing distinct but related proteins.

As well as controlling the insertion or exclusion of specific functional domains, alternative splicing can also regulate the balance between functional and non-functional mRNAs. For example, the inclusion of a particular exon in the mRNA can result in it containing a premature stop codon whereas exclusion of this exon results in a functional mRNA (**Figure 7.24**). As described in Chapter 6 (Section 6.7), such a premature stop codon will result in the degradation of the RNA by nonsense-mediated decay processes. In this manner, alternative splicing can be used to control the level of a particular protein, in a manner similar to but distinct from the processing/discard decisions which occur, for example, in *Drosophila*, as described above (compare Figure 7.1 and Figure 7.24).

This method of controlling the amount of a functional protein by alternative splicing is frequently used to regulate the amount of alternative splicing factors themselves. For example, as discussed in Chapter 10 (Section 10.2), functional nPTB protein is produced only in mammalian neuronal cells where it regulates neuron-specific alternative splicing events. However, the *nPTB* gene is also transcribed in non-neuronal cells. Alternative splicing of the nPTB primary transcript produces an mRNA containing a premature stop codon in non-neuronal cells, but produces a functional mRNA in neuronal cells.

Interestingly, all types of alternative splicing are particularly common in non-dividing cells such as muscle cells and nerve cells. As discussed in Chapter 3 (Section 3.2), the reprogramming of cellular commitment and gene transcription often requires a cell-division event allowing alterations in the chromatin structure of the DNA and its associated proteins to occur. Alternative splicing may therefore represent a means of conveniently

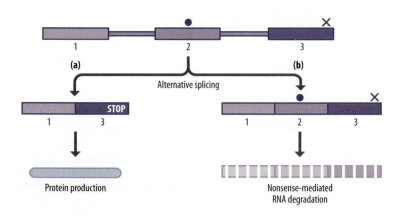

Figure 7.24
An RNA can undergo alternative splicing to produce either a functional mRNA containing only a correctly located stop codon (X), which will be translated into protein (a), or an mRNA containing a premature stop codon (solid circle), which will be degraded (b).

changing the pattern of protein production in cells where such reprogramming cannot be achieved by cell division.

7.2 RNA EDITING

Some cases of RNA editing involve a change from a C to a U residue

The finding that two different protein products can be produced from the same RNA by alternative splicing has been supplemented by the observation that a similar result can be achieved by a post-transcriptional sequence change in the mRNA.

This is seen in the case of the apolipoprotein B protein, which plays an important role in lipid transport and is known to exist in two closely related forms. A large protein of 512 kDa, known as apo-B100, is synthesized by the liver while a smaller protein apo-B48 is made by the intestine. The smaller protein is identical to the N-terminal portion of the larger protein. Analysis of the mRNA encoding these proteins revealed a 14.5 kb RNA in both tissues. These two RNAs were identical with the exception of a single base at position 6666 which is a cytosine in the liver transcript and a uracil in the intestinal transcript (**Figure 7.25**). This change has the effect of replacing a CAA codon, which directs the insertion of a glutamine residue, with a UAA stop codon which causes termination of translation of the intestine RNA and hence results in the smaller protein being made.

Only one gene encoding these proteins is present in the genome and it is not alternatively spliced. In both intestinal and liver DNA, this gene has a cytosine residue at position 6666. Hence the uracil in the intestinal transcript must be introduced by a post-transcriptional RNA-editing mechanism.

Indeed, a cytidine deaminase enzyme has been identified and shown to bind to the apoB mRNA at sequences adjacent to the edited site. This enzyme removes an amine (NH_2) group from the cytosine residue, generating a uracil. The editing enzyme is expressed in the intestine, where editing of the apoB mRNA occurs, but not in the liver, where editing does not occur. Interestingly, however, it is also present in other tissues such as the testis, ovary, and spleen where apoB mRNA is not expressed. This indicates that this enzyme is likely to edit other transcripts expressed in these tissues. Indeed, several other transcripts which undergo C-to-U editing have now been identified and a number of different cytidine deaminase enzymes capable of carrying out this form of editing have been characterized.

Other cases of RNA editing involve a change from an A to an I residue

As well as C-to-U editing by cytidine deaminase, an adenosine deaminase enzyme also exists in mammalian cells. This enzyme removes an amino residue from adenine to produce an inosine base that is read as a guanine residue by the translation apparatus. This form of editing is particularly prevalent in humans where it is believed to affect over 1000 genes. It was first identified in the gene encoding a receptor for the excitatory amino acid glutamate which is expressed in neuronal cells. The editing of an A residue

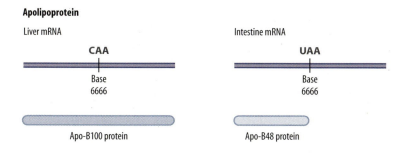

Figure 7.25
RNA editing of the apolipoprotein B transcript in the intestine produces an mRNA encoding the truncated protein apo-B48.

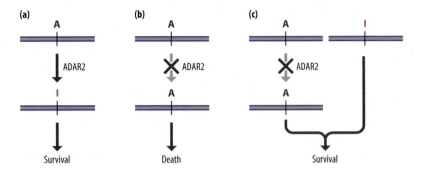

Figure 7.26
Mice in which the glutamate receptor gene cannot undergo the normal A-to-I editing do not survive (compare panels a and b). However, they can be rescued by adding another copy of the glutamate receptor gene which has been edited at the DNA level (c).

to an I (G) residue in the transcript of this gene results in it encoding an arginine rather than the glutamine found in other related receptors and alters its properties so that it is permeable to calcium. Interestingly, this editing of the glutamate receptor mRNA is essential for the survival of the animal. Mice lacking the adenosine deaminase enzyme (ADAR2) which carries out this editing are prone to seizures and die young. However, their survival can be restored to normal by inserting a glutamate receptor gene in which the alteration produced by editing has been carried out at the DNA level, removing the need for editing at the RNA level (**Figure 7.26**).

Although the glutamate receptor is therefore a critical target for **RNA editing**, editing by adenosine deamination is prevalent especially in the nervous system and also occurs in other neuronally expressed receptors such as the kainate receptors and the serotonin receptor. Indeed, a recent comprehensive search for edited transcripts in *Drosophila* identified 16 previously unidentified targets for A-to-I editing. Interestingly, all of these were involved in electrical or chemical neurotransmission and the editing event targeted a functionally important residue. Hence, A-to-I editing plays a critical role in the nervous system in producing different functional variants of molecules involved in neuronal signaling.

As well as affecting the nature of the protein produced from a particular mRNA, A-to-I editing by adenosine deaminase enzymes can also affect the splicing of the edited RNA. This is seen in the case of the gene encoding the ADAR2 adenosine deaminase enzyme itself. The ADAR2 transcript contains an AA sequence 47 bases upstream of the AG sequence which is normally used as a 3′ splice site. Editing of the AA sequence to AI (which mimics AG) creates a new 3′ splice site which is used for splicing, resulting in the 47 base sequence being retained in the mRNA (**Figure 7.27**).

This case is of particular interest in that the editing event results in a non-functional ADAR2 protein being produced due to the inclusion in the mRNA of a premature translational stop codon, leading to nonsense-mediated decay of the mRNA (see Sections 6.7 and 7.1). It therefore appears to represent a mechanism for ADAR2 to autoregulate its own production, with editing of its own mRNA by the enzyme limiting the amount of ADAR2 protein which is made. Moreover, this editing event is regulated so that it occurs frequently in the brain and lung where ADAR2 transcripts predominantly contain the 47-base extra exon and much less frequently in the heart where

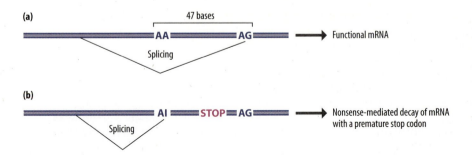

Figure 7.27
In the ADAR2 RNA, splicing normally occurs using an AG sequence as the 3′ splice site, producing a functional mRNA (a). RNA editing alters an AA sequence 47 bases upstream to AI. As AI resembles AG, this sequence is used as the 3′ splice site resulting in an extra 47 bases being retained in the final mRNA. This results in the presence in the mRNA of a premature stop codon, leading to nonsense-mediated decay of the mRNA (b).

this exon is therefore absent from the mRNA. This effect also demonstrates that editing by adenosine deaminase can affect RNA splicing as well as altering the encoded protein and that it can affect the RNA encoding the adenosine deaminase itself, as well as RNAs encoding other proteins such as the glutamate receptor.

A-to-I editing can therefore target protein-coding mRNAs to alter the encoded protein or the splicing pattern. In addition, however, the A-to-I editing process can also target the microRNAs (miRNAs) which were discussed in Chapter 1 (Section 1.5) and which play a key role in the regulation of gene expression. In this case, A-to-I editing alters the sequence of the miRNA and therefore results in its binding (by complementary base pairing) to a different set of target mRNAs.

Moreover, this effect is a tissue-specific one. For example, the miRNA miR376-a undergoes A-to-I editing in the mouse kidney but not in the liver. This alters the target mRNAs to which it can bind since the A residue in the unedited miRNA needs to base-pair with a U residue in the target mRNA, whereas the I in the edited miRNA will base-pair with a C residue (**Figure 7.28**). Sequence analysis suggests that the unedited miRNA can bind to 78 target genes whereas the edited miRNA binds to 82 target genes, with only two target genes being bound by both the unedited and edited forms. Since binding of a miRNA to its target mRNAs results in inhibition of gene expression (see Section 7.6), A-to-I editing will result in miR376-a inhibiting a different set of genes in the kidney compared to the liver.

The existence of two distinct types of editing enzymes (adenosine deaminase and cytidine deaminase) in mammalian cells each with multiple substrates indicates that this represents a widely used mechanism which like alternative splicing can produce related but distinct proteins from the same transcript. As with alternative splicing, however, this is likely to act as a supplement to transcriptional control. The *apoB* gene, for example, is transcribed in the liver and the intestine but not in other tissues whereas many of the targets for A-to-I editing are transcribed only in the nervous system. Interestingly, as well as these similarities in the role/outcome of RNA editing and alternative splicing, there is evidence that the two processes may occur in parallel. For example, it has been shown that the adenosine deaminase enzyme is found in large ribonucleoprotein particles which also contain Sm and SR splicing factors.

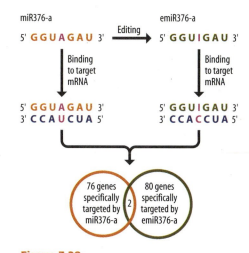

Figure 7.28
A-to-I editing of the miR376-a miRNA alters the mRNAs to which it can bind by complementary base pairing.

7.3 REGULATION OF RNA TRANSPORT

Specific proteins can regulate the transport of individual mRNAs from nucleus to cytoplasm

The process of RNA splicing takes place within the nucleus whereas the machinery for translating the spliced RNA is found in the cytoplasm. The spliced mRNA must therefore be transported to the cytoplasm if it is to direct protein synthesis (see Section 6.5). The first example of regulation at this level was described in the human immunodeficiency virus (HIV-1).

Early in infection of cells with HIV, the virus produces a high level of very short non-functional transcripts and a small amount of full-length functional RNA. The full-length RNA is spliced with the removal of two introns so that the predominant transcript which appears in the cytoplasm is a fully spliced mRNA (**Figure 7.29**). This transcript encodes the regulatory proteins Tat and Rev. As discussed in Chapter 5 (Section 5.4), the Tat protein acts to greatly increase the rate of transcription of the viral genome as well as promoting transcriptional elongation so that predominantly full-length RNAs are produced in the nucleus. However, at this second stage of infection most of the mRNAs that appear in the cytoplasm are either unspliced or have had only the first intron removed (see Figure 7.29). As these transcripts encode the viral structural proteins, this allows the high-level production of viral particles which is necessary late in infection.

This change in the nature of the HIV RNA in the cytoplasm is dependent on the action of the Rev protein which is made early in infection. Most

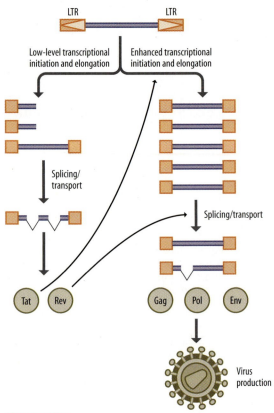

Figure 7.29
Regulation of HIV gene expression. Early in infection a low level of transcriptional initiation and elongation produces a small amount of the fully spliced viral RNA, which encodes the viral Tat and Rev proteins. When these proteins are produced, the Tat protein enhances both the initiation and elongation of transcription, resulting in an increase in the production of the viral RNA. The Rev protein then acts post-transcriptionally to promote splicing/transport so the unspliced and singly spliced RNAs encoding the viral structural proteins (Gag, env) and the reverse transcriptase (pol) accumulate. LTR, long terminal repeat.

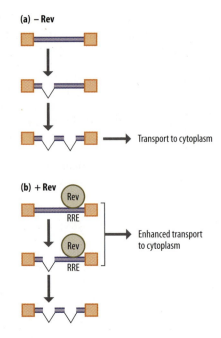

Figure 7.30
Binding of Rev to its response element (RRE) in the HIV-1 RNA promotes the transport of unspliced or singly spliced RNAs, which contain the Rev-binding site, at the expense of doubly spliced RNA, which lacks the binding site and therefore cannot bind Rev.

interestingly, Rev does not affect the amounts of the different RNAs in the nucleus. Rather it is an RNA-binding protein and binds to a specific site (the Rev response element) in the second intron of the HIV transcripts. This binding promotes the transport of these RNAs to the cytoplasm (**Figure 7.30**). As the fully spliced RNA lacks the Rev-binding site, its transport is not accelerated and so the proportion of unspliced or singly spliced RNA in the cytoplasm increases. Hence, Rev acts at the level of RNA transport and was the first regulatory protein identified as acting at this stage.

The Rev protein resembles several cellular RNA transport proteins in containing a **nuclear export signal** (NES) rich in leucine residues, which promotes transport of the RNA to which it has bound from the nucleus to the cytoplasm. Moreover, Rev has been shown to bind to a cellular nuclear export protein which is localized in the pores of the nuclear membrane and this interaction is likely to play a key role in its ability to mediate export of the HIV RNA (**Figure 7.31**). These findings indicate that Rev mediates RNA

Figure 7.31
The Rev protein binds to the HIV RNA by means of its RNA-binding domain and its nuclear-export signal (NES) then binds to cellular proteins (Exp) in the nuclear pore and promotes the export of the RNA–protein complex to the cytoplasm.

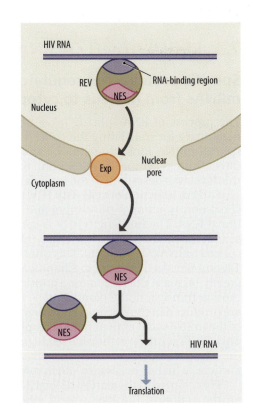

export via a cellular pathway, involving the NES signal which Rev shares with cellular RNA-transport proteins.

It is likely therefore that as well as the examples of constitutive RNA transport discussed in Chapter 6 (Section 6.5), there exist also situations in which this system is used to regulate the transport of specific cellular mRNAs in different tissues or, in response to specific stimuli. This would for example, provide a convenient explanation of the existence of RNA species that are confined to the nucleus in specific tissues of the sea urchin but are transported to the cytoplasm in other tissues (see Introduction to this chapter). Similarly, a mechanism which prevented the transport of unspliced RNA in genes regulated by processing/discard decisions (see Section 7.1) would represent a means of preventing wasteful translation of RNA species containing interruptions in the protein-coding sequence.

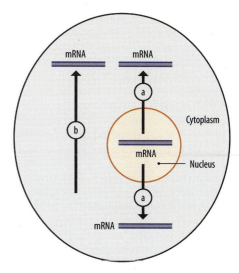

Figure 7.32
Transport of mRNA within the cell can involve both its export from the nucleus to the cytoplasm (a) as well as in some cases its preferential localization within a particular region of the cell cytoplasm (b).

RNA transport processes can also regulate the location of individual mRNAs within the cytoplasm

As well as controlling the transport of mRNA from the nucleus to the cytoplasm (**Figure 7.32a**) it is also possible for regulatory processes to control the location of an mRNA species within the cytoplasm of a polarized cell, resulting in the corresponding protein being made only in certain parts of the cell (**Figure 7.32b**). For example, this process is observed within the oocyte (egg), resulting in different proteins being localized in different regions of the egg. Following fertilization, this differential distribution of proteins in turn controls embryonic development so that different regions of the embryo develop from different regions of the fertilized egg. For example, in *Drosophila* the bicoid mRNA is located at the anterior end of the egg while the nanos mRNA is localized to the posterior end. The resulting opposite gradients of the Bicoid and Nanos proteins in turn control the polarity of the head, thorax, and abdomen of the embryo so that different regions of the body are produced depending on the relative concentrations of the two proteins (**Figure 7.33**) (see Section 9.2 for further discussion of the processes controlling embryonic development in *Drosophila*).

The specific localization of the bicoid and nanos mRNAs in the fertilized *Drosophila* egg is not an isolated example. In a recent study of more than 3000 mRNA species in the *Drosophila* embryo, over 70% showed specific subcellular localizations rather than exhibiting a uniform distribution within the cell. Similarly, in mammals, approximately 400 mRNAs are targeted to the multiple dendritic processes of neuronal cells.

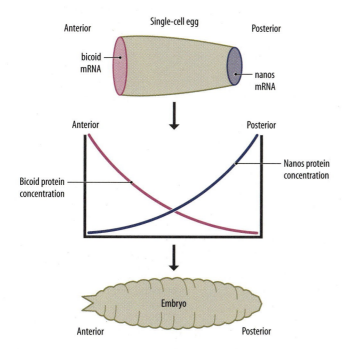

Figure 7.33
Localization of the bicoid and nanos mRNAs at opposite ends of the *Drosophila* egg sets up opposite gradients in the concentrations of their corresponding proteins, which in turn specify the anterior/posterior polarity of the resulting embryo.

In some cases, such as the bicoid mRNA, specific sequences have been localized in the 3′ untranslated region of the mRNA which are essential for correct localization and which can confer the bicoid pattern of localization on a non-localized mRNA. These sequences in the 3′ untranslated region of the bicoid mRNA are essential for it to bind both the staufen protein and a protein complex known as ESCRT-II to form a ribonucleoprotein complex. In turn, recruitment of these proteins is essential for movement of the bicoid mRNA to the anterior end of the egg (**Figure 7.34a**). Interestingly, the staufen protein is also expressed in mammalian neurons and appears to play a critical role in directing specific mRNAs to the multiple dendritic processes of the neuron and excluding them from the single **axon** (**Figure 7.34b**). Hence, staufen plays a key role in directing specific mRNAs to particular parts of the cell in very different cell types and in widely different organisms.

As with the bicoid mRNA, initial studies also identified the 3′ untranslated region of the oskar mRNA as being important for its localization in the egg, although the oskar mRNA localizes to the posterior rather than the anterior region. However, a more recent study has shown that correct localization of the oskar mRNA in the cytoplasm is also dependent on the splicing of the first two exons of the oskar transcript in the nucleus. An artificial oskar RNA with the first two exons already joined together (so that splicing is not required) is not correctly localized even though the mRNA is indistinguishable from one produced by splicing and contains the 3′ untranslated region (**Figure 7.35**).

This effect is dependent upon the linkage between RNA splicing and transport to the cytoplasm which was discussed in Chapter 6. During splicing, a protein complex known as the exon junction complex (EJC) binds to each exon just upstream of the junction between two adjacent exons and is transported to the cytoplasm bound to the mRNA (see Section 6.7). In the case of the oskar RNA the binding of the EJC to the first exon recruits another protein known as Barentsz which is responsible for the correct localization of the oskar mRNA (**Figure 7.36**), so explaining how the splicing process is necessary for correct mRNA localization.

Hence, regulatory proteins such as Staufen and Barentz play a key role in localizing specific mRNAs to particular locations within the cell. Evidently, the aim of this process is to ensure that the protein corresponding to the mRNA is made only in the appropriate subcellular location where it is required. Translation of the mRNA must therefore be prevented whilst it is being transported through the cytoplasm to the correct place in the cell. For example, if production of the Bicoid protein occurred as soon as its mRNA entered the cytoplasm, it would induce inappropriate anterior structures in the wrong place.

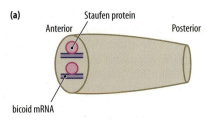

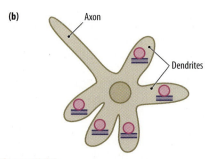

Figure 7.34
The Staufen protein plays a key role in localizing the bicoid mRNA to the anterior end of the *Drosophila* embryo (a) and in directing specific mRNAs to the dendrites and not the axon of mammalian neurons (b).

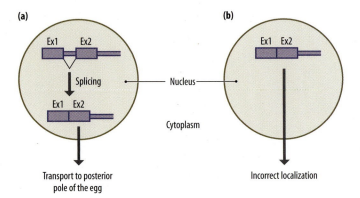

Figure 7.35
Correct localization of the oskar mRNA in the cytoplasm is dependent on the splicing of exons (Ex) 1 and 2 in the nucleus (a). If exons 1 and 2 are joined together in an artificial RNA transcript so that splicing is not required, the mRNA does not correctly localize in the cytoplasm (b).

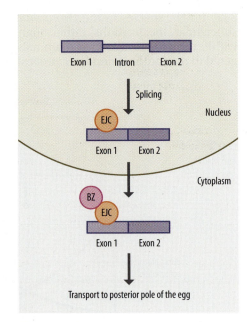

Figure 7.36
Splicing of exons 1 and 2 in the oskar mRNA results in recruitment of the exon junction complex (EJC) to exon 1. In turn, this allows recruitment of the Barentz (BZ) protein and correct localization of the mRNA to the posterior pole of the egg.

An example of how this problem is solved is seen in the β-actin mRNA which localizes to cellular projections where the β-actin protein is required to maintain and extend the projections. The transport of the β-actin mRNA to the projections is controlled by binding of the ZBP1 protein to the mRNA. Most importantly, ZBP1 also represses translation of the β-actin mRNA, preventing it being translated while it is being transported.

When the β-actin mRNA arrives at the tip of the cell, ZBP1 is phosphorylated by the Src **tyrosine kinase** which is located at the cell membrane. This results in ZBP1 no longer binding the β-actin mRNA which can now be translated in the appropriate cellular location (**Figure 7.37**). ZBP1 both directs the correct localization of the β-actin mRNA within the cell and also prevents its translation until it reaches this location.

Interestingly, in the case of the oskar mRNA discussed above, it has been shown that the **PTB** protein represses its translation until it reaches the correct point in the cell. The PTB protein is also involved in the regulation of alternative splicing (see Section 9.2), producing a further link between RNA splicing and RNA transport.

Although a number of mRNAs, such as that encoding β-actin are directed by ZBP1 to cellular projections as described above, a second system for doing this also exists. Thus, it has been shown that at least 50 other mRNAs are transported to cellular projections by a process involving the **adenomatous polyposis coli (APC) protein** which plays a key role in regulating normal cellular growth and in cancer (see Section 11.4). Hence, multiple systems exist for promoting the localization of mRNAs to cellular projections.

7.4 REGULATION OF RNA STABILITY

Gene regulation can involve alterations in RNA stability

Once the mRNA has entered the cytoplasm, the number of times that it is translated and hence the amount of protein it produces will be determined by its stability. The more rapidly degraded an RNA is, the less protein it will produce. Indeed, it is now clear that the rate of turnover of an mRNA plays an important role in determining its level in the cell (see Section 6.7).

An effective means of gene regulation could be achieved by changing the stability of an RNA species in response to a regulatory signal. A number of situations where the stability of a specific RNA species is changed in this way have been described. For example, the mRNA for the milk protein casein turns over with a half-life of around 1 h in untreated mammary gland cells. Following stimulation with the hormone prolactin, the half-life increases to over 40 h, resulting in increased accumulation of casein mRNA and protein production in response to the hormone (**Figure 7.38**). Similarly, the increased production of the DNA-associated histone proteins in the S phase (DNA synthesis) of the cell cycle is regulated in part by a fivefold increase in histone mRNA stability that occurs in this phase of the cell cycle (see Section 6.2).

Such regulation of RNA stability is not confined to isolated cases. In over 50% of cases where the abundance of an mRNA transcript is altered in human cells in response to stress, at least a part of the change was due to altered RNA stability. Interestingly, a similar genome-wide survey of all cellular mRNAs in yeast demonstrated that regulation at the level of mRNA stability was frequently observed for mRNAs whose corresponding proteins are involved in ribosomal RNA synthesis and ribosome production. Hence, this form of regulation may be particularly frequent for genes encoding proteins involved in the process of protein synthesis itself. A representative selection of cases where mRNA stability is altered in a particular situation is given in **Table 7.2**.

Specific sequences in the mRNA are involved in the regulation of its stability

The first stage in defining the mechanism of changes in RNA stability is to identify the sequences within the RNA that are involved in mediating the observed alterations. This can be achieved by transferring parts of the gene

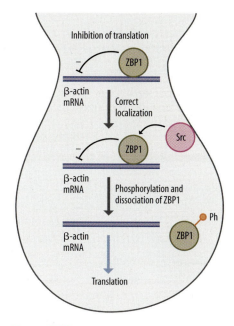

Figure 7.37
The ZBP1 protein represses translation of the β-actin mRNA and directs its transport to the tip of the cell. When the mRNA reaches its correct location in the cell, ZBP1 is phosphorylated by the membrane-associated Src protein. This results in ZBP1 dissociating from the β-actin mRNA, allowing the mRNA to be translated.

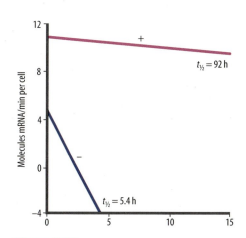

Figure 7.38
Difference in stability of the casein mRNA in the presence (+) or absence (−) of prolactin. From Guyette WA, Matusik RJ & Rosen JM (1979) *Cell* 17, 1013–1023. With permission from Elsevier.

TABLE 7.2
SOME EXAMPLES OF REGULATION OF RNA STABILITY

mRNA	CELL TYPE	REGULATORY EVENT	INCREASE OR DECREASE IN HALF-LIFE
Cellular oncogene c-*myc*	Friend erythroleukemia cells	Differentiation in response to dimethyl sulfoxide	Decrease from 35 min to less than 10 min
c-*myc*	B cells	Interferon treatment	Decrease
c-*myc*	Chinese hamster lung fibroblasts	Growth stimulation	Increase
Epidermal growth factor receptor	Epidermal carcinoma cells	Epidermal growth factor	Increase
Casein	Mammary gland	Prolactin	Increase from 1 to 40 h
Vitellogenin	Liver	Estrogen	Increase 30-fold
Type I pro-collagen	Skin fibroblasts	Cortisol	Decrease
Type I pro-collagen	Skin fibroblasts	Transforming growth factor β	Increase
Histone	HeLa	Cessation of DNA synthesis	Decrease from 40 to 80 min
Tubulin	Mammals	Accumulation of free tubulin subunits	Decrease 10-fold

encoding the RNA under study to another gene and observing the effect on the stability of the RNA expressed from the resulting hybrid gene. In a number of cases, short regions have been identified which can confer the pattern of stability regulation of the donor gene upon a recipient gene that is not normally regulated in this manner.

In many cases such regions are located in the 3′ untranslated region of the mRNA, downstream of the stop codon that terminates production of the protein. For example, the cell-cycle-dependent regulation of histone H3 mRNA stability is controlled by a 30-nucleotide sequence at the extreme 3′ end of the molecule. Similarly, the destabilization of the mRNA encoding the transferrin receptor in response to the presence of iron can be abolished by deletion of a 60-nucleotide sequence within the 3′ untranslated region. Interestingly, both of these sequences have the potential to form stem-loop structures by intra-molecular base-pairing (**Figure 7.39**), suggesting that

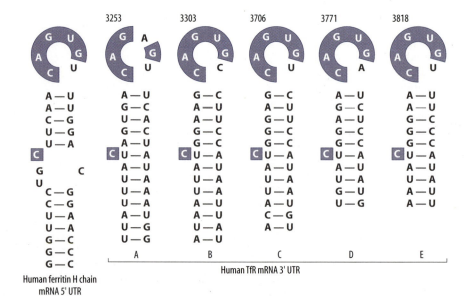

Figure 7.39
Similar stem-loop structures in the human ferritin and transferrin (TfR) receptor mRNAs. Note the boxed conserved sequences in the unpaired loops and the absolute conservation of the boxed C residue, found within the stem, 5 bases 5′ of the loop. UTR, untranslated region. From Casey JL, Hentze MW, Koeller DM et al. (1988) *Science* 240, 924–928. With permission from The American Association for the Advancement of Science.

changes in stability are likely to be brought about by alterations in the folding of this region of the RNA in response to a specific signal (see Section 6.2).

The localization of sequences involved in the regulated degradation of specific mRNA species to the 3′ untranslated region is in agreement with the important role of this region in determining the differences in stability observed between different RNA species, for example, by promoting loss of the polyA tail catalyzed by specific de-adenylation enzymes. In turn, this would open up the RNA to exonucleolytic attack via its free 3′ end (see Section 6.7). This suggests that differences in RNA stability, whether between different RNA species or in a single RNA in different situations, may be controlled primarily by this region.

Despite this, cases where other regions of the RNA mediate the observed alterations in stability have also been described. The most extensively studied of such cases concerns the auto-regulation of the mRNA encoding the microtubule protein, β-tubulin, in response to the presence of free tubulin monomers. This auto-regulation prevents the wasteful synthesis of tubulin when excess free tubulin not polymerized into microtubules is present and is caused by a destabilization of the tubulin mRNA. A short sequence only 13 bases in length from the 5′ end of the β-tubulin mRNA is responsible for this destabilization and can confer the response on an unrelated mRNA.

Most interestingly, these bases actually encode the first four amino acids of the tubulin protein. This raises the possibility that the trigger for degradation of the tubulin mRNA might be the recognition of these amino acids in the tubulin protein rather than the corresponding nucleotides in the tubulin RNA. In an elegant series of studies, Cleveland and colleagues showed that this was indeed the case. They demonstrated that changing the translational **reading frame** of this region such that the identical nucleotide sequence encoded a different amino acid sequence, abolished the auto-regulatory response (**Figure 7.40a**). In contrast, changing the nucleotide sequence in a manner which did not alter the encoded amino acids (due to the degeneracy of the genetic code) left the response intact (**Figure 7.40b**). Hence, stability of the RNA transcript can be regulated by sequences in different parts of the RNA which can be recognized at the level of the RNA itself or of the protein that it encodes.

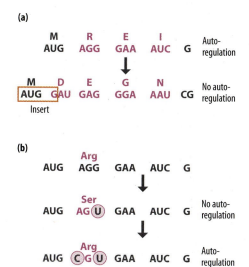

Figure 7.40
Effect of changes in the β-tubulin sequence on auto-regulation of tubulin mRNA stability.

RNA stability changes supplement transcriptional control in cases where a rapid response is required

A consideration of the situations where changes in the stability of a particular RNA occur (see Table 7.2) suggests that the majority have two features in common. First, changes in stability of a particular mRNA are very often accompanied by parallel alterations in the transcription rate of the corresponding gene. For example, prolactin treatment of mammary gland cells results in a two- to fourfold increase in casein gene transcription. Similarly, the increased stability of histone mRNA in the S phase of the cell cycle is accompanied by a three- to fivefold increase in transcription of the histone genes.

Secondly, cases where RNA stability is regulated are very often those where a rapid and transient change in the synthesis of a particular protein is required. For example, synthesis of the histone proteins is necessary only at one particular phase of the cell cycle, when DNA is being synthesized. Following cessation of DNA synthesis, a rapid shut-off in the synthesis of unnecessary histone proteins is required. Similarly, following the cessation of hormonal stimulation it would be highly wasteful to continue the synthesis of hormonally dependent proteins, such as casein or vitellogenin.

In the case of the cellular oncogene c-*myc*, whose RNA stability is transiently increased when cells are stimulated to grow, such continued synthesis would not only be highly wasteful but is also potentially dangerous to the cell. This growth-regulatory protein is only required for a short period when cells are entering the growth phase and its continued inappropriate synthesis at other times carries the risk of disrupting cellular growth-regulatory mechanisms, possibly resulting in transition to a cancerous state (see Sections 11.1 and 11.2).

These considerations suggest that alterations in RNA stability are used as a significant supplement to transcriptional control in cases where rapid changes in the synthesis of a particular protein are required. Thus, if transcription is shut off in response to withdrawal of a particular signal, inappropriate and metabolically expensive protein synthesis will continue for some time from pre-existing mRNA unless that RNA is degraded rapidly. Similarly, rapid onset of the expression of a particular gene can be achieved by having a relatively high basal level of transcription with high RNA turnover in the absence of stimulation, allowing rapid onset of translation from pre-existing RNA following stimulation. Hence as with alternative RNA processing, cases where RNA stability is regulated represent an adaptation to the requirements of a particular situation and do not affect the conclusion that regulation of gene expression occurs primarily at the level of transcription.

7.5 REGULATION OF TRANSLATION

Translational control occurs in specific situations such as fertilization

The final stage in the expression of a gene is the translation of its mRNA into protein (see Section 6.6). In theory, therefore, the regulation of gene expression could be achieved by producing all possible mRNA species in every cell and selecting which were translated into protein in each individual cell type. The evidence that different cell types have very different cytoplasmic RNA populations (see Section 1.2) indicates, however, that this extreme model is incorrect. Nonetheless, the regulation of translation such that a particular mRNA is translated into protein in one situation and not another does occur in some cases.

The most prominent of such cases is that of fertilization. In the unfertilized egg, protein synthesis is slow but upon fertilization of the egg by a sperm a tremendous increase in the rate of protein synthesis occurs. This increase does not require the production of new mRNAs after fertilization. Rather, it is mediated by pre-existing maternal RNAs which are present in the unfertilized egg but are only translated after fertilization.

Although in many species such translational control produces only quantitative changes in protein synthesis, in others it can affect the nature as well as the quantity of the proteins being made before and after fertilization. For example, in the clam *Spisula solidissima* some new proteins appear after fertilization while others which are synthesized in large amounts before fertilization are repressed thereafter. However, the RNA populations present before and after fertilization are identical (**Figure 7.41**), indicating that translational control processes are operating. Translational control also operates in the egg to ensure that mRNAs which are localized to particular regions of the egg cytoplasm (see Section 7.3) are not translated prior to their arrival at the correct region of the cell (see the Introduction to Chapter 9 for further discussion of gene regulation before and after fertilization).

As well as producing parallel changes in the translation of many RNA species, translational control processes may also operate on individual RNAs in a particular cell type. For example, the rate of translation of the globin RNA in reticulocytes is regulated in response to the availability of the **heme** co-factor which is required for the production of hemoglobin. Similarly, the translation of the RNA encoding the iron-binding protein ferritin is regulated in response to the availability of iron.

Translational control can involve either modifications in the cellular translational apparatus or specific proteins which recognize sequences in the target RNA

In principle, translational regulation could operate via modifications in the cellular translational apparatus affecting the efficiency of translation of particular RNAs. Alternatively, it could involve proteins which recognize

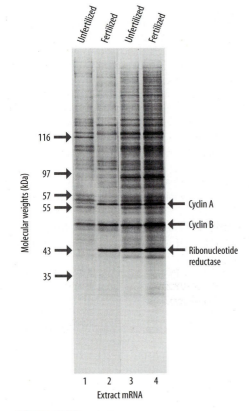

Figure 7.41
Translational control in the clam *Spisula solidissima*. Different proteins are synthesized *in vivo* before fertilization (track 1) and after fertilization (track 2). If, however, RNA is isolated either before fertilization (track 3) or after fertilization (track 4) and translated *in vitro* in a cell-free system, identical patterns of proteins are produced. The difference in the proteins produced *in vivo* from identical RNA populations must therefore be due to translational control. Courtesy of Nancy Standart and Tim Hunt, from Standart N (1992) *Sem. Dev. Biol.* 3, 367–380. With permission from Elsevier.

specific sequences in the RNA itself and which affect its translation. Evidence is available indicating that both these types of mechanism are used in different cases and they will therefore be discussed in turn.

Translational control can be produced by modifications in the cellular translation apparatus

In the absence of heme, a cellular protein kinase in the reticulocyte becomes active and phosphorylates the protein initiation factor eIF2 (see Section 6.6), resulting in its inactivation. Since this factor is required for the initiation of protein synthesis, translation of the globin RNA ceases until heme is available. However, the use of such a mechanism in which total inactivation of the cellular translational apparatus is used to regulate the translation of a single RNA is possible only in the reticulocyte, where the globin protein constitutes virtually the only translation product.

In other cell types, where a large number of different RNA species are expressed, such a mechanism is normally used only where large-scale repression of many different RNA species occurs. For example, phosphorylation of eIF2 (at the same serine amino acid but produced by a different kinase to that activated by the absence of heme) also occurs following exposure of cells to stress when the translation of most cellular mRNAs is repressed (for further discussion of the kinases which phosphorylate eIF2 see Section 8.5).

The mechanism by which phosphorylation of eIF2 inhibits translation involves its interaction with another factor, eIF2B. Thus, eIF2 is a GTP-binding protein which, as described in Chapter 6 (Section 6.6), binds to the 40 S subunit of the ribosome and migrates with it down the mRNA until the AUG **start codon** is encountered. The eIF2 bound to the 40S subunit is also bound to **GTP**. However, when the AUG start codon is reached, the eIF2–GTP complex is hydrolyzed and eIF2–GDP is released (**Figure 7.42a**). This release of eIF2 is an essential step to allow the large ribosomal subunit to bind and translation to begin.

However, before eIF2 can be recycled to participate once again in the process of translation, it must be reconverted from the eIF2–GDP form into the eIF2–GTP form which can bind to the 40S subunit. This is achieved by the binding to eIF2–GDP of the eIF2B protein. This protein is a **guanine nucleotide exchange factor** which catalyzes the exchange of GDP for GTP, resulting in the regeneration of eIF2–GTP (see Figure 7.42a).

When eIF2 is phosphorylated following exposure to heme or stress, it binds much more tightly to eIF2B. This sequesters eIF2B and prevents it from binding to other molecules of eIF2–GDP and regenerating eIF2–GTP. This therefore blocks translation by reducing the amount of available eIF2–GTP (**Figure 7.42b**). In addition, it has recently been shown that eIF2B itself can

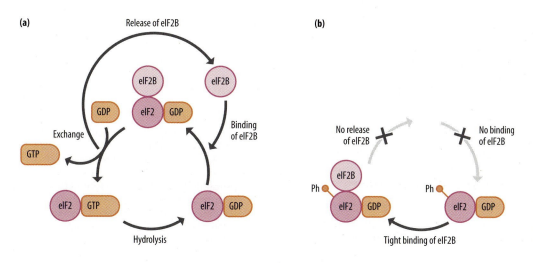

Figure 7.42
The guanine exchange factor eIF2B binds to eIF2–GDP and promotes GDP/GTP exchange, leading to the formation of eIF2–GTP. The eIF2B then dissociates and can bind to further molecules of eIF2–GDP (a). Following phosphorylation of eIF2, it binds much more tightly to eIF2B. This sequesters the eIF2B and prevents it binding to further molecules of eIF2–GDP and promoting GDP/GTP exchange (b).

also be a target for phosphorylation which inhibits its activity, providing a further mechanism for regulating protein synthesis via the eIF2/eIF2B pathway.

Interestingly, following exposure to specific stresses, some mRNAs continue to be translated since their protein products are required to mediate response to the stress. For example, following exposure of cells to elevated temperature or other stresses, the genes encoding the **heat-shock proteins** show enhanced transcription (see Sections 3.5 and 4.3). Under these conditions, it is evidently necessary to translate the mRNAs encoding these proteins so that they can produce a protective effect against the damaging effects of stress. This evidently raises the question as to how such translation can occur when general translation initiation factors have been inactivated.

A mechanism by which this is achieved has been elucidated in the case of the cationic amino acid transporter gene (Cat-1), the mRNA of which continues to be translated following the stress of amino acid starvation in yeast. The Cat-1 mRNA contains an **internal ribosome-entry site (IRES)** which allows the ribosome to bind to the mRNA internally and translate it rather than binding to the 5′ cap as normally occurs (see Section 6.1). Such IRES-mediated translation is activated after amino acid starvation, allowing the Cat-1 protein to carry out its function of mediating enhanced amino acid uptake (**Figure 7.43**).

The mechanism by which translation via the IRES is induced after amino acid starvation has been determined. Prior to amino acid starvation, the folded structure of the Cat-1 mRNA does not allow the ribosome access to the IRES. However, following amino acid starvation, a short upstream **open reading frame** is translated into a small peptide of 48 amino acids and the ribosome reading through this region has the effect of changing the structure of the RNA, thereby exposing the IRES (**Figure 7.44**). Moreover, unlike cap-dependent translation, the IRES in the Cat-1 mRNA actually requires phosphorylated eIF2 for its activity and is therefore stimulated by the

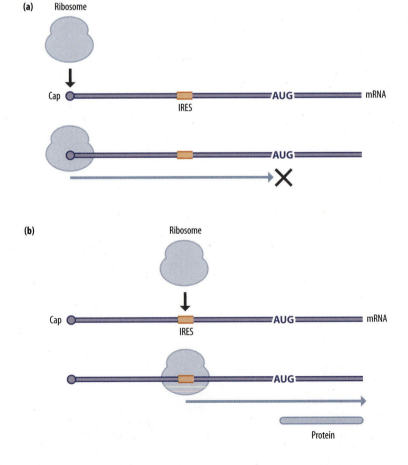

Figure 7.43
Following amino acid starvation, protein synthesis by ribosomes binding to the 5′ cap is decreased due to inactivation of specific initiation factors (a). However, translation initiated by ribosomes binding to an internal ribosome-entry site (IRES) continues, so allowing specific proteins to be made even though most protein synthesis is repressed (b).

phosphorylation of this factor during stresses such as amino acid starvation (see Figure 7.44).

IRES sequences are not confined to yeast, and have been found in a variety of viral and cellular mRNAs, including a number of mammalian mRNAs. It appears that in many of these mRNAs, as in the Cat-1 case, the IRES may have a role in allowing the RNA to continue to be translated when cap-dependent translation is inhibited, for example during mitosis or following stress.

In contrast to the Cat-1 case, however, many of these cases involve the fact that IRES-dependent translation will have a reduced requirement for eIF4G, which normally recognizes eIF4E bound to the cap and is therefore essential for cap-dependent translation (see Sections 6.1 and 6.6). When cells are exposed to stimuli which induce programmed cell death (apoptosis), eIF4G is cleaved by caspase proteases, inactivating it. In turn, this inhibits cap-dependent translation. However, translation of some proteins involved in apoptosis continues, since they have IRES elements which allow cap-independent translation (**Figure 7.45**).

As well as widespread repression of translation, regulation of translation via initiation factors can also produce translational control of specific mRNAs by acting in conjunction with sequences within the mRNA itself, which are often located within the 5′ untranslated region upstream of the translational start site. This is seen in the case of specific mRNAs whose translation is activated by treatment of cells with insulin or several growth factors. In this case, in the absence of insulin or growth factor, the cap-binding translational initiation factor eIF4E (see Section 6.6) is associated with another protein, eIF4E-binding protein (eIF4Ebp). This association prevents the other initiation factors eIF4G and eIF4A from binding to the RNA and results in a general decrease in translation.

Binding of eIF4G and eIF4A is particularly necessary for the translation of these insulin-regulated mRNAs, since they have a high degree of secondary structure in their 5′ region which must be unwound by the helicase activity of eIF4A before translation can begin. Following insulin or growth factor treatment, eIF4Ebp is phosphorylated, releasing it from eIF4E and allowing the other factors to bind and unwind the mRNA so that translation can occur (**Figure 7.46**).

It is likely that the phosphorylation of eIF4Ebp plays a key role in regulating the translation of a number of different mRNAs in different situations. For example, it has recently been shown that mice lacking functional eIF4Ebp show altered responses to viral infection, indicating a key role for eIF4Ebp in the immune response (for further discussion of translational regulation by the eIF4E/eIF4Ebp system see Section 8.5).

Although most protein synthesis is shut down when eIF4Ebp is bound to eIF4E, the translation of IRES-containing mRNAs will continue since they obviously do not require the cap-binding complex. This effect has been shown to occur in *Drosophila* for the insulin-like receptor (ILR), translation of which increases in insulin-treated cells, since it contains IRES sequences. As the cap-dependent translation of most mRNAs is inhibited by the lack of

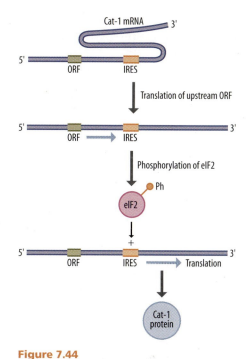

Figure 7.44
Prior to amino acid starvation, the IRES in the Cat-1 mRNA is inaccessible. Following amino acid starvation however, a small upstream open reading frame (ORF) is translated into protein and this changes the structure of the mRNA so exposing the IRES. Moreover, the activity of the IRES is stimulated by the phosphorylation of eIF2, which also occurs after stresses such as amino acid starvation.

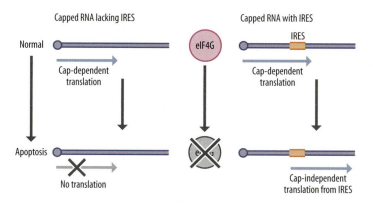

Figure 7.45
Destruction of eIF4G during apoptosis blocks cap-dependent translation but not IRES-mediated cap-independent translation.

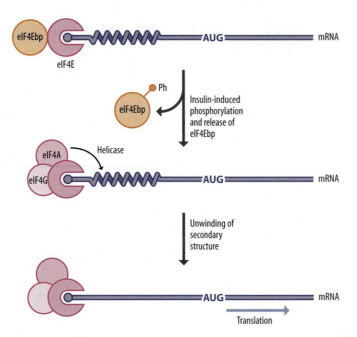

Figure 7.46
In the absence of insulin, the translation initiation factor eIF4E is bound by eIF4Ebp and translation does not occur due to the presence of secondary structure in the mRNA. Following exposure to insulin, eIF4Ebp is phosphorylated leading to its release from eIF4E. eIF4A and G then bind to eIF4E and the helicase activity of eIF4A unwinds the RNA allowing translation to occur.

eIF4Ebp phosphorylation under these conditions, the ribosomes move onto the ILR mRNA, resulting in its enhanced translation (**Figure 7.47**).

Interestingly, this process is supplemented at the transcriptional level. In the presence of insulin, the FOXO transcription factor is dephosphorylated and this allows it to induce transcription of the ILR gene resulting in more ILR mRNA. In this case, transcriptional and translational mechanisms are combined to enhance the production of the ILR protein (**Figure 7.48**).

Translational control can be produced by proteins binding to specific sequences in the RNA itself

Several other cases where translational regulation of particular mRNA species is mediated by sequences in their 5′ untranslated region have been defined. For example, the enhanced translation of the ferritin mRNA in response to iron is mediated by a sequence in this region which can fold into a stem-loop structure. The structure of this stem loop is very similar to that found in the 3′ untranslated region of the transferrin receptor mRNA whose stability is negatively regulated by the presence of iron (see Figure 7.39 and **Table 7.3**). This has led to the suggestion that such loops may represent functionally equivalent **iron-response elements** (IREs) whose opposite effects on gene expression are dependent upon their position (5′ or 3′) within the RNA molecule. This idea was confirmed by transferring the

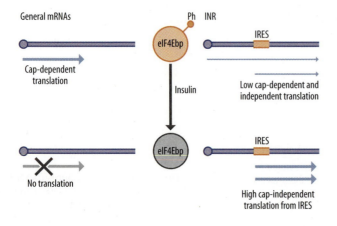

Figure 7.47
Dephosphorylation of eIF4Ebp blocks cap-dependent translation but allows IRES-dependent translation of the insulin-like receptor (INR) to continue. INR translation is therefore increased as ribosomes move off other mRNAs and onto the INR mRNA. Ph, phosphorylation.

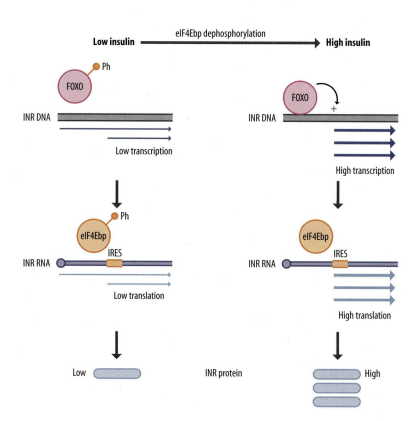

Low insulin ⎯⎯⎯ eIF4Ebp dephosphorylation ⎯⎯⎯→ High insulin

FOXO — Ph

INR DNA

Low transcription

FOXO +

INR DNA

High transcription

eIF4Ebp — Ph

INR RNA — IRES

Low translation

eIF4Ebp

INR RNA — IRES

High translation

Low   INR protein   High

transferrin receptor stem loop to the 5′ end of an unrelated RNA resulting in the iron-dependent enhancement of its translation. The identical structure is therefore capable of mediating opposite effects on RNA stability and translation depending on its position within the RNA molecule.

Such an apparent paradox can be explained if it is assumed that the stem-loop structure is an IRE which unfolds in the presence of iron (**Figure 7.49**). In the ferritin mRNA where the element is at the 5′ end, this will allow the binding and movement of the 40S ribosomal subunit along the mRNA until it reaches the initiation codon and translation begins. In contrast, in the transferrin receptor mRNA, where this element is in the 3′ untranslated region, such unfolding renders the RNA susceptible to nuclease degradation at an increased rate. In agreement with this model, an IRE-binding protein known as aconitase has been identified which binds to the stem-loop element in both the ferritin and transferrin receptor mRNAs. The RNA-binding activity of aconitase increases dramatically in cells that have been deprived of iron, suggesting that its binding normally stabilizes the stem-loop structure.

Not all cases of translational regulation mediated by sequences in the 5′ untranslated region operate via such stem-loop structures, however. Increased expression of the yeast transcriptional regulatory protein GCN4 in response to amino acid starvation is caused by increased translation of its RNA. The translational regulation of this molecule is mediated by short sequences within the 5′ untranslated region of the RNA, upstream of the

TABLE 7.3			
REGULATION OF THE TRANSFERRIN RECEPTOR AND FERRITIN GENES			
	EFFECT OF IRON ON PROTEIN PRODUCTION	**MECHANISM**	**POSITION OF STEM-LOOP STRUCTURE**
Ferritin	Increased	Increased mRNA translation	5′ untranslated region
Transferrin receptor	Decreased	Decreased mRNA stability	3′ untranslated region

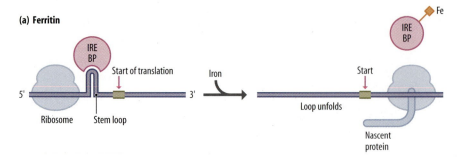

(a) Ferritin

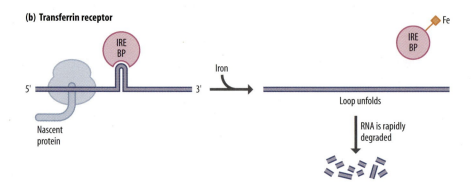

(b) Transferrin receptor

Figure 7.49
Role of iron-induced unfolding of the stem-loop structure in producing increased translation of the ferritin mRNA (a) and increased degradation of the transferrin receptor mRNA (b). In each case the unfolding of the stem loop is dependent on the dissociation of a binding protein (IRE-BP) whose ability to bind to the stem loop decreases dramatically in the presence of iron.

start point for translation of the GCN4 protein. As in the case of the upstream open reading frame in the Cat-1 mRNA (see above), such sequences are capable of being translated to produce peptides, although in this case only of two or three amino acids (**Figure 7.50**). Unlike the Cat-1 case, however, translational initiation at the second, third, or fourth of these upstream sequences in GCN4 to produce these small peptides means that the ribosome fails to reinitiate at the translational start point for GCN4 production and the protein is not synthesized.

Following amino acid starvation, the production of these small peptides is suppressed and production of GCN4 correspondingly increased. Once again, this switch in initiation also involves the phosphorylation of the eIF2 factor which is involved in translational regulation in response to heme or stress (see above). Following amino acid starvation, transfer RNA molecules lacking a bound amino acid accumulate and this activates an enzyme which phosphorylates eIF2. This decreased activity of eIF2 caused by its phosphorylation then results in the ribosome not reinitiating at three of the upstream sites in the 5′ untranslated region following translation of the first small open reading frame. This failure allows increased initiation at the downstream site

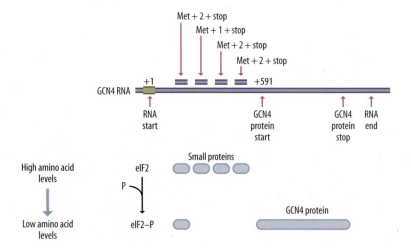

Figure 7.50
Presence of short open reading frames capable of producing small peptides in the 5′ untranslated region of the yeast GCN4 RNA. Translation of the RNA to produce these small proteins suppresses translation of the GCN4 protein. The position of the methionine residue beginning each of the small peptides is indicated together with the number of additional amino acids incorporated before a stop codon is reached. When high amino acid levels are present, the small proteins are made and the production of GCN4 is therefore suppressed. When amino acid levels fall, the eIF2 translational initiation factor is phosphorylated. Under these conditions, following translation of the first small peptide, the ribosome does not reinitiate to translate the next three small open reading frames. This favors reinitiation at the start codon for GCN4 and thereby promotes GCN4 production.

leading to enhanced GCN4 production (see Figure 7.50). (See also Section 8.5 for further discussion of the kinases which phosphorylate eIF2).

Interestingly alternative initiation codons are also found in the genes encoding two mammalian liver transcription factors, C/EBPα and C/EBPβ. However, in this case the two initiation codons are used to produce two different forms of each of the proteins. One of these is a long form, containing an N-terminal activation domain which allows it to activate transcription (see Section 5.2 for discussion of transcriptional activation domains). In contrast, the shorter form does not have this activation domain and so lacks this ability (**Figure 7.51**). It can therefore interfere with the stimulatory activity of the activating form by binding to DNA via its DNA-binding region and blocking binding of the activator (see Section 5.3 for discussion of this mechanism of transcriptional repression). Such alternative translation of the same mRNA parallels the use of alternative splicing to produce different protein products from the same gene.

Study of the cases of translational control where the mechanisms have been defined thus makes it clear that although sequences in the 5' untranslated region of particular mRNAs are frequently involved in the translational regulation of their expression, the mechanism by which they do so may differ dramatically in different cases.

Although the 5' untranslated region is an obvious location for sequences involved in mediating translational control, cases where sequences in the 3' untranslated region play a role in the regulation of translation have been reported. For example, sequences in this region are involved in modulating the efficiency of translation of specific mRNAs which occurs upon fertilization of the egg, as well as in mediating the increased translation of the mRNA encoding lipoxygenase, which occurs during erythroid differentiation.

In the lipoxygenase case, a sequence in the 3' untranslated region of the lipoxygenase mRNA directly controls the ability of the 60S ribosomal subunit to bind to the mRNA. Early in erythroid differentiation, the 40S ribosomal subunit can bind to the mRNA and, as normally occurs, it moves to the AUG translation initiation codon (see Section 6.6). However, a silencing complex, bound within the 3' untranslated region prevents binding of the 60S ribosomal subunit (**Figure 7.52a**). Hence, translation only occurs later in erythroid differentiation when this block is relieved and the 60S subunit can bind and initiate translation (**Figure 7.52b**).

A completely different mechanism involving sequences in the 3' untranslated region regulating translation is seen in the case of *Xenopus* oocytes. In this case, a number of mRNAs are maintained in a non-translated state in dormant oocytes which are arrested early in meiosis. When the oocyte is

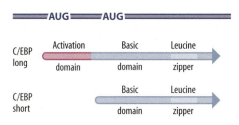

Figure 7.51
Alternative use of different translational initiation codons results in the generation of either long or short forms of the C/EBP transcription factors. The long form has a transcriptional activation domain and can therefore activate transcription, whereas the short form lacks this domain and therefore inhibits transcriptional activation by the short form.

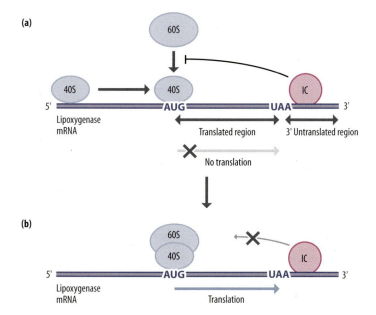

Figure 7.52
During erythroid differentiation, translation of the lipoxygenase mRNA is regulated by an inhibitory complex (IC) which binds to the 3' untranslated region of the mRNA. Early in erythroid differentiation (a), the 40S ribosomal subunit can bind to the mRNA and move to the AUG initiation codon. However, the inhibitory complex prevents the 60S ribosomal subunit from binding and so translation is inhibited. Later in the differentiation process (b), the negative effect of the inhibitory complex is blocked and translation occurs.

In dormant oocytes, the CPEB protein is bound near the 3' end of the mRNA and recruits the Maskin protein. In turn, Maskin binds to eIF4E bound to the 5' cap (blue circle) of the mRNA and thereby blocks binding of eIF4G and translation. When the oocyte is activated, CPEB is phosphorylated. In turn, this allows recruitment of the CPSF polyadenylation factor leading to polyadenylation. This results in recruitment of the polyA-binding protein PAB1. PAB1 then recruits eIF4G which binds to eIF4E, displacing Maskin and allowing translation to occur.

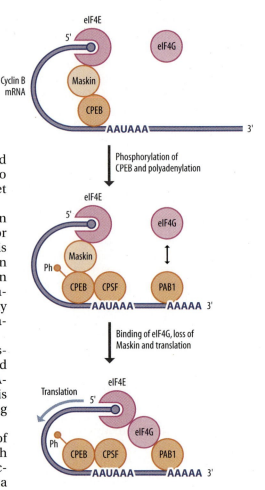

stimulated to undergo meiosis the polyA tail of these mRNAs is lengthened and the mRNA is translated. Hence, in this case a connection appears to exist between the extension of the polyA tail to its full length and the onset of translation.

In the case of one of these mRNAs encoding cyclin B, this has been shown to occur because, in the dormant oocytes, the polyadenylation factor CPSF (see Section 6.2) is not bound to the AAUAAA sequence which is involved in polyadenylation. However, another factor CPEB is bound to an adjacent region of the RNA together with a further protein known as Maskin (**Figure 7.53**). Maskin in turn appears to interact with the translation initiation factor eIF4E and prevent binding of eIF4G. Maskin therefore acts exactly like the eIF4E-binding protein described above, in that it prevents translation by inhibiting binding of eIF4G.

When the oocyte is stimulated to pass through meiosis, CPEB is phosphorylated. This allows it to recruit CPSF to the AAUAAA sequence and polyadenylation occurs. As described in Chapter 6 (Section 6.6), a polyA-binding protein, which binds to the polyA tail, can also bind eIF4G. This results in subsequent binding of eIF4G to eIF4E and in Maskin dissociating from eIF4E, allowing translation to begin (see Figure 7.53).

A further mechanism involving eIF4E and the 3' untranslated region of an mRNA, occurs in the case of gene regulation by the bicoid protein, which as described in Section 7.3, plays a key role in specifying the anterior structure in *Drosophila* (see also Section 9.2). Although the Bicoid protein is a homeodomain-containing transcription factor (see Section 5.1) able to affect the transcription of specific genes, it also acts at the translational level to repress the transcription of the caudal mRNA. Thus, although the caudal mRNA is equally distributed throughout the embryo, the caudal protein is only produced at the posterior end of the embryo where Bicoid is low or absent (for further discussion of translational regulation by Bicoid see Section 9.2).

This is achieved by Bicoid binding to the 3' untranslated region of the caudal mRNA. However, rather than directly blocking eIF4G recruitment by binding to eIF4E, bicoid recruits an eIF4E-related protein, known as 4E-HP, which binds to the cap and prevents binding of eIF4E. As 4E-HP cannot bind eIF4G, this prevents translation of the caudal mRNA (**Figure 7.54**).

Mechanisms involving regulation of translation by eIF4E can therefore involve either a protein (eIF4Ebp), which binds only to eIF4E, or one (Maskin) which is recruited by an RNA-bound protein (CPEB) and then binds to eIF4E or one (4E-HP) which binds to the mRNA and prevents eIF4E binding. Similarly, regulation of translation via sequences located in the 3' untranslated region can involve different mechanisms just as occurs for processes involving sequences near the 5' end of the mRNA.

In summary therefore, it is clear that cases of translational control can be mediated by sequences in various parts of the RNA and can involve secondary structure, use of different translation initiation codons, regulation of ribosome binding, or the regulation of polyadenylation.

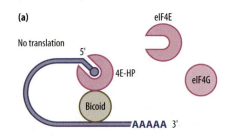

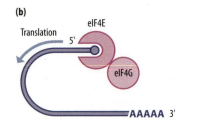

Binding of the Bicoid protein to the 3' end of the caudal mRNA results in recruitment of the 4E-HP protein to the 5' cap (blue circle). This blocks binding of eIF4E to the cap and so prevents translation (a). In the absence of bicoid, eIF4E binds to the 5' cap and translation occurs (b).

Translational control frequently occurs when a rapid response is required but also occurs for the genes encoding some transcription factors

Many cases of translational control occur in situations where very rapid responses are required. For example following fertilization, a very rapid activation of cellular growth processes is required. Similarly, following heat shock or other stresses, it is necessary to shut down rapidly the synthesis of most enzymes and structural proteins and begin to synthesize the protective heat-shock proteins. Such regulation can be achieved rapidly by translational control, supplemented by increased transcription of the genes encoding the heat-shock proteins. Once again as with other cases of post-transcriptional regulation, translational control can be viewed as supplementing the regulation of transcription in order to meet the requirements of particular specialized cases. In the case of the heat-shock proteins, this combination of transcriptional and translational control is further supplemented by an increased stability of the heat-shock mRNAs following exposure to elevated temperature providing a further means of producing a rapid and effective response.

The case of the yeast GCN4 protein provides a different aspect to translational control, however. This protein is a transcriptional regulator which increases the transcription of several genes encoding the enzymes of amino acid biosynthesis in response to a lack of one or more amino acids. In this case the synthesis of a transcriptional regulatory protein is regulated by translational control. As discussed in Chapter 10 (Section 10.2) a similar mechanism regulates the translation of the mRNA encoding the ATF4 mammalian transcription factor, indicating that this effect is not confined to yeast cells. When taken together with the increasing evidence that many mammalian transcriptional regulatory molecules may pre-exist in an inactive form and be activated by protein modifications (see Chapter 8), this suggests that it may be necessary for the cell to control the expression of some of its transcriptional regulatory molecules at levels other than that of gene transcription. Translational control is one means by which such regulation is achieved.

7.6 POST-TRANSCRIPTIONAL INHIBITION OF GENE EXPRESSION BY SMALL RNAs

Small RNAs can inhibit gene expression post-transcriptionally

Throughout this chapter, we have discussed cases where particular sequences in individual RNAs are targeted by specific proteins to produce alternative splicing, altered transport, stability, or translation, or to edit the RNA. As noted in Chapter 1 (Section 1.5), however, it has become clear that regulation of gene expression can also be achieved by very small RNA molecules of between 20 and 30 bases in length, which inhibit the expression of specific genes.

As discussed in Chapter 3 (Section 3.4), such small RNAs can produce their inhibitory effect at the level of transcription by modulating chromatin structure. However, in the majority of cases defined so far, they achieve their effect by acting post-transcriptionally either by promoting mRNA degradation or by blocking translation. These effects will therefore be discussed in this section.

A key parameter affecting whether a particular small RNA will affect mRNA degradation or translation is its complementarity to its target mRNA. As discussed in Chapter 1 (Section 1.5), small RNAs achieve their effects by binding to their target to form a partially double-stranded RNA. If there is a perfect match between the small RNA and its target then this results in endonucleolytic cleavage of the mRNA (Figure 7.55).

This effect is predominantly observed in the small interfering RNA (siRNA) class of small RNAs. These siRNAs are produced by cleavage of a

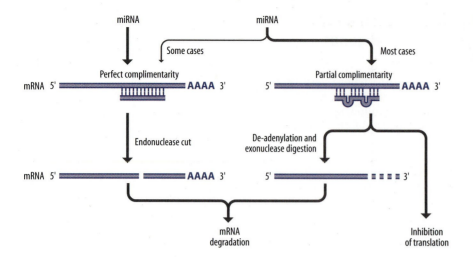

Figure 7.55
Some miRNAs and all siRNAs have perfect complementarity to their target mRNA sequence and induce cleavage of the mRNA. However, most miRNAs have only partial complementarity to their mRNA targets and produce either de-adenylation of the mRNA and its subsequent degradation or blockage of its translation.

double-stranded RNA molecule and then target further single-stranded RNA copies of the same sequence. They therefore normally have perfect complementarity to their target (see Section 1.5).

In contrast, the other major class of small RNAs, the miRNAs, are cleaved from a specific precursor RNA which is a single-stranded molecule that has folded to form a double-stranded structure. They then hybridize to distinct mRNA targets (see Section 1.5). In some cases in mammals and very often in plants, these miRNAs have perfect complementarity to their target mRNAs and will therefore induce endonucleolytic cleavage, as for the siRNAs. In the majority of cases in humans and other animals, however, they will have only partial complementarity to their target mRNAs. In these situations, they induce either de-adenylation of the mRNA leading to its exonucleolytic cleavage or block its translation (see Figure 7.55). The effects of miRNAs and/or siRNAs on mRNA degradation or translation will now be discussed in detail.

Small RNAs can induce mRNA degradation

After processing of the precursor siRNA or miRNA to the mature form is complete (see Section 1.5), the mature small RNA binds a multi-protein complex, known as the RNA-induced silencing complex (RISC). This complex contains one or other of the Argonaute family of proteins of which there are four in mammalian cells and which are RNA-binding proteins (**Figure 7.56**). The small RNA in the RISC complex then binds to a target mRNA. If there is a perfect base-pair match between the small RNA and its target mRNA, as normally occurs with siRNA and in some miRNA cases, the Argonaute protein will cleave the mRNA between the two nucleotides that are base-paired with nucleotides 10 and 11 of the small RNA (see Figure 7.56).

This will result in the mRNA being cut into two pieces, each of which has one end that is unprotected either by a 5′ cap or a 3′ polyA tail. The initial endonuclease cut within the RNA is therefore followed by further exonuclease digestion from the free end of each mRNA fragment. Interestingly, the siRNA or miRNA itself is not cut in this process but remains intact. It can therefore bind to further copies of its target mRNA and induce their degradation making the process of gene repression by small RNAs highly efficient.

As noted above in most cases involving miRNAs, the miRNA will not have perfect complementarity to its target sequence. Indeed, an individual miRNA can hybridize to many different mRNAs to which it has significant but not total homology and thereby repress a number of different genes.

Such cases of non-perfect homology can also result in degradation of the target mRNA but via a different mechanism to that discussed above. Although these cases also involve the RISC complex and hybridization to the target mRNA, the Argonaute protein in the RISC complex does not cleave the target mRNA in the region of homology to the miRNA. Rather,

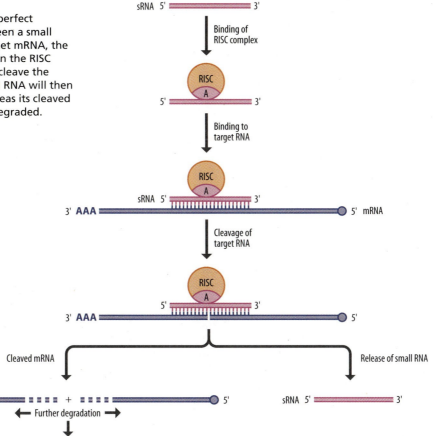

Figure 7.56
In cases where there is perfect complementarity between a small RNA (sRNA) and its target mRNA, the Argonaute (A) protein in the RISC complex will internally cleave the target mRNA. The small RNA will then be released intact whereas its cleaved target will be further degraded.

binding of the miRNA to the mRNA will induce its de-adenylation (**Figure 7.57**). As noted in Chapter 6 (Section 6.7), such de-adenylation results in rapid degradation of the mRNA involving its transport to P bodies and decapping at the 5′ end. This decapping together with the initial de-adenylation, renders the mRNA susceptible to exonuclease attack at both its 5′ and 3′ ends. As before, the miRNA is released intact allowing it to bind to a new target mRNA and catalyze repeated cycles of mRNA degradation.

Small RNAs can repress mRNA translation

Evidently, the small RNA-induced degradation of mRNA produced by internal cleavage or de-adenylation will prevent its translation into protein. In addition, however, it is also clear that miRNAs can directly repress the translation of target mRNAs, independent of any effect on mRNA degradation. As with de-adenylation, this occurs when the miRNA is not perfectly complementary to its target mRNA. Such inhibition of translation can occur either at the level of initiation of protein synthesis by the ribosome (**Figure 7.58a**) or at the level of translational elongation in which amino acids are joined together to form the protein (**Figure 7.58b**). These effects will be discussed in turn.

The inhibition of translational initiation mediated by miRNAs appears to be particularly strong in situations where the translational apparatus recognizes the 5′ cap of the mRNA compared to those where it utilizes an IRES (see Section 7.5). This has led to the suggestion that miRNAs can inhibit translational initiation via the 5′ cap on the mRNA. In agreement with this idea, it has been shown that some members of the Argonaute protein family have homology to the eIF4E translation initiation factor. When such an Argonaute protein is recruited to the mRNA with the miRNA, it could therefore bind to the cap and prevent eIF4E from binding, thereby blocking the binding of other translational initiation factors and the 40S ribosome itself (**Figure 7.59**).

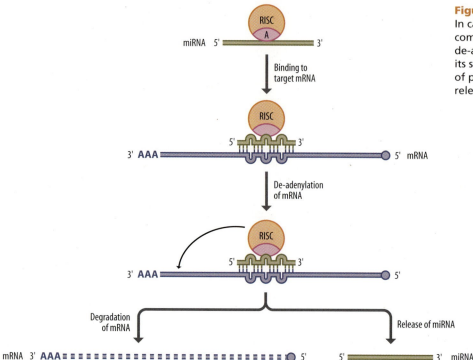

Figure 7.57
In cases where the miRNA is only partially complementary to its target, it can induce de-adenylation of the mRNA, leading to its subsequent degradation. As in cases of perfect complementarity, the miRNA is released intact.

As described in Chapter 6 (Section 6.6), binding of the 40S subunit of the ribosome is followed by its migration down the mRNA until it encounters an appropriate AUG initiation codon. At this point, the 60S ribosomal subunit binds to the 40S subunit and translation begins. Interestingly, there is evidence that miRNAs can also inhibit this stage of translational initiation.

It has been shown that the miRNA–RISC complex can associate with the translational initiation factor eIF6, which has a key role in preventing premature association of the large and small ribosomal subunits. Hence, its recruitment to the mRNA by the miRNA–RISC complex could block binding of the 60S ribosomal subunit and thereby block initiation of translation (**Figure 7.60**). Such regulation via eIF6 is of particular interest since this factor has recently been shown to be present at relatively low abundance in the cell and to act as a rate limiting factor in determining the rate of translation.

These two mechanisms for inhibiting translational initiation are not mutually exclusive. Indeed, blocking of eIF4E binding and 60S ribosomal subunit binding may act together to produce strong repression of cap-dependent translation. Similarly, blockage of 60S ribosomal subunit binding may account for the cases in which blockage of IRES-mediated translational initiation by miRNAs has been shown to occur.

Figure 7.58
Binding of a miRNA to its target mRNA can block its translation by the ribosome at either translational initiation or translational elongation.

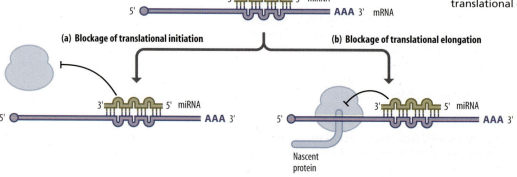

(a)

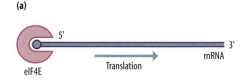

(b)

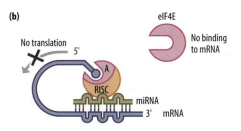

Figure 7.59
Binding of the Argonaute (A) protein in the RISC complex to the 5′ cap of the mRNA can block its translation by preventing binding of eIF4E to the cap.

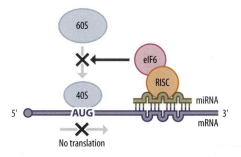

Figure 7.60
Binding of eIF6 to the RISC complex prevents the binding of the 60S ribosomal subunit to the mRNA-bound 40S ribosomal subunit, resulting in blockage of translation.

Although inhibition of translational initiation is likely therefore to be a key event in the miRNA-mediated repression of gene expression, it cannot be the only mechanism. This is seen in the case of the *lin-4* miRNA, which was the first miRNA to be characterized (see Section 1.5). When this miRNA inhibits its target *lin-14*, the *lin-14* mRNA has multiple ribosomes bound to it but *Lin-14* protein production is blocked. Thus, in this and a number of other cases where the same effect occurs, inhibition must occur after translational initiation.

Currently, however, the mechanism of this post-initiation effect is unclear. It could involve a direct effect of the RISC complex on the ability of the ribosome to translate the mRNA and produce the protein (**Figure 7.61a**). Alternatively, it could involve a protease enzyme which is recruited by the RISC complex and then degrades the nascent protein, as the ribosome produces it (**Figure 7.61b**).

Multiple mechanisms therefore mediate the effect of miRNAs on translation, acting at the level of translational initiation or elongation. It is possible that different mechanisms of translational inhibition operate for different miRNAs or that in some cases they combine together to produce a very strong inhibitory effect on gene expression.

Interestingly, recent data suggest that miRNAs can also stimulate translation under some circumstances. For example in growing cells, binding of a specific miRNA (miR369-3) to the **tumor necrosis factor** α **(TNFα)** mRNA results in inhibition of its translation. The opposite effect occurs, however, when the miR369-3 binds the TNFα mRNA under conditions where the cells are deprived of serum and have growth arrested. Under these conditions, binding of miR-369 promotes translation. Although the precise mechanism of this effect is unclear, it involves the RISC complex recruiting an additional activator protein in serum starved cells, which promotes translation of the mRNA (**Figure 7.62**).

In the case of the miRNA miR-10a, its effect on translation appears to depend on where it binds to specific mRNAs. This miRNA binds to the 5′ untranslated region of a number of different mRNAs, including several which encode ribosomal proteins and such binding stimulates their translation (**Figure 7.63a**). In contrast, it binds to the 3′ untranslated region of other mRNAs and such binding inhibits the translation of these mRNAs (**Figure 7.63b**).

Hence, miRNAs can have different effects on one specific mRNA under different conditions (as in the case of miR-369-3) or on different mRNAs, depending on the miRNA-binding site in the mRNA (as in the case of miR-10a).

(a) Blockage of translational elongation

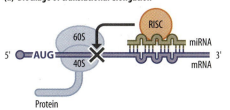

(b) Degradation of nascent protein

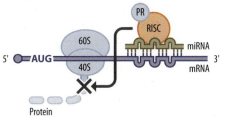

Figure 7.61
An miRNA could block translation after initiation either by directly blocking the translational elongation process (a) or by inducing degradation of the nascent protein, via a protease (PR) associated with the RISC complex (b).

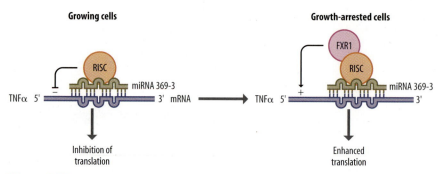

Figure 7.62
In growing cells, the binding of the miRNA 369-3 to the TNFα mRNA blocks its translation. However, in growth-arrested cells an additional protein, FXR1, is recruited to the RISC complex resulting in enhanced translation of the TNFα mRNA, following miRNA 369-3 binding.

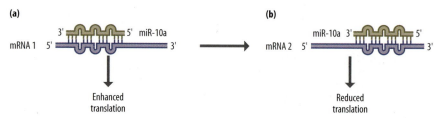

Figure 7.63
Binding of the miR-10a to the 5′ untranslated region of specific mRNAs (mRNA 1) enhances their translation (a). In contrast, binding of miR-10a to the 3′ untranslated region of other mRNAs (mRNA 2) reduces their translation (b).

miRNAs regulate gene expression at multiple levels

In this section, we have discussed the effect of miRNAs on post-transcriptional processes, such as mRNA degradation and mRNA translation. These effects, together with the effect in some cases on transcription (see Section 3.4) allow miRNAs to play a key role in regulating gene expression. Indeed, it is likely that approximately 1000 different miRNAs exist in human cells and each of these is likely to bind to and repress a number of different target mRNAs.

Although many of these mRNAs will encode structural proteins involved directly in different cellular processes, it appears that a significant number of miRNA targets are themselves regulatory proteins, such as transcription factors and regulators of alternative splicing. This reinforces our earlier conclusion (see Section 7.5) that the genes encoding transcription factors are often regulated post-transcriptionally and extends it to alternative splicing factors.

Clearly, these effects allow individual miRNAs to have a very widespread effect on gene expression. By affecting the expression of transcription factors and alternative splicing factors, they can indirectly affect the expression of many other genes in addition to those whose expression they can directly repress. Moreover, by inhibiting the expression of negatively acting regulatory molecules, an miRNA can achieve positive as well as negative effects on downstream gene expression (**Figure 7.64**).

An excellent example of this is seen in the case of miR-124 which is expressed in neuronal cells and directly represses a number of target genes

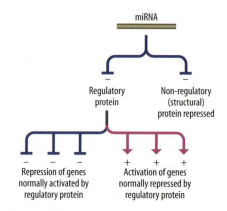

Figure 7.64
By down-regulating the expression of a regulatory protein, an miRNA can enhance the expression of any downstream genes which are normally inhibited by the regulatory protein and/or reduce the expression of any downstream genes which are normally activated by the regulatory protein.

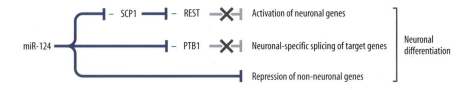

Figure 7.65
The miR-124 miRNA stimulates neuronal differentiation by repressing the expression of the SCP1 component of the REST transcriptional complex thereby preventing it from repressing neuron-specific genes. Similarly miR-124 represses expression of the PTB1 alternative splicing factor thereby preventing it from repressing neuron-specific alternative splicing. These effects are in addition to the ability of miR-124 to directly repress the expression of non-neuronal genes.

that are expressed in non-neuronal cells. In addition, however, it also represses the expression of the SCP1 transcription regulatory protein and of the PTB1 alternative splicing regulator. SCP1 is part of the **REST** transcriptional repressor complex which inhibits the expression of neuronal genes in non-neuronal cells. Hence, its repression results in the activation of many neuron-specific genes. Similarly, PTB1 represses neuron-specific splicing patterns of its target genes. Therefore repression of PTB1 expression by miR-124 results in neuron-specific alternative splicing of downstream genes (**Figure 7.65**) (for further discussion of the role of miR-124 in the regulation of gene expression in neuronal cells, see Section 10.2).

CONCLUSIONS

A wide variety of cases exist in which gene expression can be regulated at levels other than transcription. In some lower organisms, such post-transcriptional regulation may constitute the predominant form of gene control. In mammals, however, it appears to represent an adaptation to particular situations. These include for example, the need to respond rapidly to the withdrawal of hormonal stimulation or stress by the regulation of RNA stability or translation. Similarly, post-transcriptional regulation may be the predominant form of regulation for the proteins which themselves regulate the transcription or splicing of other genes.

Such post-transcriptional regulation of gene expression frequently involves small inhibitory RNAs, such as miRNAs. An individual miRNA can inhibit genes for regulatory proteins, as well as those encoding non-regulatory structural proteins. In turn, the consequent effect on the ability of the regulatory protein to control the transcription or splicing of its target genes allows an individual miRNA to regulate gene expression, indirectly as well as directly and positively as well as negatively.

As well as these post-transcriptional effects on the level of gene expression, whole-genome studies using both gene arrays (see Section 1.2) and the very large amount of DNA sequence data which are now available have confirmed early studies on individual genes and shown the importance of post-transcriptional processes, such as alternative splicing or RNA editing in generating multiple related proteins from one gene. Thus, over 90% of human genes with multiple exons are subject to alternative splicing, whereas multiple new targets for RNA editing have been identified in this way.

Overall, therefore, post-transcriptional processes play quantitative and qualitative roles in the regulation of gene expression, acting as a significant supplement to transcriptional regulation either where a rapid response is required or where multiple related proteins are produced from a single gene.

KEY CONCEPTS

- **Although transcriptional control is the predominant means of regulating gene expression, post-transcriptional control processes also occur often as a supplement to transcriptional control.**

- **Such post-transcriptional control can occur at a number of different points between initial transcription and protein synthesis.**

- Alternative splicing and RNA editing are post-transcriptional processes which are often used as a supplement to transcriptional control to produce distinct but related proteins from a single gene.

- Regulation of RNA stability or of translation are often used as supplements to transcriptional control in situations where a rapid change in protein levels is required.

- Small RNAs play a key role in the post-transcriptional regulation of gene expression by inducing degradation of their target mRNAs or inhibiting their translation.

- Post-transcriptional regulation frequently occurs for the genes encoding transcription factors ensuring that the proteins which regulate transcription are not themselves regulated at the transcriptional level.

FURTHER READING

7.1   Alternative RNA splicing

David CJ & Manley JL (2008) The search for alternative splicing regulators: new approaches offer a path to a splicing code. *Genes Dev.* 22, 279–285.

Editorial (2008) Multitudes of messages. *Nat. Genet.* 40, 1385.

Hertel KJ (2008) Combinatorial control of exon recognition. *J. Biol. Chem.* 283, 1211–1215.

Kornblihtt AR (2005) Promoter usage and alternative splicing. *Curr. Opin. Cell Biol.* 17, 262–268.

Matlin AJ, Clark F & Smith CWJ (2005) Understanding alternative splicing: towards a cellular code. *Nat. Rev. Mol. Cell Biol.* 6, 386–398.

McGlincy NJ & Smith CW (2008) Alternative splicing resulting in nonsense-mediated mRNA decay: what is the meaning of nonsense? *Trends Biochem. Sci.* 33, 385–393.

7.2   RNA editing

Samuel CE (2003) RNA editing minireview series. *J. Biol. Chem.* 278, 1389–1390.

7.3   Regulation of RNA transport

Besse F & Ephrussi A (2008) Translational control of localized mRNAs: restricting protein synthesis in space and time. *Nat. Rev. Mol. Cell Biol.* 9, 971–980.

Martin KC & Ephrussi A (2009) mRNA localization: gene expression in the spatial dimension. *Cell* 136, 719–730.

Mili S & Macara IG (2009) RNA localization and polarity: from A(PC) to Z(BP). *Trends Cell Biol.* 19, 156–164.

Sandri-Goldin RM (2004) Viral regulation of mRNA export. *J. Virol.* 78, 4389–4396.

St Johnston D (2005) Moving messages: the intracellular localization of mRNAs. *Nat. Rev. Mol. Cell Biol.* 6, 363–375.

Wharton RP (2009) A splicer that represses (translation). *Genes Dev.* 23, 133–137.

7.4   Regulation of RNA stability

Goldstrohm AC & Wickens M (2008) Multifunctional deadenylase complexes diversify mRNA control. *Nat. Rev. Mol. Cell Biol.* 9, 337–344.

Mata J, Marguerat S & Bahler J (2005) Post-transcriptional control of gene expression: a genome-wide perspective. *Trends Biochem. Sci.* 30, 506–514.

Wilusz CJ & Wilusz J (2004) Bringing the role of mRNA decay in the control of gene expression into focus. *Trends Genet.* 20, 491–497.

7.5   Regulation of translation

Komar AA & Hatzoglou M (2005) Internal ribosome entry sites in cellular mRNAs: mystery of their existence. *J. Biol. Chem.* 280, 23425–23428.

Richter JD & Sonenberg N (2005) Regulation of cap-dependent translation by eIF4E inhibitory proteins. *Nature* 433, 477–480.

Sonenberg N & Hinnebusch AG (2009) Regulation of translation initiation in eukaryotes: mechanisms and biological targets. *Cell* 136, 731–745.

7.6   Post-transcriptional inhibition of gene expression by small RNAs

Bartel DP (2009) MicroRNAs: target recognition and regulatory functions. *Cell* 136, 215–233.

Eulalio A, Huntzinger E & Izaurralde E (2008) Getting to the root of miRNA-mediated gene silencing. *Cell* 132, 9–14.

Makeyev EV & Maniatis T (2008) Multilevel regulation of gene expression by microRNAs. *Science* 319, 1789–1790.

Pillai RS, Bhattacharyya SN & Filipowicz W (2007) Repression of protein synthesis by miRNAs: how many mechanisms? *Trends Cell Biol.* 17, 118–126.

Wu L & Belasco JG (2008) Let me count the ways: mechanisms of gene regulation by miRNAs and siRNAs. *Mol. Cell* 29, 1–7.

Gene Control and Cellular Signaling Pathways

8

INTRODUCTION

Transcription factors can be regulated by controlling their synthesis or by controlling their activity

As described in Chapter 5, transcription factors are essential for the process of transcription, binding to specific DNA sequences and then acting to either activate or repress transcription of the target gene. Clearly, such transcription factors play a central role in regulating gene transcription. As described in Chapter 1 (Section 1.4), transcription is the key stage at which gene expression is regulated to produce different cell types and tissues or to allow cells to respond to specific stimuli.

To achieve this, however, it is necessary for transcription factors themselves to be differentially active in different cell types or to have their activity modulated by specific stimuli. This allows them in turn to switch their target genes on or off in the appropriate cell types or in response to a specific signal, so producing the appropriate alterations in cellular phenotype.

In general, such regulation of transcription factors themselves can be achieved in two ways (**Figure 8.1**), both of which are actually used in different cases. In the first method (**Figure 8.1a**) gene control is mediated by a transcription factor being synthesized only in certain tissues or cell types. In the tissue or cell type where it is present the factor activates the gene, whereas in its absence no activation occurs.

This mechanism is widely used in situations where it is necessary to maintain a particular pattern of gene expression over a considerable period of time and this can most readily be achieved by synthesizing and maintaining the expression of a particular transcription factor over that period. This method is therefore frequently used for transcription factors that control cell-type-specific or developmentally regulated genes. Thus, the gene encoding the transcription factor will be transcribed only in a specific cell type or at a specific stage of development, resulting in cell-type-specific or developmentally regulated synthesis of the transcription factor itself. This method will be discussed extensively in Chapters 9 and 10 as part of our discussion of developmental and cell-type-specific gene control.

Although such regulation of transcription factor synthesis by controlling the transcription of its corresponding gene represents a simple means of controlling target gene expression, it cannot be the only mechanism used to regulate transcription factors. Thus, regulation of the gene encoding a transcription factor at the level of transcription will require at least one other transcription factor to regulate the gene encoding the first transcription factor. In turn, if the second transcription factor is regulated at the level of its own transcription, then a further factor will be required, and so on (**Figure 8.2**). Hence, the exclusive use of transcriptional regulation would require a potentially endless hierarchy of transcription factor genes, each requiring another transcription factor to regulate its transcription.

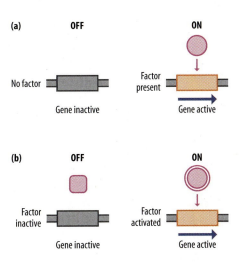

Figure 8.1
Gene activation can be mediated by either the synthesis (a) or the activation (b) of a transcription factor only in a specific cell type or in response to a specific signal.

Figure 8.2
The transcriptional regulation of the gene encoding transcription factor A allows its synthesis in specific situations resulting in the transcriptional regulation of its target genes, leading to a change in cellular phenotype. However, such transcriptional regulation of gene A requires the regulation of at least one other transcription factor (B) which could occur either by regulating the transcription of its corresponding gene or by some other means (these alternative possibilities are indicated by the dashed lines).

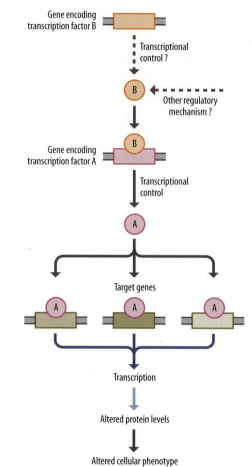

One solution to this problem would be to regulate the expression of the transcription factor gene at a post-transcriptional level. This is indeed observed in some cases such as the yeast transcription factor GCN4, whose synthesis is regulated at the level of mRNA translation in response to amino acid availability (see Section 7.5).

However, as well as regulating transcription factor synthesis at transcriptional or post-transcriptional levels, it is also possible for transcription factors to be regulated at the level of their activity (**Figure 8.1b**). In this mechanism, the transcription factor pre-exists in an inactive form in situations where its target genes are not being expressed. It can then be activated post-translationally in a particular situation to activate its target genes. Hence, in these cases the transcription factor is not regulated at the level of its synthesis but by controlling the activity of pre-existing transcription factor protein.

An example of this is provided by the heat-shock factor (HSF) which, as described in Chapter 4 (Section 4.3), binds to the **heat-shock element** (HSE) in the DNA of stress-inducible genes and plays a key role in their induction in response to elevated temperature or other stresses.

Interestingly, HSF is present in cells prior to exposure to heat shock and can activate the heat-shock genes following exposure to elevated temperature or other stresses, even in the presence of protein-synthesis inhibitors preventing its synthesis *de novo*. It is clear therefore, that upon heat shock the previously inactive HSF is activated to an active form by a post-translational modification involving an alteration of the pre-existing protein. In agreement with this, such activation can be produced in an isolated cell-free nuclear extract exposed to elevated temperature under conditions when no new protein synthesis could occur. The mechanism of this effect is discussed further in Section 8.1. Clearly therefore, the post-translational activation of transcription factors represents an important mechanism in gene regulation (see Figure 8.1b).

Both the regulation of transcription factor synthesis and the regulation of transcription factor activity are therefore used in different situations. In particular, the activation of a pre-existing transcription factor offers the opportunity for a much more rapid response than a process which requires the synthesis *de novo* of the factor via gene transcription, RNA processing, translation, etc. (**Figure 8.3**). It is therefore used frequently in the response

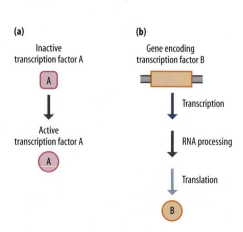

Figure 8.3
The post-translational activation of transcription factor A (a) allows a much more rapid response than the *de novo* transcription of the gene encoding transcription factor B (b), which requires multiple stages before transcription factor protein is produced.

of cells to cellular signaling pathways, where in contrast to cell-type-specific or developmentally regulated gene expression a rapid change in cellular gene expression in response to the signal is required. Thus, exposure of a cell to a specific signal will result in a change in transcription factor activity which in turn will produce changes in the expression of the target genes for the transcription factor. The altered levels of the proteins encoded by these target genes will then produce the appropriate change in cellular characteristics in response to the signal, so as to allow the signal to have a biological effect (**Figure 8.4**). The mechanisms by which cellular signaling pathways regulate transcription factor activity are considered in this chapter.

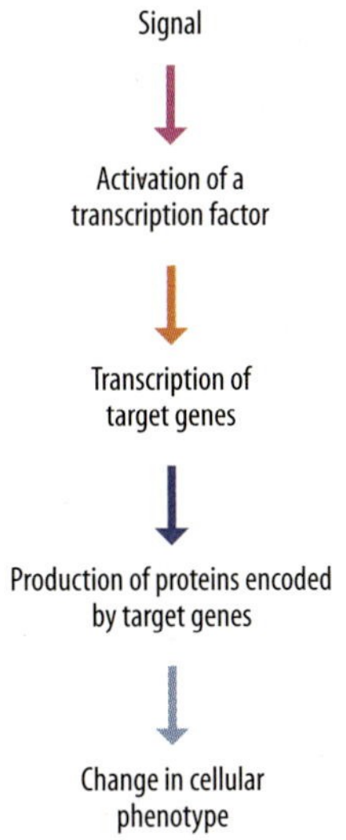

Multiple mechanisms regulate transcription factor activity

As indicated in **Figure 8.5**, a number of different mechanisms allow the regulation of transcription factor activity. In the case of signaling molecules that can enter the cell, direct binding of the ligand to the transcription factor is a frequently used mechanism for transcription factor activation. The regulation of transcription factors by such intracellular ligands is therefore discussed in Section 8.1 (**Figure 8.5a**). Conversely, signaling molecules which bind to cell-surface receptors and cannot enter the cell frequently act by inducing enzymes which can post-translationally modify proteins, for example by phosphorylation (**Figure 8.5b**). The post-translational modification of transcription factors by phosphorylation is therefore discussed in Section 8.2 and other post-translational modifications which regulate transcription factor activity are discussed in Section 8.3.

In contrast to such post-translational modification of transcription factors, other signaling molecules which bind to cell-surface receptors act by inducing the cleavage of a large inactive precursor to produce an active transcription factor molecule (**Figure 8.5c**) and this is discussed in Section 8.4. As we shall discuss, many such cases of activation by ligand binding, post-translational modification or precursor cleavage also involve a change in the interaction of the resulting protein with other proteins (**Figure 8.5d**)

Signal

Activation of a transcription factor

Transcription of target genes

Production of proteins encoded by target genes

Change in cellular phenotype

Figure 8.4
Production of a biological effect by a specific signal can occur by the signal activating a transcription factor. The transcription factor then activates its target genes, resulting in enhanced levels of the proteins they encode, thereby producing the appropriate change in cellular phenotype.

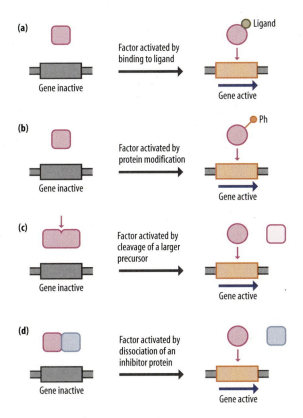

(a) Factor activated by binding to ligand — Gene inactive → Ligand — Gene active

(b) Factor activated by protein modification — Gene inactive → Ph — Gene active

(c) Factor activated by cleavage of a larger precursor — Gene inactive → Gene active

(d) Factor activated by dissociation of an inhibitor protein — Gene inactive → Gene active

Figure 8.5
Mechanisms by which transcription factors can be activated from an inactive (square) to an active (circle) form by post-translational changes.

and this will be discussed when appropriate cases are considered in Sections 8.1–8.4.

Finally, it is important to note that in addition to affecting transcription factors, signaling pathways can also regulate the post-transcriptional processes which, as described in Chapters 6 and 7, play an important role in gene control. The effect of signaling pathways at these stages of gene expression will therefore be discussed in Section 8.5.

8.1 REGULATION OF TRANSCRIPTION FACTOR ACTIVITY BY LIGANDS WHICH ENTER THE CELL

Transcription factors can be activated by direct binding of ligands which enter the cell

In the case of a signaling molecule which enters the cell, it can potentially bind to a transcription factor and directly regulate its activity, for example, by inducing a conformational change in the transcription factor (see Figure 8.5a). An example of such a protein–ligand interaction which activates a transcription factor is the yeast ACE1 factor, which mediates the induction of the metallothionein gene in response to copper. This transcription factor has been shown to undergo a major conformational change upon binding of copper. This conformational change allows it to bind to DNA-binding sites in the regulatory region of the metallothionein gene and induce transcription. Hence the activity of this transcription factor is directly modulated by copper, allowing it to activate gene expression in response to the metal (**Figure 8.6**).

An interesting variant of this direct regulation of transcription factor activity by a specific stimulus is seen in the yeast Yap1 factor, which is active under conditions of high oxygen and regulates the expression of antioxidant genes. This factor contains **disulfide bonds** between cysteine amino acids which create a structure for the protein that masks its nuclear export signal and it is therefore retained in the nucleus where it can regulate its target genes. Under conditions of low oxygen the disulfide bonds are reduced, leading to them break, and the protein refolds so that the nuclear export signal is exposed. This results in export of Yap1 to the cytoplasm where it can no longer regulate transcription (**Figure 8.7**).

Members of the nuclear receptor family of transcription factors are activated by binding of the appropriate ligand

As the examples above indicate, direct regulation of transcription factor activity by an inducing stimulus is widely used in yeast cells which are in close contact with their environment. However, it also occurs in multicellular organisms where the best-characterized example is the nuclear receptor family of transcription factors (see Section 5.1), which includes the receptors for steroid hormones and thyroid hormone.

A simple example of direct regulation by ligand is seen in the thyroid hormone receptor member of the nuclear receptor family. As discussed in Section 5.3, this receptor binds to DNA in the absence of thyroid hormone and by recruiting co-repressor molecules represses transcription. Upon binding of hormone, however, the DNA-bound transcription factor undergoes a conformational change which allows it to bind co-activator rather than co-repressor molecules and therefore activate transcription of its target genes. This allows thyroid hormone to produce its biological effect of enhancing cellular metabolism by regulating specific target genes. This case is similar to that of ACE1 in terms of direct binding of the ligand to the transcription factor. It differs, however, in that thyroid hormone produces a conformational change in the DNA-bound receptor, which contrasts with the ability of copper to produce a conformational change in ACE1 that allows it to bind to DNA.

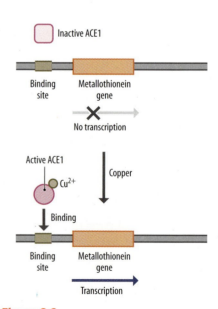

Figure 8.6
Activation of the ACE1 factor in response to copper results in transcription of the metallothionein gene.

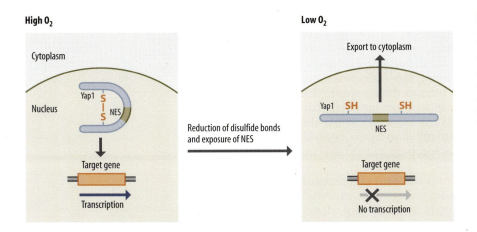

Figure 8.7
Under conditions of high oxygen, the Yap1 protein contains disulfide bonds which mask its nuclear-export signal (NES) and allow it to remain in the nucleus and activate its target genes. Under conditions of low oxygen, the disulfide bonds are reduced, exposing the NES. This results in export of Yap1 to the cytoplasm, preventing it from activating its target genes.

In contrast to the situation with the thyroid hormone receptor, several other members of the nuclear receptor family do not bind to DNA in the absence of the appropriate hormone. The glucocorticoid receptor, for example, is normally located in the cytoplasm and only binds to DNA and activates transcription when the hormone is added. Originally it was thought that, as with ACE1, binding of hormone to the receptor activated its ability to bind to DNA and switch on transcription of hormone-responsive genes. However, it has been shown that, although the receptor binds to DNA only in the presence of glucocorticoid hormone in the cell, in the test tube it will bind to DNA even when no hormone is present.

This has led to the idea that, in the cell, the receptor is prevented from binding to DNA by its association with another protein and that the hormone acts to release it from this association and allow it to fulfil its inherent ability to bind to DNA. In agreement with this idea, the glucocorticoid receptor has been shown to be associated in the cytoplasm with a 90 kDa heat-inducible protein (HSP90). Upon steroid binding, the receptor dissociates from HSP90 and forms a receptor dimer. The loss of HSP90 exposes the **nuclear localization signal** of the receptor, allowing it to interact with nuclear-import proteins. This promotes its movement to the nucleus where it regulates gene transcription (**Figure 8.8**).

Formation of a complex with HSP90 is not unique to the glucocorticoid receptor and has also been reported for other steroid receptors, such as the estrogen receptor and the progesterone receptor. Hence, transcription activated by these hormone receptors involves an interaction with the ligand which disrupts a protein–protein interaction that otherwise inhibits the inherent DNA-binding ability of the receptor by anchoring it in the cytoplasm.

Although the hormone-induced dissociation of HSP90 from the receptors is sufficient for their binding to DNA, it is not sufficient for transcriptional activation. Thus, such transcriptional activation requires a ligand-induced structural change in the receptor which exposes its C-terminal transcription activation domain (see Section 5.2), allowing the receptor to activate transcription in a hormone-dependent manner.

Structural analysis of the receptors has shown that the hormone induces a conformational change in the receptor which results in the re-alignment of the ligand-binding domain so that it forms a lid over the bound ligand. As well as preventing the ligand from dissociating from the receptor, this

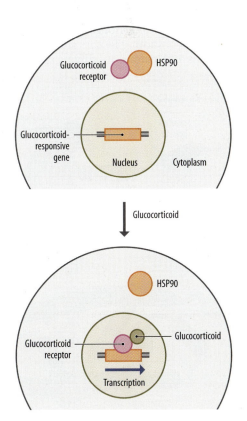

Figure 8.8
Binding of glucocorticoid to the glucocorticoid receptor results in its dissociation from HSP90 and movement of the hormone–receptor complex to the nucleus where it activates the transcription of glucocorticoid-responsive genes.

configuration allows the activation domain of the receptor to interact with transcriptional co-activator molecules (see Section 5.2), thereby allowing transcriptional activation to occur (**Figure 8.9**).

Interestingly, although estrogen itself is able to induce this change in the estrogen receptor, so activating transcription, the estrogen antagonist raloxifene does not do so, although it does bind to the receptor. This provides further evidence for the key role of this ligand-induced structural change in transcriptional activation. Moreover, it also explains the ability of raloxifene to antagonize activation of gene expression by estrogen since it competes with estrogen for binding to the receptor but cannot activate transcription.

Steroid receptors such as the glucocorticoid and estrogen receptors are therefore activated in a two-stage process involving, first, unmasking of their DNA-binding ability by dissociation from HSP90 and, second, a change in their transcriptional activation ability (**Figure 8.10a**). This combines the mechanisms of ligand-induced conformational change and dissociation from an inhibitory protein, which are illustrated respectively in Figure 8.5a and Figure 8.5d.

Comparison of this two-stage activation with the activation of the thyroid hormone receptor (see above and Section 5.3) indicates that the thyroid hormone receptor uses only the second of these two stages in which a conformational change is induced by the hormone and this allows the activation domain of the receptor to bind co-activator molecules (**Figure 8.10b**).

Hence, in all nuclear receptors, activation by ligand involves structural changes in the C-terminal activation domain which allows it to bind co-activators. However, in the case of receptors for steroids such as glucocorticoid and estrogen, this is preceded by an earlier step involving disruption of the HSP90–receptor interaction.

In all the nuclear receptors a critical role in transcription activation is played by co-activators which are recruited by the receptor in a hormone-dependent manner. A number of co-activators which are able to interact with nuclear receptors have been defined and include TIF1, TIF2, SRC-1, SRC-3, and CREB-binding protein (CBP). As discussed in Chapter 5 (Section 5.2), co-activators can activate transcription both by interacting with the basal transcriptional complex to stimulate its activity and by altering chromatin structure to allow transcriptional activation to occur. Indeed, it is clear that many co-activators exist as part of multi-protein complexes with the complex containing factors such as ATP-dependent chromatin-remodeling proteins and histone-modifying enzymes (such as histone acetyltransferases

Figure 8.9
Schematic diagrams (a, b) and structural model (c) of nuclear receptors in the presence or absence of ligand. Note that binding of ligand produces a conformational change so that the C-terminal domain (pink) can form a lid over the ligand and the activation domain can interact with co-activator molecules.

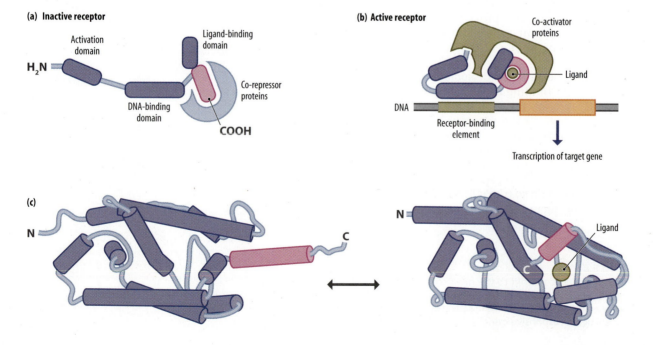

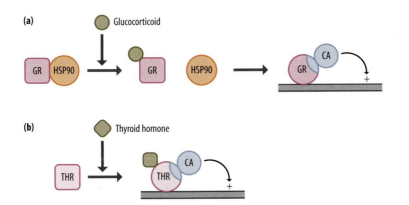

Figure 8.10
Activation of receptors such as the glucocorticoid receptor (GR) by binding of glucocorticoid involves both dissociation from HSP90 and a change in the structure of the receptor, allowing it to bind co-activator (CA) molecules (a). In contrast, activation of the thyroid hormone receptor by binding of thyroid hormone involves only the second of these stages (b).

and histone methyltransferases) which together can produce an open chromatin structure compatible with transcription (see Chapter 3).

Following ligand-mediated activation, the glucocorticoid receptor can repress as well as activate gene transcription

Interestingly, the glucocorticoid receptor can also have a negative as well as a positive effect on transcription. Unlike the thyroid hormone receptor, however, inhibition of gene transcription by the glucocorticoid receptor requires binding of glucocorticoid. Hence, glucocorticoid hormone can produce both negative and positive effects on transcription via the glucocorticoid receptor.

The glucocorticoid receptor has been shown to repress transcription by three different mechanisms. In the first of these, the hormone–receptor complex binds to a target DNA sequence in the target gene. However, this target sequence (known as the **nGRE**) is distinct from (although related to) the glucocorticoid-response element (GRE), which is present in genes that are activated by binding of the glucocorticoid receptor. In the case of the *POMC* gene, the presence of an nGRE results in the receptor binding as a trimer of three receptor molecules rather than the dimeric form which binds to the standard GRE (see Section 5.1).

Unlike the glucocorticoid receptor dimer, the trimer form of the receptor is unable to activate transcription itself and appears to repress transcription by preventing the binding of positively acting transcription factors to adjacent or overlapping sites in the gene-regulatory region (**Figure 8.11**). For example, in the human glycoprotein hormone α-subunit gene, which contains a negative GRE, the nGRE overlaps the positively acting cyclic AMP-response element (**CRE**) and only inhibits gene expression when the CRE is intact. In this situation therefore, the glucocorticoid receptor acts as an indirect repressor which inhibits transcription by binding to DNA and preventing the binding of a positively acting transcription factor (see Section 5.3 and Figure 5.54b).

In contrast to this method of transcriptional repression, the second method of transcriptional repression by the glucocorticoid receptor does not involve DNA binding. This second mechanism relies on the fact that, as described in Section 5.2, very many different transcriptional activators interact with the CBP co-activator and require it to produce transcriptional activation.

As the cell contains limiting amounts of CBP, this results in different transcription factors competing with one another for the limited amount of CBP. Thus, like the glucocorticoid receptor, the Fos–Jun complex which was described in Chapter 5 (Section 5.1; see also Section 11.2), requires CBP to produce transcriptional activation. The glucocorticoid receptor therefore competes with the Fos–Jun complex for the limiting amounts of CBP in the cell (**Figure 8.12**). This results for example, in hormone-mediated activation of the glucocorticoid receptor producing repression of the collagenase gene

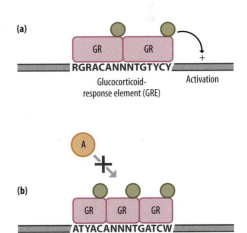

Figure 8.11
Binding of a dimer of the glucocorticoid receptor (GR) to the glucocorticoid-response element (GRE) activates transcription (a). In contrast, the receptor binds as a trimer to the related but distinct nGRE sequence. In this conformation it cannot activate transcription itself and blocks the binding of activating factors (A), so repressing transcription (b).

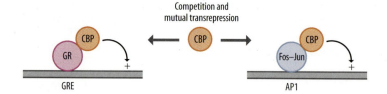

which is dependent upon Fos–Jun for its activation. This occurs via competition for CBP rather than by binding of the glucocorticoid receptor to the collagenase gene which does not contain positively or negatively acting GREs. Clearly, this effect will be a reciprocal one, so that activation of the Fos–Jun pathway, for example by phorbol esters (see Section 11.2), will result in repression of glucocorticoid receptor-dependent genes as CBP binds to Fos–Jun.

Hence, the ligand-activated glucocorticoid receptor can repress transcriptional initiation both by binding to DNA and preventing a positively acting factor from binding and by competing with other positively acting transcription factors for an essential co-activator. Interestingly, the final mechanism by which the glucocorticoid receptor can repress transcription involves the repression of transcriptional elongation. The activated glucocorticoid receptor can prevent the recruitment of the pTEF-b kinase to the promoter of the interleukin-8 gene. As noted in Chapter 4 (Section 4.2), such recruitment is required for the phosphorylation of the C-terminal domain of RNA polymerase II on serine 2 which in turn is necessary for transcriptional elongation to occur. Hence, by blocking the recruitment of pTEF-b to the promoter, the activated glucocorticoid receptor represses transcription of the interleukin-8 gene (**Figure 8.13**).

Hence, the glucocorticoid receptor illustrates three different mechanisms of transcriptional repression involving, blocking the binding of a transcriptional activator, competing with other activators for a co-activator, and inhibiting transcriptional elongation (**Figure 8.14a**) (see Section 5.3 for further discussion of transcriptional repression mechanisms). As with activation of gene expression by the receptor, all these inhibitory

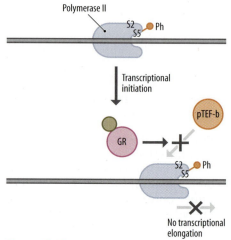

Figure 8.13
Following activation by glucocorticoid hormone, the glucocorticoid receptor (GR) can inhibit transcriptional elongation by blocking recruitment of the pTEF-b kinase.

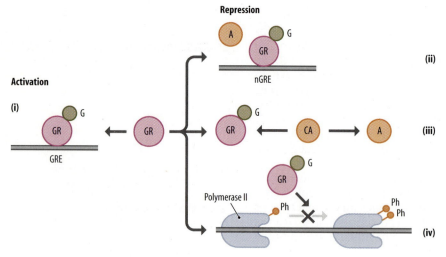

(a) Glucocorticoid receptor

(b) Thyroid hormone receptor

Figure 8.14
(a) Following binding of glucocorticoid (G) the glucocorticoid receptor (GR) can activate transcription by binding to a GRE (i). It can also repress the transcription of other target genes by binding to an nGRE and blocking binding of an activator (A) (ii), by competing with other activators for binding of a co-activator (CA) (iii), or by blocking transcriptional elongation (iv). (b) In contrast, the thyroid hormone receptor (TR) inhibits transcription in the absence of thyroid hormone and activates the same genes in the presence of thyroid hormone.

mechanisms require binding of glucocorticoid to the receptor. This contrasts with the ability of the thyroid hormone receptor to directly repress transcription in the absence of thyroid hormone while activating transcription of the same genes in the presence of thyroid hormone (**Figure 8.14b**).

The HSF is activated by stressful stimuli and induces the transcription of genes encoding protective proteins

The HSP90 protein also plays a critical role in the activation of HSF-1, which stimulates the transcription of genes induced by heat and other stresses. Genes inducible in this way include those encoding HSP70 (see Section 4.3) and HSP90 itself. As noted above, HSF can be activated in cultured cells or in cellular extracts under conditions when new protein synthesis cannot occur. It must therefore be activated by modification of a pre-existing inactive form.

In fact, prior to exposure of cells to heat or other stresses, HSF exists as a monomer associated with HSP90. Under these conditions, the molecule folds in such a way that its DNA-binding domain is masked and it cannot therefore bind to DNA. Following exposure to elevated temperature or other stresses, HSF-1 is converted to a trimer of three HSF-1 molecules that can bind to DNA. As with the steroid hormone receptors, the activation of HSF-1 involves the dissociation of HSP90 which binds to HSF-1 in unstressed cells and maintains it in an inactive form that cannot bind to DNA (**Figure 8.15**).

Interestingly, HSP90, like many of the proteins inducible by heat or other stresses, functions as a so-called "**chaperone**" protein promoting the folding of unfolded or poorly folded proteins and thereby protecting the cell against this effect of stressful stimuli. Clearly, heat or other stresses will increase the amount of such unfolded proteins with HSP90 being "called away" to deal with them, thereby releasing HSF-1. Hence, this system neatly couples the effect of the inducing stimulus in producing unfolded proteins with the activation of HSF-1.

In addition to dissociation of HSP90, two other factors are required for trimerization of HSF-1. One of these is eEF1A, a factor which is normally involved in the translation of mRNAs into protein (see Section 6.6). As heat shock or other stresses shut down most cellular translation, eEF1A is released from the translational apparatus and can participate in the activation of HSF-1. As well as dissociation of HSP90 and association with eEF1A, trimerization is also stimulated by a small RNA molecule of approximately 600 bases in size, known as HSR1. HSR1 is therefore a small RNA which is able to participate in the activation of gene expression, as opposed to the more widespread small inhibitory RNAs discussed in previous chapters. Hence, trimerization involves the dissociation of one protein (HSP90) from HSF-1 and the association of another protein (eEF1A) and a small RNA (HSR1) (**Figure 8.16**).

Trimerization of HSF-1 allows it to bind the HSE. However, this is not sufficient to promote transcriptional activation. For this to occur, HSF-1 must be phosphorylated on the serine amino acid at position 230 (see Figure 8.15). HSF-1 therefore combines activation mechanisms involving dissociation of an inhibitory protein (see Figure 8.5d) with the activation of a transcription factor by phosphorylation see (see Figure 8.5b). Such activation of transcription factors by phosphorylation is predominantly used however by factors which bind to cell-surface receptors and then trigger further events inside the cell, culminating in phosphorylation or other modifications of specific transcription factors. Such signaling pathways are discussed further in the next section.

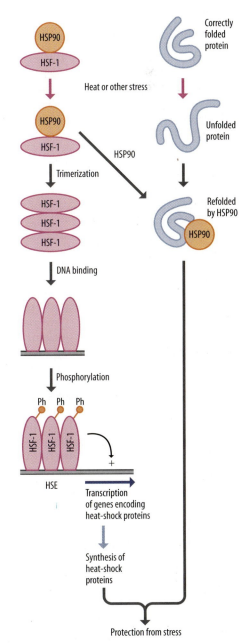

Figure 8.15

In non-stressed cells, HSF-1 exists in a non-DNA-binding form associated with HSP90. In response to stresses such as elevated temperatures, HSP90 dissociates from HSF-1 and binds to unfolded proteins produced as a result of the stress. This allows HSF-1 to form a trimer which can bind to DNA but which needs to be further modified by phosphorylation (Ph; orange dots) before it can activate transcription of the genes encoding the heat-shock proteins.

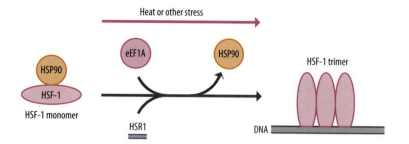

Figure 8.16
In response to heat or other stress, HSP90 dissociates from HSF-1. The eEF1A protein and the small RNA, HSR1 associate with HSP90. This allows HSF-1 to trimerize, exposing its DNA-binding domain and allowing it to bind to DNA.

8.2 REGULATION OF TRANSCRIPTION FACTOR ACTIVITY BY PHOSPHORYLATION INDUCED BY EXTRACELLULAR SIGNALING MOLECULES

Transcription factors can be phosphorylated by receptor-associated kinases

Molecules such as the steroid hormones and thyroid hormone can cross the cell membrane and therefore regulate transcription factors by binding to them directly within the cell. This mechanism is not possible for signaling molecules such as the many proteins which act at the cell surface and do not penetrate the cell. Such **extracellular signaling molecules** must therefore act by binding to cell-surface receptors which in turn send a signal into the cell that ultimately results in modification of specific transcription factors (**Figure 8.17**).

As will be discussed later in this section, many such intracellular signaling systems involve multiple steps before the modification of the transcription factor itself occurs. However, a much simpler system is used in the case of the cytokines, a family of small proteins which play a key role in regulating the growth and differentiation of specific cell types and in the response to viral infection.

These molecules bind to specific receptors which span the cell membrane. Following binding of the cytokine to the extracellular portion of the receptor, a signal is transmitted to a JAK protein which is associated with the intracellular region of the receptor (**Figure 8.18**). This signal activates the kinase activity of the JAK protein and allows it to phosphorylate members of the **STAT** (signal transducers and activators of transcription) family of transcription factors on specific tyrosine residues. In unstimulated cells, the STAT factors reside in the cytoplasm. However, following stimulation of the cell and consequent phosphorylation of the STAT factors, they form STAT–STAT dimers and migrate to the nucleus where they can bind to specific DNA sequences and switch on target gene expression. This signaling pathway involves JAK protein kinases phosphorylating STAT transcription factors and is therefore known as the **JAK/STAT pathway** (see Figure 8.18).

This simple mechanism allows the receptor to transmit a signal from an extracellular molecule to a kinase located within the cell which in turn phosphorylates a specific transcription factor. A similar mechanism is also used by a different family of receptors which bind the transforming growth factor β (TGFβ) family of proteins, leading to the activation of another family of transcription factors called **Smads**.

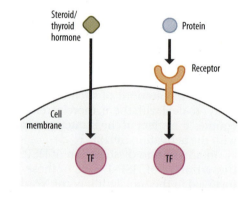

Figure 8.17
A signal (such as a steroid hormone or thyroid hormone) which can enter the cell is able to activate a transcription factor (TF) by binding directly to it. In contrast, a protein which cannot enter the cell needs to bind to a cell-surface receptor which then transduces the signal to the transcription factor either directly or via intermediate signaling molecules.

Transcription factors can be phosphorylated by kinases activated by specific intracellular second messengers such as cyclic AMP

The cases discussed above are clearly examples involving the activation of a transcription factor by phosphorylation, as illustrated in Figure 8.5b. Such modification of transcription factors by phosphorylation is a very widespread mechanism of controlling transcription factor activity. In many cases,

Figure 8.18
Binding of cytokines to the extracellular region of their specific receptors activates JAK proteins which are associated with the intracellular region of the receptor. This activates the kinase activity of the JAK protein, allowing it to phosphorylate STAT transcription factors. In turn, phosphorylation allows the STAT factor to dimerize, enter the nucleus, and bind to its specific DNA-binding site, resulting in transcription of its target genes.

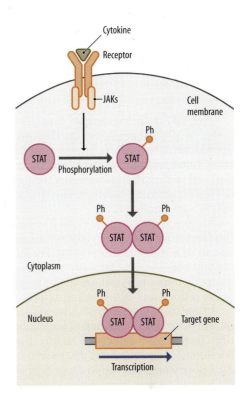

however, the signaling pathway involved is much more complex than that seen in the activation of STAT or Smad factors. In the case of the cyclic AMP-response-element-binding protein (CREB) transcription factor, transcriptional activation occurs in response to a rise in the intracellular concentration of cyclic AMP, as implied by the name of the transcription factor itself.

The cyclic AMP/CREB pathway is initiated by the binding of extracellular signaling molecules, such as glucagon, serotonin, or epinephrine (adrenalin) to specific receptors on the cell surface. Unlike the STAT or Smad systems, in the case of the cyclic AMP/CREB pathway such receptors are not associated with kinase enzymes that directly phosphorylate transcription factors. Rather, receptors of this type are associated with trimeric proteins which are known as **G proteins** because they can bind the nucleotides GDP and GTP. Activation of these **G-protein-coupled receptors** results in a move of the associated G protein from the inactive form in which it binds GDP to the active form in which it binds GTP (**Figure 8.19**). In turn, the active GTP form of the G protein activates the enzyme **adenylate cyclase** which catalyzes the production of cyclic AMP from ATP (see Figure 8.19).

Cyclic AMP is an example of a second-messenger molecule which acts within the cell following activation by an extracellular stimulus. As a small cyclic nucleotide, it can move more rapidly in the cell than can a protein molecule. Moreover, one single receptor/active G protein can stimulate the synthesis of large numbers of cyclic AMP molecules, thereby amplifying the signal effectively compared to the direct phosphorylation of target transcription factors by receptor-associated kinases.

Following its production, cyclic AMP binds to the regulatory subunit of an enzyme known as protein kinase A and promotes the dissociation of the regulatory subunit from the catalytic subunit with which it is associated (see Figure 8.19). The free catalytic subunit then enters the nucleus, where it phosphorylates the CREB transcription factor on serine 133 of the protein.

Interestingly, such phosphorylation does not modulate the DNA-binding activity of CREB. Indeed, CREB is already bound to its cyclic AMP-response element (CRE) DNA-binding site prior to activation. Rather, protein kinase A phosphorylates CREB in a region known as the phosphorylation box, located between two glutamine-rich regions which act as transcriptional activation domains (see Section 5.2). This phosphorylation results in an enhanced ability of CREB to activate transcription following phosphorylation.

As described in Chapter 5 (Section 5.2), this effect is dependent upon specific binding of the co-activator CREB-binding protein (CBP), which binds to CREB when CREB is phosphorylated on serine 133 (**Figure 8.20**). Once CBP has bound to CREB, CBP can then activate transcription via association with components of the basal transcriptional complex and/or by altering the chromatin structure to a more open configuration (see Section 5.2).

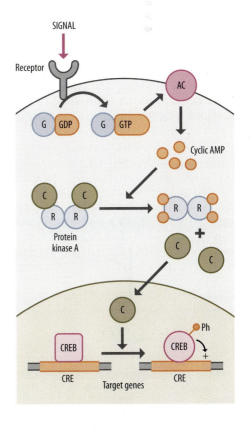

Figure 8.19
Binding of specific signal molecules to G-protein-coupled cell-surface receptors results in conversion of the associated G protein (G) from its inactive GDP-bound form to its active GTP-bound form. In turns, this results in the activation of the adenylate cyclase enzyme (AC) leading to production of cyclic AMP. Binding of cyclic AMP to the regulatory (R) subunit of protein kinase A results in the release of its catalytic subunit (C). The catalytic subunit then moves to the nucleus and phosphorylates the CREB transcription factor which is bound to the cyclic AMP-response element (CRE) in target genes.

Figure 8.20
Phosphorylation of the CREB transcription factor in response to cyclic AMP stimulates the ability of CREB to bind the CBP co-activator. It therefore results in the activation of cyclic AMP-inducible genes containing the cyclic AMP-response element (CRE) to which CREB binds.

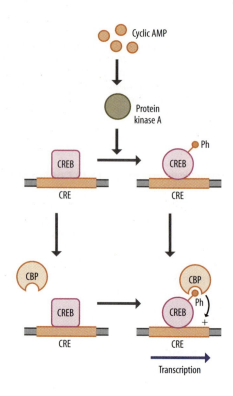

The CREB/CBP example involves the modification of a transcription factor by phosphorylation (as illustrated in Figure 8.5b) combined with an altered protein–protein interaction (as illustrated in Figure 8.5d). In this case, however, phosphorylation promotes an association between CREB and CBP rather than promoting the dissociation of an inhibitory protein, as is seen for steroid binding to the glucocorticoid receptor and dissociation of HSP90 (see Section 8.1).

Although the cases discussed so far involve the activation of a transcription factor by phosphorylation, this modification can also reduce the activation of a transcription factor. For example, kinases activated by another **second messenger**, namely an increase in cellular calcium levels, produce multiple phosphorylations of the Ets-1 transcription factor which progressively reduce its DNA-binding activity (**Figure 8.21**).

Interestingly, calcium-activated kinases also phosphorylate a histone deacetylase enzyme which is normally associated with the MEF2 transcription factor and directs a tightly packed chromatin configuration (see Sections 2.3 and 3.3 for a discussion of histone acetylation/deacetylation). Phosphorylation of the histone deacetylase results in its dissociation from MEF2 and export from the nucleus, resulting in activation of gene expression by MEF2 (**Figure 8.22**). Hence, as well as targeting transcription factors and the histones themselves (see Sections 2.3 and 3.3) phosphorylation can target histone-modifying enzymes, thereby altering chromatin structure (for further discussion of MEF2, its regulation, and its biological roles see Sections 10.1 and 10.2).

Transcription factors can be phosphorylated by signaling cascades consisting of several protein kinases

In the examples described so far, a transcription factor is phosphorylated by a kinase enzyme that is either directly associated with a cell-surface receptor or is activated by second-messenger molecules such as cyclic AMP or calcium. It is also possible, however, for a signal to be transmitted from the cell surface to a specific transcription factor by a cascade of kinase enzymes. Each kinase in the cascade phosphorylates the next kinase enzyme and activates it, allowing it to phosphorylate and activate the next kinase in the pathway, and so on. Ultimately, this results in the final kinase in the pathway phosphorylating the target transcription factor.

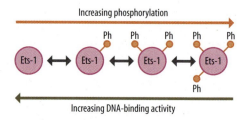

Figure 8.21
Multiple phosphorylations progressively reduce the DNA-binding activity of the Ets-1 transcription factor.

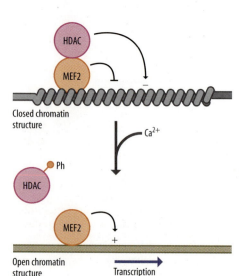

Figure 8.22
In the absence of calcium, the MEF2 transcription factor cannot activate transcription because it is bound to histone deacetylase (HDAC) enzymes which produce an inactive chromatin structure. Elevated calcium levels induce phosphorylation of the histone deacetylases, resulting in their dissociation and allowing MEF2 to activate transcription.

Figure 8.23

Activation of a receptor tyrosine kinase (RTK) by its ligand results in it phosphorylating itself on its intracellular domain. This allows it to bind an adaptor protein (A) which in turn allows the Ras protein to bind GTP, thereby activating it. Active Ras then activates the Raf kinase (shown as Raf; also known as mitogen-activating protein (MAP) kinase kinase kinase), which in turn phosphorylates MEK (also known as MAP kinase kinase), thereby activating it and allowing it to phosphorylate MAP kinase (MAPK), thereby activating it.

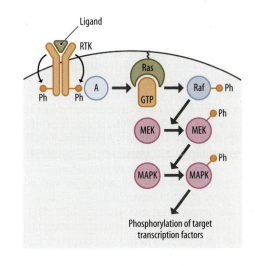

An example of this is the family of **receptor tyrosine kinases** which bind a variety of growth factors that regulate the growth and differentiation of specific cell types. Binding of the growth factor ligand activates the tyrosine activity of the receptor and allows it to phosphorylate itself. In turn, this allows the receptor to recruit adaptor proteins which bind members of the **Ras family** of proteins (Figure 8.23). Like the **trimeric G proteins** described above, **Ras proteins** can bind GDP or GTP, although they differ from trimeric proteins in that they exist as monomers. Binding of the Ras protein to the adaptor protein promotes its binding of GTP rather than GDP and therefore converts it into an active form (see Figure 8.23).

The active Ras-GTP protein then binds to the Raf kinase enzyme (also known as mitogen-activating protein (MAP) kinase kinase kinase) activating its kinase activity and setting off a kinase cascade. In this cascade, active Raf phosphorylates the MEK kinase (also known as MAP kinase kinase). MEK then phosphorylates **MAP kinase**, also known as extracellular regulated kinase (ERK) (see Figure 8.23). The activated MAP kinase then phosphorylates a number of target proteins including several transcription factors and regulates their activity so allowing target genes to be activated.

One key target for the activated MAP kinase is the transcription factor ternary complex factor (TCF), a member of the **Ets family** of transcription factors (see Section 5.1). TCF plays a key role in the activation of the c-*fos* gene (see Section 11.2) by growth factors. Active MAP kinase translocates to the nucleus where it phosphorylates TCF (Figure 8.24). This is therefore an example of the activation of a transcription by phosphorylation (see Figure 8.5b) mediated via a multiple kinase **signaling cascade**.

The situation is more complicated, however, because activation of MAP kinase also results in the phosphorylation of another transcription factor, serum-response factor (SRF). Following phosphorylation, two molecules of SRF and one of TCF form a trimeric complex which binds to the serum-response element (SRE) in the c-*fos* gene and activates its transcription (see Figure 8.24).

Interestingly unlike activation of TCF, activation of SRF by MAP kinase is indirect. Activation of MAP kinase results in the phosphorylation of another kinase p90^{RSK} in the cytoplasm. The active phosphorylated p90^{RSK} then translocates to the nucleus where it phosphorylates SRF (see Figure 8.24). Hence, in this case, two different transcription factors are activated either directly or indirectly by MAP kinase, resulting in their forming a complex and activating transcription of the c-*fos* gene.

The MAP kinase system illustrates the activation of a transcription factor by phosphorylation produced by a cascade of multiple enzymes. Hence, transcription factors in different systems can be phosphorylated either by receptor-associated kinases (Figure 8.25a), by kinases activated by second messengers such as cyclic AMP or calcium (Figure 8.25b), or by multiple kinase cascades (Figure 8.25c).

Transcription factor activity can be regulated by phosphorylation of an inhibitory protein: the NFκB/IκB system

In the cases described so far, cellular signaling processes result in the direct phosphorylation of a target transcription factor by specific kinases. It is also possible, however, for cellular signaling processes to induce the phosphorylation of an inhibitory protein which is associated with the transcription

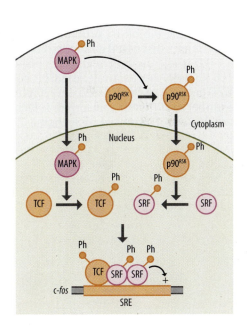

Figure 8.24

Active MAP kinase (MAPK) can translocate to the nucleus and phosphorylate the TCF transcription factor. In addition, MAPK phosphorylates p90^{RSK} in the cytoplasm. Active p90^{RSK} translocates to the nucleus where it phosphorylates the SRF transcription factor. Together one molecule of phosphorylated TCF and two molecules of phosphorylated SRF form a trimeric complex which binds to the SRE DNA sequence in the c-*fos* gene and activates its transcription.

factor. This process has been most extensively characterized in the case of the NFκB transcription factor which is activated following exposure of cells to a variety of stimuli including inflammatory or stressful stimuli. Such stimuli can result, for example, in the release of inflammatory cytokines such as tumor necrosis factor α (TNFα) or interleukin-1 which then bind to their specific cell-surface receptors. Similarly, following bacterial infection, specific bacterial components bind to receptors known as Toll-like receptors. In both cases, NFκB is activated following binding of the ligand to the appropriate receptor. Activated NFκB then plays a key role in the cellular response to inflammation or bacterial infection.

Prior to exposure to activating stimuli, the two subunits of the NFκB transcription factor (p50 and p65) are associated with an inhibitory protein, known as IκB (**Figure 8.26**). Such binding masks the nuclear localization signals on p50 and p65, thereby ensuring that NFκB remains in the cytoplasm.

Following exposure to activating stimuli and the consequent activation of the receptors described above, IκB kinases are activated and phosphorylate IκB (see Figure 8.26). This phosphorylation makes IκB a target for further modification by addition of the small protein ubiquitin. As discussed in Chapter 2 (Section 2.3), the histones are also modified by ubiquitin addition, resulting in a change in their activity. In the case of IκB and many other proteins, however, the addition of ubiquitin targets them for degradation by specific proteolytic enzymes. The degradation of ubiquitinated IκB results in the exposure of the nuclear localization signals on the NFκB molecule and allows it to move to the nucleus and activate gene transcription (see Figure 8.26).

Such activation of NFκB is similar to the dissociation of the glucocorticoid receptor from HSP90 following steroid treatment (see Section 8.1), although in this case phosphorylation rather than steroid binding dissociates the protein–protein complex. Hence, this activation of NFκB combines the disruption of a protein–protein interaction (see Figure 8.5d) with the use of phosphorylation (see Figure 8.5b), although the target for phosphorylation is the inhibitory protein rather than the transcription factor itself, as in the cases discussed earlier in this section.

IκB is not the only regulatory protein which can be phosphorylated by IκB kinases. For example, the CBP co-activator is also phosphorylated by this enzyme. As discussed in Chapter 5 (Section 5.2), NFκB is one of a number of different transcription factors which utilize CBP as a co-activator. Moreover, as noted in Section 8.1, such pathways can compete for the

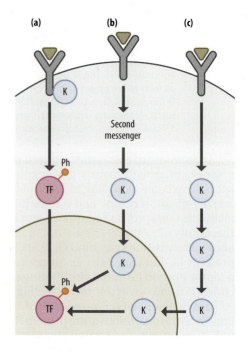

Figure 8.25
Transcription factors (TF) can be phosphorylated by receptor-associated kinases (a), by second-messenger-activated kinases (b), or by cascades of multiple kinases (c). K, kinase.

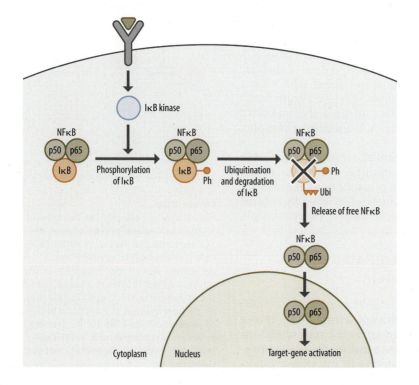

Figure 8.26
Activation of receptors, such as the TNFα receptor or the interleukin (IL)-1 receptor, activates the IκB kinase, which phosphorylates IκB. In turn, phosphorylation of IκB results in its ubiquitination and subsequent degradation. This releases free NFκB (consisting of the two subunits p50 and p65), which moves to the nucleus and activates its target genes.

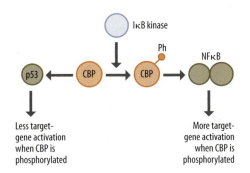

Figure 8.27
Phosphorylation of CBP by IκB kinase increases its binding to NFκB and reduces its binding to p53. It therefore enhances NFκB-dependent gene expression at the expense of p53-dependent gene expression.

limited amount of CBP in the cell. Interestingly, phosphorylation of CBP by IκB kinase enhances its binding to NFκB while reducing its binding to other transcription factors such as p53. Hence, in this case phosphorylation is used to further augment the activity of the NFκB pathway by making it a more effective competitor for the limited amounts of CBP in the cell (**Figure 8.27**). Moreover, in this case, it is CBP itself which is the target for activation by phosphorylation rather than the CREB transcription factor. This indicates that co-activators as well as DNA-binding transcriptional activators can be targets for modulation by phosphorylation.

The effectiveness of IκB can also be modulated at the level of its synthesis as well as by phosphorylation. For example, treatment with glucocorticoid hormone suppresses NFκB-mediated gene activation by stimulating the synthesis of IκB, resulting in increased levels of inactive NFκB/IκB complexes (**Figure 8.28**).

Interestingly, the synthesis of one form of IκB, IκBα, is actually stimulated by NFκB, which transcriptionally activates the gene encoding IκBα. This produces a **negative-feedback** loop in which the activation of NFκB is self-limiting since activation of NFκB activates the synthesis of IκBα which in turn inhibits NFκB (**Figure 8.29**). Such feedback pathways are often used in cellular signaling to ensure a strictly controlled response to a particular signal.

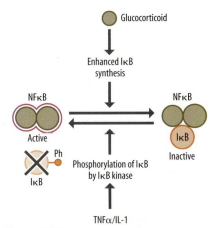

Figure 8.28
TNFα and IL-1 can activate NFκB by reducing the phosphorylation of IκB, while glucocorticoid represses NFκB-dependent gene expression by enhancing IκB synthesis.

8.3 REGULATION OF TRANSCRIPTION FACTOR ACTIVITY BY OTHER POST-TRANSLATIONAL MODIFICATIONS

Although phosphorylation is probably the most frequently used post-translational modification of transcription factors, a variety of other post-translational modifications also occur for transcription factors. This parallels the regulation of histone activity by a variety of post-translational modifications which was discussed in Chapter 2 (Section 2.3) and Chapter 3 (Section 3.3). The various post-translational modifications (other than phosphorylation) which modify transcription factor activity will be discussed in this section.

Acetylation

Although acetylation of lysine residues was originally characterized as a mechanism for regulating histone activity (see Sections 2.3 and 3.3), it has subsequently been shown to play a significant role in regulating the activity of a variety of different transcription factors. For example, acetylation occurs in the NFκB/IκB system which was discussed above (see Section 8.2). In addition to the phosphorylation of IκB, NFκB can be acetylated and this acetylation reduces its interaction with IκB. Activation of NFκB therefore requires both acetylated NFκB and phosphorylated IκB (**Figure 8.30**).

Acetylation of NFκB is carried out by CBP and the closely related p300 transcriptional co-activators which interact with NFκB (as well as interacting with other transcriptional activators; see Section 5.2). Hence, as well as acetylating histones and opening chromatin structure following recruitment to the DNA by a specific transcriptional activator, CBP/p300 can also modify the activity of the transcription factor itself by acetylation (**Figure 8.31**).

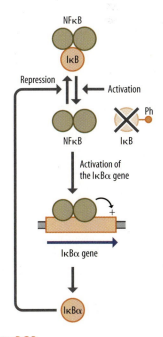

Figure 8.29
Activation of NFκB induces enhanced synthesis of IκBα which in turn represses NFκB by binding to it, so producing a negative-feedback loop.

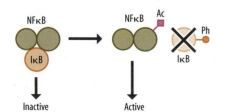

Figure 8.30
Activation of NFκB involves phosphorylation of IκB and acetylation of NFκB.

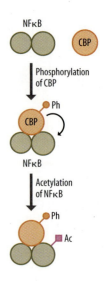

A similar acetylation of a transcription factor by p300 also occurs for the p53 anti-oncogene protein (see Section 11.3) to which CBP and p300 bind. This acetylation increases the DNA-binding activity of p53, further enhancing its ability to stimulate gene expression.

Interestingly, as noted in Section 8.2 above, the activation of IκB kinases not only phosphorylates IκB but also results in the phosphorylation of CBP increasing its affinity for NFκB and decreasing its affinity for p53. Hence, the activation of IκB kinase stimulates NFκB gene expression in multiple ways. First, it phosphorylates IκB, promoting its dissociation from NFκB. Second, it also phosphorylates CBP enhancing its binding to NFκB. In turn this will not only result in opening of the chromatin structure of NFκB-dependent genes but will also further stimulate NFκB activity by its acetylation.

In contrast, the phosphorylation of CBP will reduce its binding to p53. This will reduce histone acetylation at p53-dependent genes and also reduce the acetylation of p53 itself, thereby reducing its DNA-binding activity. This example illustrates therefore how a specific signal acting via a transcription co-activator can switch gene expression between two different pathways which both rely on the co-activator (**Figure 8.32**).

Figure 8.31
Phosphorylation of CBP enhances its ability to bind to NFκB, which it acetylates.

Methylation

As well as being modified by phosphorylation, CBP and p300 can also be modified by methylation on specific arginine residues. As with phosphorylation, methylation alters the relative affinity of these co-activators for different transcriptional activators. Thus, methylation abolishes the ability of CBP/p300 to bind to the CREB transcription factor (discussed in Section 8.2) but has no effect on the ability to bind to the nuclear receptors discussed in Section 8.1 (**Figure 8.33**).

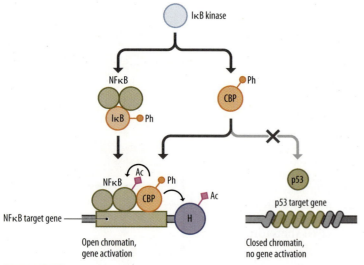

Figure 8.32
IκB kinase phosphorylates IκB, thereby releasing free NFκB. It also phosphorylates CBP, enhancing its affinity for NFκB and reducing its affinity for p53. In turn, CBP acetylates NFκB, enhancing its activity, and also acetylates histones (H), thus opening the chromatin structure of NFκB-dependent genes and allowing NFκB to activate their transcription.

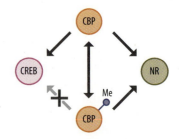

Figure 8.33
Methylation of CBP abolishes its binding to the CREB transcription factor, while having no effect on the binding of CBP to the nuclear receptors (NR).

Thus, co-activators can be modified by different post-translational modifications such as phosphorylation and methylation, which modulate their relative affinities for different transcriptional activators. This has led to the idea of a co-activator code in which a precise pattern of post-translational modifications of a particular co-activator is induced by a specific signal. In turn, this controls the activators to which co-activators bind as well as their relative affinity for each activator and thereby determines the pattern of gene activation which is produced (**Figure 8.34**). Such a co-activator code evidently parallels the histone code of post-translational modifications, which was discussed in Chapter 2 (Section 2.3) and Chapter 3 (Section 3.3) and which plays a key role in the regulation of chromatin structure.

Ubiquitination and sumoylation

As discussed in Chapter 2 (Section 2.3) and Chapter 3 (Section 3.3), histones can be modified by the addition of the small 76-amino acid protein ubiquitin and this modifies their activity resulting in either opening or closing of the chromatin structure depending on the histone that is ubiquitinated. In contrast, in a variety of other proteins including many transcription factors, the addition of ubiquitin acts as a signal for proteolytic enzymes to degrade the ubiquitinated factor. An example of this was discussed in Section 8.2, where phosphorylation of the IκB transcription factor promoted subsequent ubiquitination, leading to the degradation of IκB and the release of active NFκB.

Another example of the modification of a transcription factor resulting in its subsequent ubiquitination and degradation is the hypoxia-inducible factor **HIF-1**, which mediates gene activation in response to low oxygen levels. Under conditions of high oxygen, the HIF-1α protein is modified by the addition of a hydroxyl (OH) group to a specific proline amino acid in the protein. This modification specifically allows HIF-1α to be recognized by the von Hippel Lindau (**VHL**) anti-oncogene protein (see Section 11.3). VHL is part of a protein complex which modifies HIF-1α by ubiquitination leading to its degradation (**Figure 8.35**). Under conditions of low oxygen, such hydroxylation does not occur. HIF-1α is therefore stabilized and is free to associate with the HIF-1β protein to form a heterodimer able to activate transcription (see Figure 8.35).

In this case, therefore, the novel modification of proline hydroxylation renders a transcriptional activator a target for ubiquitination and degradation, paralleling the similar effect of phosphorylation on IκB. Interestingly, the oxygen concentration directly regulates proline hydroxylation since the enzyme prolyl hydroxylase uses oxygen as a substrate to add hydroxyl groups to proline residues in its target proteins. As the enzyme has a low affinity for oxygen, it can only do this when oxygen concentrations are high, so resulting in the oxygen-regulated hydroxylation of HIF-1α.

The role for ubiquitination in promoting the degradation of transcription factors such as HIF-1α parallels its role in achieving this effect in a variety of other proteins. Interestingly, however, in some cases, such as the yeast transcription factor GCN4, it has been shown that ubiquitination and enhanced degradation can actually stimulate the activation of gene expression by the transcriptional activator. This seemingly paradoxical effect involves the fact that after one round of transcription ubiquitination and degradation removes "spent" GCN4 from the promoter and allows it to be replaced by a fresh molecule of GCN4 able to more effectively stimulate transcription (**Figure 8.36**). Clearly, such a mechanism in which a DNA-bound transcription factor is rapidly turned over and potentially replaced by another molecule of the factor will allow cells to respond rapidly to changes in the levels of specific signaling molecules, which result in changes

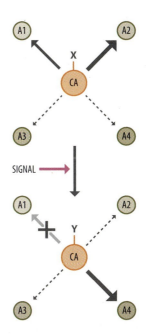

Figure 8.34
A cellular signal can alter the pattern of post-translational modifications (shown as X and Y) on a co-activator (CA) and thereby alter its relative affinity for different transcriptional activators (A1–A4). Strong binding is shown as a heavy arrow, intermediate binding as a normal arrow, and weak binding as a dashed arrow. In this case, the signal enhances co-activator binding to A4, has no effect on binding to A3, reduces binding to A2, and abolishes binding to A1. It will therefore stimulate the expression of A4-dependent genes, at the expense of those dependent on A1 and A2.

Figure 8.35
Under conditions of high oxygen, HIF-1α is modified by proline (P) hydroxylation, resulting in its subsequent ubiquitination and degradation. Under low-oxygen conditions, this proline hydroxylation does not occur. HIF-1α is therefore stabilized and can interact with HIF-1β to activate its target genes.

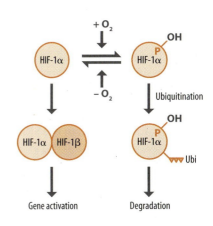

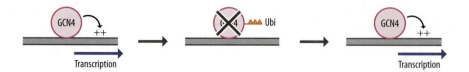

Transcription

Figure 8.36
Following a round of transcriptional activation, the GCN4 activator is ubiquitinated and degraded, allowing a new highly active molecule of GCN4 to bind and stimulate a further round of transcription.

in the degree of activation of particular transcription factors and their corresponding co-activators.

In addition to modification by ubiquitination, transcription factors have also been shown to be modified by addition of the small ubiquitin-related modifier (SUMO). As both SUMO and ubiquitin modify lysine residues, they can compete with each other in modifying a specific site on a transcription factor. Indeed, this is seen in the case of IκB, where SUMO modification protects IκB from ubiquitin modification and thereby stabilizes it (**Figure 8.37**).

Interestingly, both the N-CoR and RIP140 nuclear receptor co-repressors (see Section 5.3) are modified by sumoylation and such modification increases their transcriptional repressor activity. Hence, as with co-activators, co-repressors are also subject to post-translational modifications which can modulate their activity. Thus, the co-activator post-translational modification code discussed above is paralleled by the existence of a co-repressor code in which different post-translational modifications of co-repressors will affect their activity and their relative binding to different transcriptional repressors.

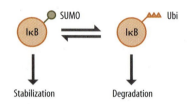

Figure 8.37
Modification of the IκB factor by sumoylation inhibits its ubiquitination and therefore stabilizes it.

8.4 REGULATION OF TRANSCRIPTION FACTOR ACTIVITY BY SIGNALS WHICH REGULATE PRECURSOR PROCESSING

Transcription factors can be activated by cleavage of a precursor which contains an inhibitory region

As discussed in Sections 8.2 and 8.3, factors such as HIF-1α and IκB can be regulated by processes which alter their rate of degradation and thereby control the level of factor which is present (**Figure 8.38a**). In addition, however, transcription factors can be regulated by processes which regulate the proteolytic cleavage of an inactive precursor molecule to generate an active transcription factor (**Figure 8.38b**), as illustrated in Figure 8.5c. An example of this is seen in the NFκB system. The NFκB-related protein, p105, is synthesized as an inactive precursor in which the active part of the molecule is inhibited by an IκB-like region of the same molecule. Following exposure to an activating stimulus, the IκB-like region is phosphorylated in the same way as for IκB itself. This results in the cleavage of the precursor molecule so that the IκB-like region is degraded, releasing the NFκB-like region which is now able to move to the nucleus and activate gene expression (**Figure 8.39**).

Transcription factors can be activated by cleavage of a membrane-bound precursor

As well as involving the removal of an inactivating region as in the p105 system, processing of an inactive precursor can involve releasing it from a cellular membrane to which it is anchored. In the case of signals which can enter the cell, the inactive precursor molecule can be anchored to an intracellular (internal) membrane, such as that of the **endoplasmic reticulum**. This is seen, for example, in the SREBP proteins which activate gene expression in response to lowered levels of **cholesterol**. In the presence of cholesterol, SREBP exists as an inactive precursor which is anchored in the membrane of the endoplasmic reticulum in association with a cholesterol-binding protein, SREBP-cleavage-activating protein (SCAP). When cholesterol levels within the cell fall, SCAP ceases to bind cholesterol and the SREBP–SCAP

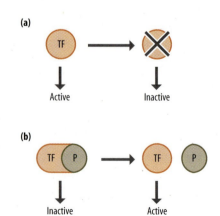

Figure 8.38
A transcription factor (TF) can be regulated by regulating its degradation (a) or its processing from an inactive precursor (TF–P) (b).

Figure 8.39

Activation mechanisms in the NFκB family. (a) Prior to exposure to an activating signal, the NFκB factor exists as an inactive form complexed to the distinct IκB inhibitory protein. In response to an activating signal the IκB factor is phosphorylated (P) resulting in the release of free NFκB which activates transcription. (b) In the p105 protein the NFκB- and IκB-like regions are part of a single large precursor which is inactive. Following exposure to an activating signal, the IκB-like region is phosphorylated and the protein is cleaved, releasing the NFκB-like region which is free to activate transcription.

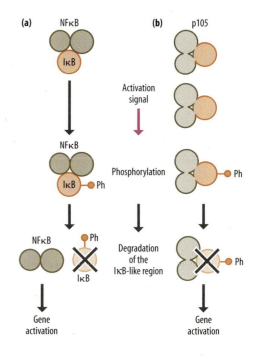

complex moves from the endoplasmic reticulum to the Golgi apparatus where SREBP is proteolytically cleaved to release its N-terminal domain. This N-terminal domain is a DNA-binding transcription factor containing a basic helix-loop-helix DNA-binding domain (see Section 5.1) Following release of the N-terminal domain, it moves to the nucleus where it can bind its target genes and activate gene transcription, thereby producing the appropriate change in cellular phenotype in response to lowered cholesterol levels (**Figure 8.40**).

This system parallels the steroid receptor system discussed in Section 8.1, in which the change in the level of a signaling molecule within the cell directly triggers activation of its target transcription factor, in this case via the cholesterol-binding protein SCAP. Similarly, such activation of SREBP results in a change in its cellular localization, as occurs for the glucocorticoid receptor (see Section 8.1) and the NFκB factor (see Section 8.2).

Proteolytic processing of a membrane-anchored transcription factor precursor is also used for signals which cannot enter the cell but act by binding to cell-surface receptors. This can involve the cell-surface receptor itself being cleaved to release an active transcription factor. An example of this is seen in the **Notch/Delta** system which mediates signaling between different cells, so that adjacent cells develop different phenotypes from one another (see Section 10.2 for discussion of the role of the Notch/Delta pathway in the development of the nervous system). The Notch protein is processed from a single polypeptide to form a heterodimeric receptor in which one chain is associated with the cell membrane and the other chain is associated with the first chain and projects into the extracellular space (**Figure 8.41**).

The Notch receptor is activated when it binds a molecule of the Delta protein which is anchored in the membrane of an adjacent cell. When this occurs the extracellular chain of the Notch protein associates with the Delta protein and dissociates from its membrane-anchored heterodimeric partner. This results in the cleavage of the membrane-bound chain releasing its intracellular region. This intracellular region of Notch is a transcription factor which migrates to the nucleus and associates with the CSL DNA-binding transcription factor. This converts CSL from a transcriptional repressor to a transcriptional activator and so activates specific Notch target genes (see Figure 8.41). Hence in this case, the active transcription factor is produced by proteolytic cleavage of the receptor itself to generate an active transcription factor which migrates to the nucleus and stimulates gene expression.

Figure 8.40

In the presence of cholesterol (C), the SREBP transcription factor exists in an inactive form in the endoplasmic reticulum, where it is associated with the SCAP protein which binds cholesterol. When cholesterol levels fall, SCAP binds less cholesterol and the SCAP–SREBP complex moves to the Golgi apparatus where SREBP is cleaved, releasing its active N-terminal domain (pink). This domain can then move to the nucleus, bind to DNA, and activate its target genes.

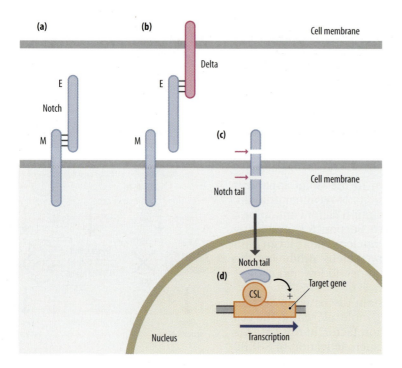

Figure 8.41
The Notch receptor is a heterodimer composed of a membrane-spanning chain (M) and an extracellular chain (E) (a). When the Notch protein binds to a Delta protein in the membrane of an adjacent cell, the extracellular chain binds to Delta and dissociates from the membrane-spanning chain (b). This results in the cleavage of the membrane-spanning chain in two places (arrows), releasing the tail region of Notch (c). This tail region then moves to the nucleus where it associates with the CSL DNA-binding protein and activates transcription (d).

Cleavage of a transcription factor can convert it from an activator to a repressor

Although in the Notch/Delta system the transcription factor is actually a fragment of the receptor, this is not always the case. Thus, a different mechanism involving cleavage of an intracellular protein, is used in the Hedgehog signaling pathway which is involved in the regulation of cellular growth and differentiation. Binding of **Hedgehog proteins** to their specific receptors inhibits phosphorylation of the cubitus interruptus (Ci) transcription factor by several different kinase enzymes, including protein kinase A (see Section 8.2). The inhibition of Ci phosphorylation stabilizes the Ci protein, allowing it to bind to DNA and recruit co-activator proteins, thereby activating specific target genes (**Figure 8.42**). In contrast, in the absence of Hedgehog receptor activation, the Ci protein is phosphorylated and proteolytically cleaved to produce a small fragment of the protein. This fragment can still bind to DNA, but rather than recruiting co-activator proteins the truncated protein recruits co-repressor molecules and thereby represses gene expression (see Figure 8.42).

Proteolytic cleavage can therefore result in the activation of transcription factors such as p105, SERBP, and Notch (as illustrated in Figure 8.5c) and can also switch a transcription factor from activator to repressor, as illustrated by the Ci transcription factor. The Ci case is also an example of an activation mechanism involving modification of protein–protein interaction (as illustrated in Figure 8.5d) since the cleaved form of Ci binds co-repressor molecules rather than the co-activator molecules bound by the full-length protein.

Cleavage of a lipid link can be used to activate a transcription factor

Interestingly, the activation of a transcription factor by cleavage can also be achieved by cleaving a lipid link rather than proteolytic cleavage of the transcription factor protein itself. This is seen in the Tubby transcription factor which binds to phospholipid in the plasma membrane and is therefore anchored in this position via a protein–lipid interaction rather than a protein–protein interaction (**Figure 8.43**). Interestingly, some G-protein-coupled receptors activate a phospholipase enzyme rather than activating

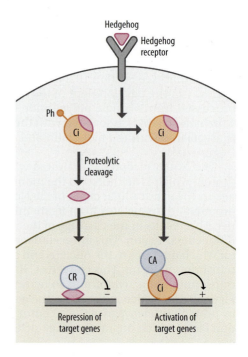

Figure 8.42
Activation of the Hedgehog receptor by its Hedgehog ligand results in the dephosphorylation of the cubitus interruptus protein (Ci). The intact Ci protein binds co-activators and therefore activates target gene expression. In the absence of Hedgehog receptor activation, the Ci protein is phosphorylated and this results in its proteolytic cleavage, producing a fragment which can still bind to DNA but binds co-repressor proteins and therefore represses transcription.

Figure 8.43
(a) The Tubby transcription factor is anchored in the cell membrane by interaction with a phospholipid (PL). (b) Following G-protein-coupled receptor activation, the phospholipase C enzyme (PLC) is activated and cleaves the phospholipid. This releases Tubby and allows it to enter the nucleus and activate transcription.

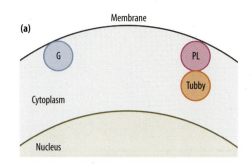

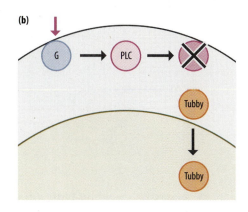

the adenylate cyclase enzyme (as described in Section 8.2). Such a phospholipase enzyme can cleave the phospholipid which anchors Tubby to the cell membrane, thereby releasing Tubby and allowing it to move to the nucleus and activate its target genes (see Figure 8.43).

8.5 REGULATION OF POST-TRANSCRIPTIONAL PROCESSES BY CELLULAR SIGNALING PATHWAYS

As discussed in Chapters 6 and 7, post-transcriptional processes play important roles in the regulation of gene expression. It is not surprising therefore that, in addition to regulating transcription factor activity, cellular signaling pathways can also regulate post-transcriptional events. Such regulatory processes are discussed in this section.

The PI3-kinase/Akt system plays a key role in regulating gene expression in response to growth factors or insulin

The binding of a number of growth factors or of the peptide hormone insulin to their specific cell-surface receptors results in the activation of the **phosphoinositide 3-kinase** (**PI3-kinase**)/**Akt** system, which plays a key role in the regulation of cellular growth and in the cellular response to insulin. Following binding of growth factors or insulin to their specific receptors, PI3-kinase is activated. This enzyme then phosphorylates the **inositol** phospholipid molecule, phosphatidylinositol 4.5-biphosphate (PIP_2) at the 3 position of the inositol ring to produce phosphatidylinositol 3,4,5-triphosphate (PIP_3) (**Figure 8.44**). PIP_3 functions as a second-messenger molecule which in particular activates **3-phosphoinositide-dependent kinase 1** (PDK1). In turn, PDK1 phosphorylates the Akt kinase (also known as protein kinase B) and activates it (see Figure 8.44).

This signal pathway thus involves a multi-kinase cascade, as in the MAP kinase system (see Section 8.2) but includes phosphorylation of lipid molecules catalyzed by a lipid kinase as well as protein phosphorylation catalyzed by protein kinases.

The activated Akt kinase can regulate both transcription and post-transcriptional events by phosphorylating specific targets. Thus like the other kinases described above, Akt can regulate transcription by phosphorylating specific transcription factors. For example, Akt phosphorylates the FOXO1 transcription factor. Unlike the phosphorylation events discussed in Section 8.2, phosphorylation of FOXO1 inhibits its ability to activate its target genes by promoting its retention in the cytoplasm (**Figure 8.45**). Akt also phosphorylates the PGC-1α protein which is a co-activator for FOXO1 and other transcription factors. Such phosphorylation of PGC-1α by Akt inhibits the activity of PGC-1α. In this case therefore insulin signaling induces phosphorylation events which inhibit the activation of

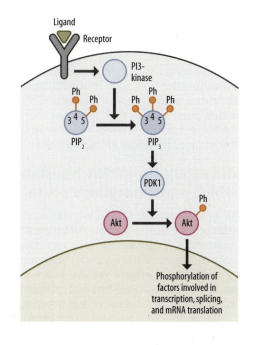

Figure 8.44
Binding of ligands, such as insulin or growth factors, to their specific receptors activates PI3-kinase which phosphorylates the inositol lipid PIP_2 on the 3 position of the inositol ring to produce PIP_3. PIP_3 activates the PDK1 kinase, which in turn phosphorylates and activates the Akt kinase. This allows Akt to promote the phosphorylation of target factors involved in RNA splicing and mRNA translation, as well as specific transcription factors.

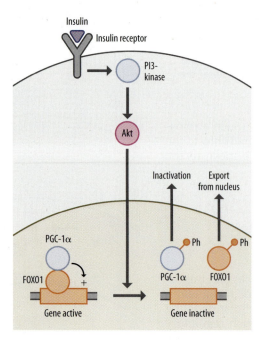

Figure 8.45
Activated Akt phosphorylates the FOXO1 transcription factor, resulting in its retention in the cytoplasm. Phosphorylation by Akt also inhibits the activity of PGC-1α, which is a co-activator for FOXO1. This dual mechanism therefore represses the transcription of FOXO1-dependent genes.

specific genes by targeting both a transcriptional activator (FOXO1) and the co-activator which it requires to activate transcription (PGC-1α) (see Figure 8.45).

Akt regulates RNA splicing by phosphorylating splicing factors

Most importantly, however, Akt also targets factors involved in post-transcriptional processes and thereby regulates gene expression at post-transcriptional levels. One example of such regulation involves the SF2 member of the SR protein family which plays a key role in determining which 5′ splice site is joined to a specific 3′ splice site (see Section 7.1). This factor is regulated by multiple phosphorylation events catalyzed by Akt. Following its synthesis in the cytoplasm, SF2 is phosphorylated within its SR domain and such phosphorylation is essential for its transport to the nucleus (**Figure 8.46**). Subsequently, further phosphorylation of SF2 takes place in the nucleus and enhances the ability of SF2 to recruit the U1 snRNP particle to the 5′ splice site. In turn, this promotes the joining of the more proximal of two 5′ splice sites to a 3′ splice site (see Section 7.1). Hence, in this case phosphorylation of SF2 enhances its activity and allows it to promote specific alternative splicing decisions.

One target for phosphorylated SF2 is the fibronectin RNA, where SF2 promotes the inclusion of the alternatively spliced EDA exon in the mRNA by stimulating the use of a proximal 5′ splice site (**Figure 8.47**). Interestingly, it has been shown that SF2 also stimulates the translation of fibronectin mRNA containing the EDA exon, indicating that it can act at the level of translation as well as at the level of splicing (see Figure 8.47).

Akt regulates mRNA translation via the TOR kinase, which phosphorylates proteins involved in translation

As well as regulating splicing, Akt can also regulate mRNA translation. Such regulation involves Akt phosphorylating and thereby activating the target of rapamycin (**TOR**) kinase, so-called because it was originally defined as the cellular target for the bacterial toxin rapamycin. Activated TOR has a number of stimulatory effects on the process of translation. Most importantly, it phosphorylates eIF4E-binding proteins, which bind to the eIF4E translational initiation factor and prevent it binding to other translation initiation factors (see Section 7.5). The interaction between eIF4E-binding proteins and eIF4E

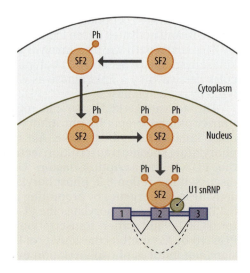

Figure 8.46
The SF2 protein is phosphorylated in the cytoplasm, facilitating its transport to the nucleus. In the nucleus, it undergoes further phosphorylation. This enhances its ability to recruit the U1 snRNP to 5′ splice sites and to promote the use of a proximal rather than a distal 5′ splice site (solid compared to dashed lines).

is inhibited by phosphorylation of eIF4E-binding proteins, so allowing eIF4E to bind the translational initiation factors eIF4G and eIF4A and thereby promoting translation (Figure 8.48a).

As noted in Chapter 7 (Section 7.5), activation of eIF4E in response to insulin or growth factors is of particular importance in enhancing the translation of specific mRNAs, which have a high degree of secondary structure in their 5′ untranslated region. Thus, the recruitment of eIF4A by eIF4E plays an important role in unwinding this secondary structure and allowing translation of these mRNAs (see Figure 7.46).

In addition to targeting eIF4E-binding proteins, TOR also stimulates translation by phosphorylating and activating the **S6 kinase** (S6K). Prior to phosphorylation and activation, S6K is associated with the translational initiation factor eIF3 in an inactive complex. Phosphorylation of S6K results in its dissociation from eIF3. S6K then phosphorylates the S6 ribosomal protein increasing its activity and also phosphorylates the eIF4B translational initiation factor. In turn, phosphorylated eIF4B now associates with the released eIF3 to form an active complex (see Figure 8.48).

By activating TOR, the PI3 kinase/Akt pathway therefore plays a key role in activating translation both by activating translational initiation and by stimulating the activity of ribosomal proteins such as S6. This therefore allows an enhanced rate of translation to occur in response to cellular growth signals or insulin treatment.

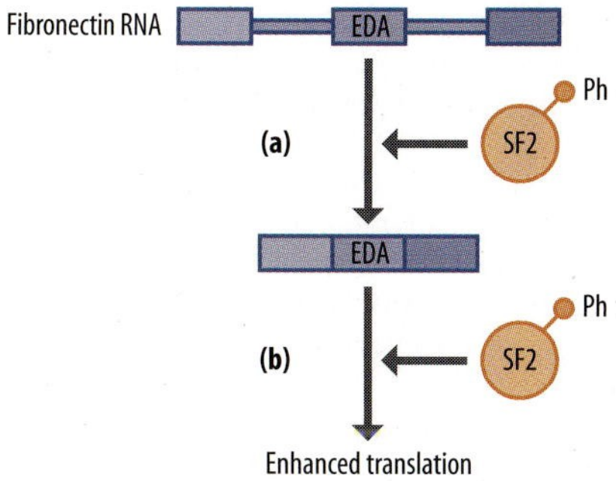

Akt/TOR can also stimulate mRNA translation by enhancing the transcription of genes encoding RNAs and proteins involved in protein synthesis

Although changes in the rate of protein synthesis are an important response to growth signals, sustained growth also requires enhanced synthesis of the ribosomal RNAs and proteins to allow more ribosomes to be produced. TOR, in addition to directly regulating translation, also stimulates transcription by all three RNA polymerases in order to do this.

In the case of RNA polymerase I, activation of TOR results in the phosphorylation of the TIF-IA transcription factor, increasing its activity. TIF-IA is bound to RNA polymerase I and, by interacting with the promoter-bound factor SLI/TIF-IB, plays a key role in recruiting RNA polymerase I to the ribosomal RNA gene promoters (see Section 4.1 for discussion of RNA polymerase I transcription). Hence, activation of TIF-IA results in enhanced

Figure 8.47
Phosphorylated SF2 acts at the level of RNA splicing to promote the inclusion of the EDA exon in the fibronectin mRNA (a). It also acts at the level of translation to promote translation of mRNA containing this exon (b).

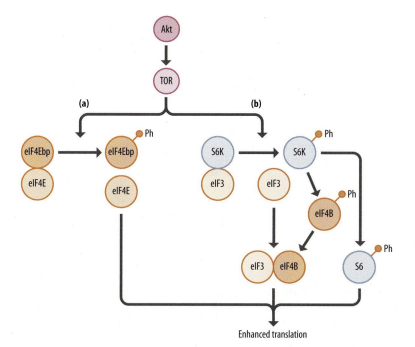

Figure 8.48
Activated Akt kinase activates the TOR kinase. In turn, TOR phosphorylates eIF4E-binding proteins (eIF4Ebp), resulting in the release of the active translational initiation factor eIF4E (a). Active TOR also phosphorylates the S6 kinase (S6K), dissociating it from eIF3 (b). Active S6K phosphorylates eIF4B, allowing it to form an active complex with free eIF3. It also phosphorylates the S6 ribosomal protein (S6), enhancing its activity.

Figure 8.49
The TOR pathway can activate the TIF-IA transcription factor, thereby activating RNA polymerase I-mediated transcription of the ribosomal genes (a). It can inhibit the CRF1 co-repressor, thereby relieving its repressive effect on ribosomal protein gene transcription by RNA polymerase II and so enhancing their transcription (b). Finally, it can inhibit the MAF1 repressor, preventing it from inhibiting transcription of the 5S and tRNA genes by RNA polymerase III, so enhancing their transcription (c).

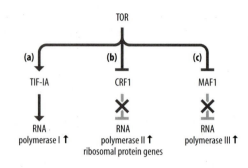

recruitment of RNA polymerase I to the promoter of the ribosomal RNA genes resulting in enhanced synthesis of the 18, 28, and 5.8S ribosomal RNAs (**Figure 8.49a**).

In contrast to RNA polymerase I, RNA polymerase II transcribes a wide range of genes. It is therefore necessary for TOR to specifically induce the transcription of the genes encoding the ribosomal proteins which are required together with ribosomal RNAs to produce new ribosomes. To achieve this, TOR inhibits the activity of the CRF1 factor, which is an essential co-repressor inhibiting the transcription of the ribosomal protein genes by RNA polymerase II. Active CRF1 enters the nucleus and acts as a co-repressor inhibiting the transcription of the ribosomal protein genes. Inactivation of CRF1 by TOR relieves this repression and activates ribosomal protein gene transcription (**Figure 8.49b**).

Finally, TOR also controls the activity of RNA polymerase III which transcribes the genes encoding the tRNAs and the 5S ribosomal RNA. This is achieved by TOR catalyzing the phosphorylation of the MAF1 transcriptional repressor. This promotes the export of MAF1 from the nucleus. In turn, this prevents MAF1 from displacing TFIIIB from RNA polymerase III promoters and thereby acting as a repressor of polymerase III gene transcription. Inactivation of MAF1 by TOR therefore activates RNA polymerase III transcription (**Figure 8.49c**).

Hence, activation of TOR can activate transcription of RNA polymerase I- and RNA polymerase III-dependent genes, as well as of RNA polymerase II-dependent genes encoding ribosomal proteins. Although these effects of TOR occur by it independently regulating three different transcription factors, one for each of the three RNA polymerases (see Figure 8.49), there is also evidence that the effects of growth signals on the three RNA polymerases can be co-ordinated with one another. For example, as well as repressing RNA polymerase III-dependent genes by blocking binding of TFIIIB to their promoters, the MAF1 repressor also binds to the promoter of the gene encoding the TBP basal transcription factor. This prevents binding of the Elk-1 transcriptional activator to the promoter and so represses TBP gene transcription. As TBP plays a critical role in the assembly of the basal transcriptional complex for all three RNA polymerases (see Section 4.1), this results in repression of transcription by all three polymerases. Such repression is correspondingly relieved by activation of TOR and consequent inhibition of MAF1 (**Figure 8.50**).

Activation of the TOR pathway by Akt therefore results in multiple effects which promote cellular growth by stimulating transcription of ribosomal components by all three RNA polymerases and by stimulating translation via a number of mechanisms including the phosphorylation of eIF4E-binding proteins. This complements and extends the roles of Akt in directly phosphorylating factors involved in transcription and RNA splicing (**Figure 8.51**).

A variety of kinases inhibit translation by phosphorylating eIF2

In contrast to the activation of translation produced by phosphorylation of eIF4E-binding proteins, phosphorylation of another translational initiation factor eIF2 is used to inhibit the translation of a wide range of cellular mRNAs during cellular stress, allowing cells to divert resources to most effectively respond to the stress itself (see Section 7.5).

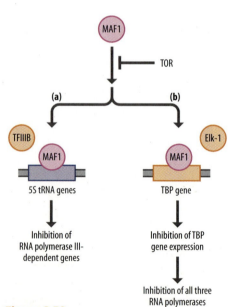

Figure 8.50
The MAF1 repressor inhibits RNA polymerase III-dependent genes by blocking the binding of the essential transcription factor TFIIIB (a). In addition however, it also blocks transcription of the gene encoding TBP by blocking binding of the Elk-1 transcriptional activator to the promoter of the gene encoding TBP (b). As TBP is required for transcription by all three RNA polymerases, this inhibits transcription by all the polymerases. These effects are blocked by TOR which promotes phosphorylation of MAF1 and inhibits its activity.

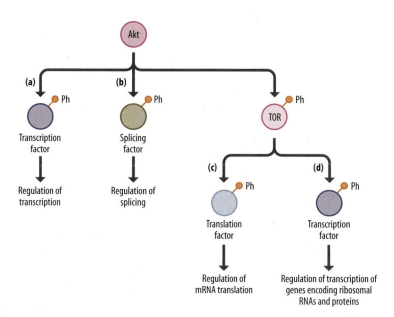

Figure 8.51
Summary of the effects of Akt. It can directly phosphorylate transcription factors and so regulate transcription (a), as well as directly phosphorylating splicing factors and so regulating RNA splicing (b). Akt can also act indirectly by activating the TOR kinase which phosphorylates factors involved in mRNA translation (c) and transcription factors, which regulate the transcription of genes encoding ribosomal RNAs and proteins (d). Both these effects of TOR will stimulate mRNA translation.

Interestingly, four different eIF2 kinases have been described in human cells, each being activated in response to different stresses (**Figure 8.52**). For example, the heme-regulated kinase (HRK) is activated in reticulocytes in response to the absence of heme as described in Chapter 7 (Section 7.5). This inhibits synthesis of globin protein (which is the predominant translation product in reticulocytes) until heme becomes available to form the hemoglobin molecule (**Figure 8.52a**).

Similarly the GCN2 eIF2 kinase is activated in the presence of free tRNA molecules which have no bound amino acid. This occurs under conditions of low amino acid concentration when it is sensible to cease protein synthesis (**Figure 8.52b**). Interestingly, in yeast, translation of the mRNA encoding the GCN4 transcription factor is actually stimulated under low-amino acid conditions when translation of most other mRNAs is repressed (see Section 7.5). Thus under these conditions, translation of short upstream open reading frames in the GCN4 mRNA is suppressed and translation of the full-length GCN4 protein is correspondingly enhanced (see Figure 7.50). This allows GCN4 to fulfil its function of activating the expression of specific genes which are required during amino acid starvation.

The other two human eIF2 kinases, PERK and PKR, are activated in the presence of unfolded proteins in the endoplasmic reticulum or following virus infection, respectively (**Figure 8.52c** and **d**). Thus, PERK suppresses protein synthesis under conditions when protein folding is proceeding incorrectly in the endoplasmic reticulum and it is inappropriate therefore to synthesize high levels of new proteins. Similarly, PKR suppresses viral protein synthesis following viral infection, allowing the cell to respond more effectively to such infection.

The examples discussed in this section have therefore shown that translation can be regulated both by phosphorylating eIF4E-binding proteins to stimulate translation and by phosphorylating eIF2 to inhibit it. Interestingly, regulation of these two factors has been shown to operate in tandem during the response of cells to the stress of hypoxia. During the initial response to hypoxia, eIF2 is phosphorylated resulting in inhibition of translation. This is supplemented later during hypoxia by enhanced association of eIF4E binding proteins with eIF4E, so providing a further mechanism of translational repression (**Figure 8.53**).

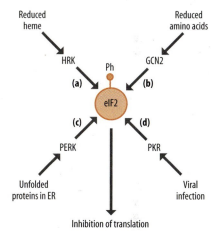

Figure 8.52
The translational initiation factor eIF2 can be phosphorylated by four different kinases: HRK, which is activated by reduced heme levels (a), GCN2, which is activated by reduced amino acid levels (b), PERK, which is activated by unfolded proteins in the endoplasmic reticulum (ER) (c), and PKR, which is activated by viral infection (d). In all cases, such phosphorylation results in inhibition of translation.

Individual kinases can produce multi-level regulation of gene expression

The findings discussed above demonstrate that post-transcriptional as well as transcriptional processes can be regulated by signaling pathways.

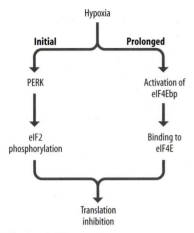

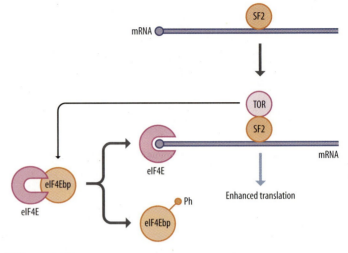

Figure 8.53
Exposure of cells to hypoxia results in an initial suppression of protein synthesis, mediated by the PERK kinase phosphorylating eIF2. Subsequently, during prolonged hypoxia eIF4E-binding protein (eIF4Ebp) is activated and binds to eIF4E, further suppressing protein synthesis.

Figure 8.54
The SF2 splicing factor can recruit the TOR kinase to an mRNA. In turn, TOR phosphorylates eIF4E-binding proteins (eIF4Ebp), so releasing eIF4E, allowing it to bind to the cap (blue circle) of the mRNA and stimulate mRNA translation.

Indeed as discussed above, the PI3-kinase/Akt kinase pathway can target transcription factors, splicing factors, and translation factors to regulate gene expression at multiple levels and produce a highly effective response to stimulation with growth factors or insulin.

Interestingly, these effects on different stages of gene expression can be linked to one another. Thus as noted above, the SF2 splicing factor is a target for Akt. In turn, binding of SF2 to an mRNA has been shown to promote recruitment of the TOR kinase, which is also an Akt target. Binding of TOR to the mRNA will enhance its translation by stimulating phosphorylation of eIF4E-binding proteins, as described above. This example therefore links a splicing factor with the recruitment of a kinase which regulates translation (**Figure 8.54**). Moreover, it provides an explanation for the enhanced translation of the alternatively spliced fibronectin mRNA produced by SF2 binding (see Figure 8.47).

Although multi-level regulation of gene expression is achieved by Akt targeting multiple regulatory factors, another kinase enzyme regulates transcription, splicing, and translation by targeting a single factor which is involved in all these processes. The Pak1 kinase is activated by the p21 growth regulatory protein (see Section 11.3) and phosphorylates the PCBP1 protein altering the effect of PCBP1 on transcription, splicing, and translation (**Figure 8.55**).

Thus, phosphorylation of PCBP1 in the cytoplasm results in its dissociation from specific mRNAs relieving its inhibitory effect on their translation (**Figure 8.55a**). In the nucleus, phosphorylated PCBP1 regulates the outcome of specific alternative splicing events (**Figure 8.55b**). Moreover, it also binds to the promoter of the gene encoding the translation initiation factor eIF4E and stimulates its transcription (**Figure 8.55c**). The resulting increased levels of eIF4E this produces will enhance translation.

PCBP1 is therefore a multi-level regulator of transcription, splicing, and translation. Its phosphorylation by Pak1 allows specific cellular signals to co-ordinately regulate multiple levels of gene expression.

CONCLUSIONS

In this chapter, we have discussed a wide variety of mechanisms by which transcription factor activity can be altered in response to specific signaling pathways so as to produce changes in gene expression and consequently in cellular characteristics (**Figure 8.56**). Where a signaling molecule such as a

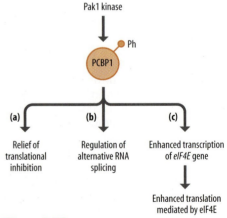

Figure 8.55
The Pak1 kinase phosphorylates the PCBP1 regulatory protein. Phosphorylated PCBP1 dissociates from mRNAs in the cytoplasm, thereby relieving inhibition of their translation (a). In the nucleus, phosphorylated PCBP1 regulates alternative splicing (b). It also stimulates transcription of the gene encoding the eIF4E translation initiation factor, thereby enhancing translation (c).

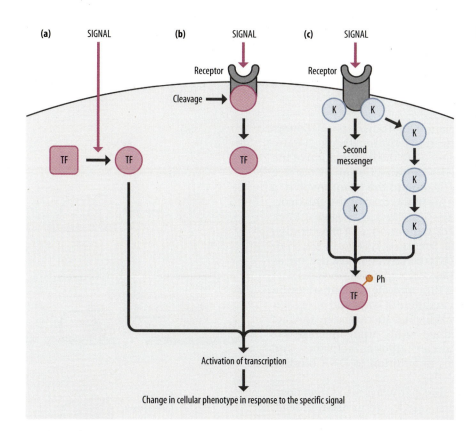

Figure 8.56
(a) A signal which can enter the cell, is able to activate a transcription factor (TF) directly. (b) A signal which cannot enter the cell can stimulate the cleavage of its receptor to generate a fragment (pink) than can activate transcription. (c) Alternatively, a signal which cannot enter the cell can bind to its receptor and stimulate phosphorylation of a transcription factor by a receptor-associated kinase, a second messenger-activated kinase, or a kinase cascade. K, kinase.

steroid hormone can enter the cell, it can interact directly with the transcription factor to alter its structure (**Figure 8.56a**). Similarly, in the case of cholesterol-regulated gene expression, cholesterol itself can bind to a transcription factor-associated protein and modulate proteolytic processing of a transcription factor that regulates cholesterol homeostasis.

In contrast, signals which cannot enter the cell need to act indirectly via cell-surface receptors. As in the case of the Notch system, this can involve a fragment of the receptor being cleaved off and moving into the nucleus where it acts as a transcription factor (**Figure 8.56b**). In the majority of cases, however, activation of a cell-surface receptor results in a post-translational modification of the transcription factor such as its phosphorylation, mediated by specific kinases that are either associated with the receptor or become activated via second messengers such as cyclic AMP or via a multiple kinase cascade (**Figure 8.56c**). These effects allow the signal to produce changes in gene expression and thereby alter the characteristics of the cell to produce appropriate biological effects.

Activation of specific transcription factors by specific signals can occur by one or more of the mechanisms illustrated in Figure 8.5. For example, the activation of steroid receptors by the appropriate hormone combines the mechanism of ligand-induced conformational change (illustrated in Figure 8.5a) with the dissociation of an inhibitory protein (as illustrated in Figure 8.5d). Similarly, the activation of the CREB factor by cyclic AMP involves phosphorylation (illustrated in Figure 8.5b) by protein kinase A, resulting in altered protein–protein interaction (as illustrated in Figure 8.5d), in this case the enhanced binding of the CBP co-activator to CREB. Such phosphorylation is also seen in the case of the p105 NFκB-like molecule, where it is combined with enhanced processing of the p105 precursor molecule (as illustrated in Figure 8.5c), which results in degradation of the IκB-like portion of the molecule and the release of the active NFκB-like portion.

Hence, different signaling mechanisms and different activation processes combine to produce the specific activation of individual transcription factors in a particular situation, thereby appropriately regulating gene expression and the cellular phenotype in response to the specific signal.

Figure 8.57
Regulatory processes can target the degradation of a transcription factor (a), its location in the cell (b), its ability to bind to DNA (c), or its ability to activate transcription following DNA binding (d).

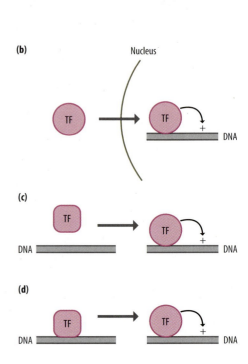

Moreover, in different cases such activation of transcription factors involves altering different properties of the transcription factor (**Figure 8.57**). For example, activation processes can regulate the degradation of the factor (**Figure 8.57a**), as in the case of the HIF-1α transcription factor (see Section 8.3). Similarly, in many cases, the activation of the factor involves a change in its location in the cell, as for example in the case of the glucocorticoid receptor (see Section 8.1) or NFκB (see Section 8.2) (**Figure 8.57b**). Moreover, activation can also involve direct regulation of DNA-binding activity, as in the case of the ACE1 factor (see Section 8.1) (**Figure 8.57c**) or regulation of the ability of a DNA-bound factor to activate transcription, as in the case of the CREB factor (see Section 8.2) (**Figure 8.57d**).

Hence, different signaling pathways use different **signal transduction** mechanisms to modify transcription factors in different ways and alter their activities at different levels. This complexity is further increased by the fact that such signaling pathways can also reduce rather than increase the activity of specific transcription factors and can target co-activators and co-repressors or histone-modifying enzymes in addition to DNA-binding transcription factors.

Similarly as discussed in Section 8.5, signaling pathways also target post-transcriptional regulatory processes as well as transcription itself. Indeed, in the Akt/TOR system or in the case of PCBP1, complementary effects often occur at transcriptional and post-transcriptional levels ensuring the most effective multi-level response to a particular signal.

Although the mechanisms used to regulate gene expression in response to specific cellular signals are highly complex, this is necessary to produce an effective response and one which can be modulated by the simultaneous activation of other signaling pathways. For example, the multiple post-translational modifications which occur for co-activators and co-repressors are likely to result in the integration of different signaling pathways which utilize a specific co-activator or co-repressor. This will allow the most appropriate pathway(s) to be activated in response to a signal with corresponding repression of other competing pathways (see Section 8.3).

Similarly, where a transcription factor is regulated at two different levels by two different stimuli, this can be used to produce the most effective response to the stimuli both individually and when they operate together. This is seen in the case of the yeast GAL4 transcription factor which is bound to DNA in the absence of galactose, but does not activate transcription because its activity is inhibited by the GAL80 protein. In the presence of galactose, GAL80 dissociates and transcription is activated (**Figure 8.58**). However in the presence of glucose, GAL4 does not bind to DNA and therefore galactose treatment has no effect (see Figure 8.58). Hence, by regulating GAL4 at the level of DNA binding by glucose and at the level of transcriptional activation by galactose the system operates so that a galactose stimulus has no effect when glucose is present.

Interestingly, transcriptional regulation by glucose and galactose has recently been shown to be supplemented in this system by post-transcriptional regulation. Thus, glucose also acts at the level of RNA stability to reduce the stability of galactose-induced mRNAs (see Section 7.4 for further discussion of the regulation of mRNA stability). Both transcriptional and post-transcriptional processes therefore produce the desired biological effect in which the presence of galactose alone activates the expression of genes encoding proteins that are necessary for the metabolism of galactose. However, this does not occur in the presence of glucose which is preferred to galactose as a metabolic substrate.

In summary, multi-level regulation of gene expression can be combined with different responses to specific stimuli when present alone or

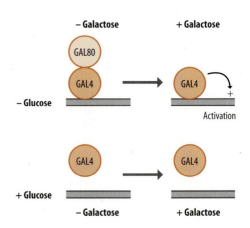

Figure 8.58
In the absence of glucose, galactose promotes the dissociation of the inhibitory GAL80 protein, allowing transcriptional activation by the GAL4 transcription factor. However, in the presence of glucose, GAL4 does not bind to DNA, so treatment with galactose has no effect.

in combination. In this manner, the cell can build up complex regulatory networks which allow it to respond effectively and appropriately to stimuli both individually and in combination. This will result in an appropriate biological response being produced in response to specific signals.

Clearly, mechanisms regulating transcription factor activity in response to specific stimuli must be paralleled by mechanisms which allow the fertilized egg to develop specific body structures during embryonic development. Moreover, mechanisms must also exist to allow individual differentiated cells in the embryo or adult organism to express the appropriate genes and proteins characteristic of that particular cell type and which are essential for its function. The regulation of gene expression during development and in specific cell types is described in the next two chapters.

KEY CONCEPTS

- The effect of transcription factors on gene expression is regulated by controlling their synthesis and/or their activity.

- Cellular signaling pathways primarily regulate transcription factor activity rather than affecting transcription factor synthesis.

- This represents a means of rapidly inducing or repressing gene expression by controlling the activity of a pre-existing transcription factor.

- Signal molecules which can enter the cell can bind to a specific transcription factor and directly regulate its activity.

- Signal molecules which bind to cell-surface molecules can alter transcription factor activity indirectly by inducing post-translational modifications mediated by specific intracellular enzymes.

- Such post-translational modification can involve phosphorylation, acetylation, methylation, ubiquitination, and sumoylation and can affect co-activators, co-repressors, and histone-modifying enzymes as well as DNA-binding transcription factors.

- Signal molecules binding to cell-surface receptors can trigger the proteolytic processing of a transcription factor precursor to produce an active factor.

- In many cases, the regulation of a transcription factor by direct binding, post-translational modification, or proteolytic cleavage produces a change in the interaction of the factor with other proteins.

- Signaling pathways can also regulate post-transcriptional events such as RNA splicing and translation into protein so producing an effective multi-level cellular response to the signal.

FURTHER READING

General

Latchman DS (2008) Eukaryotic Transcription Factors, 5th ed. Elsevier/Academic Press.

8.1 Regulation of transcription factor activity by ligands which enter the cell

Kugel JF & Goodrich JA (2006) Beating the heat: a translation factor and an RNA mobilize the heat shock transcription factor HSF1. *Mol. Cell* 22, 153–154.

Lonard DM & O'Malley BW (2007) Nuclear receptor coregulators: judges, juries, and executioners of cellular regulation. *Mol. Cell* 27, 691–700.

Morimoto RI (1998) Regulation of the heat shock transcriptional response: cross talk between a family of heat shock factors, molecular chaperones, and negative regulators. *Genes Dev.* 12, 3788–3796.

Prasanth KV & Spector DL (2007) Eukaryotic regulatory RNAs: an answer to the 'genome complexity' conundrum. *Genes Dev.* 21, 11–42.

Weatherman RV, Fletterick RJ & Scanlan TS (1999) Nuclear-receptor ligands and ligand-binding domains. *Annu. Rev. Biochem.* 68, 559–581.

Wood MJ, Storz G & Tjandra N (2004) Structural basis for redox regulation of Yap1 transcription factor localization. *Nature* 430, 917–921.

8.2 Regulation of transcription factor activity by phosphorylation induced by extracellular signaling molecules

Greer PL & Greenberg ME (2008) From synapse to nucleus: calcium-dependent gene transcription in the control of synapse development and function. *Neuron* 59, 846–860.

Hoffmann A & Baltimore D (2006) Circuitry of nuclear factor kappaB signaling. *Immunol. Rev.* 210, 171–186.

Levy DE & Darnell Jr JE (2002) Stats: transcriptional control and biological impact. *Nat. Rev. Mol. Cell Biol.* 3, 651–662.

Mayr B & Montminy M (2001) Transcriptional regulation by the phosphorylation-dependent factor CREB. *Nat. Rev. Mol. Cell Biol.* 2, 599–609.

Wietek C & O'Neill LA (2007) Diversity and regulation in the NF-kappaB system. *Trends Biochem. Sci.* 32, 311–319.

8.3 Regulation of transcription factor activity by other post-translational modifications

Bedford MT & Clarke SG (2009) Protein arginine methylation in mammals: who, what, and why. *Mol. Cell* 33, 1–13.

Calao M, Burny A, Quivy V et al. (2008) A pervasive role of histone acetyltransferases and deacetylases in an NF-kappaB-signaling code. *Trends Biochem. Sci.* 33, 339–349.

Chen ZJ & Sun LJ (2009) Nonproteolytic functions of Ubiquitin in cell signaling. *Mol. Cell* 33, 275–286.

Kaelin Jr WG & Ratcliffe PJ (2008) Oxygen sensing by metazoans: the central role of the HIF hydroxylase pathway. *Mol. Cell* 30, 393–402.

Meulmeester E & Melchior F (2008) SUMO. *Nature* 452, 709–711.

Rosenfeld MG, Lunyak VV & Glass CK (2006) Sensors and signals: a coactivator/corepressor/epigenetic code for integrating signal-dependent programs of transcriptional response. *Genes Dev.* 20, 1405–1428.

Yang XJ & Seto E (2008) Lysine acetylation: codified crosstalk with other posttranslational modifications. *Mol. Cell* 31, 449–461.

8.4 Regulation of transcription factor activity by signals which regulate precursor processing

Bray SJ (2006) Notch signalling: a simple pathway becomes complex. *Nat. Rev. Mol. Cell Biol.* 7, 678–689.

Cantley LC (2001) Translocating Tubby. *Science* 292, 2019–2021.

Kopan R & Ilagan MX (2009) The canonical Notch signaling pathway: unfolding the activation mechanism. *Cell* 137, 216–233.

Kovall RA (2008) More complicated than it looks: assembly of Notch pathway transcription complexes. *Oncogene* 27, 5099–5109.

Pomerantz JL & Baltimore D (2002) Two pathways to NF-κB. *Mol. Cell* 10, 693–695.

8.5 Regulation of post-transcriptional processes by cellular signaling pathways

Bushell M, Stoneley M, Spriggs KA & Willis AE (2008) SF2/ASF TORCs up translation. *Mol. Cell* 30, 262–263.

Franke TF (2008) PI3K/Akt: getting it right matters. *Oncogene* 27, 6473–6488.

Graveley BR (2005) Coordinated control of splicing and translation. *Nat. Struct. Mol. Biol.* 12, 1022–1023.

Ma XM & Blenis J (2009) Molecular mechanisms of mTOR-mediated translational control. *Nat. Rev. Mol. Cell Biol.* 10, 307–318.

Mayer C & Grummt I (2006) Ribosome biogenesis and cell growth: mTOR coordinates transcription by all three classes of nuclear RNA polymerases. *Oncogene* 25, 6384–6391.

Ruggero D & Sonenberg N (2005) The Akt of translational control. *Oncogene* 24, 7426–7434.

Stamm S (2008) Regulation of alternative splicing by reversible protein phosphorylation. *J. Biol. Chem.* 283, 1223–1227.

Gene Control in Embryonic Development 9

INTRODUCTION

One of the central problems of biological sciences is to understand the manner in which the single-celled fertilized egg (the zygote) develops into a multicellular organism with a vast range of different cell types, each of which forms at the appropriate time and place relative to other cells. As will be discussed in this chapter, the regulation of gene transcription by specific transcription factors plays a key role in this process. A number of these transcription factors are expressed at specific times and places during embryonic development. Such regulation of transcription factor synthesis contrasts with the regulation of transcription factor activity that occurs in response to cellular signaling pathways, as discussed in the previous chapter.

Regulation of mRNA translation occurs following fertilization

Interestingly, however, as discussed in Chapter 7 (Section 7.5), the regulation of mRNA translation plays an essential role in the very first stage of embryonic development. The unfertilized egg or oocyte contains a large number of different mRNAs which have been transcribed from the mother's genome during the process of egg formation (oogenesis). However, the mRNAs are not translated since they are associated with specific proteins in ribonucleoprotein (RNP) particles. Once the egg is fertilized by a sperm, however, the mRNAs are released from the RNP particles and are translated into protein (**Figure 9.1**).

Hence, the very first stages of embryonic development involve proteins which are produced from maternal mRNAs and which are regulated by translational control. This mechanism has the advantage that it allows the

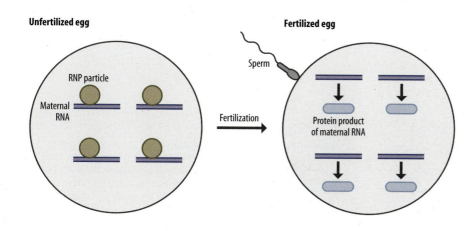

Figure 9.1
In the unfertilized egg, maternal RNAs exist in ribonucleoprotein (RNP) particles and are not translated. Upon fertilization, the maternal RNAs are released and are translated into protein.

Early embryo   **Later embryo**

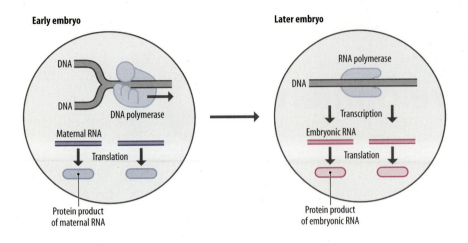

Figure 9.2
In the early embryo, the embryonic DNA is replicated by DNA polymerase, while maternal RNAs are translated into protein. In the later embryo, the embryonic DNA is transcribed by RNA polymerase to produce embryonic RNAs which are translated into protein.

rapid production of new proteins once the egg is activated by a sperm. As translation of mRNA into protein is the most energetically expensive step of gene expression, this mechanism avoids wasteful protein production in an unfertilized oocyte which may never be fertilized by a sperm while allowing rapid protein production in those oocytes which are fertilized.

Moreover, this mechanism provides a supply of protein to the fertilized egg without the need for immediate transcription of the embryonic genome, formed by the fusion of the egg and sperm. Following fertilization, the fertilized egg or zygote must divide rapidly to form two cells, then four cells, then eight cells, etc. At each stage the embryonic genome must be copied by DNA polymerase, so that each daughter cell contains a full complement of DNA. The use of pre-existing mRNA to form proteins in the very early embryo eliminates the need for extensive transcription of the embryonic genome by RNA polymerase which might interfere with rapid replication of the DNA by DNA polymerase (**Figure 9.2**).

Following fertilization and repeated cycles of cell division the single-celled mammalian zygote forms a ball of 16 cells, known as the **morula** (**Figure 9.3**). The morula develops into the blastocyst or blastula which consists of an outer layer of cells, surrounding a fluid-filled cavity and an inner cell mass (ICM) (see Figure 9.3). Subsequently, the embryo undergoes the process of **gastrulation** to form the gastrula in which the cells are organized into three distinct **germ layers**: **ectoderm**, **mesoderm**, and **endoderm** (**Figure 9.4**). These three germ layers will ultimately form the different tissues of the embryo with the ectoderm forming the skin and the nervous system, the mesoderm forming the muscle, cartilage, and bone, and the endoderm forming the gut (see Sections 10.1 and 10.2).

Transcriptional control processes activate the embryonic genome

Clearly, as embryonic development proceeds in this manner a switch will occur so that rather than simply translating pre-existing maternal mRNAs the embryo will begin to transcribe its own genome. A variety of mechanisms appear to be involved in this maternal/embryonic switch.

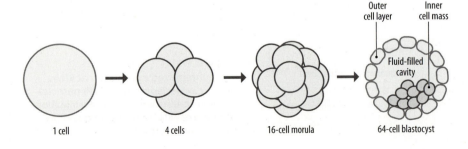

1 cell   4 cells   16-cell morula   64-cell blastocyst

Figure 9.3
Early development of the mammalian embryo, forming a ball of cells, known as the morula, which subsequently gives rise to the blastocyst containing an outer cell layer, surrounding an inner cell mass and a fluid-filled cavity.

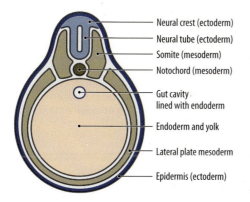

Figure 9.4
Section of amphibian gastrula-stage embryo showing the position of the three germ layers: ectoderm, mesoderm, and endoderm.

For example, during the first few cell divisions of the early embryo, the maintenance methylase Dnmt1 is excluded from the nucleus. As described in Chapter 3 (Section 3.2), this maintenance methylase plays a key role in maintaining the methylation of C residues in specific regions of the genome and thereby ensuring that the DNA in these regions is packaged into an inactive chromatin structure. The exclusion of Dnmt1 from the nucleus in the early cells of the embryo therefore allows the pattern of **DNA methylation** to be reprogrammed, so facilitating the transcriptional activation of specific genes (**Figure 9.5**).

As well as involving the regulation of chromatin structure by loss of inhibitory DNA methylation, the transition from maternal to embryonic gene expression can also involve the direct regulation of transcription by the activation of a positively acting factor which is required for transcription of the embryo genome. For example, in the nematode worm it has been shown that the TAF-4 protein which is normally part of the TFIID complex (see Section 5.2 for discussion of TAFs) is sequestered in the cytoplasm of the one- and two-cell embryo by its association with the OMA proteins. As the embryo divides further, the OMA proteins are phosphorylated. This targets them for degradation so releasing TAF-4 to join the TFIID complex (**Figure 9.6**).

Interestingly, in the zebrafish embryo the TBP component of the TFIID complex has been shown to be critical for the switching on of transcription

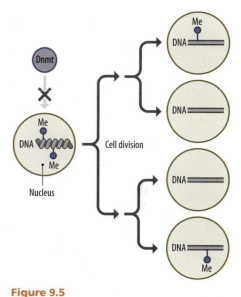

Figure 9.5
In the early embryo the maintenance methylase Dnmt1 is excluded from the nucleus. This results in progressive demethylation of the embryonic DNA as cell division occurs, allowing the DNA to move into a more open chromatin structure compatible with transcription.

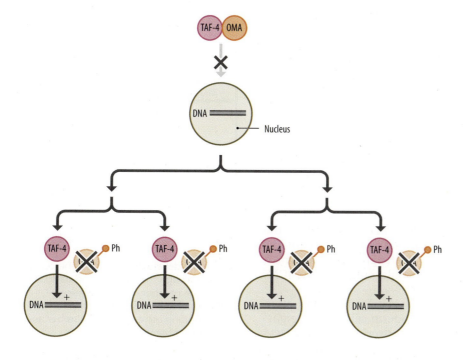

Figure 9.6
In the early nematode embryo, the TAF-4 activator protein is bound to the OMA proteins in the cytoplasm and therefore does not enter the nucleus. As development proceeds, the OMA proteins are degraded, allowing TAF-4 to enter the nucleus and activate transcription.

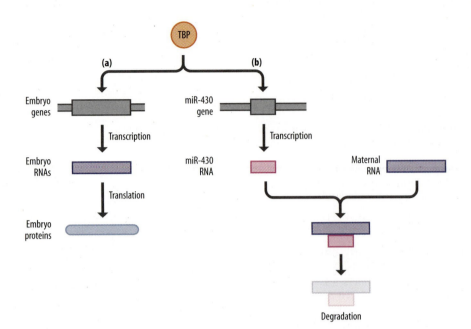

Figure 9.7
Synthesis of the TBP transcription factor allows it to activate specific protein-coding genes in the zebrafish embryo genome (a). It also activates the zebrafish embryo gene encoding the miRNA miR-430. MiR-430 binds to a variety of different maternal RNAs and induces their de-adenylation and subsequent degradation (b).

from the embryonic genome. Thus, TBP levels increase dramatically at the time that transcription of the embryonic genome begins and its inactivation prevents the transcription of a number of genes from the genome of the embryo (**Figure 9.7a**). Activation of the embryonic genome can therefore involve either TBP and/or TAF components of TFIID (see Sections 4.1 and 5.2 for further discussion of TBP and TAFs).

Hence, it is likely that several mechanisms are responsible for the progressive activation of the embryo genome during early embryonic development and these are likely to produce changes in both the chromatin structure of specific genes and their regulation by transcriptional regulatory proteins.

As well as stimulating the progressive transcription of protein-coding genes from the embryo genome, TBP also activates the expression of the gene encoding the microRNA (miRNA) miR-430 in zebrafish. Thus, in these fish, miR-430 is one of the earliest RNAs transcribed from the embryonic genome. It then binds to the 3′ untranslated region of a large number of different maternal mRNAs. This results in their de-adenylation and subsequent degradation (**Figure 9.7b**) (see Section 7.6 for further discussion of this mechanism). It is estimated that approximately 40% of maternally inherited mRNAs are degraded by an miR-430-dependent process in these fish. This process therefore allows maternal RNAs to be progressively degraded once they have fulfilled their function.

The Oct4 and Cdx2 transcription factors regulate the differentiation of ICM and trophectoderm cells

The combination of the increasing transcription of the embryonic genome and the progressive degradation of maternal mRNAs results in a progressive transition from the maternal to the embryo genome as the one-celled embryo develops through the blastula and gastrula stages (**Figure 9.8**). Clearly, as the embryo proceeds through these stages it is also necessary for individual cells to begin to differentiate into different cell types. Thus, the mammalian blastocyst, which consists of approximately 64 cells, contains two different cell types (see Figure 9.3). These are the **trophectoderm**, which will form the extra-embryonic tissues such as the placenta, and the ICM, which will give rise to the embryo itself.

Two specific transcription factors play a key role in determining whether individual cells will develop into ICM or trophectoderm cells. At the eight-cell stage, the cells express both the POU family transcription factor Oct4

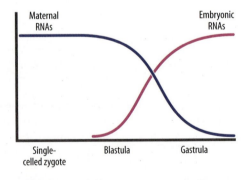

Figure 9.8
During the development of the one-cell zygote through the blastula and gastrula stages, the number and abundance of maternal RNAs declines with a corresponding increase in the number and abundance of embryonic RNAs.

(also known as Oct3/4) and the homeodomain transcription factor Cdx2 (**Figure 9.9**) (see Section 5.1 for discussion of these transcription factor families). Each of these two factors can both activate its own expression and also repress the expression of the other factor; that is, Oct4 represses Cdx2 expression and Cdx2 represses Oct4 expression (see Figure 9.9).

As embryonic development proceeds, Cdx2 expression begins to predominate in the outer cells and this represses Oct4 expression. Hence, Cdx2 levels rise further as it induces its own expression and it is relieved from the repressive effect of Oct4 (see Figure 9.9). As Cdx2 is necessary for cells to form trophectoderm, the outer cells of the embryo differentiate to form this cell type. Conversely, in the inner cells of the embryo Oct4 predominates, leading to the repression of Cdx2 and the formation of the ICM cells which requires Oct4 (see Figure 9.9).

Hence, the expression of one or other of these transcription factors by different cells in the early embryo produces the first differentiation event that occurs during embryonic development, resulting in the formation of ICM and trophectoderm cells. Moreover, the positive autoregulation of the expression of each factor and their ability to mutually inhibit each other's synthesis progressively amplifies any initial small difference in the levels of each factor between different cells.

During subsequent embryonic development, the ICM cells are able to give rise to all the different cell types in the early embryo, including derivatives of all three germ layers of the embryo: ectoderm, mesoderm, and endoderm. They are therefore referred to as being **pluripotent**. Clearly, the gene-regulatory processes which maintain these cells in a pluripotent form but also allow them to ultimately give rise to a variety of different cell types are of considerable interest in terms of the regulation of embryonic development. Fortunately, it has proved possible to isolate cell lines from the ICM of the early mammalian embryo and these are known as embryonic stem cells (ES cells). These undifferentiated cells are able to proliferate indefinitely in culture but can also, under certain conditions, differentiate into different cell types. The study of these cells has greatly aided our understanding of the processes occurring in early embryonic development and is therefore discussed in Section 9.1.

Clearly, as individual cells types are formed during embryonic development, it is necessary for regulatory processes to ensure that each cell type is formed at an appropriate time and at an appropriate place relative to other cell types. The study of the fruit fly *Drosophila* has greatly aided our understanding of these processes since it is amenable to both molecular and genetic techniques. The insights into developmental processes obtained in this organism are therefore discussed in Section 9.2. Subsequently, Section 9.3 discusses how insights obtained can be applied to the study of mammalian development.

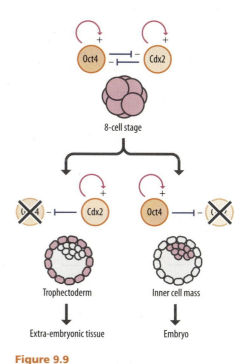

Figure 9.9
The first differentiation event in the early embryo is the formation of the ICM cells and the trophectoderm cells. Differentiation of the ICM cells is dependent upon high levels of the Oct4 transcription factor, which activates its own expression and inhibits expression of the Cdx2 transcription factor. In contrast, differentiation of the trophectoderm cells is dependent on a high level of the Cdx2 transcription factor, which induces its own expression and represses that of Oct4.

9.1 REGULATION OF GENE EXPRESSION IN PLURIPOTENT ES CELLS

ES cells can differentiate into a wide variety of cell types

As noted above, the first differentiation event in the early mammalian embryo gives rise to trophectoderm cells and to ICM cells. These ICM cells are evidently of considerable interest since they will give rise to all the cells of the embryo itself and are therefore pluripotent being able to give rise to a variety of differentiated cell types. Mechanisms must therefore exist in these cells which allow them both to maintain their pluripotency through a number of cell divisions but also allow them to differentiate into different cell types at the appropriate time. Our understanding of these processes has been greatly aided by the finding that when the ICM is disaggregated into individual cells the resulting cells can proliferate indefinitely in culture while maintaining their undifferentiated pluripotent nature (**Figure 9.10**).

The resulting ES cell lines therefore provide a source of much greater amounts of pluripotent cells than could be obtained from an early embryo.

Figure 9.10
When the ICM is isolated from a blastocyst and disaggregated, undifferentiated embryonic stem (ES) cells can be grown indefinitely in culture on a feeder layer of fibroblast cells. When the undifferentiated ES cells are grown in suspension culture, they form embryoid bodies in which an inner layer of undifferentiated cells is surrounded by an outer layer of endoderm cells. When these embryoid bodies are allowed to attach to the surface of a culture dish (in the absence of a feeder cell layer) they differentiate further to form a variety of different cell types.

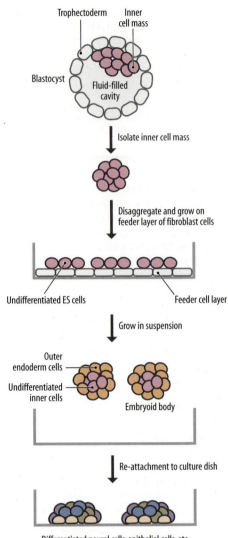

Moreover, as well as closely resembling the pluripotent cells of the early embryo in their undifferentiated nature, ES cells can also under suitable conditions differentiate to produce a wide variety of different cell types. Thus, if the ES cells are allowed to grow in suspension in the culture medium, they differentiate to form aggregates of cells known as embryoid bodies in which the inner cells retain their pluripotent nature while the outer cells differentiate to form endoderm cells (see Figure 9.10).

This is of particular interest since it closely parallels the second differentiation event in mammalian development in which cells on the surface of the ICM in the embryo differentiate to form a surface layer of endoderm cells adjacent to the fluid-filled blastocoel cavity (**Figure 9.11**).

Subsequently, if ES-cell-derived embryoid bodies in culture are allowed to re-attach to the surface of the dish in which they are grown, they can differentiate further to form a variety of different cell types including neural cells and epithelial cells (see Figure 9.10). ES cell lines therefore offer a highly convenient source of pluripotent cells which are closely related to those found in the early embryo and which, like early embryonic cells, can differentiate to form a wide variety of different cell types.

Indeed, the close equivalence of ES cell lines with pluripotent cells of the early embryo can be further demonstrated by injecting ES cells into the blastula stage of an embryo with a different **genotype** from that of the injected cells. If this chimeric blastocyst is allowed to develop, it will form a mouse in which every organ contains contributions both from the original blastocyst and from the injected ES cells. This can be demonstrated, for example, if the injected cells differ from those of the host blastocyst in the coat color which they will produce (for example, white versus black). In this case, the resulting mouse will show a chimeric coat color in which some regions are of one color, showing that they are derived from the injected ES cells, while other regions are of a different color, indicating that they are derived from the original host blastocyst (**Figure 9.12**).

ES cells, when inserted into an embryo, can therefore contribute to all the differentiated cells of the resulting adult animal. Most importantly, it has been demonstrated that they can also produce functional eggs and sperm in these animals. Thus, by breeding first-generation chimeric animals obtained as described above it is possible to produce second-generation animals which are entirely derived from the injected ES cells (see Figure 9.12).

This procedure has been extensively used to introduce genetic mutations into mice (knockout mice) by first introducing the mutation into the ES cell genome and then producing mice which carry the mutation by injecting the ES cells into a blastocyst embryo. For our purposes, however, the most important finding from these experiments is that ES cells are indeed pluripotent since as well as producing differentiated cell types they can also give rise to egg and sperm cells, allowing the production of an intact animal entirely derived from the injected ES cells.

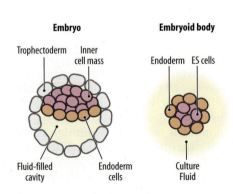

Figure 9.11
The formation of endoderm cells on the surface of an embryoid body in culture medium parallels the formation of endoderm cells on the surface of the ICM which is in contact with the fluid-filled cavity of the blastocyst embryo.

Figure 9.12

ES cells injected into the ICM of a blastocyst can contribute to all cell types in the mouse which forms from the blastocyst, when it is re-implanted into a foster mother. This can be visualized if the ES cells are of a different genotype from the recipient blastocyst; for example, having the genes required to form a "pink" coat color rather than a "gray" coat color. The injected ES cells can also contribute to the formation of sperm or egg cells in this experiment. Hence, breeding of the chimeric mice produced by ES cell injection results in offspring which are entirely derived from the injected ES cells and in this case have a "pink" rather than a "gray" coat color.

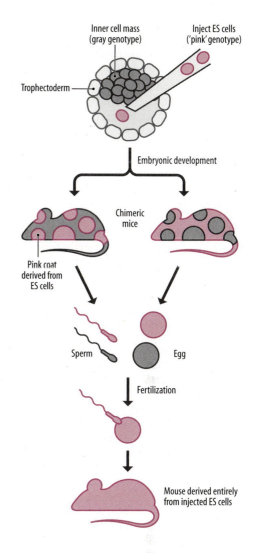

Several transcription factors are specifically expressed in ES cells and together can reprogram differentiated cells to an ES-cell-like phenotype

The ability of ES cells to remain in a pluripotent undifferentiated state through a number of cell divisions, coupled with their ability to differentiate into a variety of cell types under appropriate conditions, has led to intensive studies aimed at identifying the transcription factors which regulate these processes. A number of transcription factors expressed in pluripotent ES cells have been identified and characterized. The expression of several of these factors has been shown to be down-regulated as ES cells differentiate, suggesting that they play a key role in undifferentiated ES cells.

Most interestingly, it has been shown that the over-expression of just four of these transcription factors, Oct4, Sox2, Klf4, and c-Myc in differentiated fibroblast cells was sufficient to reprogram them to a pluripotent ES-cell-like state (**Figure 9.13**). Subsequently, this was also achieved with a different but related cocktail of regulatory factors composed of Oct4, Sox2, Nanog, and Lin28.

These experiments were primarily carried out to try and provide a source of undifferentiated stem cells for therapeutic purposes which would be more ethically acceptable than using ES cells derived from early human embryos. Thus, if differentiated cells could be reprogrammed into ES-like cells, then these could be differentiated to form specific cell types. Such cells could then be transplanted into patients suffering from degenerative diseases caused by a loss of a particular cell type.

For our purposes, however, these studies focus attention on this small group of transcription factors since they are both specifically expressed in pluripotent ES cells and can achieve the reprogramming of differentiated cells into ES-like cells. The expression of a relatively small number of transcription factors in ES cells therefore induces their pluripotent phenotype and the down-regulation of these factors allows the production of differentiated cells.

ES-cell-specific transcription factors can activate or repress the expression of their target genes

It appears that ES-cell-specific transcription factors act in part by repressing the expression of specific genes which are required for differentiation to

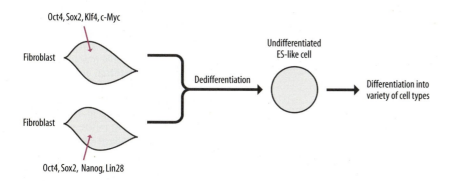

Figure 9.13

The introduction of four specific regulatory factors (Oct4 and Sox2 together with Klf4 and c-Myc or Nanog and Lin28) is sufficient to induce fibroblast cells to dedifferentiate and form undifferentiated ES-like cells, which can subsequently differentiate into many different cell types.

occur. For example, Oct4 and Nanog have been shown to interact with one another and to bind to a variety of different target genes in ES cells. In turn, such binding results in the recruitment of co-repressor complexes such as NuRD and Sin3. These complexes contain histone deacetylases which are able to induce the deacetylation of histones and thereby organize the DNA into an inactive chromatin structure, incompatible with transcription (**Figure 9.14**) (see Section 3.3 for discussion of the effect of histone acetylation on chromatin structure).

Interestingly, Oct4, Nanog, and a third factor, Sox2, co-operate together to repress the expression of the *XIST* gene in ES cells. As discussed in Chapter 3 (Section 3.6) expression of *XIST* from one of the two X chromosomes plays a key role in the inactivation of one of the two X chromosomes present in cells of female mammalian embryos, with the chromosome expressing *XIST* being the one which is inactivated. As noted in Chapter 3 (Section 3.6), X inactivation occurs randomly in the cells which form the tissues of the embryo so that different cells have inactivated either the X chromosome derived from the mother's genome or that derived from the father's genome.

However, this phase of random inactivation is preceded by an earlier phase which occurs in the cells in the early embryo prior to formation of the trophectoderm and the ICM cells. In these early undifferentiated cells, *XIST* is specifically expressed from the paternal X chromosome resulting in the inactivation of this chromosome (**Figure 9.15**). This is therefore an imprinting process (see Section 3.6) in which gene expression is regulated differently in the paternally and maternally inherited chromosomes.

Such imprinted expression of *XIST* and specific silencing of paternal X chromosome is maintained in the extra-embryonic tissues produced from the trophectoderm cells. However in the ICM cells, *XIST* expression is extinguished and the paternal X chromosome is reactivated, thereby allowing subsequent random inactivation of the maternal or paternal X chromosomes to follow during cell differentiation (see Figure 9.15).

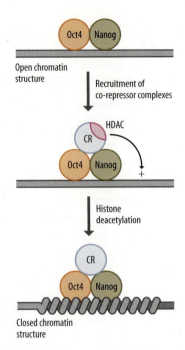

Figure 9.14
Oct4 and Nanog can interact with one another and bind to the promoters of specific genes. This results in the recruitment of co-repressor complexes (CR) with histone deacetylase (HDAC) activity, which are able to organize an inactive chromatin structure incompatible with gene transcription.

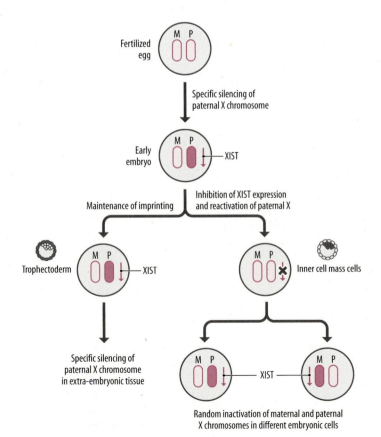

Figure 9.15
In the early embryo the *Xist* gene is specifically expressed from the paternal X chromosome, resulting in inactivation of this chromosome (dark pink). This is maintained in the trophectoderm lineage, resulting in the specific inactivation of the paternal X chromosome in extra-embryonic tissue. In contrast, in the ICM cells, the expression of *Xist* from the paternal X chromosome is specifically inactivated so allowing both the maternal and paternal X chromosomes to be active. Subsequently, *Xist* is randomly activated on the maternal or paternal X chromosome in different cells of the embryo, resulting in the embryo having a mixture of cells in which either the maternal or the paternal X chromosome has been inactivated (dark pink).

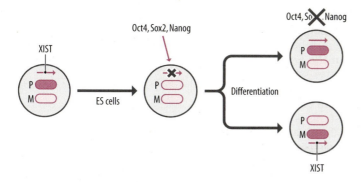

Figure 9.16
The inactivation of *Xist* gene transcription from the paternal X chromosome in ES cells is produced by the inhibitory effect of the Oct4, Sox2, and Nanog transcription factors. When the expression of these factors decreases during ES cell differentiation, *Xist* expression is reactivated on either the paternal or the maternal X chromosome, in different cells, resulting in one or other of the X chromosomes being inactivated (dark pink).

It has been demonstrated that Oct4, Nanog, and Sox2 specifically bind to the *XIST* gene on the paternal X chromosome and repress its expression, so reversing the specific inactivation of this chromosome. During differentiation when the levels of these factors fall, *XIST* expression can be reactivated. This takes place randomly on either the maternal or paternal chromosomes, as described in Chapter 3 (Section 3.6), allowing different cells to inactivate either the paternal or maternal X chromosomes (**Figure 9.16**).

Although Oct4, Sox2, and Nanog can therefore repress the expression of specific genes in ES cells, they are also able to activate the expression of other genes which are required to maintain the pluripotent nature of ES cells. Thus, it has been demonstrated that these three factors bind in combination to the regulatory regions of approximately 350 genes in the genome of ES cells and that approximately 50% of these gene loci are transcriptionally activated as opposed to repressed in ES cells.

ES-cell-specific transcription factors regulate genes encoding chromatin-modifying enzymes and miRNAs

Among the genes activated by Oct4 are the genes encoding the histone demethylases Jmjd1A and Jmjd2C. These enzymes catalyse the demethylation of histone H3 on the lysine at position 9. As described in Chapter 2 (Section 2.3) and Chapter 3 (Section 3.3) methylation of histone H3 at this position is associated with a closed chromatin structure, so that its demethylation produces an open chromatin structure, allowing specific genes to be expressed in ES cells (**Figure 9.17**).

In addition, however, Oct4 can also activate the expression of genes that promote a closed chromatin structure. Thus, Oct4 can co-operate with the STAT3 transcription factor (see Section 8.2) to activate the gene encoding the Eed protein. This protein is a component of the polycomb complex (see below) which methylates histone H3 at the lysine at position 27 and thereby produces an inactive chromatin structure (**Figure 9.18**).

Hence, Oct4 can activate the expression of genes which produce a more open chromatin structure at specific regions of the genome and also activate the expression of genes producing a closed chromatin structure, thereby indirectly repressing gene expression in other regions of genome. Such chromatin modifications will allow genes which need to be expressed in the ES cells to be organized into an open chromatin structure compatible with gene expression whereas genes which must be repressed in ES cells are organized into a closed chromatin structure incompatible with gene expression (compare Figures 9.17 and 9.18).

Interestingly, Oct4 can also indirectly repress gene expression by activating the genes encoding a variety of different miRNAs which, as described in previous chapters, have a key role in post-transcriptional inhibition of gene

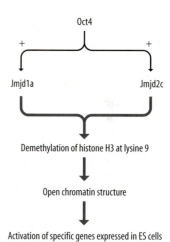

Figure 9.17
Oct4 activates the genes encoding the histone demethylases Jmjd1a and Jmjd2c. This results in the demethylation of histone H3 at lysine 9, producing an open chromatin structure and the activation of specific target genes required in ES cells.

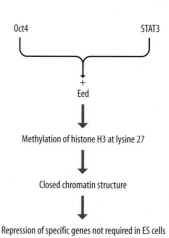

Figure 9.18
The Oct4 and STAT3 transcription factors co-operate to activate the gene encoding the Eed protein. Eed methylates lysine 27 on histone H3, producing a closed protein structure and thereby repressing the expression of specific genes not required in ES cells.

expression. For example, Oct4 co-operates with Sox2, Nanog, and another transcription factor, Tcf3, to activate the expression of the genes encoding several miRNAs which then repress the expression of a number of different target genes (**Figure 9.19**). In this case therefore, as in the Eed case, Oct4 represses gene expression indirectly by activating genes which produce inhibitory factors.

The direct and indirect mechanisms of Oct4 action are combined in the regulation of the genes encoding the DNA methyltransferases Dmnt3a and Dmnt3b, which, as described in Chapter 3 (Section 3.2), play a key role in the regulation of DNA methylation during embryonic development. Thus, Oct4, Sox2, Nanog, and Tcf3 all co-operate together to activate the expression of the *Dmnt3a* and the *Dmnt3b* genes (**Figure 9.20**). In addition, however, they also activate the expression of the genes encoding the miRNAs miR-290–295. In turn, these miRNAs repress the expression of another transcription factor Rbl2. As Rbl2 acts as a transcriptional repressor of the *Dmnt3a* and *Dmnt3b* genes, this further stimulates the expression of *Dmnt3a* and *Dmnt3b* (see Figure 9.20).

Interestingly, as well as regulatory genes encoding proteins or small miRNAs, ES-cell-specific transcription factors also regulate the transcription of large non-coding RNAs of 2 kb or greater in size which are expressed specifically in ES cells. Several such RNAs have been identified as being activated by Oct4 and Nanog. As expected from this, the expression of these RNAs declines when ES cells differentiate. Moreover, one of these large RNAs has been shown to be essential for the proliferation of undifferentiated ES cells. Hence large RNAs which do not encode proteins appear to play a key role in undifferentiated ES cells, paralleling the role of such RNAs in processes such as X-chromosome inactivation and genomic imprinting (see Section 3.6).

It is clear therefore that a small number of transcription factors maintain the pluripotency of ES cells, acting both directly by activating and repressing specific target genes and indirectly by altering the expression of genes encoding chromatin-modifying factors, miRNAs, and large non-coding RNAs. The ability of these transcription factors to regulate a very wide variety of genes, either directly or indirectly, is therefore responsible for their ability to induce differentiated cells to revert to an ES-cell-like undifferentiated state.

The REST transcription factor plays a key role in down-regulating the expression of ES-cell-specific transcription factors during differentiation

During differentiation the expression of ES-cell-specific transcription factors will be down-regulated, so inhibiting the expression of genes which are only required in the ES cells and activating the expression of genes required in specific differentiated cells. Clearly, the critical role of these factors in ES cell pluripotency and differentiation focuses attention on the manner in which they themselves are regulated.

It has been demonstrated that regulation of Oct4, Nanog, Sox2, and c-Myc involves a transcriptional repressor protein, repressor-element silencing transcription factor (REST), which also plays a key role in repressing the expression of neuronal genes in non-neuronal cells (see Section 10.2). REST is expressed at high levels in undifferentiated ES cells and its level falls during **embryoid body** formation. Moreover, REST specifically represses the expression of the gene encoding the miRNA miR-21 in ES cells. As ES cells differentiate, the level of REST expression falls and the level of miR-21 correspondingly rises. This rise in miR-21 expression results in the decreased expression of Oct4, Nanog, Sox2, and c-Myc which is required for ES-cell differentiation to occur (**Figure 9.21**).

ES cells have an unusual pattern of histone methylation

As described above, Oct4 and the other ES-cell transcription factors achieve their effects at least in part by the regulation of chromatin structure. For

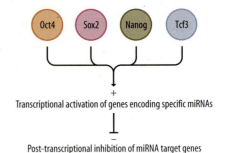

Figure 9.19
The transcription factors Oct4, Sox2, Nanog, and Tcf3 co-operate together to produce transcriptional activation of genes encoding specific miRNAs which in turn inhibit specific target genes at the post-transcriptional level.

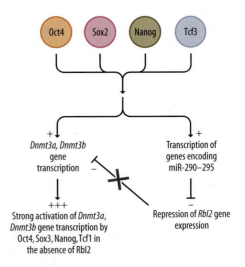

Figure 9.20
The transcription factors Oct4, Sox2, Nanog, and Tcf3 can directly activate the genes encoding the Dnmt3a and Dnmt3b methyltransferase enzymes. They also activate the genes encoding the miRNAs miR-290–295. These miRNAs repress the expression of the gene encoding the Rbl2 transcription factor. Since this factor is a repressor of the *Dnmt3a* and *Dnmt3b* genes, inhibition of Rbl2 by the miRNAs results in further activation of Dnmt3a and Dnmt3b expression.

example, Oct4 and Nanog can recruit repressor complexes containing histone deacetylases to their target genes and therefore organize a closed chromatin structure incompatible with transcription (see Figure 9.14). Similarly, Oct4 can activate the expression of genes encoding histone methylases and demethylases which can then alter the chromatin structure of other target genes to which Oct4 does not bind directly (see Figures 9.17 and 9.18).

In fact, genome-wide studies of histone modifications in undifferentiated ES cells have suggested that a number of genes in such cells may contain unique patterns of histone modification which are not normally found in other cell types. Methylation of histone H3 on the lysine at position 4 is normally associated with an open chromatin structure (see Section 3.3) and indeed this modification is present at a number of genes which are expressed in undifferentiated ES cells (**Figure 9.22a**). Similarly, methylation of histone H3 on lysine 27 is associated with a closed chromatin structure in many cell types (see Section 3.3) and is found in ES cells at gene loci which are not expressed in the undifferentiated ES cells or during their initial differentiation (**Figure 9.22b**).

Most interestingly, however, some genes in ES cells are associated with histone H3 which is methylated on both lysine 4 and lysine 27 (**Figure 9.22c**). These genes therefore have a so-called **bivalent code** in which they have both an activating histone methylation (lysine 4) and an inhibitory histone methylation (lysine 27).

When the expression pattern of these genes containing the bivalent code is studied, they are found to be repressed in ES cells but to be rapidly expressed upon differentiation. Such expression is associated with the loss of methylation at lysine 27 while methylation at lysine 4 is retained (**Figure 9.23**). Such lysine 27 demethylation is catalysed by specific demethylase enzymes such as UTX, which become active as ES cells differentiate.

It is likely therefore that the bivalent pattern of histone modification allows the ES cells to remain pluripotent by not activating these genes due to the presence of the inhibitory lysine 27 modification. However, the genes are poised for rapid activation due to the presence of methylated lysine 4. The bivalent code therefore represents a response to the unique nature of ES cells with the cells needing to retain their pluripotent nature while being able to rapidly differentiate in response to appropriate signals.

ES cells therefore maintain a number of different genes in a state which is poised for transcription but in which transcription does not normally occur until differentiation begins. Interestingly, mechanisms also exist in ES cells to prevent transcription by any RNA polymerase II complexes which bind inappropriately to such silent genes. In this case, the RNA polymerase II complex becomes associated with the **proteasome**, a protein complex

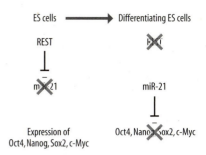

Figure 9.21
In ES cells, the REST factor represses expression of the gene encoding miR-21. As ES cells differentiate, REST expression falls, resulting in the activation of miR-21, which in turn represses the expression of the genes encoding Oct4, Nanog, Sox2, and c-Myc.

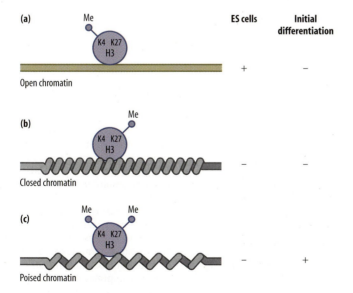

Figure 9.22
Genes which are expressed in ES cells are associated with histone H3 in which the lysine at position 4 (K4) but not the lysine at position 27 (K27) is methylated, so producing an open chromatin structure (a). In contrast, genes which are repressed in ES cells and in their initial differentiated derivatives are associated with histone H3 which is methylated on lysine 27 but not on lysine 4, so producing a closed chromatin structure (b). Genes which are repressed in ES cells but which need to be expressed immediately upon differentiation are associated with histone H3 which is methylated on both sites, producing a chromatin structure which is poised for transcriptional activation (c).

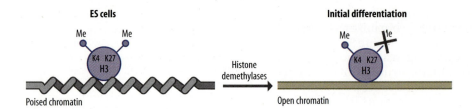

Figure 9.23
The activation of specific genes during the initial differentiation of ES cells is associated with demethylation of histone H3 at lysine 27 (K27) but not at lysine 4 (K4).

which catalyses the degradation of other proteins. The RNA polymerase II complex is therefore degraded before inappropriate transcription occurs (**Figure 9.24**).

The polycomb complex regulates histone methylation in ES cells

As discussed above, genes which are repressed in undifferentiated ES cells, but which need to be activated upon differentiation, frequently exhibit a bivalent modification of histone H3 involving methylation at both lysine 4 and lysine 27. The key role of this modification pattern in ensuring that these genes are poised for activation immediately upon differentiation focuses attention on the proteins which catalyse such methylation events.

As described in Chapter 3 (Section 3.3) the activating methylation on lysine 4 is produced by proteins of the trithorax family whereas the inhibitory modification on lysine 27 is produced by polycomb protein complexes. The trithorax and polycomb proteins were originally identified in *Drosophila* where their inactivation by mutation resulted in flies with different abnormal body patterns. This is because these factors regulate the expression of the genes encoding the homeodomain transcription factors, which themselves play a key role in specifying the identity of different cell types (see Sections 5.1 and 9.2). Thus, the trithorax proteins activate specific homeodomain-containing genes in particular cells by promoting an open chromatin structure. Inactivation of the trithorax genes by mutation therefore results in abnormal flies due to the lack of specific homeodomain gene expression in particular cells (**Figure 9.25**).

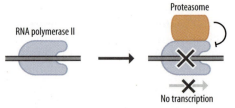

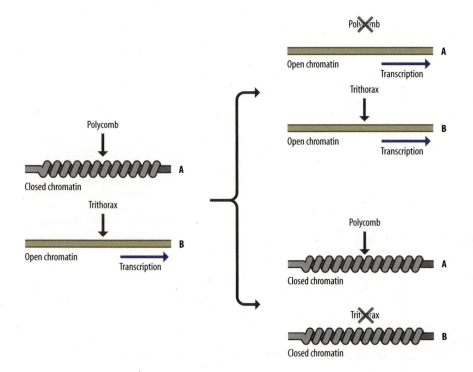

Figure 9.24
Inappropriate binding of RNA polymerase II to specific genes that should not be expressed in ES cells is followed by binding of the proteasome which degrades the RNA polymerase II complex.

Figure 9.25
The polycomb complex acts to maintain specific homeodomain-containing genes (A) in a closed chromatin structure incompatible with transcription, whereas the trithorax complex acts to produce an open chromatin structure of specific homeodomain genes (B), allowing transcription to occur. Mutation of the polycomb complex therefore results in inappropriate expression of homeodomain-containing genes, whereas mutation of the trithorax complex results in a lack of expression of specific homeodomain-containing genes.

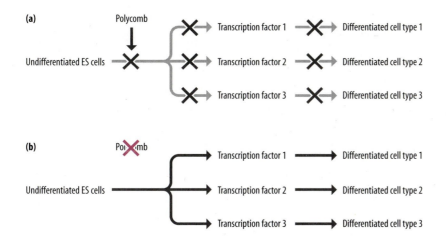

Figure 9.26
The polycomb complex maintains ES cells in an undifferentiated state by preventing the transcription of genes encoding proteins such as transcription factors which are required for differentiation into specific cell types (a). Inactivation of the polycomb complex by mutation results in the inappropriate activation of these genes and leads to premature differentiation of ES cells into a variety of differentiated cell types (b).

Similarly, the polycomb proteins repress the inappropriate expression of homeodomain-containing genes in specific cell types by organizing them into an inactive chromatin structure. Inactivation of the polycomb genes by mutation results in the inappropriate expression of homeodomain-containing genes in particular cells, again producing a mutant fly with an abnormal body pattern (see Figure 9.25).

Although originally identified in *Drosophila*, both polycomb and trithorax proteins have been found in a variety of different organisms where they play a critical role in the regulation of chromatin structure. In particular, it has been shown that the polycomb proteins play a central role in maintaining the pluripotent nature of ES cells. Thus, the polycomb proteins prevent ES cells from expressing genes such as those encoding transcription factors which are required to induce the production of specific types of differentiated cells. Genes repressed by polycomb include those encoding both homeodomain-containing transcription factors and members of other transcription factor families.

Hence, the polycomb proteins prevent the inappropriate expression of differentiation-specific transcription factor genes and thus allow ES cells to maintain their pluripotent nature (**Figure 9.26a**). In agreement with this idea, inactivation of the Eed protein, which is a component of the polycomb protein complex, induces the premature activation of differentiation-specific genes in ES cells and results in their inappropriate differentiation into a variety of different cell types (**Figure 9.26b**). Interestingly, as noted above, the gene encoding Eed itself is a target for activation by the Oct4 transcription factor, providing a link between Oct4 and the polycomb proteins, both of which play a key role in regulating gene expression in pluripotent ES cells (see Figure 9.18).

Two multiprotein polycomb complexes have been described and are referred to as PRC1 and PRC2. The PRC2 complex binds to specific genes and catalyses histone H3 methylation on lysine 27. This methylation is then recognized by PRC1 which binds to the methylated histones. The PRC1 complex can both bind to methylated lysine 27 and also catalyse further lysine 27 methylation (**Figure 9.27**).

This dual activity of PRC1 in recognizing methylated lysine 27 and also catalysing further methylation ensures that lysine 27 methylation is propagated through multiple cell divisions. Following cell division, each daughter

Figure 9.27
The PRC2 polycomb complex can methylate (small blue circle) histone H3 on lysine 27 (K27). The PRC1 polycomb complex can both recognize such methylated histone H3 and catalyze further lysine 27 methylation.

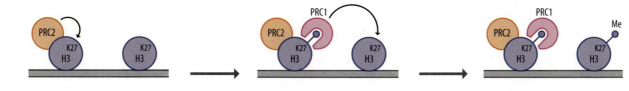

chromosome will initially have some methylated histone H3 inherited from the parental chromosome and some newly bound unmethylated histone H3. The dual activity of PRC1 will allow it to bind via the methylated histone and then methylate the unmethylated histone (**Figure 9.28**) (see Section 3.3 for discussion of a similar mechanism involving the HP1 protein). Such maintenance of histone 27 methylation through cell division is obviously of vital importance since ES cells must maintain their pluripotent nature through multiple cell divisions before eventually differentiating.

Polycomb protein complexes regulate the expression of miRNA genes in ES cells

The polycomb protein complexes have been shown by chromatin immuno-precipitation (ChIP) analysis (see Methods Box 4.3) to be associated with the regulatory regions of several hundred genes in ES cells. As described above, many of these will be genes encoding specific transcription factors which play a key role in differentiation and which therefore must be maintained in an inactive state poised for activation in ES cells. Interestingly, however, the polycomb repressor complexes also associate in ES cells with the regulatory regions of a number of genes encoding miRNAs. Thus, as described above, the genes encoding a number of different miRNAs are activated in ES cells due to the binding of the transcription factors Oct4, Sox2, Nanog, and Tcf3 (see Figure 9.19).

However, a number of other genes encoding different miRNAs also bind Oct4, Sox2, Nanog, and Tcf3. This second class of miRNA genes is not expressed in ES cells due to the simultaneous binding of polycomb repressor complexes which prevents the expression of these genes in pluripotent ES cells (**Figure 9.29**). As ES cells differentiate into various cell types, polycomb repression is relieved and specific miRNAs are expressed. For example, the gene encoding miR-132 is silenced by polycomb in ES cells but becomes active when they differentiate into muscle cells. Similarly, the gene encoding miR-124 is silenced in ES cells by polycomb complexes but becomes active as ES cells differentiate into neural precursor cells (see Sections 10.1 and 10.2 for discussion of the role of miRNAs in regulating gene expression in muscle and nerve cells respectively). Hence, the role of polycomb complexes in regulating histone methylation and thereby ensuring that differentiation-specific genes are repressed but poised for activation in ES cells is not confined to protein-coding genes but also occurs for genes which produce miRNAs (**Figure 9.30**).

Chromatin structure in ES cells is regulated by multiple effects on histones

As discussed above, the regulation of histone methylation plays a key role in modulating chromatin structure in ES cells. It has been demonstrated, however, that a number of the other effects on histones discussed in Chapter 2 (Section 2.3) and Chapter 3 (Section 3.3) also play important roles in ES cells. For example, it has been shown that the histone H2 variant H2AZ (see Section 2.3) co-operates with polycomb proteins to regulate chromatin structure in ES cells. Similarly, a specific enzyme known as SCNY, which removes ubiquitin from histone H2B, is necessary for the repression of

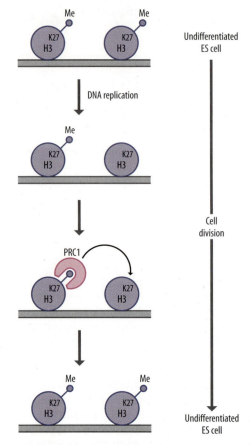

Figure 9.28
The dual activity of PRC1 allows histone methylation on lysine 27 (K27) to be maintained through cell division, thereby maintaining the pluripotent nature of ES cells. Following DNA replication, each chromosome will have some methylated histone H3 inherited from the parental chromosome and some newly deposited unmethylated histone H3 (only one daughter chromosome is shown for simplicity). PRC1 can both bind to the methylated histone H3 and catalyze methylation of the unmethylated histone H3, thereby restoring the full methylation pattern.

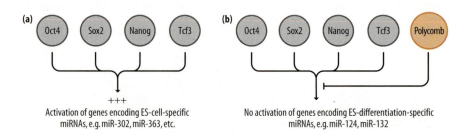

Figure 9.29
The transcription factors Oct4, Sox2, Nanog, and Tcf3 can activate the expression of genes encoding ES-cell-specific miRNAs (a). These transcription factors also bind to the regulatory regions of genes encoding differentiation-specific miRNAs. However, the expression of these miRNAs in ES cells is prevented by simultaneous binding of polycomb repressor complexes (b).

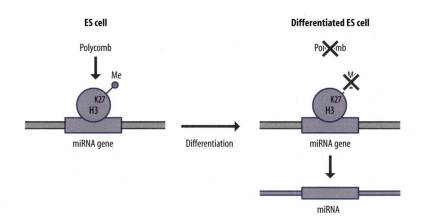

ES cell

Polycomb

Differentiated ES cell

Polycomb

Differentiation

miRNA gene

miRNA gene

miRNA

Figure 9.30
The polycomb complex produces methylation of histone H3 on lysine 27 in the regulatory regions of genes encoding miRNAs which are not expressed in ES cells. Upon differentiation lysine 27 is demethylated and the miRNA genes are expressed.

specific genes in undifferentiated ES cells. Moreover, the inactivation of this enzyme not only increases ubiquitination of histone H2B but also results in the premature differentiation of the ES cells (**Figure 9.31**).

During the normal differentiation of ES cells SCNY is down-regulated, producing enhanced ubiquitination of histone H2B. As described in Chapter 3 (Section 3.3) this has the effect of enhancing methylation of histone H3 on lysines 4 and 79, providing an example of the modification of one histone influencing the modification of another histone. Overall these changes in ubiquitination and methylation will have the effect of opening up the chromatin structure of specific genes which must be transcribed in differentiated ES cells but not in the undifferentiated cells (see Figure 9.31).

As well as these effects on the modification of individual amino acids in histones, a novel mechanism has recently been described in ES cells. This is based on the fact that the majority of the regulatory modifications of histones such as acetylation or methylation are clustered at the N-terminus of the molecule. Thus, it has been shown that during ES cell differentiation some of the histone H3 in the cell is cleaved after the alanine at position 21, so removing many of the modified residues (**Figure 9.32**). Although the role of this cleavage in the process of ES cell differentiation remains uncertain, this process is not confined to mammalian cells since cleavage of histone H3 has also been reported in yeast, suggesting that it may represent a widely used regulatory mechanism.

Regulation of chromatin structure by polycomb complexes and other mechanisms therefore plays a key role in regulating gene expression in ES cells and their differentiated derivatives. When taken together with the effects of ES-cell-specific transcription factors such as Oct4, these effects on transcription and chromatin structure allow ES cells to retain their pluripotent nature while being poised for differentiation along a variety of different pathways.

As will be discussed in Chapter 10, a variety of mechanisms exist to ensure that different cell types, such as muscle or nerve cells, exhibit the appropriate pattern of cell-type-specific gene expression. However, during embryonic development it is necessary to ensure that such cellular differentiation and the associated cell-type-specific patterns of gene expression occur at the appropriate time and place relative to the formation of other cell types. Our understanding of these processes has been greatly aided by analysis of the fruit fly *Drosophila* in which the techniques of molecular biology can readily be combined with genetic analysis of mutant flies having a particular phenotype. The insights obtained in this system are discussed in the next section.

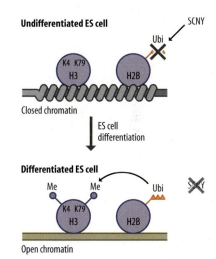

Undifferentiated ES cell

SCNY

Ubi

K4 K79
H3   H2B

Closed chromatin

ES cell differentiation

Differentiated ES cell

Me   Me   Ubi   SCNY

K4 K79
H3   H2B

Open chromatin

Figure 9.31
In undifferentiated ES cells, the enzyme SCNY removes ubiquitin (Ubi) from histone H2B. As ES cells differentiate, SCNY is down-regulated, resulting in enhanced levels of ubiquitinated histone H2B. In turn, this promotes methylation of histone H3 on lysines (K) 4 and 79, producing a more open chromatin structure for specific genes whose expression is required in differentiated ES cells.

Figure 9.32
N-terminus of histone H3 showing the various post-translational modifications (as illustrated in Figure 2.28) together with the site at which histone H3 is cleaved during ES cell differentiation (arrow). Note that this cleavage removes many of the residues whose modification regulates chromatin structure.

H3 NH₂—A R T K Q T A R K S T G G K A P R K Q L A T K A A R K S A P

Acetylation
Methylation
Phosphorylation

9.2 ROLE OF GENE REGULATION IN THE DEVELOPMENT OF *DROSOPHILA MELANOGASTER*

A gradient in expression of the Bicoid transcription factor defines the anterior–posterior axis in the early *Drosophila* embryo

The fruit fly *D. melanogaster* has been extremely well characterized genetically. In particular, a number of gene mutations have been identified, each of which can produce a fly with a specific abnormal body pattern. Analysis of these genes has shown that the majority of them encode transcription factors including homeodomain-containing transcription factors such as Bicoid and Antennapedia, as well as zinc-finger-containing transcription factors such as Kruppel and Hunchback (see Section 5.1 for discussion of these classes of transcription factors). As will be discussed in this section, the combination of molecular and genetic techniques available in *Drosophila* has provided considerable information on the key role of gene-control processes in *Drosophila* development. Moreover, as will be discussed in Section 9.3, such studies have also provided insights into the processes regulating mammalian development.

Interestingly, however, the development of *Drosophila* begins in a way very different from the early mammalian developmental processes which were discussed above. Following fertilization, 13 successive rounds of nuclear division occur without any division of the cytoplasm, so producing a single syncytial structure. In this syncytial structure, a single mass of cytoplasm surrounds approximately 1500 nuclei which progressively migrate to the periphery of the embryo (**Figure 9.33**). Only then does cell division occur to produce individual cells containing only one nucleus, around the periphery of the embryo. Clearly, to differentiate into different cell types these cells need to respond in some manner to their position within the embryo; that is, whether they are anterior or posterior or whether they are dorsal or ventral (see Figure 9.33).

A key role in this process is played by the Bicoid homeodomain-containing transcription factor. As described in Chapter 7 (Section 7.3), sequences in the 3′ untranslated region of the Bicoid mRNA result in its preferential localization at the anterior end of the embryo. A gradient in Bicoid expression is therefore created within the embryo in which the concentration of Bicoid is highest at the anterior end and then declines progressively towards the posterior end. This gradient in Bicoid expression is absolutely critical for defining the anterior–posterior axis of *Drosophila* since mutation in the Bicoid gene results in the absence of the head and thoracic structures of the animal.

Importantly, genes which are activated by the Bicoid transcription factor contain binding sites in their promoters which have either high affinity or low affinity for the Bicoid protein. The presence of high- or low-affinity binding sites (or a combination of the two) in different promoters results in the different genes being activated at different concentration levels of Bicoid.

Thus, genes with high-affinity binding sites will be activated at both low and high concentrations of Bicoid, whereas genes with low-affinity sites will be activated only at high Bicoid concentrations. Hence, the gradient in Bicoid expression in the egg is converted into a pattern of gene expression

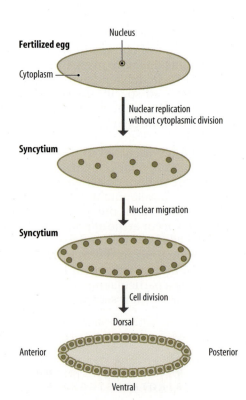

Figure 9.33
Early development in the *D. melanogaster* embryo involves nuclear replication without cytoplasmic division to produce a syncytium containing multiple nuclei in a common cytoplasm. Subsequently, the nuclei migrate to the periphery of the syncytium and cell division then occurs. During development individual cells in the embryo will differentiate according to their position along the anterior–posterior axis or the dorsal–ventral axis.

with some genes being activated only at the most anterior end of the egg since they require a high concentration of Bicoid whereas others are activated over a greater region of the egg since they require only a lower concentration of Bicoid for activation (**Figure 9.34**).

Bicoid activates a cascade of genes encoding other transcription factors, producing a segmented pattern of Eve gene expression

Most importantly, the genes activated by Bicoid include genes encoding other regulatory transcription factors. Thus, Bicoid activates the gene encoding the zinc-finger-containing gene Hunchback in a concentration-dependent manner. This results in the gradient in Bicoid expression (**Figure 9.35a**) being converted into a gradient in Hunchback expression (**Figure 9.35b**). In turn, this results in gradients at opposite ends of the egg in the expression of the Giant and Kruppel transcription factors, which are respectively activated or repressed by Hunchback (**Figure 9.35c**).

All these transcription factors then act on the promoter of the gene encoding the Eve (Even-skipped) homeodomain-containing transcription factor. However, whereas Bicoid and Hunchback activate Eve expression, Giant and Kruppel repress it. This produces a narrow band or stripe of Eve gene expression, known as Eve stripe 2 (**Figure 9.35d**).

Clearly therefore, the Bicoid protein has the properties of a **morphogen** whose concentration gradient determines position in the anterior part of the embryo. As expected from this, experimental manipulation of the Bicoid gradient produces dramatic effects on the embryo. Cells which contain artificially increased levels of Bicoid assume a phenotype which is normally characteristic of more anterior cells that normally contain that level of Bicoid. In contrast, cells which have artificially lowered levels of Bicoid assume a phenotype which is normally characteristic of more posterior cells.

As would be expected mutations in the other genes which regulate Eve expression, also produce abnormal *Drosophila*. For example, mutation of the Kruppel gene produces a larval fly lacking all three thoracic segments and the first five of the eight abdominal segments. Kruppel is therefore an example of a **gap gene** whose mutation produces an absence of a significant number of adjacent segments.

As described above, the combination of the four transcription factors Bicoid, Hunchback, Giant, and Kruppel results in a defined band or stripe of Eve gene expression (stripe 2) (**Figure 9.36a**). This can be produced by linking a 450 bp region of the *Eve* gene promoter to a marker gene and then inserting this gene into the embryo (**Figure 9.36b**). This indicates that the four transcription factors act on a defined region of the *Eve* gene promoter to produce stripe 2 expression. In fact, the *Eve* gene is expressed in seven specific stripes in the *Drosophila* embryo, each of which is produced by combinations of transcription factors acting on different regions of the *Eve* gene promoter (**Figure 9.36c** and **d**). Interestingly, computer models have now been produced which can predict the specific expression pattern of target genes such as Eve on the basis of the expression patterns of their upstream regulators and the number and affinity of the binding sites for these regulators in the target gene promoter.

The pattern of *Eve* gene expression allows the Eve protein to play a key role in producing the segmented pattern of *Drosophila*. Thus, mutation of

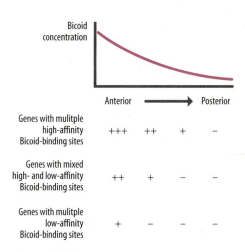

Figure 9.34
The concentration gradient of Bicoid from anterior to posterior ends of the embryo will result in differential activation of Bicoid target genes containing different-affinity binding sites. Hence, the Bicoid gradient is converted into a gradient in the expression of Bicoid-dependent genes. A plus sign indicates activation of the gene at that position in the gradient whereas a minus sign indicates that no activation will occur at that position.

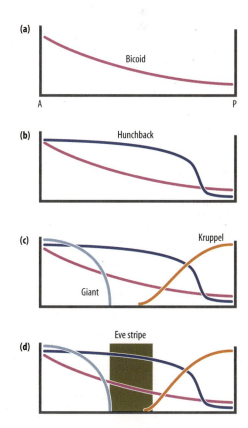

Figure 9.35
The anterior (A) to posterior (P) gradient in Bicoid expression produces a corresponding gradient in Hunchback expression, which in turn produces gradients in the expression of Giant and Kruppel. Bicoid and Hunchback activate the Eve gene whereas Giant and Kruppel repress it. The concentration gradients in expression of Bicoid, Hunchback, Giant, and Kruppel therefore produce a defined band or stripe of *Eve* gene expression.

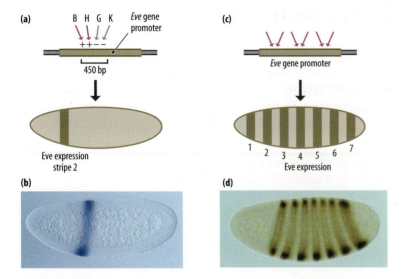

Figure 9.36
Bicoid, Hunchback, Giant, and Kruppel
(B, H, G, K) act on the 450-base pair
region of the *Eve* gene promoter to
produce a defined stripe (stripe 2) of Eve
gene expression (a shows a schematic
diagram of this effect and b shows an
actual experiment with this region of the
Eve promoter linked to a reporter gene
introduced into a *Drosophila* embryo). The
production of Eve stripe 2 is paralleled by
different transcription factors interacting
with different regions of the *Eve* gene
promoter to produce the seven strips of *Eve*
gene expression, which occur in the early
Drosophila embryo (c shows a schematic
diagram and d an actual experiment).
(b) and (d) courtesy of Stephen Small,
New York University & Michael Levene,
University of California.

the *Eve* gene produces a larva which lacks every alternate segment. Eve is thus an example of a **pair-rule gene**, whose mutation results in the larva having half the normal number of segments. Interestingly, while mutation of Eve results in the absence of the even-numbered segments (for example, abdominal segments A2, A4, A6, and A8), mutation of another pair-rule gene, Fushi-terazu, results in the absence of the odd-numbered segments, indicating that these two transcription factors have complementary roles.

The Bicoid system involves both transcriptional and post-transcriptional regulation

The Bicoid system illustrates the importance of transcription factors and transcriptional control in the early development of the *Drosophila* embryo. It should be noted, however, that this process is initiated by mechanisms which regulate RNA transport so that the Bicoid mRNA is located at the anterior end of the egg. Moreover, other post-transcriptional processes are also involved in early *Drosophila* development. For example, it has been shown that as well as regulating transcription, the Bicoid protein can also regulate mRNA translation. The anterior–posterior gradient in Bicoid expression is required to produce a posterior–anterior gradient in the expression of another protein, Caudal. Unlike the Bicoid mRNA, however, the Caudal mRNA is distributed uniformly throughout the embryo, indicating that Bicoid does not regulate *Caudal* gene transcription. Rather, Bicoid inhibits translation of the Caudal mRNA so that Caudal protein is only produced at the posterior end of the embryo where Bicoid levels are low (**Figure 9.37**). Hence, Bicoid can regulate both gene transcription and mRNA translation. Moreover, Bicoid regulates Caudal mRNA translation by binding to the Caudal mRNA, indicating that the homeodomain can interact with RNA as well as with DNA (see Section 7.5 for further discussion of translational control).

Although synthesis of Bicoid is controlled by regulating the location of its mRNA, the Bicoid system illustrates the importance of the regulated synthesis of specific transcription factors such as Hunchback, Kruppel, Giant, etc., in producing a hierarchy of transcription factor expression in the early *Drosophila* embryo. Most importantly, this can convert a simple anterior–posterior gradient in the expression of one transcription factor (Bicoid) into a defined band or stripe in the expression of another transcription factor (Eve). These processes therefore produce a segmented pattern in the early *Drosophila* embryo. Ultimately, this will result in different regions of the body developing from the different segments. Thus far, however, we have only described processes which produce a segmented pattern rather than those which lead to different segments developing different identities. This is discussed below.

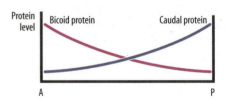

Figure 9.37
The Caudal mRNA is distributed throughout
the embryo but its translation into Caudal
protein (C) is repressed by the Bicoid
protein (B), so that Caudal protein is only
produced where Bicoid levels are low. This
produces a posterior–anterior gradient in
Caudal protein levels in the *Drosophila*
embryo, paralleling the anterior–posterior
Bicoid gradient.

Homeodomain transcription factors specify segment identity in the *Drosophila* embryo

As discussed above, a hierarchy of genes encoding specific transcription factors can convert an initial anterior–posterior gradient into a striped pattern of gene expression which defines the segmental structure of *Drosophila*. Thus, the gradient in **egg-polarity genes** such as *Bicoid* is converted into the broad expression patterns of gap genes such as *Hunchback* and *Kruppel* and then into the multiple-stripe pattern of pair-rule genes such as *Eve* (**Figures** 9.35 and **9.38**). In turn, egg-polarity genes, gap genes, and pair-rule genes co-operate together to induce expression of the **homeotic selector genes** whose protein products actually determine which structures each particular segment will form (see Figure 9.38).

As described in Chapter 5 (Section 5.1), such genes were originally identified on the basis that their mutation results in abnormal flies in which one particular structure (such as an antenna) was absent and was replaced with an additional copy of another structure (such as a leg) (see Figure 5.4). When the genes encoding these homeotic selector genes were isolated and characterized, they were found to encode homeodomain-containing transcription factors identifying this family of transcription factors for the first time (see Section 5.1).

The genes encoding these homeodomain-containing transcription factors are located on chromosome 3 in *Drosophila* and are organized into two clusters, the Antennapedia cluster and the Bithorax cluster (**Figure 9.39**). Most interestingly, the order of the genes in the clusters corresponds to the anterior–posterior order of the segments whose identity they control. Thus, the first gene in the Antennapedia cluster is *labial palps*, which controls the formation of the most anterior segment of the fly whereas the last gene in the Bithorax cluster is *Abdominal B*, which specifies the identity of the most posterior segment (see Figure 9.39).

Clearly, a complex pattern of transcriptional regulatory proteins will be required to ensure that each homeotic selector gene is expressed in the appropriate segment. Such transcriptional regulatory proteins acting on the promoters of these genes will evidently include the products of egg-polarity genes, gap genes, and pair-rule genes (as illustrated in Figure 9.38), which themselves are distributed in complex patterns in the *Drosophila* embryo (**Figure 9.40b**). Similarly, the homeotic selector genes will be regulated by modulators of chromatin structure, such as the polycomb and trithorax proteins described in Chapter 3 (Section 3.3) and in Section 9.1 (**Figure 9.40a**). The polycomb proteins will act to repress inappropriate expression of homeodomain genes in the wrong segments while the trithorax proteins will ensure that the appropriate homeodomain genes are maintained in an open chromatin structure and can therefore be activated in the appropriate segment (see also Figure 9.25).

Similarly, as described in Chapter 5 (Section 5.1) some homeodomain-containing transcription factors (such as Ubx) are able to induce their own expression, so ensuring that their expression is maintained in a particular segment once it has been switched on (**Figure 9.40c**) (see also Figure 5.7).

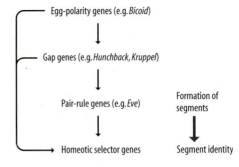

Figure 9.38
The products of egg-polarity genes, gap genes, and pair-rule genes co-operate together to regulate the expression of homeotic selector genes, which specify the identity of the different body segments.

Drosophila chromosome 3

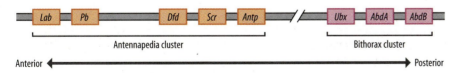

Figure 9.39
Drosophila homeodomain genes located on chromosome 3 are organized into two clusters. The order of the genes along the chromosome reflects their role in specifying the identity of more anterior or more posterior segments respectively.

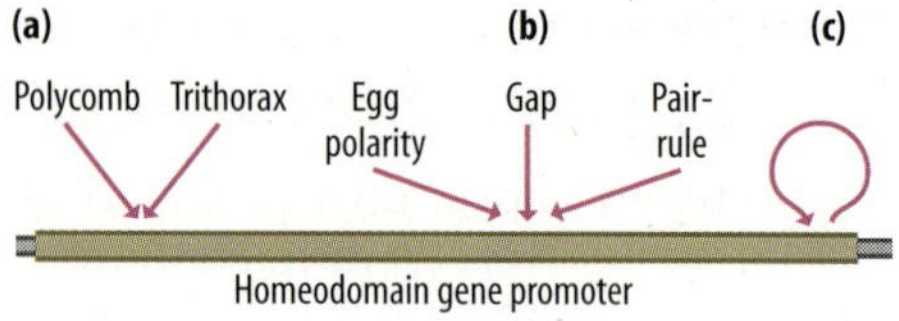

Figure 9.40
The expression of individual homeodomain genes is controlled by chromatin-regulatory proteins such as polycomb and trithorax (a), egg-polarity, gap- and pair-rule proteins (b), and by the homeodomain genes regulating their own expression (c).

Protein–protein interactions control the effect of homeodomain-containing transcription factors on gene expression

Clearly, as described in Chapter 5 (Section 5.1) the homeotic selector genes encode transcription factors which bind to specific DNA sequences via the helix-turn-helix motif in the homeodomain. Each homeotic selector gene will therefore activate and/or repress the expression of specific target genes required in particular segments. However, this will not occur simply by each homeotic selector gene being expressed only in a particular segment and modulating the expression of its target genes in that segment. Rather, the effect of a particular homeodomain transcription factor will be affected by the presence or absence of other homeodomain or non-homeodomain transcription factors.

Thus, several homeodomain proteins such as Ubx and Antp appear to have the same DNA-binding target site when tested *in vitro*. However in the intact cell, only Ubx can bind to the promoter of the *Dpp* gene and regulate its expression whereas Antp cannot do so. This difference correlates with the finding that inactivation of Ubx by mutation produces a completely different mutant fly than that which is produced by mutation of Antp.

These effects are explained by the finding that the Dpp promoter contains both a binding site for Ubx and an adjacent binding site for another homeodomain-containing protein, known as extradenticle (Exd). The Exd and Ubx proteins interact with one another on the Dpp promoter so allowing Ubx to bind strongly to the promoter and regulate its expression (**Figure 9.41**). This effect does not occur with Antp because Antp and Exd do not interact.

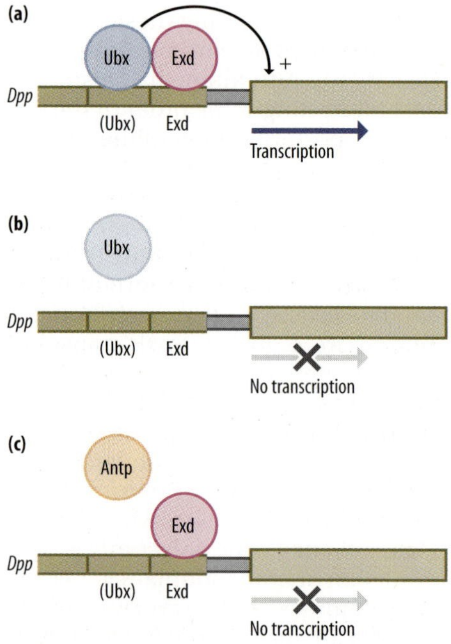

This example therefore indicates how different homeodomain-containing transcription factors can have different effects on gene expression depending on whether or not they interact with another transcription factor. Moreover, this *Drosophila* example of a homeodomain-containing transcription factor having its DNA-binding specificity modified by an interaction with another factor is paralleled in the yeast mating-type system (see Section 10.3). In this system, the DNA-binding specificity of the α2 homeodomain protein is modified differently by interaction with either the homeodomain-containing **a**1 factor or the non-homeodomain transcription factor MCM1 (see Figure 10.54).

This parallel between the yeast and *Drosophila* systems can be extended further. Thus, the key role of the *Drosophila* homeodomain transcription factors in controlling the identity of specific segments is paralleled by the role of the yeast **a** and α homeodomain factors in producing the **a** and α mating types (see Section 10.3). This vital role of homeodomain-containing transcription factors is also observed in vertebrates such as mammals, as discussed in the next section.

Figure 9.41
The Ubx protein interacts with the Exd transcription factor, allowing Ubx to bind to its binding site in the Dpp gene promoter (a). In the absence of Exd, Ubx does not bind to the Dpp promoter (b). Similarly, Antp cannot bind the Ubx promoter because it does not interact with Exd (c).

9.3 ROLE OF HOMEODOMAIN FACTORS IN MAMMALIAN DEVELOPMENT

Homeodomain transcription factors are also found in mammals

Following the findings that a number of genes involved in *Drosophila* development encode homeodomain-containing transcription factors, Southern-blot hybridization (see Methods Box 1.1) was used to identify homologous genes in other organisms including mammals. Subsequently,

Drosophila chromosome 3

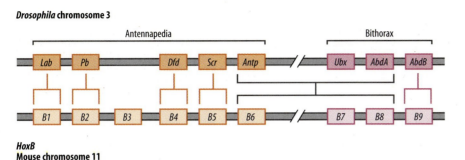

HoxB
Mouse chromosome 11

Figure 9.42
Comparison between the *Drosophila* homeodomain gene cluster on chromosome 3 and the *HoxB* gene cluster on chromosome 11 in the mouse. Each gene in the *Drosophila* complex is most homologous to the corresponding gene in the mouse complex. The *Drosophila Antp*, *Ubx*, and *AbdA* genes are too closely related to one another to be related individually to a particular mouse gene but are most closely related to the *B6*, *B7*, and *B8* genes in the mouse cluster.

these genes were characterized and it was confirmed that they do indeed encode homeodomain-containing transcription factors.

Homeodomain-containing genes therefore exist in vertebrates, including mammals, as well as in yeast and *Drosophila*. Indeed, detailed analysis of mammalian genes indicated that mammals contain individual genes homologous to specific *Drosophila* homeodomain genes such as *Engrailed* and *Deformed*, with the homology between the *Drosophila* and mammalian proteins extending beyond the homeodomain into other regions of the proteins.

The mammalian homeodomain-containing genes are organized into four gene clusters known as the *HoxA*, *HoxB*, *HoxC*, and *HoxD* gene clusters, paralleling the existence of homeodomain-gene clusters in *Drosophila* (see Figure 9.39). Moreover, the order of the genes in the mammalian clusters parallels the order of their *Drosophila* counterparts. For example in the *HoxB* cluster on mouse chromosome 11, the first gene in the complex, *HoxB1*, is most homologous to the first gene in the *Drosophila* cluster, *labial palps* (**Figure 9.42**). This pattern continues across the clusters with each gene in the mouse cluster being most homologous to the corresponding gene in the *Drosophila* cluster until the *HoxB9* gene, which is most homologous to the last gene in the *Drosophila* cluster, *Abdominal B* (see Figure 9.42).

Mammalian *Hox* genes are expressed in specific regions of the developing embryo

As noted in Section 9.2, above, the order of the genes in the *Drosophila* homeodomain gene cluster on chromosome 3 correlates with their functional roles: genes at one end of the cluster are critical for the specification of anterior structures while those at the other end of the cluster specify the posterior structures. This functional correlation is also observed in the mammalian system. The genes in the *HoxB* cluster are all expressed in the developing central nervous system and play a key role in its formation. In the central nervous system, the expression pattern of each gene is related to its position in the cluster. Thus, the *HoxB1* gene is the first gene in the cluster to be expressed and shows the most anterior boundary of gene expression (**Figure 9.43**), while each successive gene in the cluster is expressed progressively later during embryonic development and progressively less anteriorly (**Figures** 9.43 and **9.44**).

Figure 9.43
Genes in the *HoxB* cluster on mouse chromosome 11 show different expression times, expression patterns, and retinoic acid-responsiveness, which correlate with their position in the cluster. The *B1* gene is expressed most early in development, has the most anterior boundary of gene expression, and the highest retinoic acid-responsiveness, whereas the *B9* gene is expressed at the latest point in embryonic development, its expression is restricted to the posterior region, and it shows low retinoic acid-responsiveness.

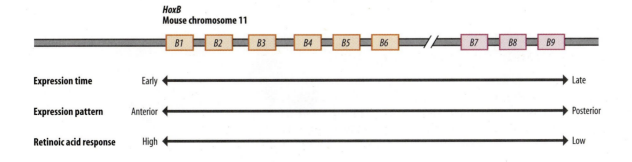

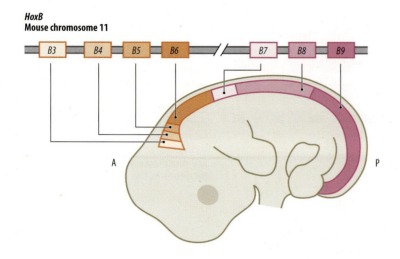

HoxB
Mouse chromosome 11

Figure 9.44
Anterior boundary of expression of *HoxB* genes related to the general structure of a 12.5-day mouse embryo and compared to the position of the gene in the *HoxB* cluster. Note the progressively more anterior (A) boundary of expression from *HoxB3* to *HoxB9*; *HoxB1* and *HoxB2* are not shown in this schematic diagram. P, posterior.

Transcription of individual *Hox* genes is regulated by gene-specific regulatory regions

Interestingly, when the regulatory regions of different *HoxB* genes are linked to a **reporter gene** they direct different expression patterns when introduced into a mouse embryo (**Figure 9.45**). This indicates that the differences in expression of the different *HoxB* genes in the embryo are produced at least in part by transcriptional regulatory processes acting on the control regions of each gene.

One possible mechanism for this effect is provided by the finding that the genes in the *HoxB* cluster are differentially sensitive to induction with retinoic acid (see Figure 9.43). As described in Chapter 5 (Section 5.1), retinoic acid regulates gene expression by binding to its receptor (which is a member of the nuclear receptor transcription factor family) and thereby regulates expression of its target genes. In this manner, retinoic acid plays a key role in vertebrate development and particularly in the development of the nervous system.

There is considerable evidence that gradients in the levels of retinoic acid exist in the embryo. Such an anterior–posterior gradient in retinoic acid levels could account for the pattern of expression of the *HoxB* genes in the developing nervous system. It has been shown that the *HoxB1* gene is the most sensitive to activation by low levels of retinoic acid with sensitivity declining across the complex so that each successive gene requires higher levels of retinoic acid for its activation (see Figure 9.43). Hence a gradient in retinoic acid, with the highest level being found in the posterior part of the central nervous system and levels declining anteriorly would result in the observed pattern of gene expression with *HoxB1* being expressed most anteriorly and the other genes progressively less anteriorly (**Figure 9.46**).

Figure 9.45
Linkage of the regulatory regions of either the *HoxB2* gene or the *HoxB4* gene to a reporter gene produces different patterns of gene expression when introduced into a mouse embryo. This indicates that the natural differences in the expression of these genes are produced, at least in part, by transcriptional regulation. Courtesy of Robb Krumlauf, Stowers Institute for Medical Research.

HoxB2

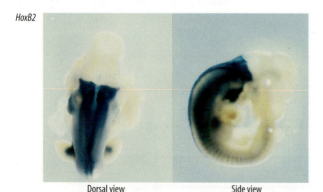

Dorsal view   Side view

HoxB4

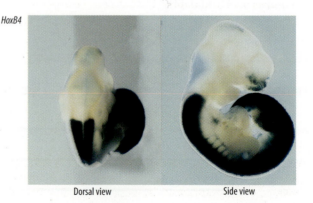

Dorsal view   Side view

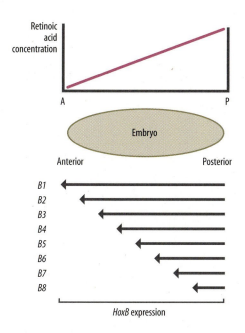

Figure 9.46
Model in which the anterior boundary of gene expression in the *HoxB* cluster is controlled by a retinoic acid gradient. Since the *HoxB1* gene is most sensitive to activation by retinoic acid, it will be expressed most anteriorly whilst the other genes will have progressively less anterior boundaries of expression due to the progressively reduced sensitivity to activation by retinoic acid.

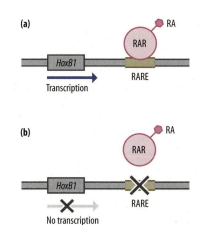

Figure 9.47
The specific transcription pattern of the *HoxB1* gene is controlled in part by the binding of the retinoic acid receptor (RAR) to its response element (RARE) in the 3′ regulatory region of the *HoxB1* gene (a). Inactivation of the RARE by mutation prevents binding of the receptor and affects the transcription of the *HoxB1* gene (b).

Regulation of the **Hox genes** by retinoic acid represents an example of the expression of genes encoding one set of transcription factors, namely the Hox proteins, being regulated by another family of transcription factors, namely the retinoic acid receptors. In agreement with this, the *HoxB1* gene has been shown to contain a retinoic acid response element in its 3′ regulatory region, which can bind the retinoic acid receptor. Moreover, inactivation of this element by mutation so that it can no longer bind the receptor abolishes expression of *HoxB1* in the neuroectoderm of the early embryo, so directly linking regulation by retinoic acid receptors to the normal pattern of *HoxB1* gene expression (**Figure 9.47**).

Hox gene transcription is also dependent on the position of the gene in the *Hox* gene cluster

Specific sequences adjacent to or within individual *Hox* genes can therefore result in their regulation at the transcriptional level, for example, by retinoic acid receptors. Such regulatory processes are supplemented by processes which regulate the transcription of *Hox* genes according to their position in the gene cluster. Thus, if individual *Hox* genes (and their adjacent regulatory elements) are moved to a different position within the gene cluster, their pattern of expression is altered so that it resembles that of the gene normally located at that position in the cluster, for example in terms of the time at which the gene is switched on during development (**Figure 9.48**).

In the case of the *HoxD* gene cluster, the gene located at one end of the cluster *HoxD13* is expressed most anteriorly and at the highest level with each successive gene being expressed at lower levels and more posteriorly. If *HoxD13* is deleted, the next gene in the cluster *HoxD12* is expressed in the manner typical of *HoxD13* even though it remains in its normal position (**Figure 9.49**).

These studies indicate therefore that, in addition to gene-specific regulatory sequences, *Hox* genes are regulated according to their position in the gene cluster by processes which operate on all genes in the cluster. In the case of the *HoxD* gene cluster, this appears to be dependent on a distant enhancer element known as the general control region (GCR) which is located at least 100,000 bases away from the gene cluster. The *HoxD* genes compete to interact with this GCR enhancer element so that the closest gene interacts most strongly and is expressed in a particular pattern, and so on (see Figure 9.49). This effect is evidently similar to that which occurs in the β-globin gene cluster (see Section 2.5) in which individual β-globin-like genes are expressed in a specific order during development

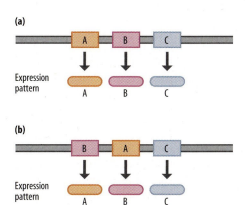

Figure 9.48
Each gene within a *Hox* gene cluster has a specific pattern of gene expression (a). Altering the position of a gene in the cluster alters the pattern of its expression to that characteristic of the gene which is normally present at this position in the cluster (b).

that is determined by their position relative to a single control element, the locus-control region.

Hox gene expression is also regulated at the post-transcriptional level by specific miRNAs

The findings discussed above indicate that *Hox* gene expression is regulated at least in part at the transcriptional level, both by sequences adjacent to the gene itself and by its position in the gene cluster. In addition, however, there is considerable evidence that regulation of *Hox* gene expression also occurs at the post-transcriptional level. It has been shown that specific miRNAs are encoded within the *Hox* gene clusters and can inhibit the expression of specific genes in that cluster and in other *Hox* gene clusters. For example, the miRNA miR-10a is produced from a region of DNA between the *HoxB4* and *B5* genes and can bind to a sequence in the 3′ untranslated region of the *HoxB3* mRNA, thereby inhibiting *HoxB3* expression (**Figure 9.50**).

Similarly, the miRNA, miR-196a-1 is produced from a region downstream of the *HoxB9* gene and can inhibit the expression of the *HoxB1*, *B6*, *B7*, and *B8* genes (see Figure 9.50). Moreover, such miRNAs do not act solely on genes in the cluster from which they are produced. For example, miR-10a also inhibits genes in other *Hox* clusters, such as the *HoxA1*, *A3*, *A5*, and *A7* genes, as well as the *HoxD10* gene. Similarly, miR-196a-1 can inhibit expression of the *HoxA5*, *A7*, and *A9* genes as well as of the *HoxC8* and *D8* genes.

Hence, a combination of sequences within and adjacent to the gene itself and more distant locus-wide regulatory regions produces a specific pattern of expression of each *Hox* gene by regulating it at both transcriptional and post-transcriptional levels. Indeed, as discussed below, the mechanisms regulating the expression of individual *Hox* genes are of critical importance in producing a particular expression pattern which in turn allows the resulting Hox protein to play a particular role during development.

Differential regulation of different *Hox* genes by Sonic Hedgehog controls the differentiation of cells in the neural tube

Although the transcriptional and post-transcriptional regulation of *Hox* gene expression may appear highly complex, this is necessary to produce specific expression patterns of individual Hox proteins which then interact to produce different body structures. This is illustrated in the processes which control the differentiation of specific neuronal cells within the **neural tube**. This structure which will form the nervous system, is located beneath an epidermal layer of ectoderm cells and above the **notochord** (see Figure 9.4 for the position of these structures in the gastrula-stage embryo). The cells in the neural tube respond to signals coming from both the epidermis and the notochord, so that sensory neurons form in the dorsal part of the neural tube, interneurons form centrally, and motor neurons form in the ventral portion of the neural tube (**Figure 9.51**).

In particular, BMP proteins, which are members of the transforming growth factor β (TGFβ) family of signaling proteins (see Section 8.2), are secreted by the ectoderm cells, overlying the neural tube. In contrast, the Sonic Hedgehog signaling molecule is secreted by the underlying notochord (see Section 8.4 for a discussion of signaling by proteins of the Hedgehog family and Section 10.1 for the role of notochord-derived Sonic Hedgehog in muscle formation). In the neural tube, these effects result in a

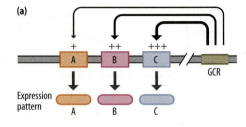

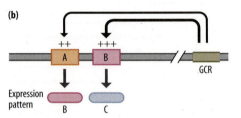

Figure 9.49
A distant enhancer element, known as the general control region (GCR), regulates the expression of the genes in the *HoxD* gene cluster, acting more strongly on the more adjacent genes (a). The deletion of one gene in the cluster (gene C) results in another gene (gene B) being the closest to the enhancer. It is therefore expressed in the normal pattern of gene C even though its physical location is unchanged (b).

HoxB
Mouse chromosome 11

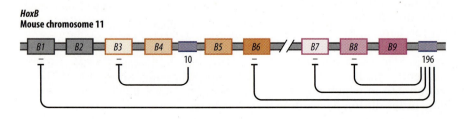

Figure 9.50
The miRNAs miR-10 and miR-196 are produced within the *HoxB* gene cluster and inhibit the expression of specific genes within the cluster.

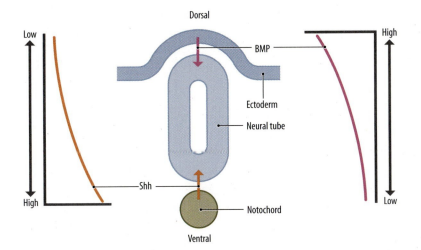

Figure 9.51
The neural tube is located between the ectoderm which produces BMP proteins and the notochord which produces Sonic Hedgehog (Shh). It is therefore subjected to opposite concentration gradients in the levels of Sonic Hedgehog (which is highest ventrally) and BMP (which is highest dorsally).

dorsal–ventral gradient of BMP expression in which levels of BMP are highest dorsally and lowest ventrally. An opposite ventral–dorsal gradient of Sonic Hedgehog expression exists, with levels being highest ventrally and lowest dorsally (see Figure 9.51).

These gradients differentially affect the expression of different homeodomain-containing genes so that the gradients are converted into unique patterns of homeodomain gene expression along the dorsal–ventral axis of the neural tube. Although all the genes are activated by BMP proteins, they differ in their response to Sonic Hedgehog. Thus, expression of the homeodomain-containing genes encoding Pax6, Pax7, Dbx1, Dbx2, and Irx3 is repressed by Sonic Hedgehog. However, their sensitivity to such repression differs so that Pax7 is the most sensitive to repression followed by Dbx1, Dbx2, Irx3, and finally Pax6, which is the least sensitive to repression. Hence, Pax6 is repressed only at much higher levels of Sonic Hedgehog than Pax7 and is therefore expressed more ventrally in the neural tube (**Figure 9.52**).

In contrast, the genes encoding the homeodomain factors Nkx6.1 and Nkx2.2 are activated by Sonic Hedgehog, with Nkx6.1 being more sensitive to activation and so being expressed at a lower level of Sonic Hedgehog than Nkx2.2 (see Figure 9.52). These different effects on the expression of different homeodomain-containing proteins result in each region having a unique pattern of expression of the different homeodomain genes, which in turn results in the formation of different neuronal cell types in the different regions (see Figure 9.52).

Interestingly, the boundaries of expression between the different regions are sharpened by mutually antagonistic effects of individual transcription

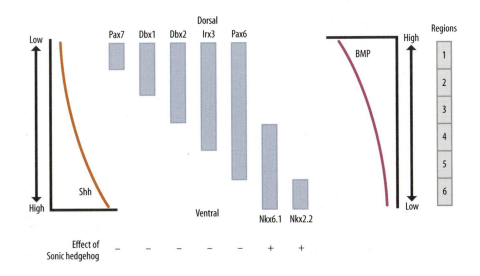

Figure 9.52
The gradients in Sonic Hedgehog and BMP expression in the neural tube produce different expression patterns of the homeodomain-containing genes *Pax7*, *Dbx1*, *Dbx2*, *Inx3*, and *Pax6*, which are all repressed by Sonic Hedgehog but differ in their sensitivity to repression. This is complemented by the different expression patterns of the homeodomain-factors Nkx6.1 and Nkx2.2 which are activated by Sonic Hedgehog but at different concentrations. In turn, these effects result in each region of the neural tube having a unique pattern of transcription factor expression.

Figure 9.53
The transcription factors Nkx2.2 and Pax6 mutually repress each other's expression (a). A similar transrepression is observed between the Nkx6.1 and Dbx2 factors (b).

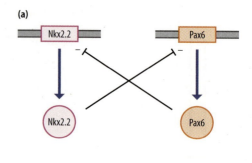

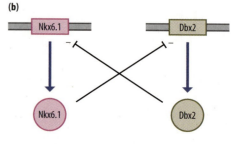

factors on expression of each other. Thus, Nkx2.2 represses the expression of Pax6 and vice versa, so producing a tight boundary of expression between regions five and six. Similarly, Nkx6.1 and Dbx2 each repress expression of the other, so producing a tight boundary between regions three and four (**Figure 9.53**).

Regulation of *Hox* gene expression by Sonic Hedgehog is also involved in limb formation

Sonic Hedgehog also plays a critical role in the development of the limbs. Each limb initially forms as a limb bud which grows out of the side of the embryo. The *sonic hedgehog* gene is activated in the posterior region of the bud by the growth factor FGF8. Such specific activation is dependent on an enhancer element which is located approximately 1 million bases away from the *sonic hedgehog* transcription unit. Although enhancers are frequently located at great distances from the genes they regulate (see Section 4.4), this is the most distant enhancer yet identified. Indeed, it was only located on the basis that mutations in this region were identified in human families with inherited polydactyly (multiple digits) (**Figure 9.54**). These findings therefore indicate the importance of the correct regulation of *sonic hedgehog* gene expression in limb development, as well as indicating the vast distances over which enhancers can act.

Within the developing limb bud, the Sonic Hedgehog protein activates expression of the *HoxD11, D12,* and *D13* genes which in turn stimulate further transcription of the *sonic hedgehog* gene. This results in high levels of expression of *sonic hedgehog* and the *HoxD11–13* genes in the posterior and not the anterior part of the limb bud (**Figure 9.55**).

Sonic Hedgehog produces its effects via the GCR which regulates the *HoxD* cluster (see above, Figure 9.49). It therefore activates the gene closest to the GCR (*HoxD13*) most strongly, and so on. Hence, *HoxD13* is expressed most strongly and most anteriorly in the limb bud, followed by *HoxD12*, and so on (**Figure 9.56**). As in other examples we have discussed, this effect therefore produces distinct expression patterns of the different *HoxD* genes, allowing them to contribute to the development of the limb and its associated digits.

Interestingly, the miRNAs which are produced from within *Hox* gene clusters, as described above (see Figure 9.50), appear to play a key role in ensuring the correct pattern of *Hox* gene expression in the developing limbs. Thus, it has been shown that retinoic acid normally induces expression of the *HoxB8* gene in the chick forelimb and in turn *HoxB8* then activates the *sonic hedgehog* gene. This process does not occur in the hind limb, even if excess retinoic acid is added. This is because the miRNA miR-196 acts to repress *HoxB8* expression in the hind limb even if it is artificially stimulated with retinoic acid (**Figure 9.57**). This suggests that miRNAs can

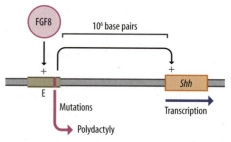

Figure 9.54
The activation of *sonic hedgehog* (*Shh*) gene expression in the limb bud by FGF8 is produced by an enhancer element (E) located approximately 1 million bases from the *sonic hedgehog* transcription unit. Mutations in this enhancer result in polydactyly (multiple digits) in humans.

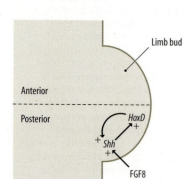

Figure 9.55
FGF8 activates *sonic hedgehog* (*Shh*) gene expression in the posterior region of the limb bud. Sonic Hedgehog then activates the expression of the *HoxD* genes which in turn activate *sonic hedgehog* transcription, so producing a positive-feedback loop in the posterior region of the limb bud.

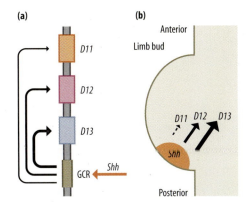

Figure 9.56

Activation of *HoxD* gene expression in the limb by Sonic Hedgehog occurs via the general control region (GCR). This results in the strength of gene activation being related to the location of the *Hox* gene relative to the GCR (a). In turn, this results in expression of the gene closest to the GCR (*HoxD13*) being the strongest and extending most anteriorly in the limb bud (b).

act as a fail-safe mechanism, blocking inappropriate *Hox* gene transcription by repressing gene expression at a post-transcriptional level. This is clearly of particular importance due to the dramatic effects of the Hox proteins, so rendering any inappropriate expression in the wrong place within the embryo, highly damaging and likely to produce an abnormal embryo.

Taken together, therefore, the studies described in this section indicate that in vertebrates such as mammals, as in *Drosophila*, homeodomain-containing transcription factors play a key role in the development of many different structures within the embryo. A variety of transcriptional and post-transcriptional processes regulate the expression of these genes, ensuring that the corresponding proteins exert their powerful effects on target gene expression only at the appropriate place and appropriate time during embryonic development.

CONCLUSIONS

In this chapter we have discussed how a range of gene-regulatory mechanisms operate during development to ensure that a wide variety of different body structures are formed at the correct place and at the correct time. Some of these processes involve post-transcriptional regulation, such as the translational control of maternal mRNA which operates before and after fertilization, or the processes which regulate the localization of the Bicoid mRNA in the early *Drosophila* embryo. Similarly, as in many other systems discussed in previous chapters, miRNAs play a key role in regulating gene expression at post-transcriptional levels. In particular, such miRNAs can serve as a fail-safe device, ensuring that any aberrant transcription of a gene encoding a regulatory protein does not result in the production of that protein in the wrong place or at the wrong time, which would produce undesirable effects on embryonic development.

Despite these examples of post-transcriptional control, however, it is clear that specific transcription factors play a key role in all stages of embryonic development by regulating their downstream target genes at the transcriptional level. For example, as discussed in Section 9.1, transcription factors such as Oct4 are specifically expressed in pluripotent ES cells and the expression of a cocktail of these factors in differentiated fibroblast cells can convert them into a pluripotent ES-like state. These factors play a key role in allowing ES cells to maintain their pluripotent nature through many cell divisions while also being poised to differentiate into specific cell types.

Individually or in combination, these factors can activate the expression of target genes whose protein products are required in undifferentiated ES cells while repressing target genes whose protein products are only required in differentiated cells. The effect of these factors is enhanced, however, by their ability to modulate the expression of genes encoding regulatory proteins (such as those which modulate chromatin structure) and of the genes encoding miRNAs. In these cases the transcription factors modulate the synthesis of proteins and RNAs which themselves regulate the expression of a number of other genes.

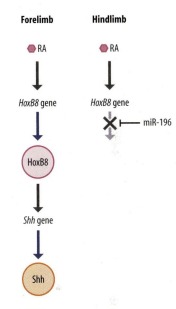

Figure 9.57

In the forelimb, retinoic acid (RA) induces expression of the *HoxB8* gene whose protein product in turn induces expression of the *sonic hedgehog* (*Shh*) gene. In the hind limb retinoic acid can induce *HoxB8* gene transcription but the production of HoxB8 protein is blocked by the miRNA miR-196.

In many instances, Oct4 and the other ES-cell-specific transcription factors act together to produce specific effects on gene expression. This combinatorial effect of transcription factors is also seen later in embryonic development. As described in Section 9.2, a cascade of transcription factor regulation is initiated by the gradient of the Bicoid protein in the syncytial *Drosophila* embryo. This ultimately results in different levels of the transcription factors Bicoid, Hunchback, Giant, and Kruppel within different regions of the developing embryo. These factors then act either positively or negatively on the transcription of the *Eve* gene, thereby converting a simple gradient in Bicoid protein levels into a band or stripe of *Eve* gene expression which defines a specific segment of the developing *Drosophila*.

Similarly, as discussed in Section 9.3, Sonic Hedgehog has different effects on the expression of specific homeodomain-containing transcription factors with different concentrations of Sonic Hedgehog being required to activate or repress the expression of different factors. In turn, this results in the gradient of Sonic Hedgehog expression across the neural tube being translated into a code whereby each region of the neural tube has a unique pattern of transcription factor expression. In turn, this allows different types of neuronal cell to differentiate at different points in the neural tube. Hence, the regulated synthesis of transcription factors plays a critical role in embryonic development. This contrasts with the critical role of transcription factor activation in the response to cellular signaling pathways which was discussed in Chapter 8. Moreover, during development simple linear gradients in the expression of one regulatory protein can be translated into a complex pattern of transcription factor gene expression and in turn allow the production of specific morphological structures during embryonic development. In turn, such structures will contain specific cell types which will perform particular functions in both the embryo and the adult. The processes regulating cell-type-specific gene expression which allow these different cell types to fulfil their different functions will be discussed in the next chapter.

KEY CONCEPTS

- The unfertilized egg contains a number of maternal mRNAs which are translated into protein only following fertilization.

- During embryonic development, these maternal mRNAs are progressively degraded and replaced by mRNAs transcribed from the embryo genome.

- The first differentiation event in the early mouse embryo is controlled by expression of the transcription factor Oct4 in the ICM cells and of the Cdx2 transcription factor in the trophectoderm cells.

- Transcription factors such as Oct4, Sox2, and Nanog are specifically expressed in ES cells which are derived from embryonic ICM cells.

- These transcription factors regulate gene expression so as to allow ES cells to maintain their pluripotent undifferentiated state while being poised to differentiate into particular cell types.

- In both *Drosophila* and vertebrates such as mammals, a variety of different classes of transcription factors, including homeodomain-containing transcription factors, play a key role in regulating embryonic development.

- The processes regulating the expression of genes encoding different transcription factors are of vital importance in allowing these factors to fulfil their different roles in embryonic development.

- Specific inducers can regulate the expression of different transcription factors either positively or negatively and with a different concentration dependence.

- These effects can convert a simple linear gradient in the expression of an inducer into a complex pattern of transcription factor expression which allows specific structures to form in the embryo.

- miRNAs can post-transcriptionally regulate the expression of specific genes including those encoding transcription factors, during embryonic development.

- The combination of transcriptional and post-transcriptional controls allows the single-celled fertilized egg to develop into the complex multicellular organism.

FURTHER READING

Introduction

Blackwell TK & Walker AK (2008) OMA-Gosh, where's that TAF? *Cell* 135, 18–20.

Cohen SM & Brennecke J (2006) Mixed messages in early development. *Science* 312, 65–66.

Schier AF (2007) The maternal-zygotic transition: death and birth of RNAs. *Science* 316, 406–407.

9.1 Regulation of gene expression in pluripotent ES cells

Baylin SB & Schuebel KE (2007) The epigenomic era opens. *Nature* 448, 548–549.

Chi AS & Bernstein BE (2009) Pluripotent chromatin state. *Science* 323, 220–221.

Gangaraju VK & Lin H (2009) MicroRNAs: key regulators of stem cells. *Nat. Rev. Mol. Cell Biol.* 10, 116–125.

Kohler C & Villar CB (2008) Programming of gene expression by Polycomb group proteins. *Trends Cell. Biol.* 18, 236–243.

Muers M (2008) Pluripotency factors flick the switch. *Nat. Rev. Genet.* 9, 817.

Niwa H (2007) Open conformation chromatin and pluripotency. *Genes Dev.* 21, 2671–2676.

Osley MA (2008) How to lose a tail. *Nature* 456, 885–886.

Schuettengruber B, Chourrout D, Vervoort M et al. (2007) Genome regulation by polycomb and trithorax proteins. *Cell* 128, 735–745.

Smith E & Shilatifard A (2009) Histone cross-talk in stem cells. *Science* 323, 221–222.

Swigut T & Wysocka J (2007) H3K27 demethylases, at long last. *Cell* 131, 29–32.

Takahashi K & Yamanaka S (2006) Induction of pluripotent stem cells from mouse embryonic and adult fibroblast cultures by defined factors. *Cell* 126, 663–676.

Zwaka TP (2006) Keeping the noise down in ES cells. *Cell* 127, 1301–1302.

9.2 Role of gene regulation in the development of *Drosophila melanogaster*

Ephrussi A & St Johnston D (2004) Seeing is believing: the bicoid morphogen gradient matures. *Cell* 116, 143–152.

Gehring WJ, Affolter M & Burglin T (1994) Homeodomain proteins. *Annu. Rev. Biochem.* 63, 487–526.

Heinrichs A (2007) Bicoid gradient tried and tested. *Nat. Rev. Mol. Cell Biol.* 8, 673.

Lawrence PA & Morata G (1994) Homeodomain genes: their function in *Drosophila* segmentation and pattern formation. *Cell* 78, 181–189.

Segal E, Raveh-Sadka T, Schroeder M et al. (2008) Predicting expression patterns from regulatory sequence in *Drosophila* segmentation. *Nature* 451, 535–540.

9.3 Role of homeodomain factors in mammalian development

Briscoe J (2009) Making a grade: Sonic Hedgehog signalling and the control of neural cell fate. *EMBO J.* 28, 457–465.

Deschamps J (2004) *Hox* genes in the limb: a play in two acts. *Science* 304, 1610–1611.

Deschamps J (2007) Ancestral and recently recruited global control of the *Hox* genes in development. *Curr. Opin. Genet. Dev.* 17, 422–427.

Duester G (2008) Retinoic acid synthesis and signaling during early organogenesis. *Cell* 134, 921–931.

Jessell TM (2000) Neuronal specification in the spinal cord: inductive signals and transcriptional codes. *Nat. Rev. Genet.* 1, 20–29.

Maden M (2007) Retinoic acid in the development, regeneration and maintenance of the nervous system. *Nat. Rev. Neurosci.* 8, 755–765.

Varjosalo M & Taipale J (2008) Hedgehog: functions and mechanisms. *Genes Dev.* 22, 2454–2472.

Yekta S, Tabin CJ & Bartel DP (2008) MicroRNAs in the Hox network: an apparent link to posterior prevalence. *Nat. Rev. Genet.* 9, 789–796.

Zeller R & Deschamps J (2002) First come, first served. *Nature* 420, 138–139.

Control of Cell-type-specific Gene Expression

10

INTRODUCTION

As discussed in Chapter 9, the regulation of gene expression plays a key role in the early development of organisms as diverse as flies and mammals. In many cases, such regulation involves transcription factors whose synthesis is regulated so that they are only produced at a particular time during development or in a particular region of the embryo. For example, the Oct4 factor is synthesized specifically in pluripotent cells of the early embryo and is important in maintaining their pluripotent nature (see Section 9.1). Similarly, in both *Drosophila* and mammals, homeodomain-containing transcription factors are synthesized in particular regions of the embryo and control the production of different embryonic structures (see Sections 9.2 and 9.3).

Ultimately, embryonic development results in the production of the different tissues and organs of the adult organism each of which contains specific differentiated cell types. As is the case in early development, the production and maintenance of such differentiated cells is controlled by specific transcription factors which are synthesized specifically in that cell type. Many of these differentiated cells must function in the body for considerable periods of time and in some cases individual differentiated cells may persist for the entire life time of the organism. This clearly requires the transcription factors involved to regulate transcription over prolonged periods. This explains why such cell-type-specific transcription factors are primarily regulated by controlling their synthesis rather than by the regulation of their activity, which as described in Chapter 8 is primarily used to provide a rapid response to cellular signaling pathways.

Regulation of transcription factor synthesis is used, for example, in the case of the C/EBPα transcription factor which is produced primarily in the liver. In this case, the actual transcription of the C/EBPα gene occurs in a small number of tissues such as the liver, so that C/EBPα is not synthesized in other tissues. This allows C/EBPα to activate the expression of a number of tissue-specific genes, such as transthyretin and α-1-antitrypsin. The regulated transcription of the C/EBPα gene controls the production of the corresponding protein which in turn directly controls the tissue-specific transcription of other genes. Interestingly, once regulated transcription has occurred, the use of alternative translational initiation codons then allows the production of different forms of C/EBPα with different activities to occur (see Section 7.5).

Similarly, in the case of B lymphocytes cell-type-specific gene expression involves the synthesis of the Oct2 transcription factor which is only produced in B lymphocytes and not in most other cell types. This factor regulates the expression of a number of B-cell-specific genes such as CD36 and the cysteine-rich secretory protein-3 (CRISP-3). Moreover, inactivation of the gene encoding Oct2 in mice results in defects in B-lymphocyte maturation indicating the essential role of this transcription factor in the production of mature B cells.

Interestingly however, B-cell-specific gene expression also involves the regulation of transcription factor activity as well as of transcription factor synthesis. Thus, the transcription factor NFκB regulates the B-cell-specific expression of the immunoglobulin κ light-chain gene. Immunoglobulin κ gene expression is activated when **pre-B cells** differentiate into mature B lymphocytes, which produce immunoglobulin. During B-cell differentiation, NFκB is activated by the dissociation of the inhibitory IκB protein from the NFκB–IκB complex, releasing mature NFκB which activates immunoglobulin κ light-chain gene expression. This parallels the mechanism used to activate NFκB in a wide variety of cell types following exposure to signals such as tumor necrosis factor α or interleukin-1 (see Section 8.2). Of course, the immunoglobulin κ light-chain gene is not activated in these other cell types since it will be in a tightly packed chromatin configuration and will not have undergone the DNA rearrangement which occurs in the B-cell lineage (see Section 1.3).

Hence, the activation of a specific transcription factor, such as NFκB from a pre-existing inactive form, can play a role in cell-type-specific gene expression as well as in the activation of gene expression in response to specific cellular signals (**Figure 10.1**). Moreover, gene expression in B lymphocytes combines the cell-type-specific synthesis of the Oct2 transcription factor with the cell-type-specific activation of the NFκB transcription factor (**Figure 10.2**).

In this chapter, we will discuss in detail three examples of the cell-type-specific regulation of gene expression and the manner in which gene expression is regulated at the level of transcription by specific transcription factors as well as by post-transcriptional regulation. Section 10.1 will discuss the regulation of cell-type-specific gene expression in skeletal muscle cells where early studies identified a **master regulatory transcription factor**. This transcription factor MyoD is a member of the basic helix-loop-helix family (see Section 5.1) and is synthesized only in skeletal muscle cells.

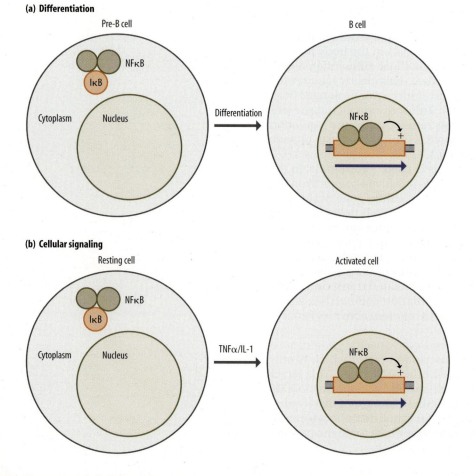

(a) Differentiation

(b) Cellular signaling

Figure 10.1
Activation of NFκB by dissociation from IκB occurs both during the differentiation of pre-B cells into mature B lymphocytes (a) and when other cell types are activated by cellular signaling molecules such as TNFα and IL-1 (b).

Similarly, Section 10.2 will discuss gene regulation at both transcriptional and post-transcriptional levels in neuronal cells. At the transcriptional level, this involves both positively acting basic helix-loop-helix transcription factors and a loss of expression of the repressor element silencing transcription factor (REST) which is expressed in non-neuronal cells and is able to repress the expression of neuron-specific genes. As will be discussed, REST expression in neuronal cells is inhibited by specific microRNAs (miRNAs) which also regulate the expression of factors involved in regulating alternative splicing in neuronal and non-neuronal cells.

Lastly, Section 10.3 will discuss the regulation of the yeast mating type system where transcription factors control the production of two mating types, **a** and **α**. This system is not only of interest in itself as a well-analyzed example of cell-type-specific gene expression in a unicellular eukaryote but also provides insights into the complex problem of gene regulation in cellular differentiation in multicellular organisms.

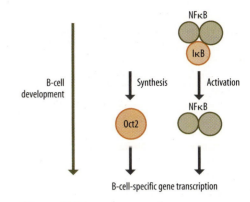

Figure 10.2
The activation of gene expression in mature B lymphocytes involves both the cell-type-specific synthesis of the Oct2 transcription factor and the activation of NFκB from a pre-existing inactive form.

10.1 REGULATION OF GENE EXPRESSION IN SKELETAL MUSCLE CELLS

The MyoD protein can induce muscle cell differentiation

As described in the Introduction to Chapter 9, the process of gastrulation results in the formation of three layers in the embryo: the inner endoderm, the central mesoderm, and the outer ectoderm. These three germ layers will each give rise to specific tissues of the later embryo and ultimately of the adult organism.

This section will consider the formation of skeletal muscle, which is formed from undifferentiated mesoderm cells in a two-stage process (**Figure 10.3**). In the first stage, undifferentiated mesoderm cells form muscle precursor cells, known as myoblasts. The myoblasts continue to proliferate but eventually a number of myoblasts fuse together to form a multinucleate non-dividing differentiated muscle cell which is known as a myotube (see Figure 10.3).

A key role in elucidating how this process is regulated was played by the observation (discussed in Section 3.2) that the fibroblast 10T½ cell line could be induced to differentiate into muscle cells by treatment with the **nucleoside** 5-azacytidine. As discussed in Chapter 3 (Section 3.2) 5-azacytidine induces the demethylation of C residues in the DNA, resulting in the activation of gene expression. This experiment therefore provided important evidence for the role of methylation of C residues in the regulation of chromatin structure and therefore gene transcription.

On the basis of this experiment, it was proposed that 5-azacytidine induces the demethylation of one or more key regulatory genes in the 10T½ cells, thereby activating their expression and in turn activating numerous other genes whose expression is dependent upon expression of the key regulatory gene(s) (**Figure 10.4**).

Evidently, such a master regulatory gene should show greatly increased expression when 10T½ cells are induced to differentiate into myoblast cells. To try and isolate the gene encoding this regulatory factor, a **subtractive hybridization** experiment was therefore performed. In this procedure (**Figure 10.5**) total mRNA prepared from the muscle cells was exposed to subtractive hybridization to remove all the mRNAs which were also expressed in undifferentiated 10T½ cells, therefore leaving only the small number of mRNAs which were specifically expressed in the differentiated cells. Further subtractive hybridization steps were then performed to remove RNAs such as that encoding myosin which were characteristic of terminally differentiated muscle cells and others which are induced by 5-azacytidine treatment in all cell types and therefore cannot include the regulatory gene which specifically induces muscle differentiation.

The remaining sample, which should contain the mRNA derived from the regulatory gene locus, was then used to screen a cDNA library prepared from 10T½ cells. This procedure resulted in the isolation of three genes,

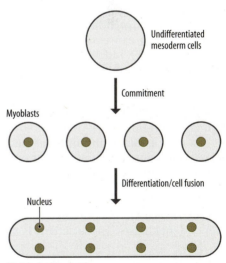

Figure 10.3
Two-stage process of skeletal muscle cell differentiation. Undifferentiated mesoderm cells first form mononucleate myoblasts which continue to proliferate but which are committed to the skeletal muscle cell lineage. Eventually however, myoblasts fuse together to form multinucleate, non-proliferating, differentiated myotubes.

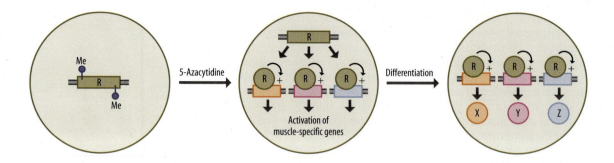

named *MyoA*, *MyoD*, and *MyoH*, whose expression was specifically activated when 10T½ cells were induced to form muscle cells by treatment with 5-azacytidine. These genes therefore represent potential candidates for the master regulatory gene. When these genes were introduced individually into untreated 10T½ cells to determine whether their over-expression could actually induce muscle cell differentiation, no muscle cell differentiation was observed upon over-expression of either *MyoA* or *MyoH*. However, over-expression of *MyoD* was able to induce muscle cell differentiation in 10T½ cells exactly as occurs upon differentiation with 5-azacytidine. This identified MyoD as a master regulatory factor which can induce muscle cell differentiation when it is artificially over-expressed or where expression of the endogenous gene encoding MyoD is induced by treating 10T½ cells with 5-azacytidine.

MyoD is a basic helix-loop-helix transcription factor which is able to regulate gene expression

The over-expression of MyoD is able to activate a number of muscle-specific genes such as those encoding myosin heavy and light chains, M-cadherin, and muscle creatine kinase (**Figure 10.6a**). In addition, MyoD over-expression in 10T½ cells induces the expression of *MyoA* and *MyoH* (**Figure 10.6b**), the other two genes whose expression was shown to be specifically induced by 5-azacytidine in 10T½ cells but which did not themselves induce differentiation when over-expressed (see above).

These findings indicated that MyoD was likely to be a transcription factor whose over-expression induces the expression of muscle-specific genes (see Figure 10.6). Indeed, analysis of the structure of MyoD revealed it to be a basic helix-loop-helix transcription factor containing a basic DNA-binding domain and adjacent helix-loop-helix dimerization domain (see Section 5.1 for a discussion of these motifs) together with an N-terminal activation domain (see Section 5.2 for discussion of transcription factor activation domains) (**Figure 10.7**).

As well as activating muscle-specific genes, MyoD can also modulate gene expression to inhibit cellular proliferation thereby producing the non-dividing phenotype characteristic of differentiated muscle cells. For example, MyoD activates the gene encoding the p21 inhibitor of **cyclin-dependent kinases** (**Figure 10.6c**) (see Section 11.3 for a discussion of p21). This will therefore result in the enhanced expression of p21, so inhibiting the cyclin-dependent kinases whose activity is necessary for cellular proliferation.

Figure 10.4
Model in which a master regulatory gene (R) is methylated in undifferentiated 10T½ cells. Treatment with 5-azacytidine demethylates and activates the regulatory gene which produces the corresponding transcription factor (R) and this factor then activates muscle-specific genes (X, Y, and Z) and produces muscle cell differentiation.

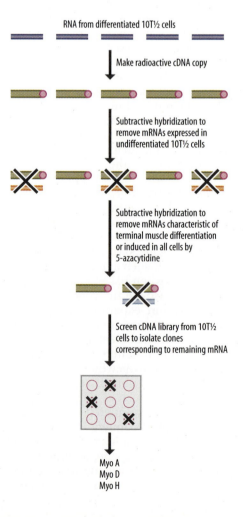

Figure 10.5
Experimental procedure used to isolate the gene encoding the master regulatory protein. RNA prepared from differentiated 10T½ cells was converted to cDNA and subtractive hybridization used to remove mRNAs expressed in undifferentiated 10T½ cells or those characteristic of terminal muscle differentiation or those which are induced in all cells by 5-azacytidine. The remaining cDNA was used to screen a cDNA library to isolate genes which are specifically induced in 10T½ cells by 5-azacytidine. This resulted in the isolation of cDNA clones derived from the genes encoding MyoA, MyoD, and MyoH.

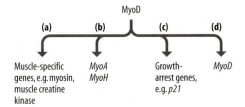

MyoD

(a)   (b)   (c)   (d)

Muscle-specific   MyoA   Growth-   MyoD
genes, e.g. myosin,   MyoH   arrest genes,
muscle creatine   e.g. p21
kinase

Figure 10.6
MyoD can induce expression of a variety of different genes including muscle-specific genes (a), *MyoA* and *MyoH*, the other two genes isolated as being specifically induced by 5-azacytidine in 10T½ cells (b), growth-arrest genes such as *p21* (c), and the gene encoding MyoD itself (d).

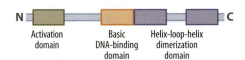

Figure 10.7
Structure of the MyoD transcription factor with an N-terminal activation domain and a basic DNA-binding domain and adjacent helix-loop-helix dimerization domain.

Interestingly, MyoD also activates expression of its own gene so that for example, when MyoD is artificially over-expressed in 10T½ cells, it activates the endogenous *MyoD* gene, further enhancing MyoD levels (**Figure 10.6d**). This indicates that a positive-feedback loop exists which will maintain high levels of MyoD expression and promote stable commitment to the muscle cell lineage once MyoD is initially expressed (**Figure 10.8**).

As a DNA-binding transcription factor, MyoD achieves its effect by binding to specific target sequences known as E boxes in the regulatory regions of muscle-specific genes, activating their expression. High-affinity binding requires MyoD to form a heterodimer with other members of the basic helix-loop-helix family, such as the **E12** and **E47** factors, which are ubiquitously expressed in all cell types. Hence, the DNA-binding heterodimer consists of one molecule of the muscle-specific MyoD and one molecule of the ubiquitously expressed basic helix-loop-helix E proteins.

As is the case for many other DNA-binding transcription factors, MyoD achieves its effects on gene expression by interacting with co-activator molecules which can then interact with components of the basal transcription complex (see Section 5.2). One such target for MyoD is the TBP-associated factor 3 (TAF3) co-activator. Interestingly however, in differentiated muscle cells rather than interacting with TBP itself, TAF3 interacts with TRF3, a TBP-like factor that substitutes for TBP in such cells (see Section 4.1). Hence, differentiated muscle cells will have a specific gene-activation complex containing MyoD, TAF3, and TRF3 (**Figure 10.9**).

As well as interacting with TAF3, MyoD can also bind the p300 transcriptional co-activator (see Section 5.2). Like its close relative CBP, p300 has histone acetyltransferase activity allowing it to open up the chromatin of target genes, as well as interacting with the basal transcriptional complex to stimulate their transcription (see Section 5.2 for discussion of the effects of co-activators on gene expression).

Hence, MyoD, via p300, can influence both chromatin structure and gene transcription (**Figure 10.10**). Moreover, MyoD can also bind the SWI–SNF chromatin-remodeling complex (see Section 3.5) allowing MyoD to open the chromatin by this means as well as via histone acetylation.

These chromatin-remodeling activities are of particular importance since they will allow MyoD to stimulate its target genes and induce the development of muscle cells from undifferentiated precursors in which muscle-specific genes may be in the tightly packed condensed chromatin structure characteristic of inactive genes.

MyoD is regulated by controlling both its synthesis and its activity

Taken together therefore, the findings discussed above indicate that MyoD is a master regulator of muscle-specific gene expression and differentiation whose expression results in the production of differentiated muscle cells. In agreement with this, MyoD mRNA and protein are present in skeletal muscle tissue taken from a number of different sites in the body but are absent in all other tissues. Hence, MyoD represents a transcription factor which is

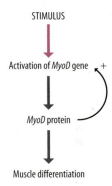

Figure 10.8
Activation of the *MyoD* gene results in increased MyoD protein which not only induces muscle differentiation but also activates expression of its own gene, producing a positive-feedback loop and maintaining differentiation.

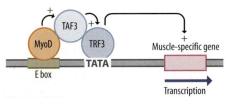

Figure 10.9
MyoD bound to its E box DNA-binding site in a muscle-specific gene can interact with the TAF3 co-activator, which in turn interacts with the TBP-like factor TRF3 bound to the TATA box, stimulating gene transcription.

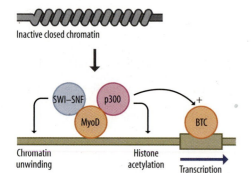

Inactive closed chromatin

Chromatin unwinding

Histone acetylation

Transcription

Figure 10.10
The MyoD transcription factor associates both with the SWI–SNF chromatin-unwinding complex and with the p300 transcriptional co-activator, which has histone acetyltransferase activity and can also stimulate activity of the basal transcriptional complex. Binding of MyoD therefore promotes the transition of inactive closed chromatin to the active open state and also enhances transcription of its target genes by the basal transcriptional complex.

controlled at the level of its synthesis with such regulation playing a key role in muscle cell differentiation.

Interestingly however, MyoD is not regulated solely at the level of synthesis. Over-expression of MyoD in undifferentiated 10T½ cells or its activation with 5-azacytidine converts the fibroblast cells into myoblasts rather than into fully differentiated multinucleate myotubes. Production of fully differentiated myotubes from the myoblasts in cell culture systems also requires the removal of serum (**Figure 10.11**).

Paradoxically MyoD levels do not increase in the transition from myoblasts to myotubes. The explanation of this paradox is provided by the existence of an inhibitory molecule, known as Id, which binds to MyoD to form a MyoD–Id heterodimer (see Section 5.3 and Figure 5.32). As Id has a helix-loop-helix domain but no basic DNA-binding domain, the MyoD–Id heterodimer cannot bind to DNA.

Upon removal of serum, Id levels decrease and MyoD is free to form a DNA-binding heterodimer with proteins such as E12 or E47, which have both a helix-loop-helix motif and a functional DNA-binding domain. This results in the activation of MyoD target genes and full muscle differentiation (see Figure 10.11). Thus, MyoD is a transcription factor which is controlled both at the level of synthesis and at the level of regulation of its activity by interaction with an inhibitory protein. Together, these mechanisms allow MyoD to play a key role in the differentiation of skeletal muscle cells and the activation of the genes expressed specifically in this cell type.

Other muscle-specific transcription factors can induce muscle cell differentiation

From the data discussed above, MyoD fulfils all the requirements of a master regulator of skeletal muscle cell differentiation. It is expressed specifically in skeletal muscle, it can induce the expression of muscle-specific genes and most importantly, its over-expression in non-muscle cells in culture can induce them to form differentiated muscle cells.

If MyoD were indeed the sole master regulator of skeletal muscle differentiation, the inactivation of the gene encoding MyoD in knockout mice should result in the absence of skeletal muscle. In fact, however, this is not the case: mice lacking MyoD exhibit generally normal muscle development and are viable and fertile, although there is some delay in the formation of some muscles such as the early limb muscles (**Figure 10.12**).

This effect occurs because there are three other genes encoding muscle-regulatory factors in addition to MyoD. The three genes encode the proteins Myf5, myogenin, and Mrf4. These proteins are closely related to MyoD and are all basic helix-loop-helix transcription factors. Most importantly, like *MyoD* when over-expressed individually in 10T½ cells each of these three genes is able to induce the formation of differentiated muscle cells.

When the gene encoding Myf5 is inactivated in knockout mice, the majority of skeletal muscle develops relatively normally although, as with MyoD, there are some muscle defects, in this case defects in trunk skeletal muscle (see Figure 10.12). However, when both MyoD and Myf5 are inactivated in double-knockout mice, the mice fail to form skeletal muscle

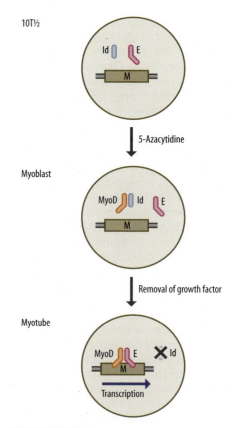

10T½

5-Azacytidine

Myoblast

Removal of growth factor

Myotube

Transcription

Figure 10.11
Undifferentiated 10T½ cells do not express MyoD. Myoblast muscle cell precursors do express MyoD but muscle-specific genes (M) are not activated because MyoD forms an inactive heterodimer with the Id factor. When cells are induced to form differentiated myotubes by removal of growth factors, Id levels decline. MyoD can then form an active DNA-binding heterodimer with ubiquitously expressed E-box-binding members of the helix-loop-helix family of transcription factors (E). Transcription of muscle-specific genes is therefore activated.

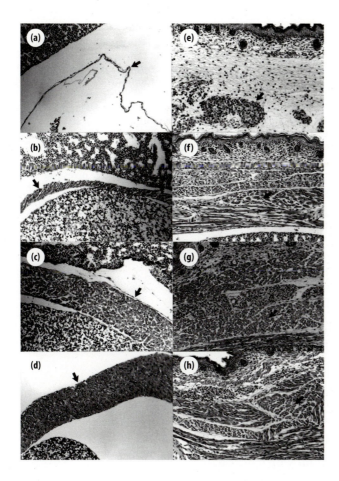

Figure 10.12
Stained sections of diaphragm muscle (a–d) and intercostal chest muscle (e–h) from *MyoD/Myf5* double-knockout mice (a and e), *MyoD* single-knockout mice (b and f), *Myf5* single-knockout mice (c and g), and wild-type mice (d and h). Note that significant loss of muscle cells occurs only in the *MyoD/Myf5* double-knockout mice. The position of the diaphragm in a–d is indicated by the arrows. Courtesy of Rudolf Jaenisch & Michael Rudnicki, from Rudnicki MA, Schnegelsberg PN, Stead RH et al. (1993) *Cell* 75, 1351–1359. With permission from Elsevier.

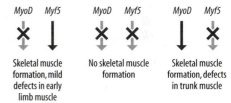

Figure 10.13
Schematic diagram of the results illustrated in Figure 10.12. Inactivation of *MyoD* alone or of *Myf5* alone produces only relatively mild defects in knockout mice and skeletal muscle is still formed. In contrast, double-knockout mice, lacking both *MyoD* and *Myf5* do not produce skeletal muscle.

myoblasts and therefore do not produce skeletal muscle (see Figure 10.12). Hence, in single-knockout mice, MyoD and Myf5 can apparently substitute for one another so that the mice produce skeletal muscle. However, when both factors are absent skeletal muscle does not form (**Figure 10.13**).

Detailed analysis of the phenotypes of knockout mice for each of the four muscle-regulatory genes has suggested that myogenin acts later than the other three factors. As noted above, in embryonic development, mesoderm cells in regions known as the **myotome** first differentiate into myoblast cells, paralleling the first stage of 10T½ cell differentiation. Subsequently, the myoblasts fuse together to form multinucleate fully differentiated myotubes. It is likely that MyoD and Myf5 act at the first stage of this process, inducing commitment to the muscle cell lineage, whereas myogenin is required for the second stage of full differentiation and Mrf4 may act at both stages (**Figure 10.14**).

In the whole animal, there is a complex relationship between the four muscle-regulatory factors with the protein product of each gene being able to activate the expression of other members of the muscle-regulatory gene family, as well as of downstream genes whose protein products are required in muscle cells. **Figure 10.15** shows the interrelationships between the four different muscle-determining genes in the myotome, which is the name given to the mesodermal regions that give rise to skeletal muscle.

As illustrated in Figure 10.15, a complex relationship exists between the four different muscle-determining genes and this relationship differs in different parts of the embryonic myotome which give rise to different skeletal muscles in the adult organism. Interestingly, as shown in Figure 10.15, *Myf5*

Figure 10.14
The four muscle-regulatory factors act at different stages of the process of myoblast and myotube formation.

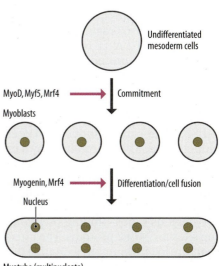

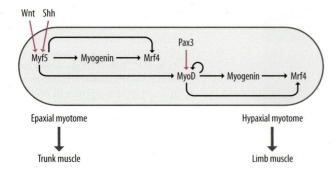

Figure 10.15
Regulatory relationships of the four muscle-determining genes, *MyoD*, *Myf5*, *myogenin*, and *Mrf4* in epaxial myotome and hypaxial myotome. Arrows indicate activation of the gene encoding one member of the family by another family member. Note that Myf5 can induce myogenin and Mrf4 in epaxial myotome, which eventually gives rise to trunk muscle, whereas MyoD can activate myogenin and Mrf4 in hypaxial myotome, which eventually gives rise to limb muscle. In turn, Myf5 is activated by Wnt and Shh signaling from the neural tube and notochord respectively, whereas MyoD expression is activated by the Pax3 paired-homeodomain family transcription factor.

is the first gene in the hierarchy of expression in the epaxial myotome and can induce the expression of myogenin and Mrf4 in this region. This correlates with the defects in trunk muscle observed in *Myf5* single-knockout mice since these muscles form from the epaxial myotome. In contrast, MyoD is responsible for inducing myogenin and Mrf4 expression in hypaxial myotome. Since this region of the myotome gives rise to the limb muscles, this correlates with the defect in limb muscle formation observed in *MyoD* single-knockout mice.

Evidently, the cascade of expression of multiple muscle-regulatory genes illustrated in Figure 10.15 must be initiated by other factors which activate expression of the first gene in the hierarchy. In agreement with this, it has been shown that MyoD expression can be activated by the Pax3 transcription factor which contains a homeodomain as well as another DNA-binding domain, known as the paired domain (see Section 5.1). Similarly, expression of Myf5 in the epaxial myotome is initiated by the signaling molecules **Wnt** (see Section 11.4) and Sonic Hedgehog (which is a member of the Hedgehog signaling molecule family; see Sections 8.4 and 9.3). These factors are derived respectively from the neural tube and notochord which are adjacent to the myotome (see Figure 10.15).

The studies described in this section therefore build on the initial identification of *MyoD* as a master regulatory gene for skeletal muscle cell and differentiation and show that in fact four such regulatory genes exist which interact with one another to activate skeletal muscle-specific gene expression in the whole animal.

MEF2 is a downstream regulator of muscle-cell specific gene transcription

The four regulatory proteins described above, MyoD, Myf5, myogenin, and Mrf4, are all capable of activating the expression of a variety of different genes which encode protein products required specifically in muscle cells. In addition however, each of these four muscle-regulatory genes is able to activate the expression of genes encoding the myocyte enhancer factor 2 (MEF2) family of regulatory proteins Four different MEF2 proteins exist, known as MEF2 A–D. In contrast to MyoD and the other three factors, the MEF2 proteins are not basic helix-loop-helix family transcription factors but belong to the **MADS family** of transcription factors which is named after its founder members, MCM1 (see Section 10.3), agamous, deficiens, and serum response factor (SRF; see Section 8.2).

Moreover, unlike MyoD, Myf5, myogenin, and Mrf4, over-expression of MEF2 alone cannot induce 10T½ cells to differentiate into muscle cells. Nonetheless, MEF2 proteins play a critical role in muscle cell-specific gene expression. The muscle master regulatory genes induce expression of MEF2 proteins which then co-operate with the master regulatory gene products to induce the expression of genes characteristic of differentiated muscle cells (**Figure 10.16**). Indeed, many muscle-specific genes contain in their regulatory regions E-box-binding sites for factors such as MyoD, adjacent to binding sites for MEF2 proteins. The muscle-regulatory proteins and MEF2 proteins are likely therefore to bind to these adjacent sites and interact with one another to stimulate muscle-specific gene expression (**Figure 10.17**).

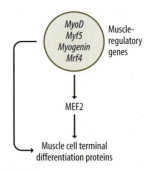

Figure 10.16
The muscle-regulatory genes, *MyoD*, *Myf5*, *myogenin*, and *Mrf4* can induce the genes encoding myocyte enhancer factor 2 (MEF2). In turn, the muscle-regulatory gene products and MEF2 co-operate to induce the expression of proteins characteristic of terminally differentiated muscle cells.

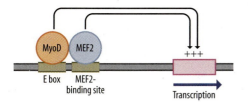

Figure 10.17
MyoD and other muscle-regulatory gene products can co-operate with MEF2 by binding to adjacent DNA-binding sites in muscle-specific genes, interacting with one another and strongly activating transcription.

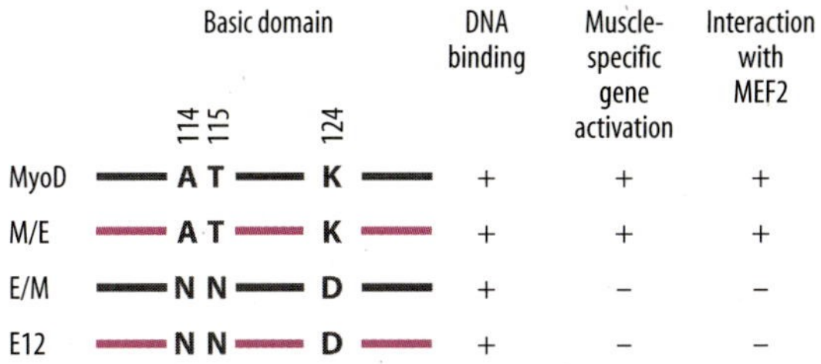

Indeed, MyoD interacts directly with MEF2 via a protein–protein interaction and this interaction appears to be of importance in allowing MyoD to activate muscle-specific gene expression. As well as mediating DNA binding, the basic domain of MyoD also mediates its interaction with MEF2 and three amino acids in this domain have been shown to be particularly critical for this interaction. Alteration of these three amino acids to their equivalents in the basic domain of the E12 ubiquitously expressed transcription factor results in a hybrid protein which cannot interact with MEF2 or activate muscle-specific gene expression even though it can still bind to the E box DNA-binding sites (**Figure 10.18**). In contrast, a hybrid basic domain containing the three key amino acids from MyoD but with the remaining portion derived from E12 can both bind to MEF2 and activate muscle-specific gene expression (see Figure 10.18).

As well as co-operating to induce genes involved in the terminal differentiation of muscle cells, MyoD and MEF2 also co-operate to activate the genes encoding two miRNAs, miR-1 and miR-133 (**Figure 10.19**). As with the other miRNAs discussed in previous chapters, miR-1 and miR-133 inhibit gene expression, turning off genes not required in muscle cells. Interestingly, however, these miRNAs also target genes encoding regulatory proteins. For example, miR-1 inhibits the expression of the gene encoding histone deacetylase 4 (HDAC4). HDAC4 can associate with MEF2, and prevent it from activating its target genes by maintaining them in an inactive chromatin structure. Inhibiting HDAC4 expression therefore results in activation of MEF2 in a manner paralleling the activation of MEF2 by phosphorylation of HDACs which was discussed in Chapters 3 (Section 3.3) and 8 (Section 8.2). Hence, the ability of MEF2 to activate gene expression in muscle differentiation can be regulated both by inhibiting the synthesis of HDACs that associate with MEF2 or by HDAC phosphorylation promoting dissociation from MEF2 (**Figure 10.20**).

In parallel with these effects, miR-1 and miR-133 also reduce the expression of the alternative splicing factor neuronal polypyrimidine tract-binding protein (nPTB; also known as PTB2). As its name suggests, this protein is

	Basic domain			DNA binding	Muscle-specific gene activation	Interaction with MEF2
	114 115		124			
MyoD	A T		K	+	+	+
M/E	A T		K	+	+	+
E/M	N N		D	+	–	–
E12	N N		D	+	–	–

Figure 10.18
The basic DNA-binding domain of MyoD can mediate DNA binding, muscle-specific gene activation and interaction with MEF2. In contrast, the basic domain of the E12 factor can bind to DNA but does not produce muscle-specific gene activation or interact with MEF2. These differences are dependent on three amino acids at positions 114, 115, and 124 in MyoD. Thus a hybrid protein (M/E) with these three amino acids from MyoD but the remaining basic domain from E12 can activate muscle-specific gene expression and interact with MEF2. In contrast, this is not the case for a hybrid protein (E/M) which contains these amino acids from E12 with the remaining basic domain derived from MyoD.

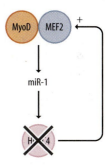

Figure 10.19
Together, MyoD and MEF2 induce expression of the small RNA miR-1. In turn, miR-1 inhibits the expression of genes whose protein products are not required in muscle cells. It also inhibits expression of the gene encoding histone deacetylase 4 (HDAC4). Loss of HDAC4 enhances the ability of MEF2 to activate transcription.

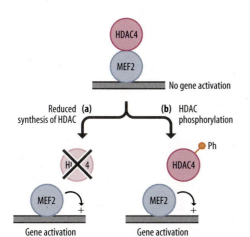

Figure 10.20
Association with histone deacetylase 4 (HDAC4) blocks the ability of MEF2 to activate gene expression. Such inhibition can be relieved either by inhibiting the synthesis of HDAC4 (a) or by phosphorylating it, so promoting its dissociation from MEF2 and nuclear export (b).

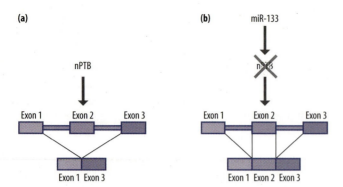

(a)

nPTB

Exon 1   Exon 2   Exon 3

Exon 1 Exon 3

(b)   miR-133

nPTB

Exon 1   Exon 2   Exon 3

Exon 1 Exon 2 Exon 3

Figure 10.21
The alternative splicing factor nPTB promotes the exclusion of muscle-specific exons (such as exon 2 in the diagram). The inhibition of its expression by miR-133 therefore promotes the inclusion of these muscle-specific exons.

highly expressed in neuronal cells (see Section 10.2) but its expression is decreased in muscle cell differentiation by the action of miR-133. This promotes the inclusion of muscle-specific exons in a variety of transcripts where their inclusion is normally inhibited by high levels of nPTB (**Figure 10.21**).

Hence, *MyoD* and the other muscle-regulatory genes can co-operate with MEF2 proteins both to regulate directly the expression of genes encoding muscle-specific proteins and to regulate the expression of miRNAs. In turn these miRNAs regulate the expression of factors involved in transcription and alternative splicing, as well of genes whose protein products are not required in muscle cells (**Figure 10.22**).

Hence MEF2 proteins, acting together with muscle-regulatory proteins such as MyoD, play a key role in muscle-specific gene expression. Unlike *MyoD*, or the other muscle-regulatory genes, however, over-expression of MEF2 alone does not induce muscle differentiation in 10T½ cells since it needs to co-operate with one or other of the muscle-regulatory proteins to induce muscle-specific gene expression. In agreement with this, MEF2 proteins are expressed in a variety of different cell types rather than being specifically expressed in muscle cells as occurs for MyoD and the other muscle-regulatory factors. Rather than being primarily regulated at the level of its synthesis, MEF2 is in fact primarily regulated by controlling its activity. As discussed above, this can involve targeting HDACs that associate with MEF2 either by regulating their synthesis or by phosphorylating them so promoting their dissociation from MEF2 and their export from the nucleus (see Figure 10.20).

Interestingly, MEF2 can also be regulated directly by post-translational modifications which alter its activity. For example the protein kinase A enzyme, which is activated by an increase in intracellular cyclic AMP (see Section 8.2) phosphorylates MEF2 converting it from a transcriptional activator to a transcriptional repressor thereby inhibiting the formation of skeletal muscle. A similar conversion of MEF2 to a transcriptional repressor is observed when the protein is modified by addition of the small protein SUMO, whereas acetylation of MEF2 promotes its ability to act as a transcriptional activator (**Figure 10.23**) (see Section 8.3, for discussion of these post-translational modifications of transcription factors).

Muscle differentiation therefore provides examples of the regulation of transcription factor synthesis and of transcription factor activity. The muscle-regulatory genes, *MyoD*, *Myf5*, *myogenin*, and *Mrf4* are synthesized only in skeletal muscle cells whereas MEF2 is synthesized ubiquitously and has its activity regulated via post-translational modification and association with other proteins such as HDACs. Together these processes regulate the levels and activity of these critical transcription factors and allow them to induce skeletal muscle cell-type-specific gene expression at the appropriate time and place.

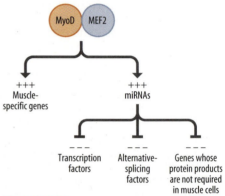

MyoD MEF2

+++
Muscle-specific genes

+++
miRNAs

Transcription factors

Alternative-splicing factors

Genes whose protein products are not required in muscle cells

Figure 10.22
MyoD and other muscle-regulatory genes co-operate with MEF2 to induce muscle-specific gene expression. They also co-operate to enhance the expression of muscle-specific miRNAs. In turn, these miRNAs inhibit the synthesis of specific transcription factors, alternative splicing factors, and genes whose protein products are not required in muscle cells.

Figure 10.23
The activity of the MEF2 transcription factor can be regulated by post-translational modification with acetylation promoting gene activation by MEF2, whereas phosphorylation and sumoylation promote its ability to repress transcription.

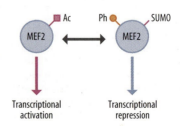

Ac   Ph   SUMO

MEF2   MEF2

Transcriptional activation

Transcriptional repression

10.2 REGULATION OF GENE EXPRESSION IN NEURONAL CELLS

Basic helix-loop-helix transcription factors are also involved in neuronal differentiation

Clearly, the brain is the most complex organ in an animal, containing for example many different types of neurons with different functions. Understanding its differentiation therefore poses unique problems. Moreover, processes such as learning and memory will require an understanding of how the brain responds to complex stimuli in a manner which affects subsequent responses. Unfortunately, no simple parallel to the 10T½ system exists for studying neuronal cell differentiation.

For this reason initial studies of genes encoding transcription factors involved in neuronal differentiation have used different approaches. One such approach involved the use of the fruit fly *Drosophila* which, as discussed in Chapter 9 (Section 9.2), is genetically well characterized. Initially, strains of the fly carrying genetic mutations which affected neuronal cell differentiation were isolated. Subsequently, the proteins encoded by the genes that were mutated in these strains of the fly were identified and characterized.

Unlike muscle cells, neuronal cells arise from ectoderm cells rather than from mesoderm. Studies in *Drosophila* have identified several basic helix-loop-helix transcription factors involved in inducing undifferentiated ectodermal cells to differentiate into neural precursor cells. Similarly, other basic helix-loop-helix proteins are involved in inducing the neuronal precursors to differentiate fully into mature neuronal cells (**Figure 10.24**). Thus for example, the Achaete and Scute proteins play a key role in the formation of neuronal progenitor cells from the initial ectoderm cells while the Asense protein is required for the precursor cells to subsequently differentiate into mature neurons (see Figure 10.24).

This system evidently has important parallels with muscle differentiation, discussed in Section 10.1. Thus in the muscle system, Myf5 and MyoD act to induce undifferentiated mesoderm cells to form myoblasts while myogenin induces the myoblasts to form myotubes. Although the basic helix-loop-helix proteins involved in the muscle system differ from those involved in neuronal cells, the same two-stage process operates with different basic helix-loop-helix proteins being involved at each of the two stages (**Figure 10.25**).

The parallels between the neuronal and muscle systems can be extended further. Activation of gene expression by Achaete and Scute involves heterodimerization with the ubiquitously expressed Da protein (see Figure 10.25), paralleling the role of the ubiquitously expressed E12/E47 proteins in the muscle system. Similarly, activation of gene expression by Achaete and Scute is inhibited by a protein known as Emc which contains a helix-loop-helix motif but lacks the basic binding domain. It therefore acts in the same negative manner as the Id protein discussed in Section 10.1 (see Figure 10.25).

Interestingly, the ectodermal cells that will eventually form neurons in *Drosophila* exist as a single layer of cells in which some cells will differentiate

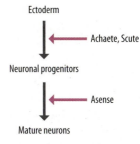

Figure 10.24
Differentiation of neuronal cells in *Drosophila* involves the Achaete and Scute basic helix-loop-helix proteins which induce the transition from undifferentiated ectoderm into neuronal precursors. Subsequently the basic helix-loop-helix protein Asense induces differentiation of the neuronal precursors to differentiated neuronal cells.

Figure 10.25
Parallels between the vertebrate muscle differentiation pathway and the *Drosophila* neuronal differentiation pathway. The role of the muscle-specific basic helix-loop-helix domain proteins Myf5, MyoD, and myogenin in different stages of muscle differentiation is paralleled by the roles of Achaete (Ac), Scute (Sc), and Asense in different stages of neuronal differentiation. In both situations the cell-type-specific proteins heterodimerize with ubiquitously expressed basic helix-loop-helix proteins: E12/E47 in the vertebrate muscle system and Da in the *Drosophila* nerve cell system. Similarly, in each case, DNA binding is inhibited by a helix-loop-helix protein which lacks a DNA-binding domain: Id in the case of the vertebrate muscle system and Emc in the *Drosophila* neuronal system.

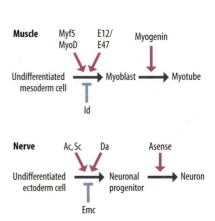

to produce nerve cells whereas other adjacent cells will differentiate to produce epidermal cells. This process is controlled by the Notch/Delta signaling pathway which, as discussed in Chapter 8 (Section 8.4), mediates signaling between adjacent cells.

Initially, all cells in the ectodermal layer will express low levels of the Achaete and Scute proteins. Differentiation is initiated when an individual cell (cell 1 in **Figure 10.26**) in the ectodermal layer randomly begins to produce more Achaete and Scute proteins than an adjacent cell (cell 2 in Figure 10.26). Increased expression of these proteins results in transcriptional activation of the gene encoding the Delta protein leading to increased levels of this protein in the membrane of cell 1 (see Figure 10.26). In turn, as described in Chapter 8 (Section 8.4). the Delta protein activates the Notch receptor on adjacent cells, such as cell 2 in Figure 10.26. Interestingly, such activation results in down-regulation of the genes encoding Achaete and Scute in cell 2, enhancing the initial difference in expression between the two cells. In turn, the reduced Achaete/Scute expression in cell 2 results in decreased levels of Delta expression in cell 2 (see Figure 10.26). This has the effect of preventing Delta in cell 2 from activating Notch in cell 1 and so prevents any inhibitory effect of Notch on the increasing levels of Achaete and Scute in cell 1 (see Figure 10.26).

The overall process therefore results in cell 1 having a high level of Achaete, Scute, and Delta but a low level of Notch activation. In contrast, cell 2 has a low level of Achaete, Scute, and Delta but a high level of Notch activation. This results in cell 1 developing into a neuronal progenitor with target genes activated by Achaete and Scute whereas cell 2 differentiates into an epidermal cell due to the low levels of Achaete and Scute. This

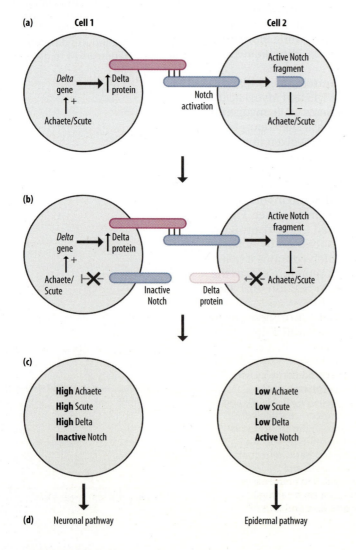

Figure 10.26
A small increase in the level of Achaete and/or Scute in cell 1 compared to cell 2 will result in Achaete/Scute activating the *Delta* gene in cell 1. In turn, the resulting Delta protein will signal to the Notch receptor in cell 2, which will inhibit Achaete/Scute expression in cell 2 (a). This results in an enhanced difference in Achaete/Scute levels between cell 1 and cell 2, which is further enhanced since the reduced levels of Achaete/Scute in cell 2 will down-regulate the Delta protein in cell 2 (light pink). In turn, this will result in the Notch receptor not being activated in cell 1, so preventing any repression of Achaete/Scute expression in this cell (b). This feedback loop will result in cell 1 having high levels of Achaete, Scute, and Delta together with inactive Notch whereas cell 2 will have low levels of Achaete, Scute, and Delta together with active Notch (c). In turn, this will result in cell 1 differentiating along the neuronal pathway whereas the adjacent cell 2 will differentiate along the epidermal pathway (d).

Figure 10.27

In a group of ectodermal cells, when one cell develops the gene-expression pattern characteristic of cell 1 in Figure 10.26, the cells adjacent to it will develop the different expression pattern of cell 2 in Figure 10.26. This process of lateral inhibition results in the central cell differentiating into a neuronal cell type (N), whereas the adjacent cells become epidermal cells (E).

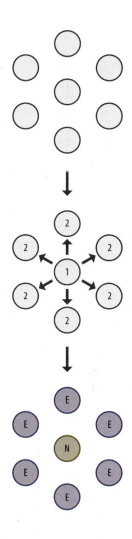

process is known as **lateral inhibition** in which one cell not only differentiates into a particular cell type but also prevents its neighbor from doing so.

In this manner, it is possible to see how regulatory processes can not only produce the pattern of gene expression characteristic of a particular cell but ensure that this occurs in the appropriate place relative to other differentiated cells (**Figure 10.27**). A similar process operates in the first differentiation event in the mammalian embryo which produces trophectoderm and inner cell mass cells. As discussed in Chapter 9 (Introduction) this involves elevated levels of the Oct4 transcription factor in the inner cell mass cells repressing expression of the Cdx2 transcription factor whereas the opposite effect occurs in the trophectoderm cells with Cdx2 repressing Oct4 expression.

Although the role of basic helix-loop-helix transcription factors in neurogenesis was first worked out in *Drosophila* as discussed above, it is clear that related mechanisms also operate in vertebrates. For example, a basic helix-loop-helix protein known as neurogenin is involved in inducing the formation of **neuroblast** precursor cells in vertebrates including mammals while another basic helix-loop-helix protein, NeuroD is involved in subsequent neuronal differentiation. Similarly, a member of the Id protein family, known as Id2, has an inhibitory role in preventing neuron-specific gene expression in mammals and is degraded during neuronal differentiation. This parallels the role of the inhibitory Emc protein in *Drosophila*, as well as the role of the original Id protein in muscle differentiation (see Section 10.1). Hence, neuronal differentiation in both *Drosophila* and mammals involves basic helix-loop-helix transcription factors, paralleling the role of homeodomain-transcription factors in the early development of both flies and mammals (see Sections 9.2 and 9.3).

Thus, as in muscle cell-specific expression (see Section 10.1) activator and repressor members of the helix-loop-helix family play a key role in regulating neuronal progenitor differentiation and neuron-specific gene expression in both mammals and *Drosophila*.

The REST transcription factor represses the expression of neuronal genes

In previous sections of this chapter, we have discussed how specific basic helix-loop-helix proteins are expressed in neuronal or muscle precursor cells and can then activate the expression of genes required in neurons or muscle cells respectively (**Figure 10.28a**). Clearly, cell-type-specific gene expression could also involve the release of transcriptional repression of cell-type-specific genes by a decrease in the abundance of a transcriptional repressor protein during cellular differentiation (**Figure 10.28b**).

This has been observed in neuronal cell differentiation in the case of repressor element silencing transcription factor (REST) (see Figure 10.28b). The REST protein is a zinc-finger transcription factor (see Section 5.1, for discussion of this family of transcription factors) in which the central DNA-binding domain is flanked by N- and C-terminal transcription-repression domains.

Figure 10.28

Activation of different neuronal specific genes (NSG) in neuronal cell differentiation can involve stimulation of transcription by activating basic helix-loop-helix (bHLH) transcription factors (a) or release of transcriptional repression by the REST transcription factor (b).

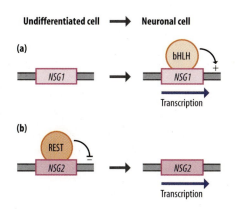

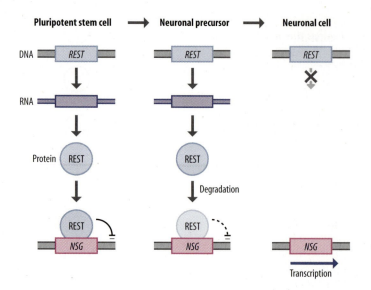

Figure 10.29
When pluripotent stem cells produce neuronal precursor cells, the REST protein is degraded at an increased rate. This reduces REST levels (pale blue). Neuronal specific genes (NSG) continue to be repressed but are poised for activation. When neuronal precursor cells subsequently differentiate into mature neuronal cells, the gene encoding REST is transcriptionally repressed thereby further reducing REST levels and neuron-specific genes are activated.

As discussed in Chapter 9 (Section 9.1) REST is expressed at high levels in pluripotent stem cells which can differentiate into a wide variety of cell types. This results in the expression of neuronal genes in these cells being repressed (**Figure 10.29**). When these stem cells differentiate to neuronal precursor cells, the transcription of the REST gene continues and its mRNA level remains the same. However, the REST protein is rapidly degraded resulting in lower levels of REST protein. The lowered level of REST is still sufficient to repress at least some neuron-specific genes but leaves them poised for rapid activation. Subsequently, when neuronal progenitors are induced to differentiate into neurons, the gene encoding REST itself is transcriptionally repressed allowing full activation of neuron-specific genes (see Figure 10.29).

REST represses the expression of neuron-specific genes by binding to specific DNA-binding sites via its central zinc-finger domain. Each of the two repressor domains can then recruit co-repressor complexes, which induce a tightly packed chromatin structure preventing gene transcription (**Figure 10.30**). The N-terminal inhibitory domain recruits the mSIN3 co-repressor complex which has HDAC activity. Similarly, the C-terminal repressor domain recruits the co-REST complex. This co-REST complex contains a variety of different proteins that can induce a tightly packed chromatin structure. These include the bromodomain protein BRG1 which is an ATP-dependent chromatin-remodeling enzyme, HDACs and a histone demethylase (which produce histone modifications characteristic of inactive tightly packed chromatin), and the MeCP2 protein (which binds specifically to methylated DNA) (see Figure 10.30) (see Chapters 2 and 3 for discussion of the effects of these factors on chromatin structure).

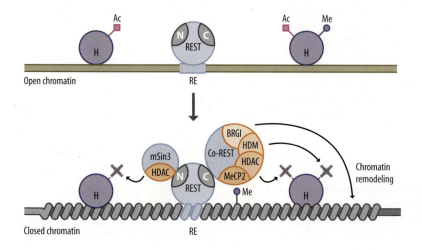

Figure 10.30
Following binding of the REST repressor to its binding site (RE), its N-terminal inhibitory domain recruits the mSin3 co-repressor complex which can deacetylase histones (H). The C-terminal inhibitory domain of REST recruits the co-REST co-repressor complex which contains the BRG1 chromatin-remodeling protein, histone deacetylases (HDAC) and histone demethylases (HDM), and the MeCP2 protein which binds to methylated DNA. Together these proteins organize an inactive tightly packed chromatin structure which does not allow gene transcription to occur.

The REST factor can therefore recruit a number of factors which together organize a tightly packed chromatin structure incompatible with gene expression. The decrease in REST synthesis due to transcriptional and post-transcriptional regulation therefore results in the derepression of a number of neuron-specific genes during neuronal differentiation.

When taken together with the studies on basic helix-loop-helix factors in neuronal differentiation discussed above, these findings indicate that the activation of neuron-specific gene expression as neuronal cells differentiate involves both activation by positively acting transcription factors and the release of repression by negatively acting transcription factors.

Neuronal cells express specific alternative splicing factors

The regulation of neuronal cell-specific gene expression at the level of transcription discussed above is significantly supplemented by post-transcriptional regulation of gene expression. In particular, a very large number of genes expressed in the nervous system are subject to alternative splicing of the primary transcript (see Section 7.1, for a general discussion of alternative splicing and its regulation). Neuronally expressed genes which are subject to alternative splicing include those involved in axon pathfinding, the formation and function of synapses and cell adhesion.

Clearly, the nervous system is highly complex and alternative splicing (as discussed in Section 7.1) represents a convenient means of generating multiple related forms of a protein with slightly different functions without requiring the existence of different genes. Furthermore, in terminally differentiated neuronal cells which do not divide, it is not possible to utilize certain mechanisms for controlling chromatin structure and transcription which require cell division (see Section 3.2). This results in a greater need for post-transcriptional mechanisms such as alternative splicing, to re-program gene expression in particular situations.

A key role in the regulation of alternative splicing in neuronal cells is played by the nPTB splicing factor (also known as PTB2) whose down-regulation in muscle cell differentiation was discussed in Section 10.1. Thus, nPTB is expressed at high levels in neuronal cells but not in other cell types whereas the related PTB protein (also known as PTB1) is expressed in other cell types but is down-regulated in neuronal cells.

The balance between PTB and nPTB plays a key role in regulating specific splicing decisions (**Figure 10.31**). For example, in the presence of a high level of PTB, the N1 exon of the *Src* gene (see Section 11.1) is not included in the mRNA because PTB represses inclusion of this exon. In contrast, nPTB promotes the inclusion of this exon and the N1 exon is therefore included in the Src mRNA in neuronal cells (see Figure 10.31).

PTB and nPTB expression is controlled by a combination of transcriptional and post-transcriptional regulation. The gene encoding PTB is transcribed in non-neuronal cells but not in differentiated neuronal cells, producing the appropriate pattern of expression of PTB. In contrast, the gene encoding nPTB is transcribed both in neuronal and non-neuronal cells (**Figure 10.32**). However in non-neuronal cells, PTB acts on the nPTB primary transcript to prevent exon 10 from being included in the mRNA (see Figure 10.32). This changes the translational reading frame of the nPTB mRNA introducing a premature translational stop codon. This results in the nPTB mRNA being degraded by the nonsense-mediated decay processes which target non-functional mRNAs (see Section 6.7, for discussion of nonsense-mediated mRNA decay).

In neuronal cells where PTB is not expressed, exon 10 of the nPTB RNA is included in the mRNA and a functional nPTB protein is therefore produced (see Figure 10.32). This combination of transcriptional and post-transcriptional control results in PTB being produced primarily in non-neuronal cells whereas nPTB is primarily produced in neuronal cells. This allows PTB and nPTB to differentially regulate specific splicing decisions in neuronal compared to non-neuronal cells.

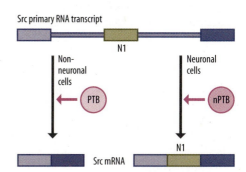

Figure 10.31

In non-neuronal cells, the PTB alternative splicing factor promotes the exclusion of the N1 exon from the Src mRNA, whereas in neuronal cells the nPTB factor promotes the inclusion of this exon in the Src RNA. It is therefore included in the mRNA in a neuron-specific manner.

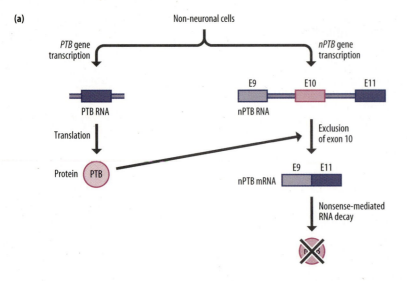

The role of PTB/nPTB in regulating alternative splicing in neuronal cells has been explored by artificially altering the expression of PTB or nPTB and then analyzing the effect on the splicing of a wide variety transcripts using microarray analysis (see Figure 7.23). Interestingly, such studies have indicated that this system regulates up to 25% of all neuron-specific alternative splicing decisions, indicating a very critical role for PTB/nPTB in this process. One of the genes whose alternative splicing was identified by such screening procedures as being regulated by PTB/nPTB was that encoding the MEF2 transcription factor. Thus, in addition to the important role of MEF2 proteins in regulating gene expression in muscle cells (see Section 10.1), these transcription factors are also expressed in other cell types. In particular they play an important role in regulating gene expression in neuronal cells, in conjunction with other transcription factors such as the basic helix-loop-helix proteins and the REST protein discussed above.

The *MEF2* gene contains a specific exon known as β, which encodes an additional transcriptional activation domain. Inclusion of the β exon in the mRNA therefore greatly enhances the ability of the resulting protein to activate transcription. It has been shown that PTB represses the inclusion of this exon whereas nPTB allows its inclusion in the mRNA (**Figure 10.33**). Hence, neuronal cells produce a very active form of MEF2 which can strongly stimulate gene expression (see Figure 10.33).

The nPTB protein therefore plays a critical role in neuron-specific alternative splicing events and is expressed widely in neurons throughout the brain. Clearly however, there will be situations where particular types of neurons differ in their pattern of alternative splicing from that observed in neurons in other regions of the brain. It is therefore of interest that other

Figure 10.32
In non-neuronal cells (a) the gene encoding PTB and the gene encoding nPTB are both transcribed. The resulting PTB protein promotes the exclusion of exon 10 from the nPTB mRNA. This results in a non-functional nPTB mRNA containing a premature translational stop codon which is degraded by nonsense-mediated decay, so preventing production of the nPTB protein. In contrast, in mature neuronal cells (b) the *PTB* gene is not transcribed whereas *nPTB* gene transcription occurs. In the absence of PTB, exon 10 is included in the nPTB mRNA and a functional nPTB protein is therefore produced (b).

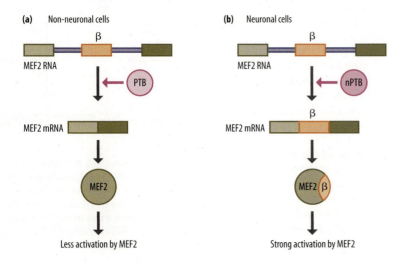

Figure 10.33
In non-neuronal cells, the PTB protein promotes the exclusion of the β exon from the RNA encoding MEF2, producing a less active form of the protein (a). In neuronal cells, the nPTB protein promotes the inclusion of the β exon, resulting in a more active form of the MEF2 transcription factor (b).

Figure 10.34

Of 35 proteins whose RNA transcript exhibited regulation of alternative splicing by Nova2, 26 exhibited a protein–protein interaction with another protein whose RNA transcript was also regulated at the level of alternative splicing by Nova2, whereas nine showed protein–protein interactions only with non-Nova2-regulated proteins. This indicates that Nova2 is likely to regulate a functional network of interacting proteins which play a key role in synaptic function.

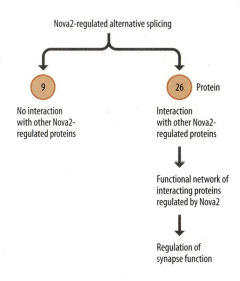

alternative splicing factors such as Nova1 and Nova2 are expressed only in certain regions of the brain. Nova1, for example, is expressed in neurons of the hindbrain and spinal cord whereas Nova2 is expressed in neurons of the cortex, hippocampus, and dorsal spinal cord. Again, these proteins regulate numerous alternative splicing decisions in the neurons which express them with, for example, approximately 7% of alternative splicing events in neuronal cells having been shown to be regulated by Nova2.

Interestingly, in an analysis of 40 proteins whose RNAs exhibited regulation of alternative splicing by Nova2, 35 had previously been shown to exhibit a protein–protein interaction with at least one other protein. Importantly, in 26 (74%) of these cases the interactions involved one or more proteins whose RNA splicing was also regulated by Nova2 (**Figure 10.34**). This suggests that Nova2 may regulate the alternative splicing of a network of genes whose protein products interact together to regulate biological processes such as synaptic function in specific neuronal cells.

Translational control plays a key role in synaptic plasticity in neuronal cells

As well as being regulated at the levels of transcription and alternative splicing, gene expression in neuronal cells can also be regulated at the level of mRNA translation into protein. Interestingly, such translational control appears to play a key role in the process of **synaptic plasticity**. In this process, repeated activation of a specific synapse produces sustained potentiation of transmission across the synapse. This is of considerable importance since such mechanisms are likely to be involved in processes such as learning and memory.

Potentiation of synaptic transmission involves an early phase that is dependent upon the modification of pre-existing proteins and a subsequent later phase, known as **long-term potentiation**, which involves the synthesis of new proteins (**Figure 10.35**). This long-term potentiation phase is regulated by translational control of protein synthesis involving the α form of the translation initiation factor eIF2.

Phosphorylation of eIF2α by the GCN2 kinase occurs in neuronal cells (**Figure 10.36**). As described in Chapters 7 (Section 7.5) and 8 (Section 8.5), such phosphorylation inhibits its ability to stimulate translation of most mRNAs. However, it enhances its ability to enhance the translation of the mRNA encoding the transcription factor ATF4 (see Figure 10.36). In turn, ATF acts as a repressor of genes which are normally activated by the CREB transcription factor (see Section 8.2).

Interestingly, activation of gene expression by CREB plays a critical role in long-term potentiation of synaptic activity. Thus, the phosphorylation of eIF2α by GCN2 will stimulate translation of ATF4 and thereby produce transcriptional inhibition of CREB target genes, leading to reduced long-term potentiation (see Figure 10.36). Hence, this system illustrates the co-operative interaction of transcriptional and translational control processes with translational control regulating the synthesis of a transcription factor.

The regulation of ATF4 mRNA translation by eIF2α in mammalian cells closely parallels the regulation of translation of the mRNA encoding the yeast GCN4 transcription factor by eIF2α, which was discussed in Chapter 7 (Section 7.5). As in the case of GCN4, phosphorylation of eIF2α regulates the balance between translation of small upstream open reading frames in the ATF4 mRNA and the translation of the ATF4 coding sequence (**Figure 10.37**). Thus, initially ribosomes translate a short three-amino acid upstream

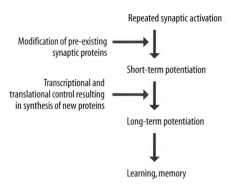

Figure 10.35

Repeated activation of a particular synapse results in short-term and subsequently long-term potentiation of its activity which is believed to be of vital importance for processes such as learning and memory. Although short-term potentiation involves the modification of pre-existing synaptic proteins, long-term potentiation involves the synthesis of new proteins which is controlled at the levels of transcription and translation.

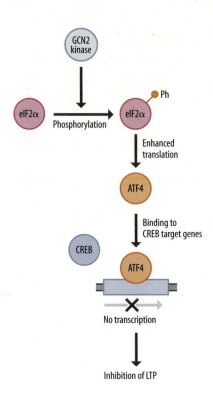

Figure 10.36
The GCN2 kinase can phosphorylate the eIF2α translation initiation factor which results in enhanced translation of the ATF4 transcription factor protein. In turn, ATF4 binds to specific genes and prevents their activation by the CREB transcription factor, so inhibiting long-term potentiation (LTP).

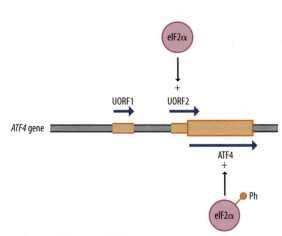

Figure 10.37
Following translation of a small upstream open reading frame (UORF1) in the *ATF4* gene, translation can reinitiate either at a second upstream frame (UORF2) or at the initiation codon for ATF4 production. Phosphorylation of eIF2α favors reinitiation at the translation start codon for ATF, so resulting in enhanced ATF4 production.

open reading frame (UORF1) in the ATF4 mRNA. In the presence of unphosphorylated eIF2α, when levels of eIF2–GTP are high (see Section 7.5) rapid translational re-initiation occurs at a second downstream open reading frame, known as UORF2. As this open reading frame overlaps with the reading frame for the production of functional ATF4, its use suppresses ATF4 protein production (see Figure 10.37). In contrast, in the presence of phosphorylated eIF2a when levels of eIF2–GTP are low, the ribosome fails to reinitiate translation at UORF2. This allows translational reinitiation to occur at the start codon for ATF4 thereby allowing functional ATF4 protein to be produced (see Figure 10.37).

This system demonstrates the importance of translational control in a major neuronal process, which is critical for the proper functioning of the brain. Moreover, it indicates how transcriptional and translational control can be combined to regulate the critical processes of long-term potentiation, leading to learning and memory formation.

miRNAs play a key role in the regulation of neuronal gene expression

As in muscle cells (Section 10.1), the expression of specific miRNAs in neuronal cells plays a key role in the regulation of gene expression. The miR-124 miRNA is expressed specifically in neuronal cells and represses the expression of a number of genes whose protein products are required in non-neuronal cells but not in neuronal cells. These include for example, proteins involved in cell proliferation. Introduction of miR-124 into non-neuronal cells results in repression of these genes whereas its inactivation in neuronal cells switches on non-neuronal genes in the neuronal cells (**Figure 10.38**).

As well as regulating a number of different **structural genes** encoding proteins not required in differentiated neurons, miR-124 also inhibits the expression of regulatory genes encoding proteins involved in the control of transcription and splicing. For example, miR-124 inhibits the expression of the gene encoding the small C-terminal **phosphatase** protein (SCP1). SCP1 is one of the proteins recruited to gene-regulatory regions by the binding of the REST transcriptional repressor discussed above. Inhibiting

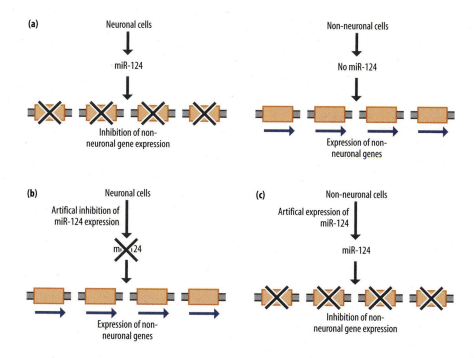

Figure 10.38
The production of miR-124 in neuronal cells results in the inhibition of a number of non-neuronal genes, whereas these genes are active in non-neuronal cells which do not produce miR-124 (a). The artificial inhibition of miR-124 expression in neuronal cells results in the expression of these neuronal genes (b). In contrast, the artificial expression of miR-124 in non-neuronal cells inhibits the normal expression of these non-neuronal genes (c).

SCP1 synthesis by miR-124 during neuronal cell differentiation therefore limits the ability of REST to repress neuronal gene expression and complements the down-regulation of REST itself which occurs during this process (**Figure 10.39**).

Similarly, miR-124 also represses the expression of the gene encoding the PTB alternative splicing factor. As discussed above, expression of PTB in non-neuronal cells prevents the production of the related nPTB protein by regulating the alternative splicing of its mRNA (**Figure 10.40**). Hence, down-regulation of PTB by miR-124 during neuronal differentiation results in upregulation of nPTB thereby promoting neuron-specific alternative splicing decisions.

Clearly, miR-124 regulates neuronal gene expression at multiple levels. As well as directly repressing the expression of structural genes encoding proteins not required in neuronal cells, it also controls the expression of regulatory proteins involved in transcription and alternative splicing. Such regulatory proteins will in turn regulate their target genes either to allow neuron-specific genes to be transcribed or to promote neuron-specific alternative splicing decisions (**Figure 10.41**).

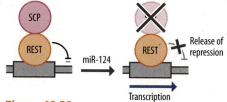

Figure 10.39
The miRNA miR-124 inhibits the expression of the small C-terminal domain phosphatase 1 (SCP) which is involved in the repression of neuron-specific genes by REST.

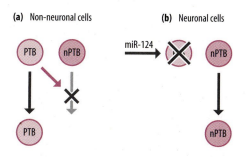

Figure 10.40
In non-neuronal cells, the PTB alternative splicing factor inhibits the production of the related nPTB protein (a). In neuronal cells PTB expression is inhibited by miR-124, resulting in the production of nPTB (b).

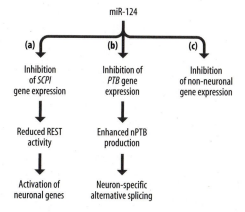

Figure 10.41
The miRNA miR-124 regulates gene expression in neuronal cells at multiple levels. By inhibiting SCP1 expression, it reduces REST activity, allowing activation of neuronal genes (a). By inhibiting PTB, it enhances nPTB production resulting in neuron-specific alternative splicing (b). Finally, it directly inhibits the expression of a number of non-neuronal genes in neuronal cells (c).

Hence, in neuronal cells, as well as in muscle cells, miRNAs play a key role in regulating gene expression. Moreover, they can act both by regulating the expression of specific genes whose protein products play a direct role in neuronal cells and by regulating the expression of proteins that themselves have a regulatory role in the process of gene expression.

10.3 REGULATION OF YEAST MATING TYPE

Yeast cells can be a or α in mating type

In previous sections of this chapter, we have discussed some of the regulatory mechanisms that control cell-type-specific gene expression in the vast array of differentiated cells found in multicellular organisms. Clearly, expression of different genes within different cells of the same organism does not occur in single-celled eukaryotes such as yeast, where the organism consists of only a single cell. Nonetheless, similar regulatory processes can operate in these organisms. Moreover, the relative simplicity of these organisms greatly enhances our ability to analyze their control processes using both genetic and molecular biology techniques. Insights obtained in this way may be relevant not only to the processes regulating cell-type-specific gene expression in multicellular organisms but also to gene regulation during embryonic development, when each cell type must be produced at an appropriate time and in an appropriate region of the embryo relative to other differentiated cells (see Chapter 9 for discussion of gene regulation in embryonic development).

In this section we will discuss the processes that regulate the production of two distinct mating types in yeast, allowing the organism to have a form of sexual reproduction. These haploid mating-type yeast cells, known as **a** and **α**, can grow and divide or mate and fuse together to create a diploid cell. The two mating types each express different genes encoding products such as pheromones or pheromone receptors which are involved in mating to produce a diploid cell. Although the **a** and **α** mating types are distinct individual organisms of different phenotypes, they can be regarded as analogous to the different types of differentiated cell in multicellular eukaryotes.

In some yeast strains, known as **heterothallic** strains, the two mating types are entirely separate as in multicellular organisms. However, we shall discuss the situation in **homothallic** yeast strains where each cell can switch its mating type from **a** to **α** or vice versa and behaves following switching exactly as any other cell with its new mating type.

The process of switching is very precise. It occurs only in the G1 phase of the cell cycle and only in a mother cell that has already produced a daughter

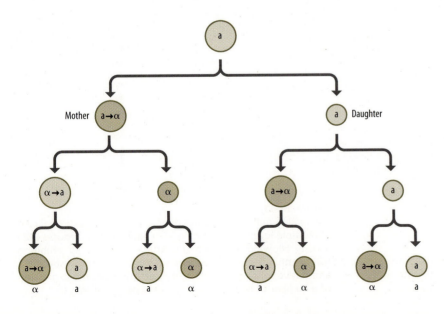

Figure 10.42
Mating-type switching in yeast. Following cell division, the larger mother cell, which has produced a small daughter cell, switches its genotype from **a** to **α** or vice versa. In contrast, the daughter cell does not switch until it has itself become a mother cell, undergone cell division, and produced a daughter cell.

by budding (**Figure 10.42**). The daughter cell itself does not switch mating type until it has grown and become a mother cell which has itself produced a daughter cell. This process is thought to be a consequence of the need to produce diploid progeny rapidly. Switching from **a** to α or vice-versa allows this to occur among the progeny of a single cell without the need for contact with another strain of different mating type.

This process of **mating-type switching** in homothallic yeast is of interest in terms of gene control from two points of view. Firstly, how does a cell switch from **a** to α and vice-versa? Secondly, how is the expression of genes expressed only in **a** or α cells regulated to produce the required pattern of cell-type-specific gene expression? These two questions will be considered in turn.

Mating-type switching is controlled by regulating the transcription of the *HO* gene

In homothallic yeast, whether a cell is **a** or α in mating type is controlled by a single gene locus on chromosome 3 known as *MAT* (mating type). If this transcriptionally active locus contains an **a** gene, the yeast cell is of **a** mating type whereas if it contains an α gene, the yeast cell is of α mating type. In addition, the yeast cell also contains transcriptionally silent copies of both the **a** and α genes elsewhere on chromosome three at the HML and HMR loci. The change in mating type occurs via a cassette mechanism in which one of the silent mating-type gene copies replaces the active gene at the *MAT* locus, so changing the mating type (**Figure 10.43**).

Hence, the switching of mating type is controlled by a DNA re-arrangement event. In turn, this process is controlled by an endonuclease enzyme which is the product of the **HO** (homothallism) gene and which makes a double-stranded cut in the DNA at the *MAT* locus, initiating switching.

At first sight, a process involving DNA cutting by an endonuclease appears to have little relevance to the study of gene control in higher organisms where, as discussed in Chapter 1 (Section 1.3) alterations in DNA such as DNA rearrangements, do not generally play a role in the regulation of gene expression. In fact, however, expression of the HO endonuclease is controlled at the transcriptional level by transcription factors which ensure that the *HO* gene is transcribed only in the G1 phase of the cell cycle and only in mother cells and not in daughter cells.

The SBF transcription factor activates HO transcription only in the G1 phase of the cell cycle

Genetic analysis has defined a number of genes, known as *SWI* (switching) genes whose protein products are necessary for *HO* gene expression. One of these, Swi5, undergoes dephosphorylation which allows it to enter the nucleus and bind to the *HO* gene promoter (**Figure 10.44**). In turn, this allows the recruitment of the SWI–SNF complex which can open the chromatin (see Section 3.5) and of the SAGA co-activator complex which can acetylate histones (see Section 5.2) (**Figure 10.45**). These changes then allow the binding of the SBF protein complex, composed of the Swi4 and Swi6 proteins, to its DNA-binding site (see Figure 10.45).

Binding of SBF to DNA is critical for transcriptional activation of the *HO* gene, since it recruits basal transcription factors, such as TFIIB and the RNA polymerase II itself (see Figure 10.45). However, the ability of SBF to do this is blocked by its association with the Whi5 inhibitory protein. In the G1 phase of the cell cycle, Whi5 is phosphorylated by cyclin-dependent protein kinases that are only active at this phase of the cell cycle. This promotes

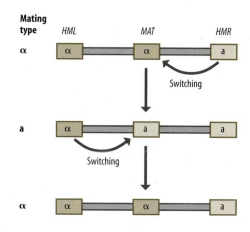

Figure 10.43
Mating-type switching involves the movement of an **a** or α gene from the inactive *HML* or *HMR* loci to the active *MAT* locus.

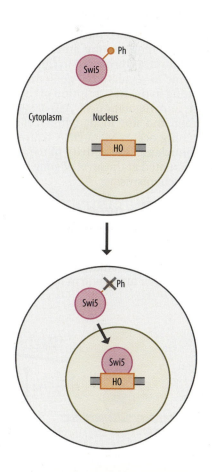

Figure 10.44
Phosphorylated Swi5 protein is located in the cytoplasm. Its dephosphorylation allows it to enter the nucleus and bind to the *HO* gene.

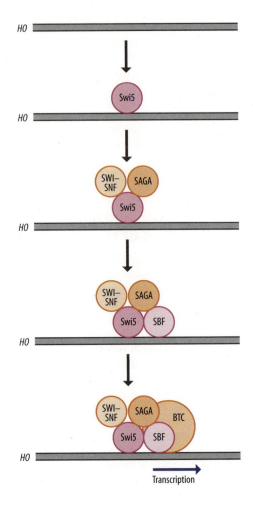

Figure 10.45
Ordered assembly of transcriptional regulatory proteins on the *HO* gene promoter involves successive binding of the Swi5 transcription factor, the SWI–SNF chromatin-remodeling complex, the SAGA co-activator complex, and the SBF transcription factor complex, which then recruits the basal transcriptional complex (BTC) including RNA polymerase II itself.

dissociation of Whi5 from SBF and the export of Whi5 from the nucleus allowing SBF to activate transcription (**Figure 10.46**).

Hence, *HO* gene transcription involves a cascade of steps which is initiated by the dephosphorylation of Swi5 and concludes with the phosphorylation of Whi5 allowing recruitment of the basal transcriptional complex and consequent transcription. This transcription occurs only in the G1 phase of the cell cycle when Whi5 is phosphorylated by cyclin-dependent kinases that are active only in this phase of the cell cycle.

The Ash-1 transcription factor represses HO transcription in daughter cells

Interestingly, as well as activating the gene encoding HO, Swi5 also activates the gene encoding the Ash-1 protein, although this occurs in a cell-cycle-independent manner. The *Ash-1* gene is transcribed in both mother and daughter cells throughout the cell cycle. However, its mRNA preferentially localizes to the daughter cell where it is translated to produce Ash-1 protein. Ash-1 is a sequence-specific DNA-binding protein related to the GATA-1 transcription factor that regulates erythroid (red blood cell) development in mammals. Ash-1 binds to the *HO* promoter in daughter cells and prevents its activation by Swi5 (**Figure 10.47**), thereby rendering *HO* gene transcription mother-cell-specific.

In the mating-type system, transcription factors such as Swi5 and SBF are activated post-translationally representing examples of control of transcription factor activity by phosphorylation and dephosphorylation events, as discussed in Chapter 8 (Section 8.2). In contrast, synthesis of the Ash-1 protein is regulated post-transcriptionally by controlling the localization of its mRNA representing an example of gene control at the level of mRNA localization, as discussed in Chapter 7 (Section 7.3). These two mechanisms combine to regulate *HO* gene transcription with regulation of SBF activity resulting in the transcription of the *HO* gene only in the G1 phase of the cell cycle and regulation of Ash-1 synthesis ensuring that it occurs only in mother and not in daughter cells. Therefore, even though mating-type switching occurs via a DNA rearrangement, mediated by the action of an endonuclease, the activity of the endonuclease and hence switching itself are regulated at the level of gene transcription by several co-operating control mechanisms.

The a and α gene products are homeodomain-containing transcription factors

The transcriptional and post-transcriptional processes described above therefore result in mating-type switching with a particular yeast cell having either an **a** or an **α** gene in the *MAT* locus and therefore producing an **a** or **α**

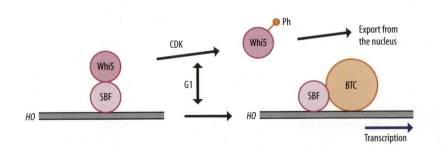

Figure 10.46
The ability of the SBF transcription factor to recruit the basal transcriptional complex (BTC) is blocked by its association with the Whi5 inhibitory protein. Phosphorylation of Whi5 by cyclin-dependent kinases occurs only in the G1 phase of the cell cycle and results in Whi5 being exported from the nucleus. In turn, this allows SBF to recruit the basal transcriptional complex so that *HO* is transcribed only in the G1 phase of the cell cycle.

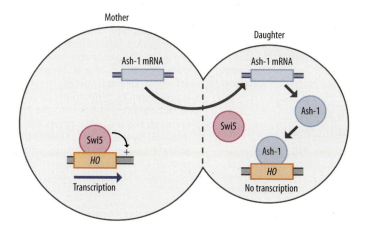

Figure 10.47
The mRNA encoding Ash-1 is produced in both mother and daughter cells. However, it is preferentially transported from the mother cell to the daughter cell, which is being produced by budding of the mother cell. In the resulting daughter cell, Ash-1 binds to the *HO* gene promoter and prevents its activation by Swi5. Hence, Swi5 activates *HO* gene transcription only in mother cells.

gene product. The presence of the **a** or α gene product results in the yeast cell being **a** or α in phenotype. Approximately 50 genes are transcribed at a significantly higher level in **a** cells compared to α cells whereas 32 genes are transcribed at a significantly higher level in α cells compared to **a** cells. This indicates that the **a** and α proteins are likely to act by regulating the transcription of a number of different target genes.

Indeed, DNA sequence analysis of the **a** and α genes suggested that they encode transcription factors. Thus, the **a** and α transcription factors have a homeodomain DNA-binding region homologous to that which is found in a wide variety of transcription factors that control critical developmental processes in higher organisms, such as *Drosophila* and mammals (see Section 5.1 and Sections 9.2 and 9.3) (**Figure 10.48**).

The α1 and α2 proteins interact with the MCM1 transcription factor to respectively activate α-specific genes and repress a-specific genes

Hence, the **a** and α gene products act to produce the corresponding mating type by regulating the expression of downstream target genes. Interestingly, however, this does not occur by a simple mechanism in which the **a** gene product activates genes which are expressed at high levels in **a** mating-type cells while the α gene product activates genes whose products are required in α mating-type cells. Rather, the products of the **a** and α loci interact together to regulate **a**- and α-specific genes (**Figure 10.49**).

In **a** mating-type cells, the **a** gene is transcribed resulting in the eventual production of the **a**1 protein. However, the **a**1 protein does not directly activate **a**-specific genes. Rather these genes are activated by a constitutively expressed protein, MCM1, which is a member of the MADS family of transcription factors that also includes the MEF2 transcription factors (see Section 10.1) and the SRF transcription factor (see Section 8.2).

In α mating-type cells, which have an α gene in the *MAT* locus, this gene is transcribed to produce two proteins, α1 and α2. The α1 protein specifically interacts with MCM1 to activate expression of the α-specific genes whereas the α2 protein interacts with MCM1 to repress expression of the **a**-specific genes (see Figure 10.49). This therefore results in activation of the

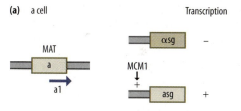

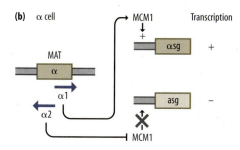

Figure 10.49
In a cells the **a** gene in the *MAT* locus is transcribed to produce the **a**1 protein. The MCM1 transcription factor activates expression of the **a**-specific genes, whilst the α-specific genes are not transcribed (a). In α cells, the α gene in the *MAT* locus is transcribed to produce α1 and α2 proteins. The α1 protein co-operates with MCM1 to activate the α-specific genes, whereas the α2 protein blocks the ability of MCM1 to activate the **a**-specific genes (b).

α2	Ser	Leu	Ser	Arg	Ile	Gln	Ile	Lys	Asn	Trp	Val	Ser	Asn	Arg	Arg	Arg	Lys	Glu
Ftz			Glu	Arg			Ile			Phe	Gln					Met		Ser

Figure 10.48
Relationship of the α2 yeast mating-type protein and the homeodomain of the *Drosophila* transcription factor Fushi-terazu (Ftz). Only amino acids in Ftz which differ from the corresponding amino acid in α2 are shown; all other amino acids are identical.

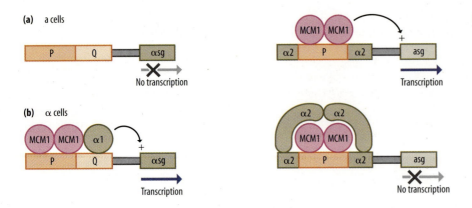

(a) a cells

| P | Q | αsg |

No transcription

(b) α cells

MCM1 MCM1 α1

| P | Q | αsg |

Transcription

MCM1 MCM1

| α2 | P | α2 | asg |

Transcription

α2 α2 / MCM1 MCM1

| α2 | P | α2 | asg |

No transcription

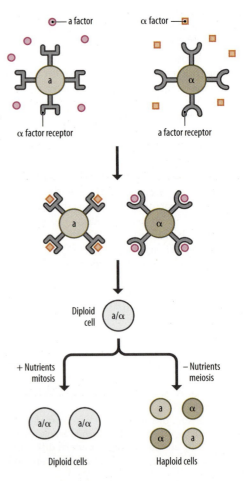

Figure 10.50
In **a** mating-type cells, a dimer of the MCM1 transcription factor binds to the P box in the **a**-specific genes and activates their transcription (a). In α mating-type cells activation of the **a**-specific genes by MCM1 is prevented by binding of the α2 protein to its binding sites flanking the P box. In contrast, the α-specific genes are activated in α cells by the MCM1 dimers binding to the P box and interacting with the α1 protein bound to the adjacent Q box (b).

a-specific genes only in **a** mating-type cells whereas the α-specific genes are activated only in α mating-type cells.

In turn, these differences in gene expression result from differences in the binding of MCM1 to the promoters of **a**- and α-specific genes (**Figure 10.50**). Thus, MCM1 binds as a dimer to a DNA sequence known as the P box in **a**-specific genes and activates their expression. However when the α2 protein is present, it binds to DNA-binding sites flanking the P box and prevents transcriptional activation by MCM1. Hence, the **a**-specific genes are transcribed only in **a** mating-type cells.

In contrast, MCM1 alone cannot activate expression of the α-specific genes. This requires not only the binding of MCM1 to the P box in the regulatory region of these genes but also the binding of the α1 protein to the adjacent Q box. Thus, the interaction of the MADS family transcription factor MCM1 with the α1 and α2 homeodomain-containing transcription factors results in the cell-type-specific regulation of the **a** and α genes.

Hence, in this case, the ability of a constitutively expressed transcription factor MCM1, to activate gene expression is modulated by cell-type-specific transcription factors with the α2 repressor preventing MCM1 from activating **a**-specific genes while the α1 activator protein co-operates with MCM1 to allow it to activate α-specific genes.

The a1 factor plays a key role in repressing haploid-specific genes in diploid cells

Although the α1 and α2 proteins have a key role in regulating gene expression in α mating-type cells, this does not appear to be the case for **a** mating-type cells and the **a**1 protein even though the **a**1 protein is specifically expressed in **a** and not in α mating-type cells (see Figure 10.50). Rather, the **a**1 protein has a key role in regulating gene expression in diploid cells which are produced by the fusion of **a** and α haploid cells.

Thus, the function of the mating-type system is to produce different mating types which, under suitable conditions, can fuse to form diploid cells providing yeast with the advantages of a sexual reproductive system (**Figure 10.51**). This process involves the secretion of different pheromones by the **a** and α haploid cells (known respectively as **a** factor and α factor) which bind to pheromone receptors on the **a** or α cell (see Figure 10.51)

Figure 10.51
Cells of **a** mating type produce both an **a**-factor pheromone and an α-factor pheromone receptor, whereas α mating-type cells produce an α-factor pheromone and a receptor for the **a**-factor pheromone. Following binding of the pheromone produced by one type of the cell to the receptor on the other type of cell, the cells fuse to produce an **a**–α diploid. Under high nutrient conditions, the diploid cell undergoes mitosis to produce more **a**–α diploid cells. However, under poor nutrient conditions it undergoes meiosis to produce **a** and α haploid spores, which can in turn develop into **a** and α mating-type haploid cells, so promoting genetic variation under poor conditions.

and induce fusion of the **a** and α haploid cells to form a diploid cell. Under high nutrient conditions this diploid cell continues to grow and divide by mitosis. However, under low-nutrient conditions the diploid cell undergoes a meiotic process to form haploid spores,which in turn can germinate to produce **a** and α haploid cells. This allows the yeast to have a sexual reproductive cycle that generates genetic variation when the organisms are exposed to poor conditions (see Figure 10.51).

Evidently, the products of certain genes will be required in both **a** and α haploid cells but not in diploid cells. These include for example, the genes encoding the **a** and α factor pheromones and their respective pheromone receptors. Similarly, diploid cells have an **a**/α genotype and do not undergo mating-type switching. Hence, they do not need to express the gene encoding the HO endonuclease.

These haploid-specific genes are repressed in diploids since only diploid cells contain both the **a**1 protein and the α2 protein. The **a**1–α2 protein complex can bind to the promoters of the haploid-specific genes and repress their expression (**Figure 10.52**). Interestingly, the **a**1–α2 protein complex also prevents the expression of the α1 protein in diploid cells. Since α1 is necessary for activation of the α-specific genes, they are not expressed in diploid cells. Similarly, the **a**-specific genes will not be expressed in diploid cells since they will be repressed by the α2 protein as described above. Hence, diploid cells do not express either **a**- or α-specific genes nor do they express haploid-specific genes such as that encoding HO itself and this effect is dependent upon the presence of the **a**1–α2 protein complex which is present only in diploid cells (see Figure 10.52).

Hence, the *HO* gene is regulated by multiple mechanisms. The regulation of SBF activity results in it being expressed only in the G1 phase of the cell cycle, the control of Ash-1 synthesis results in it being expressed only in mother cells and the presence of the **a**1–α2 complex in diploid cells results in its repression in these cells (**Figure 10.53**).

In both haploid and diploid cells, the presence of the α2 protein at a promoter will result in its repression. When present alone, the α2 repressor does not bind to DNA with high affinity. Rather, it is directed to distinct DNA-binding sites by interaction with either the MCM1 or **a**1 proteins. Interaction with MCM1 occurs via an N-terminal region of α2. This results in dimerization of α2 which then binds to sites in the **a**-specific genes and represses their expression (**Figure 10.54**). In contrast, the initial interaction of α2 with **a**1 results in a region at the C-terminal of α2 moving into an α-helical configuration (**Figure 10.55**). In turn, this facilitates the formation of an **a**1–α2 heterodimer which has a different binding specificity, leading to its binding to specific sites in the haploid-specific genes (see Figure 10.54).

The yeast mating-type system offers insights of relevance to multicellular organisms

As described in Chapter 1 (Section 1.3) changes in the DNA of cells, such as DNA rearrangements (as well as DNA loss and DNA amplification) do not appear to play a major role in the regulation of gene expression in multicellular organisms. At first sight, a process such as the regulation of yeast

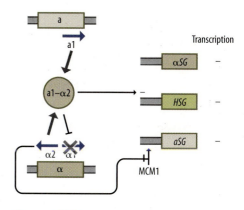

Figure 10.52

In diploid cells the presence of both the **a**1 and α2 proteins leads to the formation of an **a**1/α2 heterodimer, which binds to the haploid-specific genes (HSG) and represses their expression. This heterodimer also blocks the expression of the α1 protein. Since this protein is required for expression of the α-specific genes (αSG), they are not expressed in diploid cells. Similarly, the **a**-specific genes (aSG) are not expressed in diploid cells due to the presence of the α2 protein, which blocks the ability of MCM1 to activate these genes in the same manner as occurs in α haploid cells.

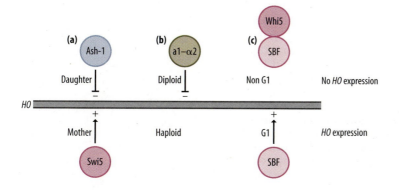

Figure 10.53

The *HO* gene is activated by Swi5 in mother cells but this is inhibited by Ash-1 in daughter cells (a). The *HO* gene is expressed in haploid cells but is repressed by **a**1–α2 in diploid cells (b). It is activated by SBF in the G1 phase of the cell cycle but this activation is blocked by Whi5 in other phases of the cell cycle (c).

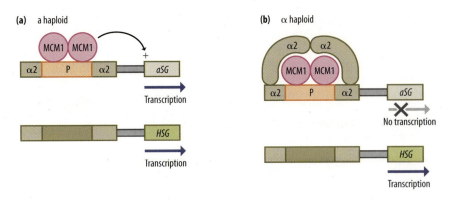

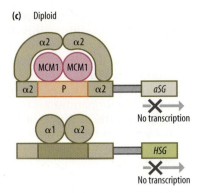

mating type which involves a DNA rearrangement event appears to be of little relevance for the regulation of gene expression in multicellular organisms. However, as discussed above, the detailed analysis of this process, which is possible in unicellular yeast, has provided a number of insights which are relevant to an understanding of gene regulation in multicellular organisms. For example, all the regulatory proteins which are involved in this process belong to families that include transcription factors that play a critical role in multicellular organisms. The **a** and α gene products for example, contain a DNA-binding homeodomain which is also found in proteins that regulate development in *Drosophila* and mammals (see Sections 9.2 and 9.3), while the Ash-1 protein is related to the GATA-1 factor which regulates gene expression in erythroid (red blood) cells. Similarly, the MCM1 protein is a member of the MADS family of transcription factors which includes the mammalian MEF2 transcription factors (see Section 10.1) and the SRF (see Section 8.2).

Moreover, the manner in which these factors interact with one another indicates how the effects of one factor can be modified in the presence of another factor. Thus, the ability of MCM1 to activate transcription is blocked in the presence of the α2 transcription factor while α2 itself is guided to different binding sites in the DNA depending on whether it interacts with MCM1 or the **a**1 transcription factor. Interestingly, the interaction between MCM1 and α2 is paralleled by the interaction between the SRF factor and proteins of the homeodomain family in mammalian and *Drosophila* cells with homologous regions of MCM1 and SRF being involved in each case.

Similarly, as described in Chapter 11 (Section 11.3), cell-cycle-specific transcription is regulated in higher eukaryotes by cyclin-dependent kinases which phosphorylate the Rb transcription factor blocking its ability to inhibit the activity of the **E2F** transcriptional activator. This evidently parallels the phosphorylation of the Whi5 protein by cyclin-dependent kinases, which blocks its inhibitory interaction with SBF.

As well as such similar interactions between different transcription factors, it is possible also to draw an analogy between the process of mating-type switching in single-celled yeast and the cell lineages which occur in multicellular organisms. Thus, the switching system illustrated in Figure 10.42 can be viewed as similar to a system in which **a** represents a stem cell and α is a differentiated cell derived from it. If it is then assumed that unlike the yeast situation switching from **a** to α is irreversible, the model lineage illustrated in **Figure 10.56** is obtained. In this model, the stem cell is continually dividing to produce both a differentiated daughter cell and another daughter cell which maintains the stem cell lineage. This type of system is commonly found in higher organisms, being used in the development of a number of different cell types and organisms (see Section 3.2 and Figure 3.15). Moreover, the specific partitioning of the Ash-1 mRNA to the

Figure 10.54
The α2 transcriptional repressor is guided to different genes by the MCM1 and **a**1 proteins. In the absence of the α2 protein in **a** haploid cells, both the **a**-specific genes and the haploid-specific genes are transcribed (a). In α-haploid cells, the α2 protein associates with MCM1 and represses the **a**-specific genes, whereas haploid-specific genes continue to be transcribed (b). In diploid cells association of MCM1 and α2 represses the **a**-specific genes, as in α cells. In addition however, the haploid-specific genes are repressed by the **a**1–α2 heterodimer (c).

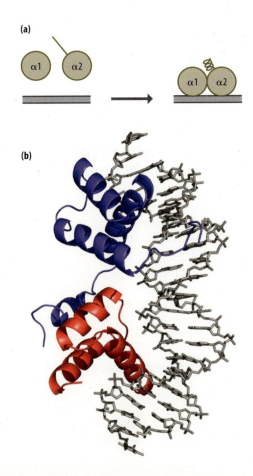

Figure 10.55
Initial interaction of the **a**1 and α2 proteins results in the unstructured region at the C-terminus of the α2 protein forming an α-helical structure. This region interacts with the **a**1 protein resulting in an **a**1–α2 heterodimer, which can bind to DNA with high affinity. (a) Schematic diagram; (b) structure of the **a**1–α2 heterodimer bound to DNA (**a**1 in red, α2 in blue).(b) Courtesy of Cynthia Wolberger, Johns Hopkins University.

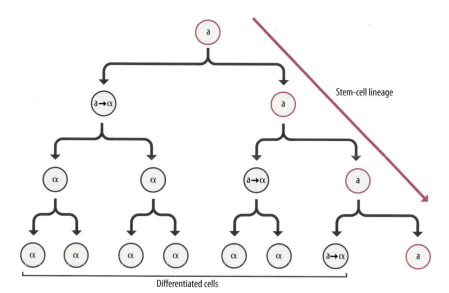

Figure 10.56
Model for the generation of a stem-cell lineage producing differentiated cells by a system based on **a**/α mating-type switching. In the model, **a** represents the stem cell and α the differentiated cell and it is assumed that, unlike the mating-type system, switching is irreversible.

daughter cell in the yeast mating-type system (see Figure 10.47) offers a simple model for how one regulatory process controlling the intracellular localization of an mRNA can result in two cells produced by cell division developing completely different phenotypes.

CONCLUSIONS

The three systems discussed in this chapter illustrate the critical role played by transcription factors in regulating cell-type-specific gene expression in different cell types and organisms. This can involve the activation of target genes by factors such as MyoD and the α1 yeast mating-type protein, or their transcriptional repression by factors such as REST or the yeast α2 mating-type protein.

In all cases, the regulation of cellular phenotype will depend on the ability of an individual transcription factor to regulate the expression of many target genes so contributing to the specific phenotype of the particular type of differentiated cell. Indeed, in some cases a single transcription factor can actually produce the differentiated phenotype when it is expressed in an undifferentiated cell. Thus, expression of the **a** or α gene controls whether a yeast cell is **a** or α in phenotype. Similarly, *MyoD* or any of the three other muscle-regulatory genes can induce a differentiated muscle phenotype in undifferentiated cells when each of them is over-expressed individually.

As described in this chapter, many transcription factors involved in the regulation of cell-type-specific gene expression are regulated at the level of their synthesis being produced only in the appropriate cell type during its differentiation. For example, *MyoD* and the other muscle-regulatory genes are expressed only in skeletal muscle cells, the REST transcriptional repressor is expressed in non-neuronal cells but not in neuronal cells and the **a** or α mating-type genes are expressed only in the appropriate mating type. In a number of these cases, the regulation of transcription factor synthesis occurs at the level of transcription itself. Thus, the **a** or α genes are only transcribed when they are present at the *MAT* locus and the presence of an **a** or α gene at this locus determines the mating type of the cell. Similarly, the REST gene is transcriptionally repressed when cells differentiate into neurons and the genes encoding muscle-regulatory proteins such as MyoD are transcribed only in skeletal muscle cells.

Nonetheless, post-transcriptional processes can also regulate the levels of factors involved in the control of cell-type-specific gene expression. For example, prior to the transcriptional down-regulation of the *REST* gene, the protein itself is destabilized resulting in its degradation in neuronal precursor cells. Similarly, the ATF4 transcriptional repressor is controlled at the level of mRNA translation while the activity of the MEF2 transcription factor can be controlled by the production of two different alternatively spliced forms with different abilities to activate transcription.

It is important to note, however, that the regulation of transcription factors at the level of their activity rather than their synthesis also plays a role in the regulation of cell-type-specific gene expression. For example, the MEF2 transcription factor is also regulated by post-translational modifications such as phosphorylation, acetylation, and sumoylation, which control whether it functions as a transcriptional activator or transcriptional repressor. Similarly, the activity of the MyoD factor is regulated by its interaction with the Id inhibitory protein. However, in general such regulation of transcription factor activity supplements the processes controlling transcription factor synthesis in the regulation of cell-type-specific gene expression, rather than having a predominant role as occurs in the ability of transcription factors to regulate gene expression in response to cellular signaling pathways (see Chapter 8).

As well as regulating transcription factor synthesis, post-transcriptional processes also play a role in the overall regulation of cell-type-specific gene expression, complementing the regulation of transcription. For example, the PTB and nPTB alternative splicing factors regulate the alternative splicing of a number of different RNAs resulting in a different pattern of exon inclusion in neuronal cells where nPTB predominates, compared to other cell types where PTB predominates.

Moreover, as described in this chapter, for both skeletal muscle cells and neuronal cells, miRNAs play an important role in regulating a number of different target genes and, as described in Chapter 7 (Section 7.6), such regulation of gene expression by miRNAs occurs predominantly at the post-transcriptional levels of RNA stability and mRNA translation.

Interestingly, the targets of miRNA-mediated regulation include not only genes encoding cell-type-specific structural proteins but also genes encoding regulatory proteins, such as transcription factors and alternative splicing factors. In turn, altering the expression of these genes will affect the many genes whose transcription or splicing are themselves regulated by these regulatory factors. This provides the basis for regulatory networks in which the regulation of a single miRNA can affect the expression of a large number of different genes either directly, or indirectly via genes encoding regulatory proteins.

These gene-regulatory processes therefore allow a particular cell to switch on a specific pattern of gene expression, as well as suppressing genes characteristic of other cell types, and to do this in a stable manner. These mechanisms provide a basis for understanding how individual differentiated cells can activate and repress the expression of a wide range of genes in order to assume a particular differentiated phenotype.

KEY CONCEPTS

- The regulation of cell-type-specific gene expression can involve positively acting transcription factors (such as MyoD) which activate genes expressed only in a particular cell type.

- It can also involve negatively acting transcription factors (such as REST) which repress genes expressed in other cell types.

- In some cases, such as MyoD, the expression of a single transcription factor is sufficient to induce the formation of a particular type of differentiated cell.

- Many transcription factors involved in the regulation of cell-type-specific gene expression are themselves regulated by controlling their synthesis.

- In addition, transcription factors involved in regulating cell-type-specific gene expression can be controlled by regulating their activity, for example, by post-translational modifications such as phosphorylation.

- As well as transcriptional regulation by specific transcription factors, cell-type-specific gene regulation can also involve the regulation of gene expression at post-transcriptional levels such as alternative splicing or mRNA translation

- miRNAs play a key role in such post-transcriptional regulation, acting both on genes encoding structural proteins and on genes encoding regulatory proteins such as transcription factors and alternative splicing factors.

- The mating-type system in unicellular yeast has provided insights into the processes regulating cell-type-specific gene expression in multicellular organisms.

FURTHER READING

10.1 Regulation of gene expression in skeletal muscle cells

Berkes CA & Tapscott SJ (2005) MyoD and the transcriptional control of myogenesis. *Semin. Cell Dev. Biol.* 16, 585–595.

Bryson-Richardson RJ & Currie PD (2008) The genetics of vertebrate myogenesis. *Nat. Rev. Genet.* 9, 632–646.

Buckingham M & Relaix F (2007) The role of Pax genes in the development of tissues and organs: Pax3 and Pax7 regulate muscle progenitor cell functions. *Annu. Rev. Cell Dev. Biol.* 23, 645–673.

Hart DO & Green MR (2008) Targeting a TAF to make muscle. *Mol. Cell* 32, 164–166.

Jones KA (2007) Transcription strategies in terminally differentiated cells: shaken to the core. *Genes Dev.* 21, 2113–2117.

Stefani G & Slack FJ (2008) Small non-coding RNAs in animal development. *Nat. Rev. Mol. Cell Biol.* 9, 219–230.

van Rooij E, Liu N & Olson EN (2008) MicroRNAs flex their muscles. *Trends Genet.* 24, 159–166.

10.2 Regulation of gene expression in neuronal cells

Bertrand N, Castro DS & Guillemot F (2002) Proneural genes and the specification of neural cell types. *Nat. Rev. Neurosci.* 3, 517–530.

Costa-Mattioli M, Sossin WS, Klann E & Sonenberg N (2009) Translational control of long-lasting synaptic plasticity and memory. *Neuron* 61, 10–26.

Coutinho-Mansfield GC, Xue Y, Zhang Y & Fu XD (2007) PTB/nPTB switch: a post-transcriptional mechanism for programming neuronal differentiation. *Genes Dev.* 21, 1573–1577.

Li Q, Lee, J-A & Black DL (2007) Neuronal regulation of alternative pre-mRNA splicing. *Nat. Rev. Neurosci.* 8, 819–831.

Makeyev EV & Maniatis T (2008) Multilevel regulation of gene expression by microRNAs. *Science* 319, 1789–1790.

Ooi L & Wood IC (2007) Chromatin crosstalk in development and disease: lessons from REST. *Nat. Rev. Genet.* 8, 544–554.

Qiu Z & Ghosh A (2008) A brief history of neuronal gene expression: regulatory mechanisms and cellular consequences. *Neuron* 60, 449–455.

Richter JD & Klann E (2009) Making synaptic plasticity and memory last: mechanisms of translational regulation. *Genes Dev.* 23, 1–11.

Stefani G & Slack FJ (2008) Small non-coding RNAs in animal development. *Nat. Rev. Mol. Cell Biol.* 9, 219–230.

10.3 Regulation of yeast mating type

Cosma MP (2002) Ordered recruitment: gene-specific mechanism of transcription activation. *Mol. Cell* 10, 227–236.

Cosma MP (2004) Daughter-specific repression of *Saccharomyces cerevisiae* HO: Ash1 is the commander. *EMBO Rep.* 5, 953–957.

Dolan JK & Fields S (1991) Cell type-specific transcription in yeast. *Biochim. Biophys. Acta* 1088, 155–169.

Herskowitz I (1989) A regulatory hierarchy for cell specialization in yeast. *Nature* 342, 749–757.

Herskowitz L (1985) Master regulatory loci in yeast and lambda. *Cold Spring Harb. Symp. Quant. Biol.* 50, 565–574.

Gene Regulation and Cancer

11

INTRODUCTION

As we have discussed in preceding chapters, the regulation of gene expression in higher eukaryotes is a highly complex process. It is not surprising therefore that this process can go wrong. Indeed, the identification of the molecular basis of many human diseases has shown some to be due to defects in gene regulation.

Although a wide variety of human diseases arise due to defects in gene regulation, the human disease that exhibits the most extensive malregulation of gene expression is cancer. Not only does cancer often result from the over-expression of certain cellular genes (known as proto-oncogenes) due to errors in their regulation, but several of these genes themselves actually encode transcription factors and cause the disease by affecting the expression of other genes. This chapter will therefore focus on the connection between cancer and the malregulation of gene expression. It will illustrate how our increasing knowledge of this connection has aided our understanding both of the disease and of the processes that regulate gene expression in normal and transformed cancer cells. Chapter 12 will focus on the role of gene regulation in other human diseases as well as the potential therapy of human diseases which may be possible by manipulating gene-regulatory processes.

11.1 GENE REGULATION AND CANCER

Oncogenes were originally identified in cancer-causing viruses

In order to provide a background to our discussion of cancer and gene regulation, it is necessary to discuss briefly the nature of cancer-causing onco-genes and the process that led to their discovery.

As long ago as 1911, Peyton Rous showed that a connective tissue cancer in the chicken was caused by an infectious agent. This agent was subsequently shown to be a virus and was named Rous sarcoma virus (RSV) after its discoverer. This cancer-causing, tumorigenic virus is a member of a class of viruses called **retroviruses** whose genome consists of RNA rather than DNA as in most other organisms.

The majority of viruses of this type do not cause cancers or kill the infected cell. Rather, they simply infect a cell and produce a persistent infection with continual production of virus by the infected cell. In the case of RSV, however, such infection also results in the conversion of the cell into a cancer cell, capable of indefinite growth and eventually killing the organism containing it.

In the case of non-tumorigenic retroviruses, the genome contains only three genes, which are known as *gag, pol,* and *env,* and which function in the normal life cycle of the virus (**Figure 11.1**). Following cellular entry, the

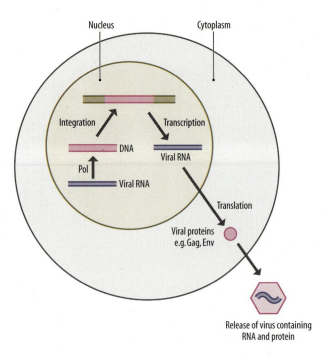

Figure 11.1
Life cycle of a typical non-tumorigenic retrovirus. Note the integration of the viral DNA (pink) into the cellular genome (green).

viral RNA is converted into DNA by the action of the Pol protein and this DNA molecule then integrates into the host chromosome. Subsequent transcription and translation of this DNA produces the viral structural proteins Gag and Env which coat the viral RNA genome, yielding viral particles that leave the cell to infect other susceptible cells.

Inspection of the RSV genome (**Figure 11.2**) reveals an additional gene, known as the *src* gene, which is absent in the other viruses. This suggests that this gene is responsible for the ability of the virus to cause cancer. This idea was confirmed subsequently by showing that if the *src* gene alone was introduced into normal cells it was able to transform them to a cancerous phenotype. This gene was therefore called an oncogene (from the Greek *onkas* for mass or tumor) or cancer-causing gene.

Following the identification of the *src* oncogene in RSV, a number of other oncogenes were identified in other oncogenic retroviruses infecting both chickens and mammals such as the mouse or rat (**Table 11.1**).

Cellular proto-oncogenes are present in the genome of normal cells

The identification of individual genes that are able to cause cancer obviously opened up many avenues of investigation for the study of this disease. From the point of view of gene regulation, however, the most exciting aspect of oncogenes was provided by the discovery that these cancer-causing genes are derived from genes present in normal cellular DNA. For example, using Southern-blotting techniques (see Methods Box 1.1), a cellular equivalent of the viral *src* gene was detected in the DNA of both normal and cancer cells. Moreover, it was shown that an mRNA capable of encoding the Src protein was produced in normal cells.

Subsequent studies identified cellular equivalents of all the retroviral oncogenes. When these cellular equivalents of the viral oncogenes were

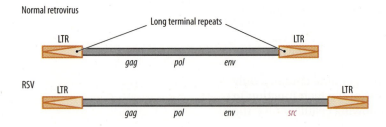

Figure 11.2
Comparison of the genome of a non-tumorigenic retrovirus with that of the tumorigenic retrovirus RSV.

TABLE 11.1
PROTO-ONCOGENES AND THEIR FUNCTIONS

ONCOGENE	SPECIES INFECTED BY VIRUS	NORMAL FUNCTION OF PROTEIN
abl	Mouse	Tyrosine kinase
erbA	Chicken	Transcription factor and hormone receptor
erbB	Chicken	Receptor for epidermal growth factor, tyrosine kinase
ets	Chicken	Transcription factor
fes	Cat	Tyrosine kinase
fgr	Cat	Tyrosine kinase
fms	Cat	Tyrosine kinase, receptor for colony-stimulating factor
fos	Mouse	Transcription factor
jun	Chicken	AP1-related transcription factor
kit	Cat	Tyrosine kinase
lck	Chicken	Tyrosine kinase
mos	Mouse	Serine/threonine kinase
myb	Chicken	Transcription factor
myc	Chicken	Transcription factor
raf	Chicken	Serine/threonine kinase
ras	Rat	GTP-binding protein
rel	Turkey	Transcription factor
ros	Chicken	Tyrosine kinase
sea	Chicken	Tyrosine kinase
sis	Monkey	Platelet-derived growth factor B chain
ski	Chicken	Nuclear protein
src	Chicken	Tyrosine kinase
yes	Chicken	Tyrosine kinase

cloned and characterized, they were shown to encode proteins identical or closely related to those present in the retroviruses. To avoid confusion, the viral oncogenes are given the prefix v, as in v-*src*, while their cellular equivalents are designated proto-oncogenes and given the prefix c, as in c-*src*. Over 20 cellular proto-oncogenes which were originally identified in this way are now known (see Table 11.1).

The presence of cellular equivalents of the retroviral oncogenes suggests that the viral genes have been picked up from normal cellular DNA following integration of a non-tumorigenic virus next to the proto-oncogene. The subsequent incorrect excision of the viral genome resulted in its picking up the gene and converted it into a tumorigenic virus (**Figure 11.3**). Hence the oncogene in the virus is derived from a cellular gene which is present in the DNA of normal cells.

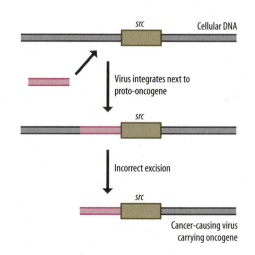

Figure 11.3
Model for the way in which a retrovirus (pink) could have picked up the cellular *src* gene by integration adjacent to the gene and subsequent incorrect excision.

Cellular proto-oncogenes can cause cancer when they are over-expressed or mutated

Paradoxically the same gene can be present in normal cells without any apparent adverse effect and yet can cause cancer when incorporated into a virus. It is now clear that this conversion of the proto-oncogene into a cancer-causing oncogene is caused by one of two processes. Either the gene is mutated in some way within the virus so that an abnormal product is formed, or alternatively the gene is expressed within the virus at much higher levels than are achieved in normal cells and this high level of the normal product results in **transformation** (Figure 11.4).

Such conversion of a proto-oncogene into a cancer-causing oncogene is not confined to viruses however. Cases of cancer which have no viral involvement in humans have now been shown to be due either to the over-expression of individual cellular proto-oncogenes or to their alteration by mutation within the cellular genome. Hence, these genes play an important role in the generation of human cancer.

The potential risk of such proto-oncogenes causing cancer raises the question of why these genes have not been deleted during evolution. In fact, proto-oncogenes have been highly conserved in evolution, equivalents to mammalian and chicken oncogenes having been found not only in other vertebrates but also in invertebrates such as *Drosophila* and even in single-celled organisms such as yeast.

The extraordinary evolutionary conservation of many of these genes despite their potential danger led to the suggestion that their products were essential for the processes regulating the growth of normal cells and that their malregulation or mutation therefore results in abnormal growth and cancer. This idea has been confirmed abundantly as more and more proto-oncogenes have been characterized and shown to encode growth factors that stimulate the growth of normal cells, cellular receptors for growth factors and other cellular proteins involved in transmitting the growth signal within the cell, either by acting as a protein kinase enzyme or by binding GTP (see Table 11.1).

Ultimately the growth regulatory pathways controlled by oncogene products end in the nucleus with the activation of genes whose corresponding proteins are required by the growing cell. It is not surprising therefore, that several proto-oncogenes have been shown to encode transcription factors that regulate the expression of genes activated in growing cells (Table 11.2).

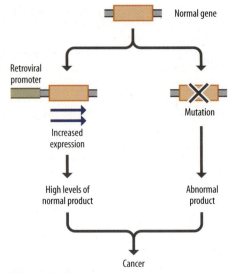

Figure 11.4
A cellular proto-oncogene can be converted into a cancer-causing oncogene by increased expression or by mutation.

TABLE 11.2
ONCOGENES ENCODING TRANSCRIPTION FACTORS

ONCOGENE	COMMENTS
erbA	Mutant form of the thyroid hormone receptor
ets	Binding site often found in association with AP1 site
fos	Binds to AP1 site as Fos–Jun dimer
jun	Can bind to AP1 site alone as Jun–Jun dimer
mdm2	Inhibits gene activation by p53
mdm4	Inhibits gene activation by p53
myb	DNA-binding transcriptional activator
myc	Requires Max protein to bind to DNA
rel	Member of the NFκB family
spi-1	Identical to PU.1 transcription factor

Hence, oncogenes present two aspects of importance from the point of view of the regulation of gene expression. First, since cancer is often caused by elevated expression of cellular oncogenes, the processes whereby this occurs are of interest both from the point of view of the etiology of cancer and for the light they throw on the mechanisms which control gene expression. This topic is discussed below. Secondly, the study of the transcription factors encoded by a few proto-oncogenes has led to a better understanding of the processes regulating gene expression in cells growing normally and in cancer cells. This is discussed in Section 11.2.

Viruses can induce elevated expression of oncogenes

The products of cellular proto-oncogenes play a critical role in cellular growth control and in many cases are synthesized only at specific times and in small amounts. It is not surprising, therefore, that transformation into a cancer cell can result when these genes are expressed at high levels in particular situations. The simplest example of such over-expression occurs in the case of retroviruses where, as we have already discussed, the oncogene comes under the influence of the strong promoter contained in the retroviral **long terminal repeat** (LTR) region and is hence expressed at a high level (see Figure 11.4).

A similar up-regulation due to the activity of a retroviral promoter is also seen in the case of avian leukosis virus (ALV) of chickens. Unlike the retroviruses described so far, however, this virus does not carry its own oncogene. Rather it transforms by integrating into cellular DNA next to a cellular oncogene, the c-*myc* gene (**Figure 11.5a**). The expression of the c-*myc* gene is brought under the control of the strong promoter in the retroviral LTR and it is hence expressed at levels 20–50 times higher than normal producing transformation. This process is known as promoter insertion.

In other cases of this type, the ALV virus has been shown to have integrated downstream rather than upstream of the c-*myc* gene. Hence, the elevated expression of the c-*myc* gene in these cases cannot be due to promoter insertion. Rather, it involves the action of an enhancer element in the viral LTR which activates the c-*myc* gene's own promoter. As discussed in Chapter 4 (Section 4.4), enhancers, unlike promoters, can act in either

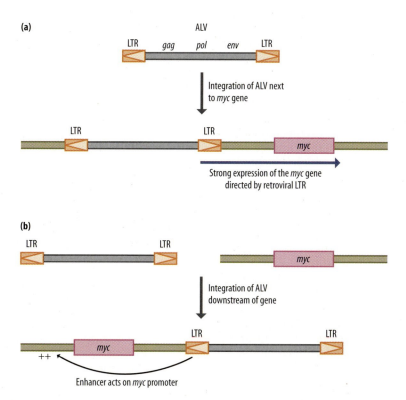

Figure 11.5
Avian leukosis virus (ALV) can increase expression of the *myc* proto-oncogene either via promoter insertion (a) or by the action of its enhancer (b).

orientation and at a distance. Hence the ALV enhancer can activate the *myc* promoter from a position downstream of the gene (**Figure 11.5b**).

Proto-oncogene expression can also be enhanced by cellular mechanisms

The cases discussed so far are of interest from the point of view of gene regulation as indicating how viral regulatory systems can subvert cellular control processes. Of potentially greater interest, however, are the cases where up-regulation of a cellular oncogene occurs through the alteration of internal cellular regulatory processes rather than through viral intervention. The best-studied example of this type concerns the increased expression of the c-*myc* oncogene which occurs in the transformation of B cells in cases of Burkitt's **lymphoma** in humans or in the similar plasmacytomas that occur in mice.

When these tumors were studied, it was note that they commonly contained very specific chromosomal translocations which involved the exchange of genetic material between chromosome 8 and chromosome 14 (**Figure 11.6**). Most interestingly, the region of chromosome 8 involved includes the c-*myc* gene and the translocation results in the *myc* gene being moved to chromosome 14 where it becomes located adjacent to the gene encoding the immunoglobulin heavy chain. This translocation results in the increased expression of the *myc* gene which is observed in the tumor cells.

Such translocation of specific oncogenes to the immunoglobulin locus resulting in increased expression is not unique to the *myc* gene but has been observed for a variety of other oncogenes in different B-cell **leukemias**. Moreover, similar translocations involving oncogene translocation to the T-cell receptor gene locus (which is highly expressed in T lymphocytes) have been observed in T-cell leukemias (see also Section 11.2). The mechanisms responsible for the observed up-regulation of oncogene expression have been best described, however, in the case of the *myc* oncogene and these will now be discussed.

A variety of cellular mechanisms mediate enhanced expression of proto-oncogenes in different cancers

Detailed study of the processes mediating increased *myc* gene expression has indicated that it is produced by different mechanisms in different lymphomas, depending on the precise break points of the translocation within the c-*myc* and immunoglobulin genes. In all cases studied, however, the break point of the translocation occurs within the immunoglobulin gene resulting in a truncated gene lacking its promoter being linked to the c-*myc* gene. This fact, together with the fact that the genes are always linked in a head-to-head orientation (**Figure 11.7**) indicates that the up-regulation of the c-*myc* gene does not occur via a simple promoter insertion mechanism in which it comes under the control of the immunoglobulin promoter. In some cases, however, the B-cell-specific enhancer element which is located between the joining and constant regions of the immunoglobulin genes (see Sections 1.3 and 4.4) is brought close to the *myc* promoter by the translocation (**Figure 11.8**). This enhancer element is highly active in B cells and can activate the *myc* promoter in a manner analogous to the enhancer of ALV (see Figure 11.5b).

Hence, in this case the c-*myc* gene is up-regulated by the action of the B-cell-specific regulatory mechanisms of the immunoglobulin gene, the immunoglobulin enhancer activating the c-*myc* promoter rather than its own promoter which has been removed by the translocation. In other cases, however, this does not appear to be the case, the break point of the translocation having removed both the immunoglobulin promoter and enhancer. This leaves the c-*myc* gene adjacent to the constant region of the immunoglobulin gene without any obvious B-cell-specific regulatory elements. In these cases it is likely that the up-regulation of the c-*myc* gene arises from its own truncation in the translocation. This results in the removal of

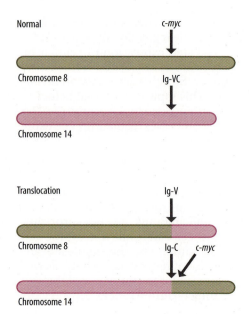

Figure 11.6
Translocation of the c-*myc* gene from chromosome 8 to the immunoglobulin heavy-chain gene locus on chromosome 14 that occurs in cases of Burkitt's lymphoma.

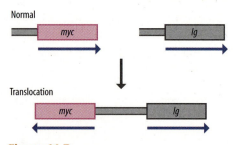

Figure 11.7
Head-to-head orientation of the translocated *myc* gene and the immunoglobulin gene.

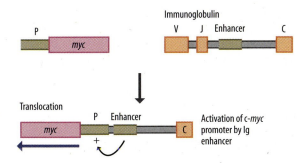

Figure 11.8
In some cases the enhancer of the immunoglobulin heavy-chain gene activates the *myc* gene promoter (P).

negative regulatory elements that normally repress its expression, such as the upstream silencer element which normally represses the c-*myc* promoter (see Section 4.4 and **Figure 11.9**).

More extensive truncation of the c-*myc* gene involving the removal of transcribed sequences rather than upstream elements has also been observed in some tumors. Frequently, this involves the removal of the first exon of the c-*myc* gene which does not contain any protein-coding information. This exon may thus fulfil a regulatory role by modulating the stability of the c-*myc* RNA or by affecting its translatability. Hence, its removal could enhance the level of c-*myc* protein by increasing the stability or the efficiency of translation of the c-*myc* RNA produced at a constant level of transcription. A similar increase in gene expression could be achieved by the removal of sequences within the first intervening sequence which inhibit the transcriptional elongation of the nascent c-*myc* transcript (see Section 5.4).

The increased expression of an oncogene produced by the removal of sequences that negatively regulate it is also seen in the case of the *lck* proto-oncogene which encodes a tyrosine kinase related to the c-*src* gene product. In this case, activation of the oncogene in tumors is accompanied by the removal of sequences within its 5′ untranslated region, upstream of the start site of translation. The removal of these sequences results in a fiftyfold increase in the initiation of translation of the *lck* mRNA into protein.

Most interestingly, the region removed contains three AUG translation initiation codons which are located upstream of the correct initiation codon for production of the Lck protein (**Figure 11.10**). The elimination of these codons results in increased translation initiation from the correct AUG, suggesting that initiation at the upstream codons inhibits correct initiation. This is exactly analogous to the regulation of translation of the GCN4 and ATF4 transcription factors which was discussed respectively in Chapter 7 (Section 7.5) and Chapter 10 (Section 10.2).

It is likely that the processes regulating the expression of oncogenes such as c-*myc* and *lck*, which are revealed by studying their over-expression in tumors, also play a role in their normal pattern of regulation during cellular growth. Hence, their further study will throw light not only on the mechanisms of tumorigenesis but also on the processes regulating gene expression in normal cells.

Other examples of up-regulation of oncogene expression in tumorigenesis may occur, however, by mechanisms that are unique to the transformed cell. For example, as discussed in Chapter 1 (Section 1.3), DNA amplification is relatively rare in normal cells. In tumors, however, it is observed

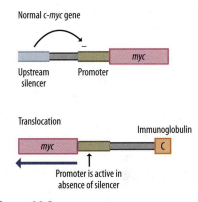

Figure 11.9
Activation of *myc* gene expression can be achieved by removal of the upstream silencer element.

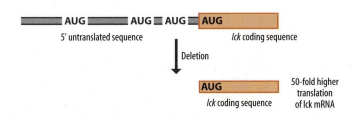

Figure 11.10
Increased translation of the lck proto-oncogene mRNA can be achieved by deletion of AUG translational start sites upstream of the AUG that initiates the coding sequence of the Lck protein.

frequently for specific oncogenes and results in the presence of regions of amplified DNA which are visible in the microscope as homogeneously staining regions or as **double-minute chromosomes**. Such amplification is especially common in human lung tumors and brain tumors and frequently involves the c-*myc*-related genes N-*myc* and L-*myc*. The expression of these genes in the tumor is increased dramatically due to the presence of up to 1000 copies of the gene in the tumor cell.

A variety of mechanisms involving both the subversion of normal control processes or abnormal events occurring in the tumor cell therefore result in the observed over-expression of oncogenes in tumor cells. In turn, such over-expression is critical for tumor formation. For example, if the high levels of c-*myc* in a mouse model of liver cancer are reduced, the tumor cells lose their cancerous phenotype and differentiate into hepatocytes. If over-expression of c-*myc* is then restored, the cells return to a cancerous phenotype. When taken together with the production of abnormal oncogene products due to mutations, it is likely that the over-expression of specific oncogenes is involved in a wide variety of human cancers.

11.2 TRANSCRIPTION FACTORS AS ONCOGENES

As described in Section 11.1, the isolation of the cellular genes encoding particular oncogene products led to the realization that they were involved in many of the processes regulating cellular growth. Ultimately the onset and continuation of cellular growth is likely to involve the activation of cellular genes that are not expressed in quiescent cells. It is not surprising, therefore, that several proto-oncogenes have been shown to encode transcription factors which regulate the transcription of genes activated in growing cells (see Table 11.2). Several of these cases will be discussed.

The Fos and Jun oncogene proteins are cellular transcription factors which can cause cancer when over-expressed

The chicken retrovirus avian sarcoma virus AS17 contains an oncogene, v-*jun*, whose equivalent cellular proto-oncogene encodes a nuclearly located DNA-binding protein. Sequence analysis of this protein revealed that it showed significant homology to the DNA-binding domain of the yeast transcription factor GCN4 suggesting that it might bind to similar DNA sequences (**Figure 11.11**). Interestingly, GCN4 itself had been shown previously to bind to similar sequences to those bound by the activator protein 1 (AP1) factor which had been detected in mammalian cell extracts by its DNA-binding activity (**Figure 11.12**).

This relationship of Jun and AP1 to the sequence and binding activity respectively of GCN4 led to the suggestion that Jun might be related to AP1. This was confirmed by the findings that antibody to Jun reacted with purified AP1 preparations and that Jun expressed in bacteria was capable of binding to AP1-binding sites in DNA. Moreover, Jun was capable of stimulating transcription from promoters containing AP1-binding sites but not from those which lacked these sites. Hence, the *jun* oncogene encodes a sequence-specific DNA-binding protein capable of stimulating transcription of genes containing its binding site, which is identical to the AP1-binding site.

Figure 11.11

Comparison of the C-terminal amino acid sequences of the chicken Jun protein and the yeast transcription factor GCN4. Boxes indicate identical residues.

Although Jun undoubtedly binds to AP1-binding sites, preparations of AP1 purified on the basis of this ability contain several other proteins in addition to c-Jun. Several of these are encoded by genes related to *jun* but another is the product of a different proto-oncogene namely c-*fos*. Although Fos is present in AP1 preparations, it does not bind to DNA when present alone but requires the product of the c-*jun* gene for DNA binding. Hence, in addition to its ability to bind to AP1 sites alone, Jun can also form a complex with Fos that binds to this site. As discussed in Chapter 5 (Section 5.1), this association takes place through the leucine-zipper domains of the two proteins. This results in a Fos–Jun heterodimer which binds to the AP1-binding site with much greater affinity than the Jun homodimer.

Both Fos and Jun which were identified originally through their association with oncogenic retroviruses are thus also cellular transcription factors. Such a finding raises the question of the normal role of these factors and how they can cause cancer. In this regard it is of obvious interest that the AP1-binding site is involved in mediating the induction of genes that contain it in response to treatment with phorbol esters, which are also capable of promoting cancer. Not only do many phorbol ester-inducible genes contain AP1-binding sites (see Table 4.3) but, in addition, transfer of AP1-binding sites to a normally non-inducible gene renders that gene inducible by phorbol esters. Increased levels of Jun and Fos are also observed in cells after treatment with phorbol esters. Hence, these substances act by increasing the levels of Fos and Jun, which in turn cause increased transcription of other genes containing AP1-binding sites that mediate induction by the Fos–Jun complex.

In the case of the *ccl2* gene, binding of Jun to multiple APl sites has been shown to produce a cascade of changes which result in activation of the gene. These include phosphorylation of histone H3 on serine 10 and histone acetylation, both of which are associated with an open chromatin structure (see Section 2.3 and Section 3.3), as well as recruitment of the NFκB transcription factor and recruitment of RNA polymerase II itself (**Figure 11.13**).

Most interestingly, increased levels of Jun and Fos are produced by treatment with serum or growth factors which stimulates the growth of quiescent cells. Hence, the transduction of the signal to grow which begins with the growth factors and their cellular receptors and continues with intracytoplasmic signal transducers such as protein kinases and **GTP-binding proteins**, ends in the nucleus with the increased level of the transcription factors Jun and Fos (**Figure 11.14**). These proteins will then activate the genes whose products are necessary for the process of growth itself.

Clearly, it is relatively easy to fit the oncogenic properties of Jun and Fos into this framework. Thus, if these proteins are normally produced in response to growth-inducing signals and activate growth, their continual abnormal synthesis will result in a cell which will be stimulated to grow continually and will not respond to growth-regulating signals. Such continuous uncontrolled growth is characteristic of the cancer cell.

In agreement with this idea, mutations in the leucine-zipper region of Fos, which abolish its ability to dimerize with Jun and induce genes containing AP1 sites, also abolish its ability to transform cells to a cancerous

DNA-binding site

```
GCN4  5'  T G A C/G T C A T  3'
AP1   5'  T G A  G  T C A G  3'
```

Figure 11.12
Relationship of the DNA-binding sites for the yeast transcription factor GCN4 and the mammalian transcription factor AP1.

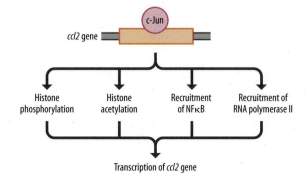

Figure 11.13
Binding of c-Jun to the *ccl2* gene initiates a number of events which together result in transcription of the *ccl2* gene.

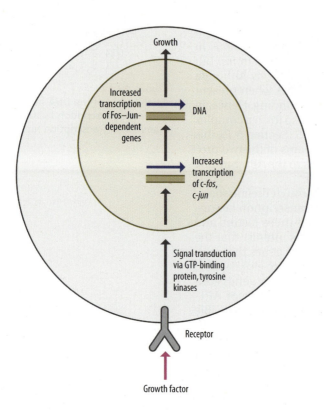

Figure 11.14
Growth factor stimulation of cells results in increased transcription of the c-*fos* and c-*jun* genes, which in turn stimulates transcription of genes that are activated by the Fos–Jun complex.

phenotype. Hence, the ability of Fos to cause cancer is directly linked to its ability to act as a transcription factor for genes containing the appropriate binding site.

Interestingly, AP1 sites in growth-regulated genes are often located close to binding sites for the Ets protein, another transcription factor which is encoded by a cellular proto-oncogene (see Table 11.2). Several different oncogenic transcription factors may co-operate therefore to produce high-level transcription of specific genes in actively growing cells.

The v-*erb*A oncogene protein is a mutant form of the cellular thyroid hormone receptor

Unlike most other retroviruses, avian erythroblastosis virus (AEV) carries two cellular oncogenes, v-*erb*A and v-*erb*B. When c-*erb*A, the cellular equivalent of v-*erb*A, was cloned it was shown to encode the cellular receptor that mediates the response to thyroid hormone. As discussed in Chapter 5 (Sections 5.1 and 5.3), this receptor is a member of the nuclear receptor family, which following binding of a particular hormone, induce the transcription of genes containing a binding site for the hormone-receptor complex. In the case of **ErbA**, the protein contains a region that can bind thyroid hormone. Following such binding, the hormone–receptor complex induces transcription of thyroid hormone-responsive genes, such as those encoding growth hormone or the heavy chain of the myosin molecule (**Figure 11.15**).

The finding that the cellular homologue of the v-*erb*A oncogene is a hormone-responsive transcription factor provides a further connection between oncogenes and cellular transcription factors. It raises the question, however, of the manner in which the transfer of the thyroid hormone receptor to a virus can result in transformation. To answer this question it is necessary to compare the protein encoded by the virus with its cellular counterpart. As shown in **Figure 11.16**, the cellular ErbA protein has the typical structure of a member of the nuclear receptor family (see also Figure 5.21), containing both DNA-binding and hormone-binding regions. The viral ErbA protein is generally similar except that it is fused to a portion of the retroviral gag protein at its N-terminus. It also contains a number of

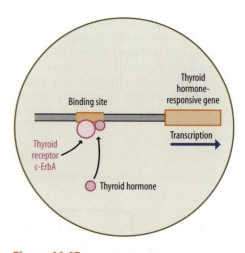

Figure 11.15
The c-*erb*A gene encodes the thyroid hormone receptor and activates transcription in response to thyroid hormone.

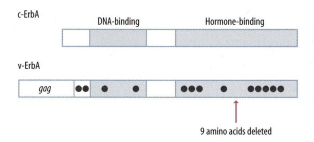

Figure 11.16
Relationship of the cellular ErbA protein and the viral protein. The dots indicate single-amino acid differences between the two proteins and the arrow indicates the region where nine amino acids are deleted in the viral protein.

mutations in both the DNA-binding and hormone-binding regions, as well as a small deletion in the hormone-binding domain.

Of these changes, it is the alterations in the hormone-binding domain which have the most significant effects on the function of the protein and which are thought to be critical for transformation. These changes abolish the ability of the protein to bind thyroid hormone and activate transcription. However, the protein retains the inhibitory domain which allows the ErbA protein to repress gene transcription in the absence of thyroid hormone (see Section 5.3). The viral protein is therefore functionally analogous to the alternatively spliced form of the c-*erb*A gene product discussed in Chapter 5 (Section 5.3), which lacks the hormone-binding domain and dominantly represses the ability of the hormone-binding receptor form to activate thyroid hormone-responsive genes.

The idea that the non-hormone-binding viral ErbA protein might also be able to do this has been confirmed by studying the effect of this oncogene on thyroid hormone-responsive genes. As expected, the v-*erb*A gene product was able to abolish the responsiveness of such genes to thyroid hormone by binding to the thyroid hormone response elements in their promoters and preventing activation by the cellular ErbA protein-thyroid hormone complex (**Figure 11.17**).

Interestingly, however, this repression by v-*erb*A does not simply involve the passive blockage of binding by c-*erb*A following thyroid hormone treatment. Thus, mutations in the inhibitory domain described in Chapter 5 (Section 5.3) can abolish the oncogenic activity of v-*erb*A. These mutations do not affect the ability of v-*erb*A to bind to DNA but prevent it from recruiting the inhibitory co-repressor which is essential for its ability to actively repress transcription. Hence the ability to actively repress transcription is essential for transformation by v-*erb*A (**Figure 11.18**).

The explanation of how such gene repression by viral ErbA can result in transformation is provided by the observation that the introduction of the viral gene into cells can repress transcription of the avian erythrocyte anion transporter gene. This gene is one of those which is switched on when chicken erythroblasts differentiate into erythrocytes. It has been known for some time that the viral ErbA protein can block this process and it is now clear that this is achieved by blocking the induction of the genes needed for this to occur. In turn, such blockage of differentiation allows the cells to continue to proliferate. When this is combined with the introduction of the v-*erb*B gene, which encodes a truncated form of the epidermal growth factor receptor and renders cell growth independent of external growth factors, transformation results (**Figure 11.19**).

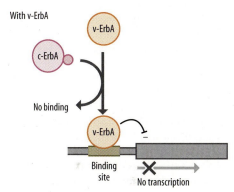

Figure 11.17
Inhibitory effect of the viral ErbA protein on gene activation by the cellular thyroid hormone receptor (c-ErbA) in response to thyroid hormone. Note the similarity to the action of the α-2 form of the c-ErbA protein, illustrated in Figure 5.58.

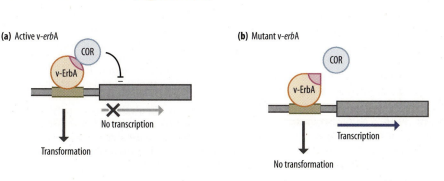

Figure 11.18
The ability of v-*erb*A to transform cells to a cancerous phenotype requires it to be able to actively inhibit transcription via its inhibitory domain (pink), which acts by recruiting an inhibitory co-repressor (COR) (a). Mutations in the v-*erb*A inhibitory domain (pink) which abolish its ability to bind the co-repressor also abolish its ability to transform cells, even though it can still bind to DNA (b).

Transformation caused by v-*erb*A thus represents an example of the activation of an oncogene by mutation resulting in this case in its losing the ability to activate transcription but retaining its ability to act as a dominant repressor of transcription. As discussed in Chapter 5 (Section 5.3), however, one alternatively spliced transcript of the c-*erb*A gene is also able to do this and, like the v-*erb*A gene product, cannot bind thyroid hormone. Hence, this repression of transcription by a non-hormone-binding form of the receptor is likely to be of importance in normal cells also.

Other transcription factor-related oncogenes are over-expressed due to chromosomal translocations

Although the *fos/jun* and *erb*A cases represent well-characterized examples of the connection between oncogenes and transcription factors, several other cellular oncogenes also encode transcription factors (see Table 11.2). One of these, the Myc protein, has been studied intensively in view of its over-expression in many human tumors due to its involvement in tumor-specific chromosomal translocations (see Section 11.1). The Myc protein clearly encodes a transcription factor which contains the leucine-zipper motif characteristic of many transcription factors including Fos and Jun, as well as a helix-loop-helix motif (see Section 5.1). Moreover, mutations in the leucine-zipper region of the protein abolish its oncogenic ability to transform normal cells, suggesting that the ability to act as a transcription factor is essential for Myc-induced transformation.

Despite all this evidence, for many years the actual role of Myc in transcriptional control remained unclear. This was because it was not possible to demonstrate the binding of Myc to a specific DNA sequence in the manner that had been shown to occur for Jun and ErbA. This problem was resolved, however, by the finding that Myc has to heterodimerize with a second factor known as Max, in order to bind to DNA and activate transcription (**Figure 11.20a**). Myc therefore resembles Fos in requiring another factor for sequence-specific binding. This finding indicates once again the importance of heterodimerization in regulating the activity of transcription factors and also illustrates how the function of a particular factor can remain obscure simply because its partner has not yet been isolated.

Interestingly as well as regulating transcription, Myc can also regulate the translation of mRNAs. This is achieved by Myc binding to the TFIIH component of the basal transcriptional complex (see Section 4.1). In turn, TFIIH interacts with the cap guanine methyltransferase enzyme and stimulates its activity. The cap guanine methyltransferase is responsible

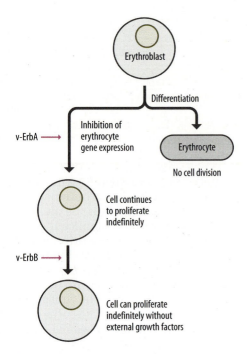

Figure 11.19
Inhibition of erythrocyte-specific gene expression by the v-ErbA protein prevents erythrocyte differentiation and allows transformation by the v-ErbB protein.

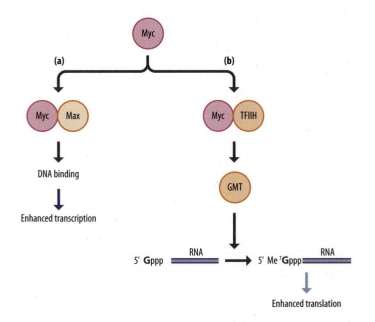

Figure 11.20
The Myc oncoprotein can activate transcription by interacting with the Max protein and binding to the DNA of its target genes (a). It can also stimulate translation by interacting with the TFIIH factor and stimulating the activity of the cap guanine methyltransferase (GMT). The GMT enzyme methylates the cap structure located at the 5′ end of mRNAs, enhancing their translation (b).

for methylating the cap at the 5′ end of the mRNA on the 7 position of the guanine base, so converting the 5′-Gppp cap to a 5′-me7Gppp cap (see Section 6.1). This methylated cap strongly promotes translation of the mRNA compared to the unmethylated cap. The stimulation of cap methylation by Myc therefore results in Myc stimulating mRNA translation (**Figure 11.20b**).

As described in Section 11.1 the *myc* gene was initially identified by studies on RNA **tumor viruses** and subsequently shown to be over-expressed in many human cancers due to chromosomal translocations. In addition, however, a number of oncogenes encoding transcription factors have been identified specifically on the basis of their involvement in the chromosomal translocations which occur in human leukemias, as well as in some solid tumors. As well as involving previously characterized oncogenes such as the *myc* gene, these translocations can involve genes not previously shown to produce cancer when over-expressed. For example, a member of the homeodomain family of transcription factors (see Section 5.1 and Section 9.3), has been shown to be involved in cases of acute childhood leukemia where its expression is activated by translocation to the T-cell receptor gene locus.

Chromosomal translocations can also produce novel oncogenic fusion proteins involving transcription factors

As well as resulting in increased expression of a proto-oncogene by translocation to a highly active immunoglobulin or T-cell receptor gene locus, chromosomal translocations can also cause leukemias or solid tumors by producing fusion genes between two genes which were previously located on separate chromosomes (**Figure 11.21**). The corresponding fusion proteins presumably produce cancer because the fusion protein has cancer-causing properties distinct from those of either intact protein alone. Thus, the fusion protein which is present in 15% of human acute myeloid leukemias involves the AML and ETO transcription factors. This AML–ETO fusion protein is able to bind to basic helix-loop-helix factors (see Section 5.1) and prevent them binding the CBP co-activator, thereby acting as a transcriptional repressor.

Such chromosomal translocation can also involve factors regulating transcriptional elongation rather than transcriptional initiation. For example, the *ELL* gene, which is involved in chromosomal translocations leading to fusion protein products in acute myeloid leukemia, is a transcription elongation factor (see Section 4.2 for discussion of transcriptional elongation and its control).

As in the case of translocations causing over-expression, such translocations can involve genes encoding transcription factors previously identified within RNA tumor viruses or those which have not previously been shown to have an oncogenic effect. For example, the c-*ets*-1 proto-oncogene (see

Figure 11.21
Activation of an oncogene through chromosomal translocation can occur if its expression is increased by translocation to the actively transcribed immunoglobulin or T-cell receptor gene loci (a) or if a fusion gene encoding a novel protein with oncogenic properties is created (b).

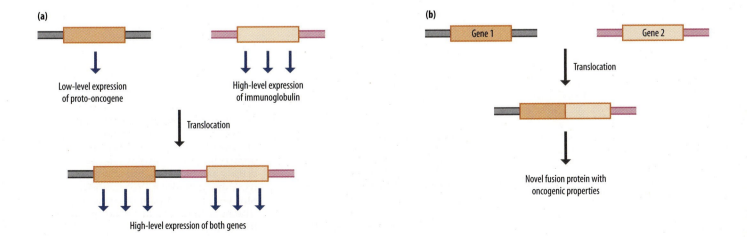

Section 5.1) was originally identified in a chicken retrovirus (see Table 11.1). Subsequently it was shown that the c-*ets*-1 gene becomes fused to the platelet-derived growth factor receptor gene in chronic myelomonocytic leukemia. In contrast, in acute promyelocytic leukemia (PML) the translocation involves the fusion of two transcription factor genes not characterized in oncogenic retroviruses, namely the retinoic acid receptor α gene (RAR; a member of the nuclear receptor family discussed in Section 5.1) and the PML transcription factor, a zinc-finger transcription factor which was originally characterized on the basis of its involvement in this disease.

As in the case of the AML–ETO fusion protein, the RAR–PML fusion protein acts as a transcriptional repressor even though the retinoic acid receptor alone can act as an activator. This inhibitory activity of RAR–PML involves its ability to recruit histone deacetylases and produce an inactive chromatin structure. Indeed, the AML–ETO fusion protein can also bind histone deacetylases (in addition to its effect on CBP binding by basic helix-loop-helix factors). This suggests that enhanced deacetylation induced by oncogenic fusion proteins may play a critical role in human leukemia (**Figure 11.22**) and may be a target for therapy (see Section 12.5).

In view of the key role of altered histone acetylation in producing changes in chromatin structure, these findings indicate that altered chromatin structure is likely to play a key role in human cancer. Such effects on chromatin structure are not confined to altered histone acetylation, however. For example, alterations in other modifications which affect chromatin structure such as DNA methylation on C residues (see Section 3.2) and histone methylation or ubiquitination (see Section 2.3 and Section 3.3) have been observed at a variety of different gene loci in human tumors compared to the corresponding normal human cells.

An example of these effects is provided by the MLL factor, which was identified on the basis of its involvement in the chromosomal translocations which occur in human mixed-lineage leukemias (hence the name, MLL). Such translocations lead to the production of fusion proteins of MLL linked to other factors. The intact MLL protein present in normal cells acts as a histone methylase which is able to methylate histone H3 on lysine 4. Moreover, MLL also protects its target genes from DNA methylation on C residues. Both of these effects result in the MLL target genes having an open chromatin structure (**Figure 11.23a**). These activities are respectively lost or reduced in the MLL fusion proteins present in mixed-lineage leukemias resulting in MLL target genes being repressed in these cancers (**Figure 11.23b**).

Interestingly, the MLL protein is a human homolog of the trithorax proteins, originally identified in *Drosophila* (see Section 3.3 and Section 5.1). Like the trithorax proteins MLL functions to open the chromatin structure of specific *Hox* genes encoding homeodomain factors during development.

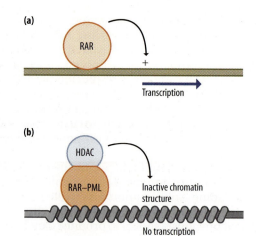

Figure 11.22
The RAR–PML fusion protein found in promyelocytic leukemia can recruit a histone deacetylase (HDAC) which produces an inactive chromatin structure. It therefore acts as a transcriptional repressor (b) unlike the normal retinoic acid receptor (RAR) (a).

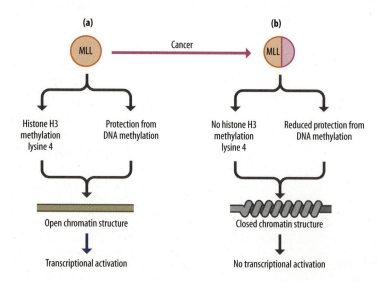

Figure 11.23
The normal MLL protein can methylate histone H3 on lysine 4 and protect its target genes from DNA methylation. This results in the target genes having an open chromatin structure and being transcribed (a). The MLL fusion protein present in cancers has lost its histone methylation activity and has a reduced ability to protect its target genes from DNA methylation. MLL target genes therefore have a closed chromatin structure, incompatible with transcription (b).

As described in Chapter 3 (Section 3.3) and Chapter 5 (Section 5.1) the chromatin-opening action of trithorax proteins is opposed by polycomb family proteins which act to produce a closed chromatin structure.

Such polycomb proteins are also involved in cancers. For example, the EZH2 protein, which is part of the PRC2 polycomb complex, is over-expressed in several different cancers and appears to repress the transcription of a number of anti-oncogenes (which would otherwise inhibit tumor formation; see Section 11.3) by methylating histone H3 on lysine 27. This modification not only itself produces a closed chromatin structure but, in cancer cells, also promotes DNA methylation, further promoting tight packing of chromatin. Hence as in normal development, trithorax and polycomb proteins have opposite effects on cancer development with the loss of MLL/trithorax function due to chromosomal translocation or the enhancement of EZH2/polycomb expression leading to cancer (**Figure 11.24**).

Another member of the polycomb family encoded by the *JJAZ1* gene is involved in a chromosomal translocation that occurs in human endometrial tumors. This translocation produces a fusion protein derived from the JJAZ1 polycomb protein and another protein encoded by the *JAZF1* gene. This fusion protein can enhance the growth of cells and protect them from stimuli which would otherwise induce cell death, explaining why it is oncogenic.

Most interestingly, however, this fusion protein is also made at low levels in normal endometrial cells. However, as described in Chapter 6 (Section 6.3) in normal cells, the fusion protein is encoded by a novel mRNA produced by trans-splicing of exons from the JAZF1 and JJAZ1 RNA transcripts. This novel finding suggests that in some cases, DNA translocation in cancer may produce a high level of fusion proteins mimicking those produced at low level in normal cells by trans-splicing at the RNA level (**Figure 11.25**). In this case therefore, cancer would be produced by a DNA translocation leading to over-expression of a fusion protein which is made at low levels by a trans-splicing process in normal cells.

It is clear, therefore, that both transcription factors and factors which alter chromatin structure can be over-expressed or altered in human cancers and that such effects play a critical role in cancer development.

In addition to the cellular oncogenes identified in RNA tumor viruses or at sites of chromosomal translocations, it is worth noting that cancer-causing DNA viruses also encode oncogenes capable of regulating cellular gene expression. In particular, both the large T oncogenes of the small DNA viruses, SV40 and polyoma, and the E1A protein of adenovirus are capable of affecting the transcription of specific cellular genes and this ability is critical for the ability of these proteins to transform cells. Unlike the oncogenes of RNA viruses, however, the genes encoding these viral proteins do not appear to have specific cellular equivalents and are likely to have evolved within the virus rather than having been picked up from the cellular genome.

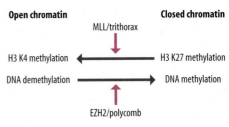

Figure 11.24
The intact MLL/trithorax family protein promotes an open chromatin structure by enhancing histone H3 methylation on lysine (K) 4 and inhibiting DNA methylation. This effect is lost when the *MLL* gene is inactivated by chromosomal translocation in specific cancers. In contrast, the EZH2 polycomb family protein promotes a closed chromatin structure by enhancing histone H3 methylation on lysine (K) 27 and promoting DNA methylation. This effect is enhanced when the *EZH2* gene is over-expressed in specific cancers.

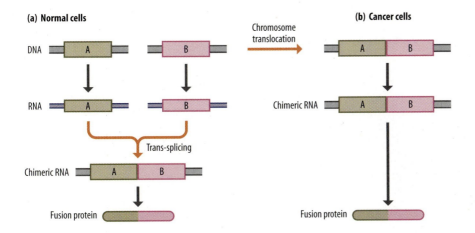

Figure 11.25
In normal cells (a) trans-splicing of two RNAs encoded by different genes (A and B) can produce a chimeric RNA encoding a fusion protein. The same fusion protein can be produced in cancer cells by a chromosomal translocation which fuses genes A and B at the DNA level (b).

The similar ability of oncogenes from DNA and RNA tumor viruses to affect cellular gene expression despite their very different origins, suggests that such modulation of cellular gene expression is critical for the transforming ability of these viruses.

11.3 ANTI-ONCOGENES

Anti-oncogenes encode proteins which restrain cellular growth

Following the discovery of cellular oncogenes it was rapidly shown that they encoded proteins which promoted cellular growth so that their activation by over-expression or by mutation resulted in abnormal growth, leading to cancer. Subsequently, however, it became clear that cancer could also result from the deletion or mutational inactivation of another group of genes. This indicated that these genes encoded products which normally restrained cellular growth so that their inactivation would result in abnormal, unregulated growth. These genes were therefore named anti-oncogenes or tumor-suppressor genes.

Clearly the inhibitory role of anti-oncogenes indicates that cancers will involve the deletion of these genes or the occurrence of mutations within them which inactivate their protein product, rather than their over-expression or mutational activation as occurs for the oncogenes (**Figure 11.26**). A number of genes of this type have been defined on the basis of their mutation or deletion in specific tumor types (**Table 11.3**). Most interestingly, three of the best-defined anti-oncogenes all encode transcription factors. Two of these, p53 and the Wilms' tumor gene product, appear to act by binding to target sites in the DNA of specific genes and regulating their expression while the third, the **retinoblastoma gene**

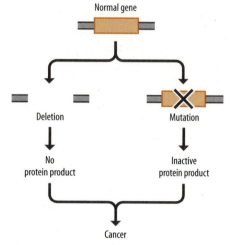

Figure 11.26
Deletion of an anti-oncogene or its inactivation by mutation can result in cancer.

TABLE 11.3 ANTI-ONCOPROTEINS AND THEIR FUNCTIONS		
ANTI-ONCOPROTEIN	**TUMORS IN WHICH GENE IS MUTATED**	**NATURE OF PROTEIN PRODUCT**
APC	Colon carcinoma	Cytoplasmic protein
BRCA-1	Breast cancer	DNA repair, transcription factor?
BRCA-2	Breast cancer	DNA repair, transcription factor?
DCC	Colon carcinoma	Cell adhesion molecule
NF1	Neurofibromatosis	Activator of Ras GTPase activity
NF2	Schwannomas, meningiomas	Cytoplasmic protein
p53	Sarcomas, breast carcinomas, leukemia, etc.	Transcription factor
PTEN	Glioblastoma, melanoma, prostate cancer	Phosphatase
RB1	Retinoblastoma, osteosarcoma, small lung cell carcinoma	Transcription factor
VHL	Pheochromocytoma, kidney carcinoma	Protein degradation
WT1	Wilms' tumor	Transcription factor

product (**Rb-1**), acts predominantly via protein–protein interactions with other regulatory transcription factors. The p53 and Rb-1 proteins will therefore be discussed as examples of these two types.

The p53 protein is a DNA-binding transcription factor

The p53 protein was originally identified as a 53 kDa protein which bound to the large T oncogene protein of the small DNA virus SV40. Subsequent studies have shown that the gene encoding this protein is mutated in a very wide variety of human tumors, especially carcinomas. In normal cells, p53 is induced in response to DNA damage and its activation results in growth arrest and/or cell death. Most interestingly, inactivation of the p53 gene in mice does not prevent the normal development of the animal. Rather, it results in an abnormally high rate of tumor formation which results in the early death of the animal. This has led to the idea that the p53 gene product normally acts to prevent cells with damaged DNA from proliferating by inducing their growth arrest or death. In the absence of p53 such cells proliferate and form tumors which occur at high frequency when p53 is inactivated by mutation.

The detailed characterization of the p53 gene product has shown that it is a transcription factor capable of binding to a specific DNA sequence and activating the expression of specific genes. The mutations which occur in human tumors result in a loss of the ability to bind to this specific DNA sequence, indicating that the ability of p53 to do this is crucial for its ability to control cellular growth and suppress cancer. The p53 protein thus functions at least in part by activating the expression of specific genes whose protein products act to inhibit cellular growth (**Figure 11.27a**). Its inactivation by gene deletion (**Figure 11.27b**) or by mutation (**Figure 11.27c**) leads to a failure to express these genes, resulting in uncontrolled growth.

Although loss of p53 function by deletion of the gene or an inactivating mutation can produce cancer, some p53 cancer-associated mutants appear to have acquired an additional function which allows them to act as an oncogene with the mutant protein actually promoting tumor formation. Such mutants can activate the expression of genes encoding growth-promoting genes even though they do not bind to the normal DNA-binding site for wild-type p53. This may involve such mutant proteins binding to a different DNA-binding site in growth-promoting genes (**Figure 11.28a**) or the mutant p53 being recruited to the gene by a protein–protein interaction with other DNA-binding transcription factors (**Figure 11.28b**).

As well as developing in the absence of wild-type p53 or in the presence of mutant p53, cancer can also occur due to a failure of p53-mediated gene activation even in the presence of functional p53 (**Figure 11.27d**). For example, many human soft-tissue sarcomas contain intact p53 protein but have amplified the cellular *mdm2* oncogene. The MDM2 oncoprotein binds to p53 and promotes its modification by the addition of ubiquitin. Interestingly, the addition of a single ubiquitin residue to p53 promotes its export from the nucleus whereas the addition of two such residues promotes its degradation. Either effect evidently prevents p53 activating its target genes in the nucleus (**Figure 11.29a**).

As well as regulating the stability of p53, MDM2 also regulates the translation of the p53 mRNA. MDM2 interacts with the L26 ribosomal protein, which normally binds to the p53 mRNA and stimulates its translation to produce p53 protein. Binding of MDM2 to L26 prevents it binding to the p53 mRNA and thereby reduces p53 protein production (**Figure 11.29b**).

MDM2 can therefore inhibit p53 both by reducing the translation of its mRNA and by inducing degradation of the p53 protein. Inhibition of p53 can also be achieved by over-expression of the *mdm*4 oncogene. which occurs in many human tumors. In this case, however, the MDM4 protein inhibits p53 activity by binding to it and masking its transcriptional activation domain, so preventing it activating transcription (see Section 5.3 for discussion of this mechanism of transcriptional repression).

The interaction of p53 with oncogenic proteins is not unique to the MDM2 and MDM4 cellular oncoproteins. As noted above, it was

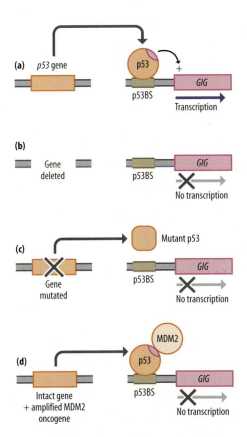

Figure 11.27
The functional p53 protein binds to its DNA-binding site (p53BS) and stimulates the transcription of genes whose protein products inhibit growth (*GIG*) (a). This effect can be prevented, however, by the deletion of the p53 gene (b) or by its inactivation by mutation (c), as well as by the MDM2 oncoprotein which binds to p53 and inactivates it (d). The activation domain of p53 is indicated in pink.

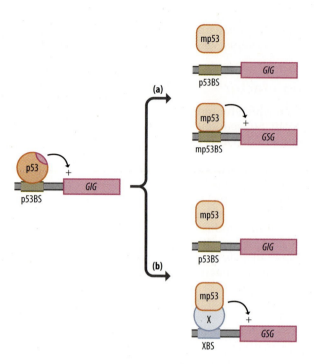

Figure 11.28
Wild-type p53 (circle) binds to its binding site (p53BS) in the regulatory regions of genes encoding growth-inhibitory proteins (*GIG*) and activates their expression. Mutant p53 (square) may activate the expression of genes encoding growth-stimulatory proteins (*GSG*) either by binding to a binding site distinct from that for wild-type p53 (mp53BS) (a) or by being recruited to the DNA by a DNA-binding transcription factor (b).

the interaction of p53 with the SV40 large T oncoprotein which led to the original identification of the p53 protein. As in the case of the MDM2 and MDM4 proteins, the interaction of p53 with either the large T protein or the transforming proteins of several other DNA viruses prevents p53 from activating its target genes. Such functional inactivation of p53 appears to play a critical role in the ability of these viruses to transform cells to a cancerous phenotype. When taken together with the action of the MDM2 and MDM4 proteins this indicates that functional interactions between oncogene and anti-oncogene products are likely to be critical in the control of cellular growth. Changes in this balance due to the over-expression of specific oncogenes or the loss of anti-oncogenes will result in cancer.

As well as being modified by ubiquitination, the activity of p53 is also regulated by multiple post-translational modifications including phosphorylation, methylation, and acetylation. For example, the acetylation of p53 at multiple lysine residues near its C-terminus prevents its interaction with MDM2. This modification therefore stabilizes the p53. Moreover, such acetylation of p53 also stimulates its ability to recruit the TAF1 component of the TFIID basal transcription factor (see Section 4.1), so enhancing the ability of p53 to stimulate transcription (**Figure 11.30**).

The acetylation of p53 at its C-terminus leading to its activation is catalyzed by the p300 and CBP transcriptional co-activators which, as discussed in Chapter 5 (Section 5.2), have histone acetyltransferase activity. In contrast, the RAR–PML fusion protein found in promyelocytic leukemia (see Section 11.2) can induce deacetylation of p53, thereby inhibiting its activity, providing an example of the activity of an anti-oncogene being targeted by an oncogenic fusion protein.

Figure 11.29
Mechanisms by which MDM2 inhibits p53: (a) MDM2 catalyzes the addition of one or two ubiquitin residues to p53 resulting in its inactivation by respectively promoting its export from the nucleus or its degradation. (b) MDM2 interacts with the L26 ribosomal protein and prevents it stimulating the translation of the p53 mRNA.

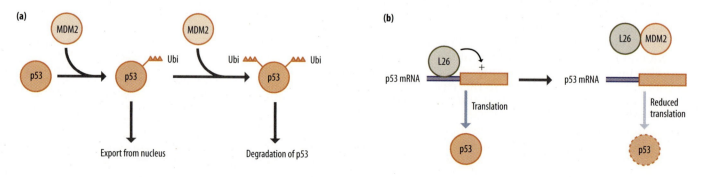

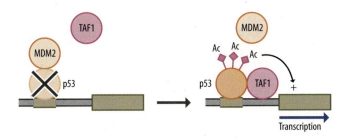

Figure 11.30
Acetylation of p53 on multiple sites at the C-terminus of the protein increases its ability to stimulate its transcription by both decreasing its ability to bind to MDM2 (thereby stabilizing p53) and by increasing its ability to recruit the TAF1 co-activator protein.

The critical role of p53 in regulating the expression of genes encoding growth-inhibitory proteins focuses attention on the nature of these growth-inhibitory genes. A number of potential target genes for p53 have now been identified. One of these encodes a protein (p21) that acts as an inhibitor of cyclin-dependent kinases. Cyclin-dependent kinases are enzymes that stimulate cells to enter cell division. Hence, the identification of a p53-regulated gene as an inhibitor of these enzymes immediately suggested that p53 acts by stimulating the expression of this inhibitory factor. In turn this would inhibit the cyclin-dependent kinases and therefore prevent cells replicating their DNA and undergoing cell division (**Figure 11.31**).

Similarly, the observation that p53 also stimulates the expression of the *bax* gene whose protein product stimulates programmed cell death or apoptosis, supports a role for p53 in promoting the death of cells which have become abnormal and can no longer divide normally.

It is likely therefore that p53 acts as a sensor for damage to cellular DNA, for example by irradiation, which could result in mutations occurring. When it is activated, growth arrest genes are stimulated and the cell ceases to divide so that it can repair the damage. If the damage to the cell is irreparable, however, genes inducing apoptosis are activated by p53 and the cell dies. Evidently in the absence of p53, cells with DNA damage or mutations will continue to replicate and if the mutations activate specific oncogenes then a cancer will result (**Figure 11.32**).

Interestingly, acetylation of p53 on the lysine at position 120 in the central DNA-binding domain enhances its ability to induce expression of the *bax* gene while having no effect on its ability to activate the *p21* gene. Hence, acetylation at this position has a distinct effect to that produced by acetylation of the C-terminal region of p53 (see above). Moreover, modifying p53 in this way may allow cells to induce the cell-death response to elevated p53 levels rather than the growth-arrest response (see Figure 11.32).

Hence, the p53 gene product plays a key role in regulating cell division and survival acting by regulating the expression of specific target genes involved in growth inhibition and cell death. Its inactivation by mutations or by specific oncogene products is likely to play a critical role in the majority of human cancers.

As well as these effects on growth inhibition and apoptosis, p53 can also regulate cellular differentiation by modulating the expression of specific

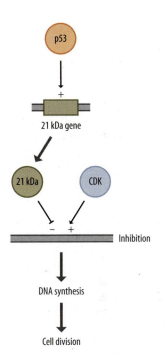

Figure 11.31
The p53 protein stimulates the transcription of the gene encoding the 21 kDa inhibitor of cyclin-dependent kinases (CDK). By inhibiting the activity of the kinases, the 21 kDa protein prevents DNA synthesis and thus cell division.

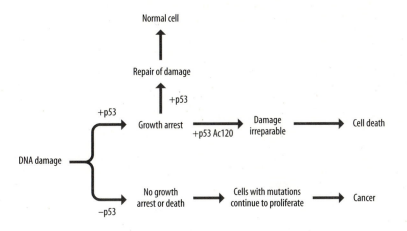

Figure 11.32
In normal cells, p53 is activated by damage to DNA. The cell is then induced to undergo growth arrest, allowing the damage to be repaired. If the damage cannot be repaired, p53 activates cell death and the damaged cell dies. In the absence of p53, the damaged cell continues to proliferate and, if it has mutations in oncogenes, will form a cancer. Acetylation of p53 on the lysine at position 120 (Ac120) promotes the cell-death response.

genes. For example, activation of p53 in undifferentiated embryonic stem cells (see Section 9.1) induces their differentiation. This occurs because p53 can repress the gene encoding the Nanog transcription factor, which as described in Chapter 9 (Section 9.1) plays a key role in maintaining the pluripotent nature of ES cells and inhibiting their differentiation. Hence, p53 can repress specific target genes as well as activating the expression of others and can modulate the expression of genes involved in regulating differentiation, as well as those involved in growth arrest or cell death (**Figure 11.33**).

Interestingly, p53 also stimulates the expression of the gene encoding MDM2, which as described above promotes p53 degradation and inhibits p53 mRNA translation. This therefore constitutes a negative-feedback loop so that, in normal cells, increases in p53 levels will induce MDM2 production which in turn will down-regulate p53 (see Figure 11.33).

Although p53 was initially believed to be a unique protein unrelated to any others, two proteins which are related to p53 have been identified and named p63 and p73 on the basis of their molecular weights. Neither p63 or p73 appears to be commonly mutated in human cancer but they do appear to be required for cellular differentiation and normal development. For example, loss or mutation of p63 leads to the human disease known as ectrodactyly, ectodermal dysplasia, and cleft lip (EEC) syndrome in which patients have limb defects and facial clefts.

The retinoblastoma protein interacts with other proteins to regulate transcription

The retinoblastoma gene (*Rb-1*) was the first anti-oncogene to be identified. This was on the basis that its inactivation results in the formation of eye tumors known as retinoblastomas. As with p53, the Rb-1 protein acts as an anti-oncogene by regulating the expression of specific target genes. Unlike p53 however, it appears to act primarily via protein–protein interactions with other transcription factors.

In particular, Rb-1 has been shown to interact with the cellular transcription factor E2F. E2F normally stimulates the transcription of several growth-promoting genes such as the cellular oncogenes c-*myc* and c-*myb* (see Sections 11.1 and 11.2) and the genes encoding DNA polymerase α and thymidine kinase. The interaction of Rb-1 with E2F does not affect the ability of E2F to bind to its target sites in the DNA but prevents it stimulating transcription. This is due to Rb-1 binding to the transcriptional activation domain of E2F and blocking its activity via a quenching-type mechanism (see Section 5.3).

In addition, however, it has been shown that Rb-1 can bind to both histone deacetylases and histone methylases and thereby promote histone deacetylation and methylation. As discussed in Chapter 3 (Section 3.3), both these modifications promote a more closed chromatin structure incompatible with transcription. Interestingly, Rb proteins have also been shown to interact with condensin proteins which play a key role in the final stages of chromatin condensation to form the mitotic chromosome (see Section 2.5). This suggests that Rb proteins may also regulate this level of chromatin structure.

Rb-1 can therefore block the action of E2F both by inhibiting its ability to activate transcription and via promoting a closed chromatin structure (**Figure 11.34**). Hence Rb-1 acts as an anti-oncogene by preventing the

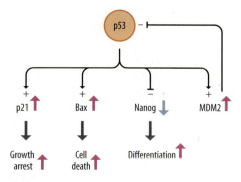

Figure 11.33
By regulating the expression of its target genes either positively or negatively, p53 can enhance cellular growth, death, and differentiation. Only one target gene is shown in each case for clarity, although p53 is known to regulate multiple genes involved in each process. In addition, p53 stimulates expression of the gene encoding MDM2, which then inhibits p53 synthesis and promotes its degradation in a negative-feedback loop.

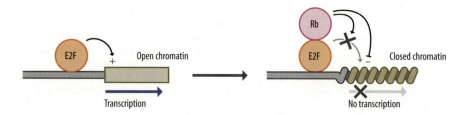

Figure 11.34
Following binding of the retinoblastoma protein (Rb) to the DNA-bound E2F transcription factor, transcription is inhibited, since Rb both blocks the ability of E2F to activate transcription and promotes the organization of a tightly packed chromatin structure (wavy line) which does not allow transcription to occur.

transcription of several growth-promoting genes, including oncogenes such as c-*myc* which themselves encode transcription factors.

During the normal cell cycle of dividing cells, the Rb-1 protein becomes phosphorylated by cyclin-dependent kinases. This prevents its interacting with E2F, allowing the E2F factor to activate the growth-promoting genes whose protein products are necessary for cell-cycle progression (**Figure 11.35a**). Such an effect can also be achieved by the deletion of the *Rb-1* gene or its inactivation by mutation which prevents the production of a functional Rb-1 protein (**Figure 11.35b**). Similarly, it is also possible for such an absence of functional Rb-1 protein to arise from transcriptional inactivation of an intact *Rb-1* gene. For example in one study, six out of 77 retinoblastomas were found to have a heavily methylated Rb-1 gene which would result in a failure of its transcription (see Section 3.2) even though the gene itself was theoretically capable of encoding a functional protein.

As with the p53 protein, the Rb-1 protein can be inactivated by protein–protein interaction with the oncogene products of **DNA tumor viruses**, such as the SV40 T antigen, the adenovirus E1A protein, or the E7 protein of human papilloma viruses. In this case, however, the association of the viral protein with Rb-1 dissociates the Rb-1–E2F complex, releasing free E2F which can then activate gene expression (**Figure 11.35c**). In contrast, the large DNA virus cytomegalovirus encodes a protein known as UL97, which phosphorylates Rb-1, thereby mimicking the manner in which it is inactivated in the normal cell cycle.

The Rb-1 protein therefore plays a critical role in cellular growth regulation, modulating the expression of specific oncogenes and acting as a target for the transforming oncoproteins of specific viruses. Interestingly, Rb-1 has also been shown to interact with the RNA polymerase I transcription factor UBF (see Section 4.1) and thereby repress the transcription of the genes encoding ribosomal RNA. In addition, it can also repress the transcription of the 5S rRNA and tRNA genes by RNA polymerase III (see Section 4.1) by interacting with TFIIIB.

Since transcription of the RNAs involved in protein synthesis is evidently necessary for cellular growth, Rb-1 can act as a global inhibitor of cellular growth by preventing the transcription of all the genes transcribed by RNA polymerases I and III as well as blocking transcription by RNA polymerase II of specific genes essential for growth (**Figure 11.36**).

Unlike the case with p53 (see above), mice lacking Rb-1 are not viable, indicating that Rb-1 function is essential for normal development. Interestingly, mice lacking Rb-1 can be rescued by also inactivating the gene encoding the Id2 factor, an inhibitory member of the helix-loop-helix family of transcription factors discussed in Chapter 5 (Section 5.1).

This indicates that Rb-1 and Id2 are likely to have antagonistic functions so that inactivating one factor minimizes the effect of inactivating the other.

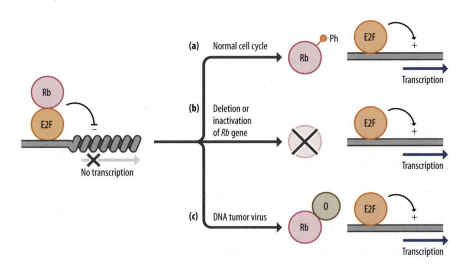

Figure 11.35
The retinoblastoma protein (Rb) binds to the E2F transcription factor and represses transcription. This inhibition can be relieved allowing E2F to activate transcription when: (a) the Rb protein is phosphorylated (orange dot) which occurs during the normal cell cycle and prevents its interacting with E2F; (b) the gene encoding Rb is deleted or inactivated by mutation; (c) the Rb protein binds to the product of a DNA tumor virus oncogene (O) which releases it from E2F.

Figure 11.36
By binding to the polymerase I transcription factor UBF, the polymerase II transcription factor E2F, and the polymerase III transcription factor TFIIIB, Rb can repress transcription of the ribosomal RNA genes by RNA polymerase I, the transcription of E2F-dependent genes by RNA polymerase II, and the transcription of the tRNA and 5S rRNA genes by RNA polymerase III.

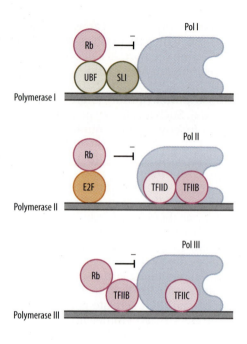

Indeed, it has been shown that Rb-1 and Id2 interact with one another, resulting in the inhibition of Rb-1 activity. In tumor cells over-expressing the N-*myc* oncogene transcription factor (see Section 11.1), the *Id2* gene is transcriptionally activated by N-*myc*. The excess Id2 binds to Rb-1 and inactivates it, allowing the tumor to grow. This therefore represents another example of oncogenes and anti-oncogenes interacting antagonistically to regulate cellular growth with the over-expression of an oncogene protein, resulting in the inactivation of an anti-oncogene protein.

Together with p53 therefore Rb-1 plays a key role in restraining cellular growth. Indeed many parallels exist between these two proteins. For example, as well as being regulated by tumor virus proteins, Rb-1 like p53 is inhibited by interaction with the MDM2 protein. Moreover, acetylation of Rb-1 enhances its interaction with MDM2 indicating that both p53 and Rb-1 can be regulated via phosphorylation and acetylation.

In addition, there is evidence that the p53 and Rb-1 pathways interact with one another. The phosphorylation of Rb-1, which regulates its activity is carried out by the cyclin-dependent kinases. Hence, the action of p53 which stimulates the p21 inhibitor of these kinases will reduce their activity and hence maintain Rb-1 in its non-phosphorylated growth-inhibitory form (**Figure 11.37**).

Interestingly, the regulation of Rb activity during the cell cycle also parallels the regulation of the Whi5 factor in the yeast cell cycle by cyclin-dependent kinases, which was discussed in Chapter 10 (Section 10.3). In turn, Whi5 inhibits the transcriptional activator SBF in yeast, paralleling the inhibition of E2F by Rb-1 in mammalian cells (**Figure 11.38**).

Other anti-oncogene proteins also regulate transcription

In addition to p53, Rb-1, and the Wilms' tumor gene product, it is clear that other anti-oncogenes are also transcription factors. For example, the products of the *BRCA-1* and *BRCA-2* genes, which are mutated in many cases of breast cancer, appear to act primarily by regulating the rate at which damaged DNA is repaired. However, it appears that they can also regulate transcription. BRCA-1 can, for example, interact with both p53 and Rb-1 to regulate their activity. Moreover, BRCA-1 has been shown to interact with the C-terminal domain of RNA polymerase II and modulate its phosphorylation, which is critical for transcriptional elongation (for discussion of the C-terminal domain of RNA polymerase II and its role in transcriptional elongation see Section 4.2).

The process of transcriptional elongation is also targeted by the von Hippel Lindau (VHL) anti-oncogene protein. VHL is part of a protein complex which adds the small protein ubiquitin to the phosphorylated form of RNA polymerase II. As ubiquitination targets proteins for degradation (see also Section 8.3), this results in degradation of the polymerase. Moreover, since the phosphorylated form of RNA polymerase II is involved in transcriptional elongation (see Section 4.2), the VHL protein specifically targets this process.

As discussed in Chapter 8 (Section 8.3), VHL is also involved in the degradation of the HIF-1α protein, which plays a key role in the induction of gene expression in response to hypoxia. Hence, VHL regulates diverse cellular processes.

Both these processes, however, are likely to play a role in the cancers which develop when VHL is inactivated. The mutant forms of VHL which are found in cancers do not inhibit transcriptional elongation, indicating that

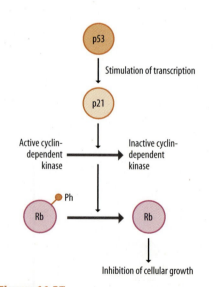

Figure 11.37
Activation of the gene encoding the p21 inhibitor of cyclin-dependent kinases by p53 results in the inhibition of these kinases which in turn maintains Rb in its non-phosphorylated growth-inhibitory form.

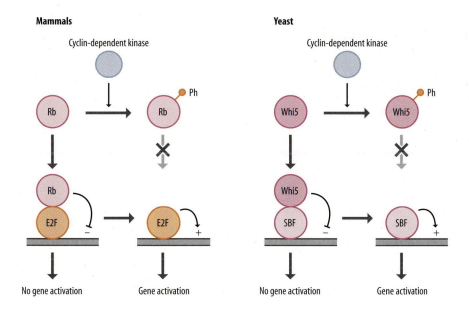

Mammals

Cyclin-dependent kinase

Yeast

Cyclin-dependent kinase

No gene activation   Gene activation   No gene activation   Gene activation

Figure 11.38
Parallels in cell-cycle gene regulation between the yeast and mammalian systems. Cyclin-dependent kinases phosphorylate the Rb protein in mammals and the Whi5 protein in yeast. In turn, this blocks their ability to interact with specific activating transcription factors (E2F in mammals and SBF in yeast) and prevent them activating gene expression.

this effect is important in cancer development. This is likely to be due to the fact that several cellular oncogenes, such as c-*myc* and c-*fos*, are regulated at the level of transcription elongation (see Section 5.4). Hence, in the absence of VHL these factors will be over-produced since the block to productive elongation of their RNA transcripts will be abolished (**Figure 11.39**).

Similarly, the enhanced HIF-1 activity in the absence of VHL will assist tumor growth since it will induce expression of genes which are normally only induced under hypoxic conditions. These include for example, genes involved in blood-vessel formation, so enhancing the blood supply to the tumor and stimulating its growth (see Figure 11.39).

The vital role of p53, Rb-1, and the other anti-oncogene products indicates that, although discovered later than the oncogenes, the anti-oncogenes are likely to play as critical a role in regulating cellular growth in general and the pattern of gene expression in particular. Hence, the precise rate of cellular growth is likely to be controlled by the balance between interacting oncogene and anti-oncogene products, with cancer resulting from a change in this balance by activation/over-expression of oncogenes or inactivation/reduced expression of anti-oncogenes.

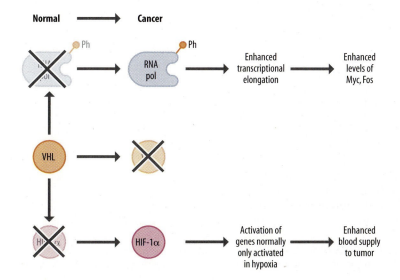

Normal → **Cancer**

Enhanced transcriptional elongation → Enhanced levels of Myc, Fos

VHL

HIF-1α

Activation of genes normally only activated in hypoxia → Enhanced blood supply to tumor

Figure 11.39
Inactivation of the *VHL* gene in specific cancers abolishes the enhanced degradation of proteins which is normally produced by functional VHL protein. In the case of the phosphorylated form of RNA polymerase, this will result in enhanced transcriptional elongation at genes such as c-*myc* and c-*fos*. In the case of HIF-1α it will result in the induction of genes whose protein products are normally only induced in response to hypoxic conditions.

11.4 REGULATION OF GENE EXPRESSION: THE RELATIONSHIP OF CANCER AND NORMAL CELLULAR FUNCTION

As discussed in the preceding sections of this chapter, cancer frequently results from alterations in gene regulation. Indeed, the study of the processes whereby increased expression or mutation of certain cellular genes or the deletion or inactivation of other genes can cause cancer has greatly increased our knowledge of the process of transformation, whereby normal cells become cancerous. Similarly, the recognition that the products of these cellular genes play a critical role in the growth regulation of normal cells has allowed insights obtained from studies of their activities in tumor cells to be applied to the study of normal cellular growth control. The relationship between cancer and normal cellular growth and their regulation by the products of oncogenes and anti-oncogenes are discussed in this section.

Oncogenes and anti-oncogenes interact to regulate the expression of genes encoding proteins which control cellular growth

The relationship between cancer and normal cellular growth is well illustrated in the case of the oncogenes and anti-oncogenes that encode cellular transcription factors. For example, the study of the Rb-1 and p53 anti-oncogene proteins has greatly enhanced our knowledge of the processes regulating the transcription of genes whose protein products enhance or inhibit cellular growth. Similarly, the isolation of the *fos* and *jun* genes in tumorigenic retroviruses has aided the study of the effects of growth factors on normal cells, while the recognition that the v-*erb*A gene product is a truncated form of the thyroid hormone receptor has allowed the elucidation of its role in transformation via the inhibition of erythroid differentiation. These two cases also illustrate the two mechanisms by which cellular genes can become oncogenic: namely mutation or over-expression. In the case of the v-*erb*A gene, mutations have rendered the protein different from the corresponding c-*erb*A gene from which it was derived. These alter the properties of the protein so it cannot bind thyroid hormone and it behaves as a dominant repressor of transcription (see Section 11.2).

Interestingly, in a situation where an oncogene product and an anti-oncogene product interact, cancer can result from mutations which enhance the activity of the oncogenic protein or decrease the activity of the anti-oncogene protein. This is seen in the case of the adenomatous polyposis coli (APC) anti-oncogene protein (see Table 11.3), which is not a transcription factor, and the β-catenin oncogene protein, which plays a key role in cell–cell adhesion but also acts as a transcription factor. These two factors normally interact and this interaction results in the export of β-catenin to the cytoplasm and its rapid degradation (**Figure 11.40a**). This prevents it from moving to the nucleus and interacting with the LEF-1 transcription factor (see Section 4.4), and stimulating the ability of LEF-1 to activate transcription. This role of APC in promoting the intracellular transport of a protein (β-catenin) is in addition to its role in promoting the intracellular transport of specific mRNAs, which was discussed in Chapter 7 (Section 7.3).

In normal cells, the LEF-1–β-catenin interaction is regulated in response to specific secreted signaling proteins known as Wnt proteins (see Section 10.1 for discussion of the role of these proteins in muscle cell differentiation). In response to **Wnt signaling**, β-catenin is stabilized and can stimulate the ability of LEF-1 to activate transcription, resulting in cellular proliferation (**Figure 11.40b**). However, such stabilization can also be achieved in the absence of the Wnt signal, either by the inactivation of APC by mutation or by mutation of β-catenin itself, resulting in its enhanced stability (**Figure 11.40c**).

Hence, normal cellular growth is controlled by interaction of the anti-oncogene product APC and the oncogene product β-catenin and can be

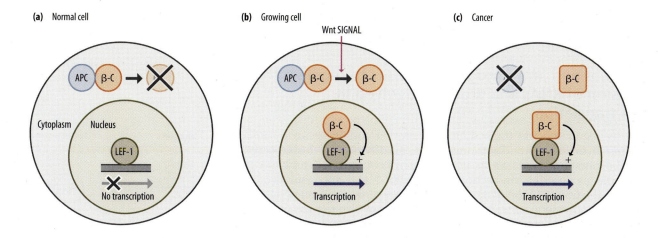

stimulated by inactivating mutations in the anti-oncogenic partner or activating mutations in the oncogenic partner.

Interestingly, β-catenin degradation can also be stimulated by the VHL anti-oncogene protein discussed in Section 11.3, indicating that multiple anti-oncogene proteins can regulate β-catenin. Similarly, although the Wnt secreted proteins which activate the β-catenin pathway play a key role in normal development (see Section 10.1) they can also act as oncogenic proteins when they are constitutively activated and inappropriately stimulate β-catenin-dependent gene expression. Hence, β-catenin is at the centre of a network of oncogenic and anti-oncogenic proteins (**Figure 11.41**).

In addition to illustrating another example of oncogene activation by mutation, β-catenin also illustrates a unique dual role, acting both as a factor critical for cell-to-cell adhesion and as an oncogenic transcription factor capable of regulating gene activity. Indeed, it is likely that this dual role allows it to transmit signals from cell-adhesion components to the nucleus resulting in changes in gene expression in response to extracellular events. Among the genes activated by β-catenin are those encoding the Myc and Jun oncogenic proteins (see Section 11.2), providing a further link between the various pathways which are involved in cancer (see Figure 11.41).

In contrast to the situation with oncogenes encoding proteins which pre-exist in an inactive form and can be activated by mutation, oncogenes whose products are made only in response to a particular growth signal and whose activity then mediates cellular growth can cause cancer simply by the normal product being made at an inappropriate time. Thus, in the case of Fos or Jun which are synthesized in normal cells in response to treatment with growth-promoting phorbol esters or growth factors, their continuous synthesis is sufficient to transform the cell (see Section 11.2). Interestingly, it has been shown that the c-Jun protein antagonizes the pro-apoptotic activity of p53 (see Section 11.3), providing a further example of proto-oncogene/anti-oncogene antagonism.

Such high-level expression of a normal oncogene product causing cancer also occurs in a number of cases without any evidence of retroviral involvement. These examples of alterations in cellular regulatory processes producing increased expression of particular genes provide another aspect to the connection between cancer and gene regulation. For example, information on the origin of the translocations of the c-*myc* gene in Burkitt's lymphoma is obviously important for the study of cancer etiology. Similarly, the fact that such translocations increase expression in some cases by

Figure 11.40
Interaction of the oncogene product β-catenin (β-C) and the anti-oncogene product APC. In normal cells APC and β-catenin are associated in the cytoplasm and this results in the rapid degradation of β-catenin (a). In response to Wnt signaling molecules, β-catenin is stabilized and it can move to the nucleus where it interacts with the LEF-1 transcription factor (green) and stimulates its ability to activate its target genes (b). This effect can also be achieved in cancer by the inactivation of APC by mutation or the activating mutation of β-catenin which converts it to a form (square) which is not susceptible to degradation. In either case the β-catenin moves to the nucleus and interacts with LEF-1, stimulating its activity (c).

Figure 11.41
Links between the β-catenin oncogenic protein and other oncogenic (pink) and anti-oncogenic (blue) proteins. The oncogenic Wnt signaling proteins activate β-catenin, whereas APC and VHL anti-oncogene proteins target it for degradation. Activated β-catenin stimulates the expression of the oncogenes encoding the Myc and Jun proteins.

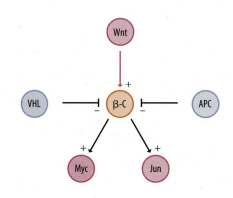

removing elements that normally inhibit c-*myc* expression allows the characterization of such negative elements and their role in regulating c-*myc* expression in normal cells (see Section 11.1).

Oncogenes and anti-oncogenes interact to regulate the expression of RNAs and proteins involved in mRNA translation

As well as regulating the transcription of protein-coding genes by RNA polymerase II, oncogenes and anti-oncogenes can also regulate the transcription of ribosomal RNA and transfer RNA genes by RNA polymerases I and III. For example, as described in Section 11.3, the Rb-1 anti-oncogene protein can repress the transcription of the ribosomal RNA genes by interacting with the UBF RNA polymerase I transcription factor. It can also repress the transcription of the tRNA and 5S ribosomal RNA genes by interacting with the TFIIIB RNA polymerase III transcription factor (see Figure 11.36) (see Section 4.1 for discussion of transcription by RNA polymerases I and III).

Interestingly, the anti-oncogene p53 also represses transcription by RNA polymerases I and III whereas in contrast the c-Myc oncogene protein stimulates transcription by both these polymerases. Hence, anti-oncogenic proteins can restrain cellular growth by limiting the production of RNAs required for protein synthesis whereas oncogenic proteins have the reverse effect and therefore stimulate cellular growth (**Figure 11.42**).

Although p53, Rb-1, and c-Myc produce their different effects on RNA polymerase III transcription by interacting with TFIIIB, they do not all interact with the same components of the multi-protein TFIIIB complex. As described in Chapter 4 (Section 4.1), TFIIIB is a multi-protein complex composed of the TBP factor and two other proteins, Bdp1 and Brf1. Rb-1 and c-Myc both interact with the Brf1 component of TFIIIB to respectively reduce or enhance its activity, whereas p53 interacts with TBP itself (see Figure 11.42).

As expected from this model, over-expression of Brf1 in cells results in their transformation into tumor cells indicating that Brf1 can act as an oncogene when over-expressed. This effect of Brf1 can be mimicked by over-expression of the initiator tRNA for methionine whose role is to insert the initial methionine amino acid with which protein synthesis begins (see Section 6.6). This demonstrates that, as well as proteins, a tRNA can act as an oncogene when over-expressed. Moreover, it indicates that the effect of Brf1 on cellular transformation is likely to be dependent upon its ability to enhance RNA polymerase III-mediated transcription of the gene encoding the initiator tRNA for methionine (**Figure 11.43**).

These findings suggest that the initiation of mRNA translation is a key regulatory point in controlling cellular growth and tumor formation. In agreement with this idea, over-expression of the translation initiation factor eIF4E (see Section 6.6) can transform normal cells into cancer cells.

As described in Chapter 7 (Section 7.5) the activity of eIF4E is regulated by the eIF4E-binding protein (eIF4Ebp) which binds to eIF4E and prevents it binding the translation initiation factors, eIF4A and eIF4G. Inhibition of eIF4E by eIF4Ebp has a particularly severe effect on the translation of mRNAs with secondary structure in their 5′ region. Thus, eIF4A is required to unfold this secondary structure and allow the ribosome to move along the mRNA from the 5′ cap to the initiator methionine codon (see Figure 7.46). RNAs of this type include a number encoding proteins involved in cellular growth, including the mRNAs encoding c-Myc and cyclin D1. For this reason, eIF4E or eIF4A, which promote translation of these mRNAs, can act as oncogenes when over-expressed whereas eIF4Ebp which inhibits this effect acts as an anti-oncogene when over-expressed.

It has been shown that eIF4Ebp is phosphorylated in growing cells and in a number of human tumors, with such phosphorylation decreasing its ability to bind to eIF4E and thereby releasing it from inhibition by eIF4Ebp. Such phosphorylation is produced by the PI3-kinase/Akt/TOR kinase cascade, which was discussed in Chapter 8 (Section 8.5) (**Figure 11.44**).

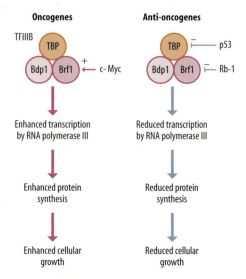

Figure 11.42
Regulation of the RNA polymerase III transcription factor TFIIIB by oncogene and anti-oncogene proteins. The anti-oncogenic proteins p53 and Rb-1 interact respectively with the TBP and Brf1 components of TFIIIB to inhibit its activity and therefore reduce RNA polymerase III transcription. In contrast the oncogenic protein c-Myc interacts with the Brf1 component of TFIIIB to enhance its activity and therefore increase RNA polymerase III transcription.

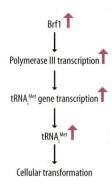

Figure 11.43
Enhanced Brf1 activity enhances RNA polymerase III-mediated transcription of the gene encoding the initiator tRNA for methionine (tRNA$_i$^{Met}). In turn, enhanced tRNA$_i$^{Met} levels lead to cellular transformation. Hence, both Brf1 and tRNA$_i$^{Met} can act as oncogenes when over-expressed.

As well as phosphorylating eIF4Ebp, the PI3-kinase/Akt/TOR pathway also phosphorylates a number of other factors so as to enhance mRNA translation. As described in Chapter 8 (Section 8.5), these include proteins involved in regulating transcription by RNA polymerases I and III, as well as the S6 protein kinase which in turn phosphorylates other factors involved in mRNA translation.

One of the factors phosphorylated by the PI3-kinase/Akt/TOR pathway is the Brf1 component of TFIIIB (see above). This promotes its association with TBP, so enhancing the activity of TFIIIB and RNA polymerase III-dependent transcription. The PI3-kinase/Akt/TOR pathway can therefore stimulate translation both by inhibiting the activity of the anti-oncogenic eIF4Ebp and by enhancing the activity of the Brf1 protein, which as described above has oncogenic activity when over-expressed (**Figure 11.45**).

In view of these effects of the PI3-kinase/Akt/TOR pathway, it is not surprising that the different components of this kinase cascade can themselves act as oncogenes when over-expressed (see Figure 11.45). Conversely, the anti-oncogenic PTEN phosphatase protein, which is mutated in a wide variety of human cancers, produces its anti-oncogenic effect by antagonizing PI3-kinase (see Figure 11.45).

Hence, the regulation of translation itself and of transcription by the RNA polymerases which produce the RNAs involved in it plays a key role in regulating cellular growth and can produce cancer when different regulatory factors are over-expressed or mutated.

Oncogenes and anti-oncogenes interact to regulate the expression of microRNAs

As discussed throughout this book, microRNAs (miRNAs) play a key role in many different cellular processes by inhibiting the expression of protein-coding genes. It is not surprising therefore that increased or decreased expression of specific miRNAs has been observed in human cancers and plays a key role in their development.

For example, B-cell chronic lymphocytic leukemia (CLL), the most frequent adult leukemia in the Western world, is caused by the deletion of a region of chromosome 13 which contains the genes encoding the miRNAs miR-15a and miR-16–1 (**Figure 11.46a**). Cancer can therefore result from the loss of genes encoding specific miRNAs which therefore fulfil the definition of anti-oncogenes.

Interestingly, in human epithelial ovarian cancer expression of a number of different miRNAs is down-regulated. Approximately, 15% of these effects are due to deletion of the gene encoding the miRNA while 36% involve the miRNA gene being packaged into a closed chromatin structure due to enhanced DNA methylation and/or decreased histone acetylation (see Sections 3.2 and 3.3 for discussion of the effect of these modifications on chromatin structure). Hence, down-regulation of miRNAs in human tumors can arise from gene deletion (see Figure 11.46a) or decreased transcription due to altered chromatin structure (**Figure 11.46b**).

As well as changes in chromatin structure, altered miRNA expression can also occur due to altered transcription of miRNA-encoding genes produced by oncogenic or anti-oncogenic transcription factors (**Figure 11.46c**). For example, it has been shown that the genes encoding the let-7 and miR-34a miRNAs are transcriptionally repressed by the c-Myc oncogene protein while p53 activates the *miR-34a* gene.

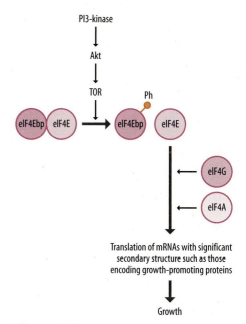

Figure 11.44
The PI3-kinase/Akt/TOR pathway phosphorylates eIF4Ebp. This results in its dissociation from eIF4E, allowing eIF4E to bind eIF4A/eIF4G and stimulate the translation of mRNAs, particularly those with secondary structure in their 5′ regions, a number of which encode growth-promoting proteins.

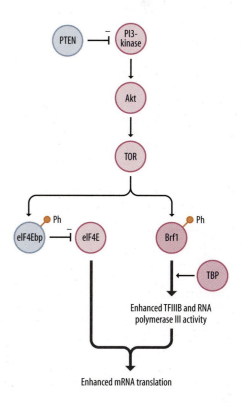

Figure 11.45
The PI3-kinase/Akt/TOR pathway phosphorylates the oncogenic Brf1 protein so enhancing its association with TBP and producing enhanced RNA polymerase III activity. It also phosphorylates the anti-oncogenic eIF4Ebp, so releasing eIF4E. Both these effects enhance mRNA translation. Due to this, components of the PI3-kinase/Akt/TOR can act as oncogenes when over-expressed or mutated, whereas the PTEN phosphatase acts as an anti-oncogene by inhibiting this pathway. Potential oncogenes are shown in pink and anti-oncogenes in blue.

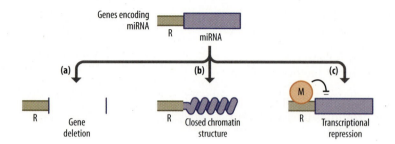

Figure 11.46
Mechanisms mediating reduced expression of the gene encoding an miRNA in human tumors include gene deletion (a), production of a tightly packed chromatin structure (b), or transcriptional repression, for example, by the Myc protein (M) binding to the regulatory region (R) controlling the transcription of the miRNA gene (c).

Multiple mechanisms therefore operate in human tumors to down-regulate the transcription of genes encoding miRNAs. In turn, such a decrease in the level of specific miRNAs will result in an increase in the expression of the protein-coding genes which they normally regulate. For example, genes which are normally down-regulated by the let-7 miRNA include the c-Myc and c-Ras oncogenes and the Hmga2 chromatin-remodeling protein (**Figure 11.47**). Similarly, deletion of the gene encoding the miR-101 miRNA produces over-expression of the oncogenic polycomb protein EZH2 (see Section 11.2).

Although down-regulation of specific miRNAs therefore plays a key role in different human tumors, other miRNAs are upregulated in human cancers. For example, the gene locus encoding the miR-17–92 cluster of miRNAs is amplified at the DNA level in several types of lymphomas and solid tumors, resulting in the over-expression of miR-17–92 miRNAs in these tumors. Interestingly, the transcription of the gene encoding miR-17–92 is also activated by both c-Myc and E2F. Hence, the over-expression of genes encoding miRNAs can be achieved by either DNA amplification or transcriptional activation (**Figure 11.48**). This parallels the multiple mechanisms producing reduced expression of other miRNA genes in human tumors. Moreover, the c-Myc oncogene protein can transcriptionally activate or repress the genes encoding different miRNAs (compare Figures 11.46 and 11.48).

Hence, the products of cellular oncogenes and anti-oncogenes can regulate the expression of miRNA genes as well as regulating the expression of protein-coding genes and the genes encoding RNAs involved in protein synthesis. The study of these effects has contributed greatly to our knowledge both of cancer and of cellular growth-regulatory processes and is likely to continue to do so in the future.

CONCLUSIONS

The malregulation of gene expression plays a central role in human cancer, paralleling the critical role that gene-control processes play in the regulation of normal cellular growth. At one level, such malregulation involves the increased expression of a cellular oncogene or the decreased expression of

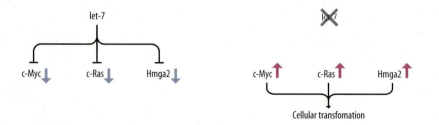

Figure 11.47
The let-7 miRNA normally inhibits expression of the c-Myc, c-Ras, and Hmga2 proteins, so that their expression is increased when let-7 expression is reduced or abolished.

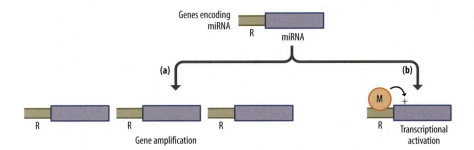

Figure 11.48
Mechanisms mediating enhanced expression of an miRNA in human tumors include gene amplification (a) or transcriptional activation, for example by the Myc protein (M) binding to the miRNA gene-regulatory region (R) (b). Compare with Figure 11.46.

an anti-oncogene encoding growth-regulatory proteins. This parallels the manner in which cancer can also be caused by mutations that produce altered forms of oncogenic or anti-oncogenic proteins.

At another level, however, some of the oncogenic or anti-oncogenic proteins which exhibit altered expression or mutation are themselves gene regulatory proteins able to control the expression of other target genes whose protein products are involved in cellular growth. Hence, their altered expression or mutation will alter the expression of a number of other genes. This can occur at the transcriptional level where the oncogene or anti-oncogene encodes a transcription factor able to regulate transcription of genes by one or more RNA polymerases. Alternatively it can occur post-transcriptionally either via oncogenic or anti-oncogenic factors encoding proteins involved in mRNA translation or via miRNAs which target other RNAs for degradation or inhibit their translation.

Overall therefore the study of gene regulation in cancer offers insights into the nature of the disease as well as the mechanisms by which gene-control processes regulate the growth of normal cells. However, alterations in gene-control processes occur in a wide variety of other human diseases apart from cancer. The involvement of malregulation of gene expression in these other human diseases is discussed in the next chapter together with the manner in which the artificial manipulation of gene expression may be of benefit in specific human diseases, including cancer.

KEY CONCEPTS

- Widespread changes in gene regulation are seen in cancer resulting from the enhanced expression/activity of oncogenes or the reduced expression/activity of anti-oncogenes.

- A number of oncogenes and anti-oncogenes encode transcription factors which control the transcription of other genes and thereby play a key role in regulating normal cellular growth processes and in cancer.

- As well as regulating the transcription of protein-coding genes by RNA polymerase II, oncogenes and anti-oncogenes can also regulate transcription by RNA polymerases I and III.

- Oncogenes and anti-oncogenes can also regulate gene expression post-transcriptionally by encoding proteins or miRNAs which can affect post-transcriptional processes.

- Study of these different factors can enhance our understanding of normal cellular gene regulation as well as of the abnormalities observed in cancer.

FURTHER READING

11.1 Gene regulation and cancer

Albertson DG (2006) Gene amplification in cancer. *Trends Genet.* 22, 447–455.

Karnoub AE & Weinberg RA (2008) Ras oncogenes: split personalities. *Nat. Rev. Mol. Cell Biol.* 9, 517–531.

Meyer N & Penn LZ (2008) Reflecting on 25 years with MYC. *Nat. Rev. Cancer* 8, 976–990.

Mitelman F, Johansson B & Mertens F (2007) The impact of translocations and gene fusions on cancer causation. *Nat. Rev. Cancer* 7, 233–245.

Pawson T & Warner N (2007) Oncogenic re-wiring of cellular signaling pathways. *Oncogene* 26, 1268–1275.

Vogelstein B & Kinzler KW (2006) Cancer genes and the pathways they control. *Nature Milestones Cancer Supplement* S33–S42.

Weinberg RA (2007) The Biology of Cancer. Garland Science Taylor & Francis Group.

11.2 Transcription factors as oncogenes

Cole MD & Cowling VH (2008) Transcription-independent functions of MYC: regulation of translation and DNA replication. *Nat. Rev. Mol. Cell Biol.* 9, 810–815.

Eilers M & Eisenman RN (2008) Myc's broad reach. *Genes Dev.* 22, 2755–2766.

Esteller M (2007) Cancer epigenomics: DNA methylomes and histone-modification maps. *Nat. Rev. Genet* 8, 286–298.

Jones PA & Baylin SB (2007) The epigenomics of cancer. *Cell* 128, 683–692.

Krivtsov AV & Armstrong SA (2007) MLL translocations, histone modifications and leukaemia stem-cell development. *Nat. Rev. Cancer* 7, 823–833.

Ozanne BW, Spence HJ, McGarry LC & Hennigan RF (2007) Transcription factors control invasion: AP-1 the first among equals. *Oncogene* 26, 1–10.

Rowley JD & Blumenthal T (2008) The cart before the horse. *Science* 321, 1302–1304.

Shaulian E & Karin M (2002) AP-1 as a regulator of cell life and death. *Nat. Cell Biol.* 4, E131–E136.

Sparmann A & van Lohuizen M (2006) Polycomb silencers control cell fate, development and cancer. *Nat. Rev. Cancer* 6, 846–856.

11.3 Anti-oncogenes

Burkhart DL & Sage J (2008) Cellular mechanisms of tumor suppression by the retinoblastoma gene. *Nat. Rev. Cancer* 8, 671–682.

DeCaprio JA (2009) How the Rb tumor suppressor structure and function was revealed by the study of Adenovirus and SV40. *Virology* 384, 274–284.

Kaelin Jr WG (2008) The von Hippel-Lindau tumor suppressor protein: O2 sensing and cancer. *Nat. Rev. Cancer* 8, 865–873.

Kruse JP & Gu W (2008) SnapShot: p53 posttranslational modifications. *Cell* 133, 930.

Polager S & Ginsberg D (2008) E2F - at the crossroads of life and death. *Trends Cell Biol.* 18, 528–535.

Riley T, Sontag E, Chen P & Levine A (2008) Transcriptional control of human p53-regulated genes. *Nat. Rev. Mol. Cell Biol.* 9, 402–412.

Salmena L & Pandolfi PP (2007) Changing venues for tumor suppression: balancing destruction and localization by monoubiquitylation. *Nat. Rev. Cancer* 7, 409–413.

Stiewe T (2007) The p53 family in differentiation and tumorigenesis. *Nat. Rev. Cancer* 7, 165–167.

Strano S, Dell'Orso S, Di Agostino S et al. (2007) Mutant p53: an oncogenic transcription factor. *Oncogene* 26, 2212–2219.

Vousden KH & Lane DP (2007) p53 in health and disease. *Nat. Rev. Mol. Cell Biol.* 8, 275–283.

11.4 Regulation of gene expression: the relationship of cancer and normal cellular function

Behrens J (2008) One hit, two outcomes for VHL-mediated tumorigenesis. *Nat. Cell Biol.* 10, 1127–1128.

Berns A (2008) A tRNA with oncogenic capacity. *Cell* 133, 29–30.

He L, He X, Lowe SW & Hannon GJ (2007) microRNAs join the p53 network--another piece in the tumor-suppression puzzle. *Nat. Rev. Cancer* 7, 819–822.

Johnson DL & Johnson SA (2008) RNA metabolism and oncogenesis. *Science* 320, 461–462.

Kent OA & Mendell JT (2006) A small piece in the cancer puzzle: microRNAs as tumor suppressors and oncogenes. *Oncogene* 25, 6188–6196.

Klaus A & Birchmeier W (2008) Wnt signalling and its impact on development and cancer. *Nat. Rev. Cancer* 8, 387–398.

Mamane Y, Petroulakis E, Lebacquer O & Sonenberg N (2006) mTOR, translation initiation and cancer. *Oncogene* 25, 6416–6422.

Marshall L & White RJ (2008) Non-coding RNA production by RNA polymerase III is implicated in cancer. *Nat. Rev. Cancer* 8, 911–914.

Mendell JT (2008) miRiad roles for the miR-17-92 cluster in development and disease. *Cell* 133, 217–222.

Mosimann C, Hausmann G & Basler K (2009) Beta-catenin hits chromatin: regulation of Wnt target gene activation. *Nat. Rev. Mol. Cell Biol.* 10, 276–286.

Nelson WJ & Nusse R (2004) Convergence of Wnt, beta-catenin, and cadherin pathways. *Science* 303, 1483–1487.

Ventura A & Jacks T (2009) MicroRNAs and cancer: short RNAs go a long way. *Cell* 136, 586–591.

White RJ (2008) RNA polymerases I and III, non-coding RNAs and cancer. *Trends Genet.* 24, 622–629.

Gene Regulation and Human Disease

12

INTRODUCTION

The central role of gene-control processes in normal cellular function which has been discussed throughout this book makes it inevitable that abnormalities in such processes will result in disease. In addition to the abnormalities which occur in cancer (see Chapter 11), it has been demonstrated that many human genetic diseases involve the inheritance of mutated genes encoding proteins which regulate gene expression. A number of different diseases have been shown to involve mutations in proteins involved in each of the three fundamental processes that regulate gene expression (see Chapters 2–7), namely the processes of transcription itself, the regulation of chromatin structure which is necessary for transcription to occur, and post-transcriptional processes. These cases will be discussed in Sections 12.1–12.3 respectively while Section 12.4 will describe alterations in gene-regulatory processes that can occur in human infectious disease. Finally, it is clear that the insights obtained by studies on gene regulation in cancer and other human diseases may lead to the development of effective therapies for manipulating gene expression in these diseases and this is discussed in Section 12.5.

12.1 TRANSCRIPTION AND HUMAN DISEASE

A number of different elements involved in the process of transcription and its control can be mutated in human disease and these will be discussed in turn.

DNA-binding transcription factors

In discussing human diseases caused by defects in specific transcription factors, it should be noted that only gene mutations which are compatible with life will manifest as human diseases, however severe these may be. Mutations in factors which are incompatible with survival, at least to birth, will evidently not be detected and it is likely that many mutations affecting gene regulation fall into this category. Nonetheless, a number of congenital diseases which are detectable at birth or shortly after involve mutations in the genes encoding specific transcription factors (**Figure 12.1a**). For example, mutations in individual members of the PAX transcription factor family (see Section 5.1), such as Pax3 and Pax 6, have been shown to be involved in a number of congenital eye disorders. Similarly, mutations in the gene encoding the Pit-1 member of the POU family of transcription factors (see Section 5.1) result in a failure of pituitary gland development and consequent dwarfism in both mice and humans.

Mutations in Pit-1 and Pax6 are both dominant, with one single copy of the mutant gene being sufficient to produce the disease even in the presence of a functional copy. However, this dominance arises for different

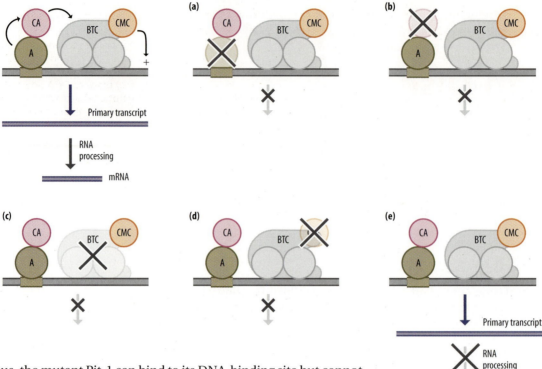

reasons. Thus, the mutant Pit-1 can bind to its DNA-binding site but cannot activate gene expression. It therefore not only fails to stimulate transcription of its target genes but can also act as a **dominant negative** factor inhibiting gene activation by preventing the wild-type protein from binding to DNA (**Figure 12.2a**) (see Section 5.3 for discussion of this mechanism of transcriptional repression).

In contrast, the dominant nature of the Pax6 mutation does not reflect any dominant negative action of the mutant protein because such mutations often involve complete deletion of the gene. Rather, it is due to **haploid insufficiency** in which the amount of protein produced by a single functional copy of the gene is not enough to allow it to activate its target genes effectively (**Figure 12.2b**).

Interestingly, although many diseases involving mutant transcription factors show very early onset of symptoms, mutations in the MEF2A transcription factor (see Section 10.1) have been shown to produce coronary artery disease in middle-aged patients. This indicates that mutations in transcription factors can produce disease later in life as well as in early development.

As well as such defects in the development or functioning of specific organs, mutations in the genes encoding the members of the nuclear receptor family (see Section 5.1) can result in a failure to respond to the specific hormone which normally binds to the receptor and thereby regulates gene expression. For example, mutations in the gene encoding the glucocorticoid receptor result in a syndrome of steroid resistance in which the patients do not respond to glucocorticoid.

Interestingly, mutations in the peroxisome proliferator-activated receptor γ (PPARγ), which is also a member of the nuclear receptor family, have been found in a few human individuals with insulin resistance, leading to type 2 diabetes. Although such cases are rare, they suggest that PPARγ plays a key role in insulin responses and that it may be a valuable therapeutic target to enhance such responses in other cases of diabetes and in obesity (see Section 12.5).

DNA-binding sites for specific transcription factors

As well as involving mutations in DNA-binding transcription factors, specific human diseases can also involve mutations in the DNA sequences to which such factors bind (**Figure 12.3**). An example of this was discussed in

Figure 12.1
Mutations can affect a number of components of the gene-expression processes, including (a) transcriptional activators (A), (b) co-activators (CA), (c) the basal transcriptional complex (BTC), (d) chromatin-modeling complexes (CMC), or (e) factors involved in RNA processing.

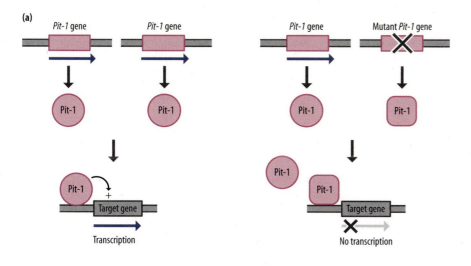

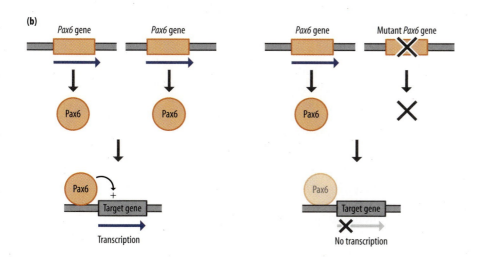

Figure 12.2
Mutations in Pit-1 (a) and Pax6 (b) are dominant and cause disease, even in the presence of a gene copy encoding functional protein. In the case of Pit-1 (a), this is because the mutant protein (square) binds to DNA and prevents transcriptional activation by the functional protein. In the case of Pax6 (b) this is because one functional gene cannot make enough protein to produce transcriptional activation.

Chapter 9 (Section 9.3) in which mutations in an enhancer element that regulates the expression of the gene encoding Sonic Hedgehog produces the human disease polydactyly (multiple digits).

Although in this case and a number of others, specific mutations have a dramatic effect on gene expression and produce severe disease, in other cases the effects of such DNA sequence changes can be more subtle, producing a quantitative change in the level of gene transcription rather than severely reducing or abolishing transcription (**Figure 12.4**). Thus, gene-chip analysis (see Section 1.2) comparing the total RNA expression patterns in different normal human individuals has revealed quantitative differences in the expression of many different RNAs which correlate with

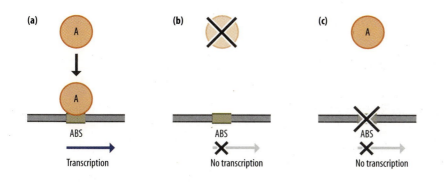

Figure 12.3
The normal binding of a transcription factor (A) to its binding site (ABS) (a) can be affected by mutation of the transcription factor (b) or by mutation of its binding site (c).

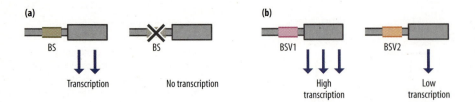

Figure 12.4
Some mutations in a binding site (BS) for a transcription factor will abolish transcription of the target gene (a). In other cases, sequence variation in the binding site (BSV1/BSV2) will produce a quantitative change in the amount of gene transcription (b).

differences in the promoter or enhancer sequences of the corresponding genes in the different individuals.

Rather than specifically causing a particular disease, such quantitative variations in gene expression are likely to enhance or reduce the risk of specific multifactorial diseases which develop due to the interaction of multiple genetic and environmental factors. For example, quantitative differences between different individuals in the expression of a number of mRNAs in adipose tissue have been shown to be dependent on sequence differences between the regulatory regions of the corresponding genes in different individuals and to correlate with the risk of their developing obesity.

Transcriptional co-activators

As well as affecting DNA-binding transcription factors or their DNA-binding sites, specific mutations can also affect co-activators which as described in Chapter 5 (Section 5.2) play a key role in the action of DNA-binding transcriptional activators (**Figure 12.1b**). In particular, the CBP co-activator is of vital importance in a number of different signaling pathways linking the transcription factor which is activated by the pathway to either the basal transcriptional complex or altering chromatin structure. It is not surprising therefore that mutations which result in individuals lacking functional CBP are incompatible with life. Indeed, even having a single inactive *CBP* gene while retaining a single functional gene results in the severe human disease Rubinstein–Taybi syndrome, which is characterized by mental retardation and physical abnormalities.

Hence, even if a single *CBP* gene remains functional, the inactivation of the second copy produces disease. This is likely to be due to the amount of CBP in normal cells being low with various transcription factors competing for it (see Sections 5.2 and 8.1). Hence, the reduction in the level of CBP due to loss of one gene copy results in disease (**Figure 12.5**). The *CBP* gene mutation in Rubinstein–Taybi syndrome is therefore dominant for the same reason as in the Pax6 case discussed above, namely the inability of one functional gene copy to produce the required amount of protein.

CBP can also be involved in disease processes even when it is intact and unmutated. It has been shown that individuals with the neurodegenerative disease Huntington's chorea produce an abnormal form of a protein known as Huntingtin. In this disease, a CAG sequence in the coding sequence of the Huntingtin protein has expanded producing a run of glutamine amino acids (encoded by the CAG triplet) in the Huntingtin protein. This abnormal form of Huntingtin binds to the normal CBP protein and prevents it from regulating transcription by depositing it in insoluble aggregates (see Figure 12.5). As

Figure 12.5
To function effectively the cell needs to have two intact genes encoding the CBP co-activator protein (a). Specific diseases result when one gene is inactive, resulting in reduced levels of CBP (b), or when CBP is intact but is sequestered by an abnormal Huntingtin protein (H) so that it cannot fulfil its co-activator role (c).

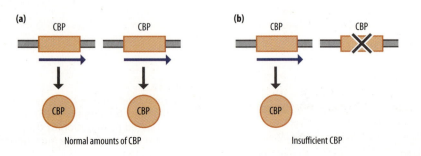

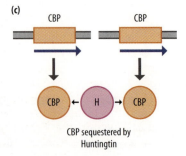

well as binding to CBP, mutant Huntingtin has also been shown to interfere with the functioning of the Sp1 transcription factor and the TBP component of the basal transcriptional complex, indicating that it can disrupt the function of different classes of transcriptional regulator (**Figure 12.6**).

Interestingly, the Huntingtin protein has recently been shown to associate with the Argonaute proteins which are involved in the processing of small inhibitory RNAs (see Section 7.6). Moreover, Huntingtin and Argonaute proteins co-localize in P bodies which, as described in Chapter 7 (Section 7.6), are involved in RNA degradation. Hence, Huntingtin may also be involved in the post-transcriptional inhibition of gene expression by small inhibitory RNAs.

Components of the basal transcriptional complex

The effect of Huntingtin on TBP, described above, indicates the ability of disease processes to disrupt the functioning of the basal transcriptional complex which is one of the major targets for activators and co-activators (see Section 5.2). Given the critical role for this complex (see Section 4.1) it is likely, however, that most mutations in components of the complex would be incompatible with survival to birth and would not therefore be observed as post-natal human diseases. However, mutations in components of TFIIH have been observed in the human disease xeroderma pigmentosum, which results in skin defects and a higher risk of cancer. Importantly, these mutations have been shown to result in defective responses to transcriptional activators and repressors rather than affecting the basal activity of TFIIH (**Figure 12.7**). This is presumably why such mutations are compatible with survival, albeit with abnormal functioning which results in disease. Hence, specific human diseases can also be caused by mutations in components of the basal transcriptional complex (**Figure 12.1c**).

Factors involved in transcription by RNA polymerases I and III

All the examples discussed above involve proteins which regulate transcription by RNA polymerase II. However, specific diseases have also been identified where the abnormal protein is involved in transcription by the other RNA polymerases. For example, in the case of RNA polymerase I, mutations in the CSB protein which forms a complex with RNA polymerase I and other proteins cause the disorder known as Cockayne's syndrome with abnormalities in the nervous system and skeleton. Similarly, the abnormal craniofacial development characteristic of Treacher Collins syndrome has been shown to result from mutations in the gene encoding a protein that interacts with the RNA polymerase I basal transcription factor UBF (see Section 4.1).

12.2 CHROMATIN STRUCTURE AND HUMAN DISEASE

As discussed in Chapters 2 and 3, changes in chromatin structure are an essential prerequisite for activation and repression of transcription. It is not surprising therefore that human diseases result when these processes are disrupted (**Figure 12.1d**). The different processes involved in chromatin structure which can be disrupted in human disease will be discussed in turn.

DNA methylation

A number of different mutations which affect DNA methylation have been shown to be the cause of specific human diseases. For example, mutations in the DNA methyltransferase enzyme Dnmt3b which normally adds methyl groups to C residues, (see Section 3.2), have been shown to cause the human ICF syndrome (immunodeficiency, centromeric instability, facial anomalies). Similarly, mutations in the MeCP2 factor which prevent

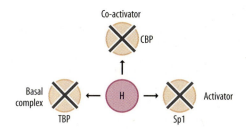

Figure 12.6
The abnormal Huntingtin protein (H) found in Huntington's disease can interfere with the functioning of a transcriptional activator, a co-activator, and a component of the basal transcriptional complex.

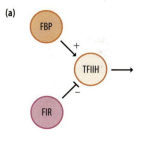

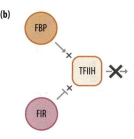

Figure 12.7
Wild-type TFIIH (circle) can be activated by the FBP protein and repressed by the FIR protein (a). In contrast, mutant TFIIH (square) does not respond appropriately to either of these proteins (b).

it from specifically recognizing methylated DNA result in the neurological disease Rett syndrome (see Section 3.2).

In addition, as well as resulting from alteration in methylation recognition proteins or DNA methylases, disease can also result from changes in methylation patterns on the DNA. For example in the human fragile-X syndrome, a triplet sequence CGG which is present in 10–50 tandem copies in the first exon of the *FMR-1* gene is amplified so that over 230 tandem copies are present in patients with this syndrome. This amplified sequence is then heavily methylated on the C residues in the triplet repeat, resulting in transcriptional silencing of the *FMR-1* gene. The protein normally produced by the *FMR-1* gene acts as a regulator of the translation of other mRNAs and its absence in turn produces the mental retardation and other symptoms characteristic of the fragile-X syndrome. This represents another mechanism by which an expansion in a three-base sequence can cause disease, in addition to the sequestration of transcription factors which was discussed above (Figure 12.8a).

Histone-modifying enzymes

As well as involving DNA methylation, **triplet-repeat diseases** can also involve alterations in histone modifications which regulate chromatin structure (see Sections 2.3 and 3.3) (Figure 12.8b). For example ataxin, the gene which is mutated in the human triplet-repeat disease SCA7 encodes a histone acetyltransferase. The mutant ataxin protein found in SCA7 patients acts as a dominant negative inhibitor of the wild-type protein and prevents its opening the chromatin via its histone acetyltransferase activity. Hence, triplet-repeat diseases can involve different effects on chromatin structure caused by altered DNA methylation or altered histone modifications (see Figure 12.8).

Altered histone modification is also observed in the human disease X-linked mental retardation. Some patients with this disease have mutations in the gene encoding the histone demethylase JARID1C. This demethylase normally functions to produce a closed chromatin structure by demethylating histone H3 on the lysine at position 4 (see Sections 2.3 and 3.3) and this activity is lost in the disease.

Interestingly, the JARID1C demethylase functions as a co-repressor for the REST transcriptional repressor which normally represses the expression of neuronal genes in non-neuronal cells (see Section 10.2). Hence, inactivation of JARID1C interferes with the ability of REST to repress gene expression (Figure 12.9a).

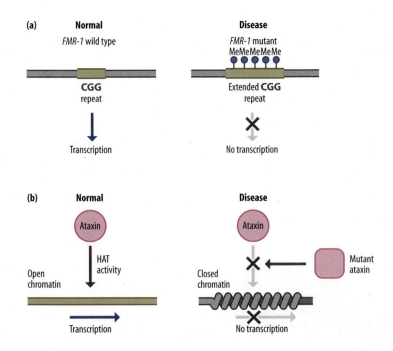

Figure 12.8
Involvement of chromatin structure in triplet-repeat diseases. The triplet-repeat expansion in the *FMR-1* gene results in its methylation on C residues (blue circles) and blocks its transcription (a). In SCA7 disease, abnormal ataxin protein encoded by a mutant ataxin gene with a triplet-repeat expansion acts as a dominant negative protein blocking the histone acetyltransferase (HAT) activity of the wild-type protein. This blocks the transition of ataxin target genes into an open chromatin structure compatible with transcription (b).

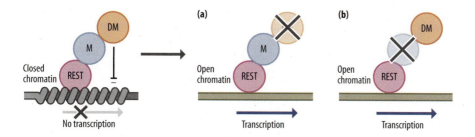

Figure 12.9
The human disease X-linked mental retardation can result from mutations which disrupt the ability of the REST factor to organize a tightly packed chromatin structure. This can involve mutation of a histone demethylase enzyme (DM) (a), or of the mediator complex (M), which links REST to the demethylase (b).

This association of REST and JARID1C appears to involve the mediator complex (see Section 5.2). Thus, X-linked mental retardation can also be produced by mutations in the gene encoding the MED12 component of the mediator complex. This MED12 mutation also impairs REST function and demethylation of histone H3 on lysine 4 (**Figure 12.9b**). Hence, X-linked mental retardation occurs in patients where the ability of REST to repress gene expression is compromised by mutations in either the mediator complex or in the histone demethylase which produces the closed chromatin structure by which REST represses its target genes (see Figure 12.9).

Chromatin-remodeling complexes

Apart from DNA methylation and histone modifications, other aspects of the regulation of chromatin structure can also be affected in human diseases. For example, the Williams syndrome transcription factor (WSTF), which is mutated in the human disease Williams syndrome, is a component of the WINAC chromatin-remodeling complex. This complex interacts with the vitamin D receptor which is a member of the nuclear receptor family of transcription factors (see Sections 4.3 and 5.1). Chromatin-remodeling by the WINAC complex is essential for recruitment of the vitamin D receptor to its target genes and this process is defective in Williams syndrome, producing mental retardation and growth deficiency.

Similarly, a mutation in the SNF2 factor which is part of the chromatin-remodeling SWI–SNF complex discussed in Chapter 3 (Section 3.5) results in a severe form of α-**thalassemia**. Patients with this form of thalassemia not only lack globin gene expression as in other thalassemias (see for example, Section 2.5) but also have other symptoms such as mental retardation. This indicates that the activity of SNF2 is necessary for opening the chromatin structure of both the α-globin genes and a number of other genes, so preventing their transcription if it is absent (**Figure 12.10**).

Interestingly, in another type of α-thalassemia changes in chromatin structure are also involved but in a completely different way. In this case, a deletion removes one of the two α-globin genes, leaving the other copy on the same chromosome intact. However, the deletion also removes the 3′ end of the LUC7L gene, which is adjacent to the α-globin genes but is

Figure 12.10
The SWI–SNF complex produces an open chromatin structure of the α-globin gene and a number of other genes (X and Y) (a). Mutation of the SNF2 component of the complex prevents this effect and the lack of expression of α-globin and other genes produces α-thalassemia and mental retardation (b).

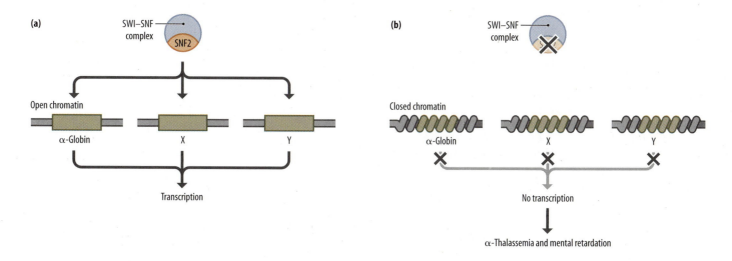

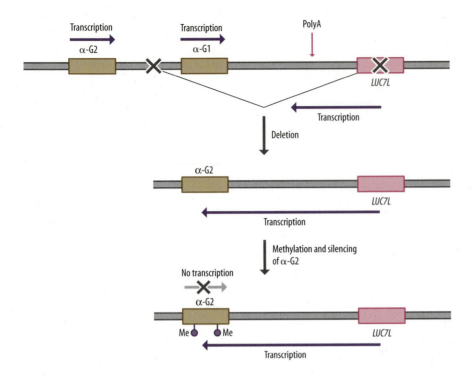

Figure 12.11
In a particular form of α-thalassemia, one of the two α-globin genes (α-G1) and part of the *LUC7L* gene, including its polyadenylation site (polyA) are deleted. This results in the LUC7L RNA transcript extending into the remaining α-globin gene (α-G2) in an antisense orientation. In turn, this results in the methylation (Me) and transcriptional silencing of the α-G2 gene.

transcribed from the opposite strand of the DNA (**Figure 12.11**). This removes the polyadenylation signal terminating the RNA of the *LUC7L* gene. In turn, this results in the LUC7L transcript continuing through the remaining α-globin gene where it is transcribed off the opposite antisense strand to that from which the α-globin gene is transcribed.

By a mechanism which is not understood but may be related to that of small interfering RNAs (see Section 3.4), this antisense RNA induces C methylation of the α-globin gene. In turn this results in transcriptional silencing of the remaining α-globin gene, leading to thalassemia even though all the regulatory elements of the gene and its coding sequence are intact (see Figure 12.11).

12.3 POST-TRANSCRIPTIONAL PROCESSES AND HUMAN DISEASE

As well as affecting transcription factors and chromatin structure, mutations can also affect gene regulation at specific post-transcriptional stages (**Figure 12.1e**) and these will be discussed in turn (see Chapters 6 and 7 for an account of these processes and their role in gene regulation).

RNA splicing

A number of mutations which affect the processes of RNA splicing or alternative RNA splicing have been described and fall into two classes (**Figure 12.12**). The first class involves a mutation in a sequence affecting the splicing of a particular gene. Such mutations can result in a failure to properly splice its RNA (**Figure 12.12a**) or a failure to produce one of the alternatively spliced mRNAs which are normally produced (**Figure 12.12c**).

An example of a failure of splicing due to a mutation in the gene is provided by the dystrophin gene. Thus, deletions of this gene producing a lack of dystrophin protein result in severe cases of muscular dystrophy. However, some mild cases in which dystrophin function is compromised but not entirely lost result from a mutation in exon 31 of the dystrophin gene. This results in abnormal splicing to produce an RNA lacking exon 31 (**Figure 12.13a**). In turn, this produces a dystrophin protein with reduced activity, resulting in a mild disease phenotype.

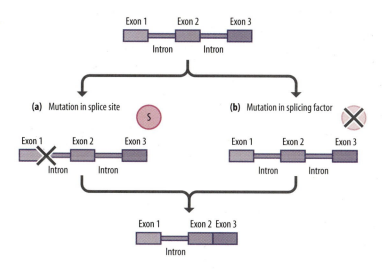

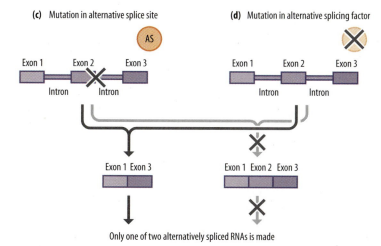

Only one of two alternatively spliced RNAs is made

Figure 12.12
The basic process of splicing can be disrupted by a mutation (X) in a sequence involved in the splicing of a particular RNA (a) or in a splicing factor (S) required for its correct splicing (b). Similarly, alternative splicing events can be affected by a mutation in a sequence involved in regulating the alternative splicing of a particular RNA (c) or in an alternative splicing factor (AS) (d).

In contrast, a mutation in exon 10 of the gene encoding the tau protein alters its alternative splicing. Normally alternative splicing results in exon 10 being included in 50% of tau mRNA molecules but being excluded from the remaining 50% of tau mRNAs. In turn, this produces equal amounts of two different forms of tau which are respectively encoded by mRNAs with or without exon 10 (**Figure 12.13b**). The mutation in exon 10 results in a greater proportion of the mRNA retaining exon 10, so resulting in an imbalance between the different forms of tau and leading to the neuropathological disease frontotemporal dementia (see Figure 12.13b).

In the second class of mutation affecting splicing, the mutation affects a splicing factor and therefore affects a number of genes whose splicing requires this factor either for the basic process of splicing itself (**Figure 12.12b**) or for a particular pattern of alternative splicing (**Figure 12.12d**). Hence, the effect of this class of mutation is not confined to a single gene.

An example of this type in which mutation of the protein affects the splicing of a number of different RNAs is seen in the *SMN1* gene. This gene was originally defined on the basis of its mutation in the disease spinal muscular atrophy (SMA). The SMN1 protein encoded by this gene has now been shown to be essential for RNA splicing, being required for assembly of the snRNP particles which catalyze RNA splicing (see Section 6.3). Its inactivation by mutation in SMA patients affects the splicing of a number of RNAs and the alternative splicing pattern observed for other RNAs, so combining the mechanisms illustrated in Figure 12.12b and Figure 12.12d.

Interestingly, alternative splicing is also affected in the muscle disease type 1 myotonic dystrophy. As with Huntington's disease and SCA7 (see

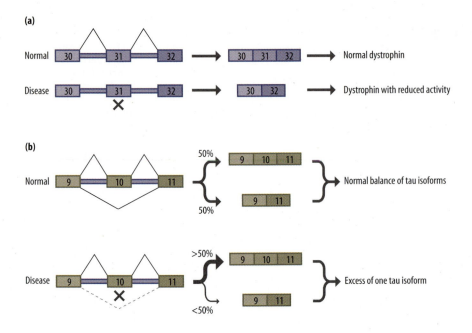

Figure 12.13
Examples of mutations (X) which affect the splicing (a) or alternative splicing (b) of a particular gene. (a) A mutation in exon 31 of the dystrophin gene results in this exon being spliced out of the RNA transcript. (b) A mutation in exon 10 of the *tau* gene results in enhanced production of the alternatively spliced RNA containing this exon at the expense of the alternatively spliced RNA which lacks exon 10.

above), this disease involves the presence of abnormal repeats of a three-base sequence, in this case CUG, in a specific gene. However, this run of multiple CUG triplets is located in the 3′ untranslated region of the DM kinase gene rather than in the protein-coding sequence, as is the case for Huntingtin and SCA7.

Moreover, this triplet repeat exerts its disease-causing effect at the RNA level. Thus, the repeated sequence binds a specific alternative splicing factor, MBNL1, and prevents it binding its normal RNA targets. As MBNL1 is required for the transition from embryonic to adult patterns of alternative splicing in a number of muscle-specific genes, this results in a failure to produce adult forms of these proteins so producing the disease.

This example is therefore similar to the Huntingtin/CBP case discussed above in that the repeat sequence sequesters a regulatory protein and prevents it carrying out its normal functions. However, it differs in that it occurs at the RNA level rather than at the protein level and involves a splicing factor rather than a transcription factor (**Figure 12.14**).

Overall the examples discussed above indicate that specific human diseases can result both from mutations in RNA splicing proteins which affect the splicing of their target RNAs and from mutations in specific RNAs which affect their splicing pattern. Moreover, in different cases both these

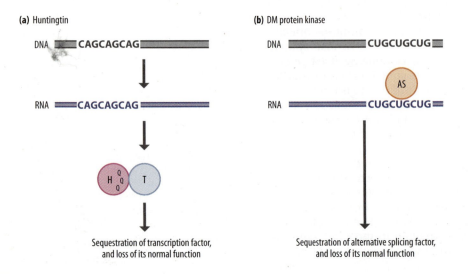

Figure 12.14
The expansion of a specific triplet sequence in the Huntingtin gene in Huntington's disease (a) or in the DM protein kinase gene in myotonic dystrophy (b) both produce disease by binding regulatory proteins and preventing them performing their normal function. In the case of Huntingtin, the repeated sequence is found in the protein-coding sequence and produces a mutant Huntington protein (H) with a run of glutamine residues (Q) which binds transcriptional regulators (T). However, in the DM protein kinase case, the repeat is in the 3′ untranslated region and is not translated into protein. Rather, it is the mRNA containing the repeat which binds a regulator of alternative splicing (AS).

effects can act at the level of the basic process of splicing and at the level of alternative splicing (see Figure 12.12).

RNA translation

As well as affecting the splicing of mRNAs, mutations involved in specific human diseases can also affect the processing of other cellular RNAs. For example, mutations in the enzymes which process transfer RNA precursors to produce the mature tRNA have been shown to cause the human neurological disorder pontocerebellar hypoplasia.

Evidently, such defects in the production of mature tRNAs will affect the translation of mRNAs by the ribosome in which tRNAs play a critical role (see Section 6.6). This is only one mechanism by which human diseases can be caused by effects on the tRNAs involved in mRNA translation. For example, mutations in the gene encoding glycyl tRNA synthase (which links the glycine amino acid to its appropriate tRNA) cause one form of Charcot Marie Tooth disease which affects the peripheral nervous system.

Effects on tRNA synthesis (**Figure 12.15a**) or tRNA binding of amino acids (**Figure 12.15b**) can therefore produce specific human diseases. Moreover, specific diseases can also be produced by mutations which target other aspects of the translation apparatus such as ribosomal proteins (**Figure 12.15c**) or translation initiation factors (**Figure 12.15d and e**) (see Section 6.6). For example, mutations that affect the S19 ribosomal protein produce the erythroid cell disease Diamond–Blackfan syndrome. Similarly, mutations in factors which regulate the activity of the translation initiation factor eIF2 also produce specific human diseases. For example, the neurological disease leukoencephalopathy with vanishing white matter is produced by mutations in the eIF2B regulator of eIF2 (see Section 7.5) whereas mutations in the PERK kinase which normally phosphorylates eIF2 produce early-onset diabetes (see Section 8.5).

As discussed in Chapter 10 (Section 10.2) one of the functions of eIF2 phosphorylation is to control the balance between translation of the ATF4 mRNA to produce functional ATF4 protein and the translation of small upstream open reading frames. In the ATF4 case, the translation of these small open reading frames prevents ATF4 protein production because the ribosome does not reinitiate at the translational start codon for ATF4 protein production. Such small open reading frames upstream of the protein-coding region are also found in the gene encoding the hairless transcription factor (**Figure 12.16**). In this case, however, the second of these upstream open reading frames appears to encode a repressor which represses the translation of the Hairless protein. Mutations of this upstream open reading frame which affect the amino acid sequence of the repressor protein result in over-production of the Hairless protein (see Figure 12.16). This results in the disease Marie Unna hypotrichosis in which the individual becomes totally bald.

Although splicing and translation represent the post-transcriptional stages of gene expression which have been most characterized in terms of disease-causing mutations, other post-transcriptional stages also appear to be affected in specific diseases. For example, a defect in RNA editing of the glutamate receptor mRNA has been suggested to be involved in motor neuron disease whereas mutations in the mRNA-export protein GLEI produce lethal motor neuron syndrome, resulting in death of the fetus before birth.

12.4 INFECTIOUS DISEASES AND CELLULAR GENE EXPRESSION

In this chapter, we have discussed the involvement of mutations in gene-regulatory proteins and DNA sequences in the causation of specific inherited human diseases. In addition however, gene-regulatory processes are also involved in human diseases caused by infection with agents such as viruses and bacteria. In particular, viruses are intracellular parasites which utilize the cellular machinery to replicate their nucleic acid and synthesize their

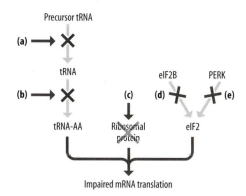

Figure 12.15
Specific human diseases can be produced by mutations in various components of the translational apparatus. These include enzymes which process precursor tRNAs to the mature tRNA (a), enzymes which link tRNAs with amino acids (b), ribosomal proteins (c), and regulators of eIF2 initiation factor activity such as eIF2B (d) and the PERK kinase (e).

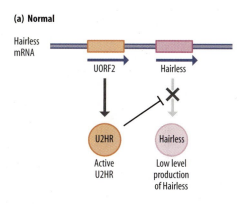

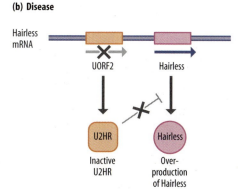

Figure 12.16
An upstream open reading frame (UORF2) in the Hairless mRNA encodes a protein (U2HR) which normally represses translation of the mRNA to produce the Hairless protein, so ensuring that the correct amount of Hairless protein is present (a). Specific mutations which inactivate U2HR (square) result in over-production of the Hairless protein, leading to disease (b).

Figure 12.17

Cellular mRNAs have a cap structure (small blue circle) which is recognized by the eIF4F complex and then by the 40S ribosomal subunit (a). The capless RNAs of picornaviruses have internal ribosome-entry sites (IRES) which are recognized by the 40S ribosomal subunit whereas the capless RNAs of hantaviruses bind the viral N protein which is recognized by the 40S ribosomal subunit (c).

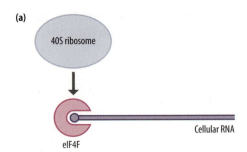

(a)

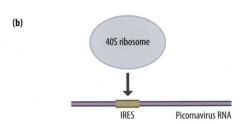

(b)

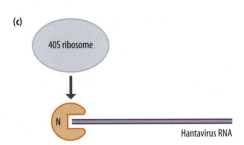

(c)

proteins. In many cases this results in the cells becoming a bag of viral particles (containing viral nucleic acid packaged in viral proteins) which eventually bursts open to release the viral particles. To convert the cell into a virus-producing factory, these viruses frequently encode proteins which inhibit cellular gene expression so as to enhance the efficiency of viral gene expression. For example, herpes simplex virus produces a protein, known as vhs (virion host shut off protein), which degrades cellular mRNAs so ensuring that cellular ribosomes are available to translate viral mRNAs.

Similarly, several groups of viruses produce a decapping enzyme which removes the 5′ cap from cellular mRNAs (see Section 6.1), thereby preventing their translation. This mechanism effectively prevents translation of cellular mRNAs but obviously raises the question of how viral mRNAs are recognized by the 40S ribosomal subunit, which normally recognizes the cap and the translational factors bound to it (**Figure 12.17a**). In the case of the picornaviruses, this is achieved by the viral RNA having internal ribosome-entry sites (IRES), so allowing the 40S ribosomal subunit to bind to the RNA in a cap-independent manner (**Figure 12.17b**) (see Section 7.5 for further discussion of IRES sequences).

In contrast, another group of viruses, the hantaviruses encode a protein (known as N) which binds to the 5′ end of the RNA and mimics the cellular eIF4F complex (containing the eIF4E-cap-binding protein) (see Section 6.6). This allows the 40S ribosomal subunit to recognize the hantavirus RNA so allowing its translation (**Figure 12.17c**).

As well as inhibiting cellular gene expression, viruses must also attempt to overcome the cellular antiviral responses which occur following viral infection. A complex interplay therefore exists between cellular defense mechanisms and the attempts of viruses to overcome these and such interplay frequently involves gene-regulatory processes.

This interplay is well illustrated by the cellular interferon system and its inhibition by various viruses. Following infection with a number of different viruses, the cellular genes encoding the antiviral proteins interferon-α and interferon-β are induced. In turn, these proteins bind to cellular receptors and induce the phosphorylation of the STAT family transcription factors, STAT-1 and STAT-2 (**Figure 12.18**) (see Section 8.2 for description of these factors and their activation by phosphorylation). The activated STAT-1 and STAT-2 then bind to each other to form a STAT-1–STAT-2 heterodimer. They then enter the nucleus where they interact with another protein, IRF9, and then bind to the DNA of their target genes, activating their expression (see Figure 12.18). These target genes encode a variety of antiviral proteins and microRNAs, which act to inhibit viral infection.

Many viruses interfere with this activation of antiviral transcriptional regulatory proteins by targeting one or more of the components of the pathway. For example, specific viruses interfere with the receptor-associated JAK kinases, which phosphorylate the STAT transcription factors (**Figure 12.18a**), while others induce the degradation of STATs (**Figure 12.18b**) or dephosphorylate them (**Figure 12.18c**). Other viruses block the movement of activated STATs to the nucleus (**Figure 12.18d**) or target the IRF9 transcription factor (**Figure 12.18e**).

As well as blocking the signaling pathway producing transcriptional activation of specific genes by interferon-α and -β, viruses can also target the transcription of the genes encoding the interferon-α and -β. For example, the NFκB transcription factor plays a key role in the activation of the gene encoding interferon-β following viral infection (see Section 4.4). Specific viruses can inhibit this effect by inhibiting either NFκB itself or the IκB kinases, (which normally phosphorylate the inhibitory IκB protein, and thereby activate NFκB) (see Section 8.2 for discussion of the NFκB/IκB system).

These examples illustrate therefore the involvement of gene-regulatory processes in infectious diseases. Moreover, as in the inherited diseases discussed in Sections 12.1–12.3, these effects can involve both transcriptional and post-transcriptional processes.

12.5 GENE REGULATION AND THERAPY OF HUMAN DISEASE

As discussed in Chapter 11 and this chapter respectively, the role of aberrant gene regulation in cancer and many other diseases is now increasingly well understood. As our knowledge of this area develops it is becoming clear that such increased understanding could lead to improved therapy for such diseases based on the manipulation of gene expression. These possibilities are discussed in this section both for cancer and for the other human diseases.

Therapy could be achieved by altering the expression of transcription factors

In some cases, therapy might involve artificially increasing the expression of a regulatory protein. For example, as described in Chapter 8 (Section 8.2) activation of the NFκB transcription factor plays a key role in the immune response. Hence, blocking its activity would be of value in treating human diseases involving damaging inflammation. This could potentially be achieved by enhancing the levels of the inhibitory IκB protein either by identifying drugs which can switch on the patient's own *IκB* gene or by using a gene-therapy procedure to deliver exogenous copies of the *IκB* gene (**Figure 12.19a**).

Alternatively, a similar therapeutic benefit could be achieved by inhibiting the synthesis of NFκB itself (**Figure 12.19b**). One potential method of doing this would be to artificially synthesize small inhibitory RNAs directed

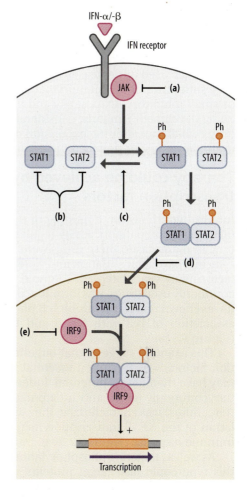

Figure 12.18
Activation of gene expression by interferons α and β involves activation of the receptor-associated JAK kinases which then phosphorylate the STAT1 and STAT2 transcription factors allowing them to heterodimerize and enter the nucleus. The STAT1–STAT2 heterodimer then interacts with the IRF9 transcription factor and the tripartite complex binds to specific target genes and activates their expression. Different viruses can interfere with this process by inhibiting the JAK kinase (a), inducing the degradation of STAT1–STAT2 (b), inducing their dephosphorylation (c), blocking their transport to the nucleus (d), or inhibiting IRF9 (e).

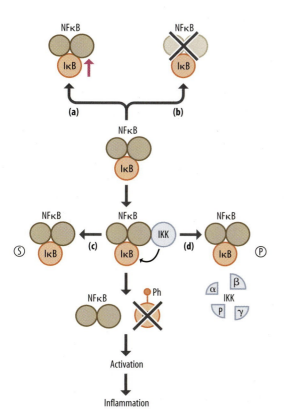

Figure 12.19
Potential therapeutic strategies for disrupting the activation of NFκB. (a) Enhanced expression of IκB, (b) reduced expression of NFκB, (c) and (d) inhibiting the phosphorylation of IκB which is essential for its activation either by directly inhibiting phosphorylation using a drug such as salicylate (S) (c) or by using a small peptide (P) to disrupt the interaction between the different proteins in the IκB kinase complex (IKK), which is responsible for phosphorylating IκB (d) (see also Figure 8.39).

against NFκB. These would act in the same way as naturally occurring small inhibitory RNAs (see Section 7.6) to block NFκB synthesis.

Although these potential therapeutic approaches are of interest, they are not specific to gene-regulatory proteins but could be used for any protein where increased or decreased expression was desired. Moreover, currently they suffer from the limitation that methods to deliver an exogenous gene or small inhibitory RNAs safely and efficiently to human patients are still being developed. Similarly, considerable effort will be required to identify drugs which can alter the expression of the endogenous gene encoding for example a transcription factor in a safe, specific, and effective manner.

Therapy could be achieved by altering the activity of transcription factors

Interestingly, however, a number of therapeutic drugs which are currently taken by many patients act by regulating the activity of transcription factors. In many cases, these drugs were introduced on the basis of their efficacy in a particular situation and were only shown to act via a transcription factor many years later. For example, salicylate (aspirin), one of the most commonly used drugs, was introduced many years ago. Much more recently, it was shown that it can inhibit IκB phosphorylation thereby promoting the association of IκB with NFκB and having an anti-inflammatory effect (**Figure 12.19c**).

This example suggests that targeting the phosphorylation of transcription factors, which is often essential for their activation (see Section 8.2) may have therapeutic potential. Indeed, when the mechanism of action of the commonly used anti-inflammatory drugs cyclosporin and FK506 (tarcolimus) was characterized, they were found to block the dephosphorylation of the NFAT transcription factor which is required for an effective immune response (**Figure 12.20a**).

As well as targeting phosphorylation of transcription factors, therapies of this type could also target proteins which regulate other stages of gene expression and which are regulated by phosphorylation. An obvious potential example of this is the PI3-kinase/Akt/TOR kinase cascade, which regulates the process of mRNA translation and which is up-regulated in various cancers (see Sections 8.5 and 11.4). Indeed, various chemical inhibitors of each of the kinases in this pathway have been developed and are being tested both alone and in combination for their therapeutic potential in cancer.

Clearly therapies could also target post-translational modifications other than phosphorylation which affect transcription factor activity (see Section 8.3). For example, the HIF-1 factor is an attractive therapeutic target since its inhibition could prevent the growth of blood vessels supplying tumors and therefore have a beneficial effect in cancer (see Section 11.3). Conversely, in patients suffering from cardiovascular disease it would be beneficial to promote blood vessel formation by stimulating HIF-1 activity. As described in Chapter 8 (Section 8.3), HIF-1 activity is regulated by hydroxylation of proline residues within the protein by a proline hydroxylase enzyme. Drugs have now been developed which target this enzyme and which could potentially be used therapeutically to enhance or reduce its activity (**Figure 12.20b**).

Similarly, the degradation of p53 induced by MDM2 (see Section 11.3) could be reduced by drugs which inhibit MDM2-mediated ubiquitination of p53. This would stabilize p53 and allow it to exert its anti-oncogenic effect, even in tumors where MDM2 is over-expressed (**Figure 12.20c**).

As well as targeting post-translational modification it is also possible to target protein–protein interactions which can also play a role in transcription factor activation (see Chapter 8). It is now relatively easy to map the region of a specific protein which interacts with another protein and to confirm this by structural studies. Once this has been achieved, small peptides can be prepared which mimic the binding region. These therefore compete for binding and can be used to disrupt the protein–protein interaction.

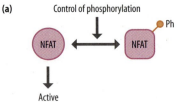

(a) Control of phosphorylation

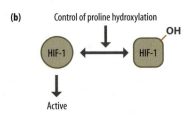

(b) Control of proline hydroxylation

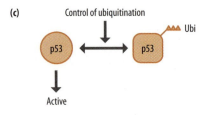

(c) Control of ubiquitination

Figure 12.20
Therapeutic drugs could target different post-translational modifications that regulate transcription factor activity such as phosphorylation (a), proline hydroxylation (b), or ubiquitination (c).

For example, a small peptide has been used to disrupt the protein–protein interactions in the IκB kinase (IKK), which is a multi-protein complex required for the phosphorylation of IκB. This peptide therefore blocked IκB phosphorylation (which is required for NFκB activation) and produced an anti-inflammatory effect *in vivo* (**Figure 12.19d**).

Hence, a variety of potential therapeutic methods for inhibiting NFκB exist with salicylate treatment being used clinically (see Figure 12.19). This potential modulation of NFκB is of particular importance in view of its key role in inflammation and more recent indications that it is involved in certain types of cancer.

Although the identification of peptides which can disrupt protein–protein interactions is of value, their potential therapeutic application is limited by the ability to deliver them efficiently to the target cells in the patient's body. Ideally, it would be preferable to develop small diffusible molecules which would be simple to deliver and would disrupt specific protein–protein interactions. This has been achieved in the case of the interaction between the anti-oncogene p53 and MDM2 (see above). Thus, structural studies indicated that part of the p53 protein inserts itself into a deep pocket in the MDM2 molecule. A chemical compound designed to fill this pocket was able to block p53/MDM2 binding and thereby activated p53 leading to reduced tumor growth both in culture and in the intact animal (**Figures 12.21** and **12.22**).

This type of drug which targets protein–protein interaction is therefore likely to be of value clinically paralleling the use of drugs such as salicylate or cyclosporin/FK506 which target phosphorylation. It should be noted that this approach and that of targeting p53 ubiquitination (see above) would be of value only in tumors with intact p53. In tumors which have a mutant p53 that acts as an oncogene (see Section 11.3), stabilizing the mutant p53 by blocking MDM2 would be counterproductive.

As well as transcription factor activity being regulated by phosphorylation or protein–protein interaction, it can also be regulated by ligand binding (see Section 8.1). This is seen particularly in the members of the nuclear receptor family of transcription factors which are activated by their appropriate ligand. This effect has been exploited therapeutically in different situations to either stimulate a receptor or inhibit it. For example, in breast cancers which are dependent on estrogen for their growth, therapy can involve the drug tamoxifen. This competes with estrogen for binding to the estrogen receptor but does not induce gene activation by the receptor following binding.

Conversely, as discussed in Section 12.1, the PPARγ member of the nuclear receptor family is mutated in a few patients with diabetes indicating its normal role in preventing the disease. In the vast majority of patients who have a normal PPARγ receptor and whose disease is due to other causes, the disease can be treated by stimulating the activity of the receptor. This is achieved by using synthetic drugs known as thiazolidinediones which bind to the receptor and stimulate its activity.

Hence, in different therapeutic applications, treatment can be carried out by either inhibiting the activity of a member of the nuclear receptor family of transcription factors (**Figure 12.23a**) or artificially stimulating it (**Figure 12.23b**).

A similar therapeutic approach has also been used in cases of promyelocytic leukemia involving the oncogenic fusion protein linking the PML protein with the retinoic acid receptor α (RARα) protein, which is a member of the nuclear receptor family. As discussed in Chapter 11 (Section 11.2), this fusion protein acts as a repressor of transcription, unlike the parental RARα protein which stimulates transcription following activation by retinoic acid.

One means of treating this form of leukemia therefore involves the administration of retinoic acid to stimulate the gene activation properties of the retinoic acid receptor portion of the fusion protein and overcome the transcriptional inhibition normally produced by the fusion protein (**Figure 12.24a**).

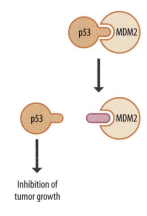

Figure 12.21
Binding of p53 to the MDM2 inhibitor involves a deep pocket in MDM2 into which p53 inserts itself. Filling this pocket with a specific synthetic chemical known as nutlin (pink) prevents binding of p53 to MDM2 and allows p53 to inhibit growth.

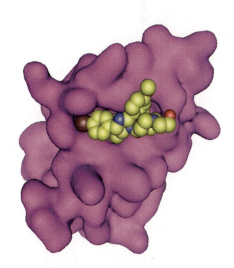

Figure 12.22
Structural model showing that the binding of p53 to a pocket in the MDM2 molecule (pink) is mimicked by the synthetic small molecule nutlin-2. Courtesy of Bradford Graves & Lyubomir Vassilev, Roche Research Center.

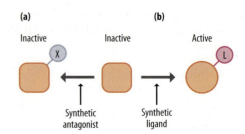

Figure 12.23
Manipulating the activity of members of the nuclear receptor family of transcription factors. (a) The activity of a receptor can be inhibited by using a synthetic antagonist (X) which competes with its normal ligand for binding but which does not activate the receptor. (b) The activity of a receptor can be stimulated by using a synthetic ligand (L) which binds to the receptor and activates it.

Therapy could be achieved by targeting proteins which alter chromatin structure

Although treatment of promyelocytic leukemia with retinoic acid is effective in some cases, especially when combined with chemotherapy, it does not produce a long-term cure in many cases. This has led to the idea of an alternative therapy based on the finding that the RAR–PML fusion protein represses gene expression by recruiting histone deacetylases which produce an inactive chromatin structure (see Section 11.2). Histone deacetylase inhibitors have been developed and may ultimately have a clinical role in leukemia caused by the RAR–PML fusion protein, acting by blocking the gene repression caused by the fusion protein (**Figure 12.24b**).

As noted in Chapter 11 (Section 11.2), histone deacetylation is involved in gene repression induced by other oncogenic fusion proteins such as AML–ETO. This approach may therefore be applicable in a number of different human leukemias. Indeed, the involvement of changes in chromatin structure, histone modification, and DNA methylation in a number of human diseases (see Sections 11.2 and 12.2) indicates that therapies involving the manipulation of chromatin structure by chemicals which alter histone acetylation or DNA methylation may have widespread applicability in the treatment of cancer and other human diseases. For example, it may be possible to treat the neurodegenerative disease SCA7 by stimulating the histone acetyltransferase activity of the wild-type ataxin protein, so as to overcome the inhibitory effect of the mutant ataxin protein which causes this disease (see Section 12.2 and Figure 12.8).

Hence, a wide range of actual and potential therapies exist involving the manipulation of chromatin structure or of transcription factor activation, whether achieved by ligand binding, post-translational modification, or protein–protein interaction. These methods are evidently targeted against a specific gene or genes which are regulated by a particular transcription factor or which show alteration in their chromatin structure in specific diseases.

Therapy could be achieved using designer zinc fingers to alter gene transcription

As well as the gene-specific methods described above, another potential therapeutic approach exists which takes advantage of a specific property of transcription factors in order to manipulate the expression of any gene in the genome. As described in Chapter 5 (Section 5.1), the two-cysteine–two-histidine zinc finger has an α-helical region which contacts the DNA. It has been shown that the nature of the amino acids at the N-terminus of this α-helix determines the exact DNA sequence to which the zinc finger binds. Moreover, it is now possible to predict the exact DNA sequence which will be bound by a finger with particular amino acids at the N-terminus of the α-helix.

On the basis of this, zinc fingers can be designed which will bind to a specific target sequence that is present in a particular gene of interest. If such a **designer zinc finger** is introduced into cells, it will bind specifically

Figure 12.24
Alternative approaches to overcoming the transcription inhibitory effect of the RAR–PML fusion protein. (a) Treatment with retinoic acid (RA) stimulates gene activation by the RAR component. (b) Treatment with histone deacetylase (HDAC) inhibitors blocks the inhibitory effect on gene expression of histone deacetylases which bind to the RAR–PML fusion protein. Compare with Figure 11.22.

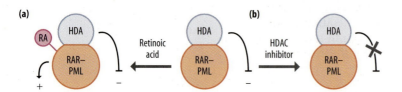

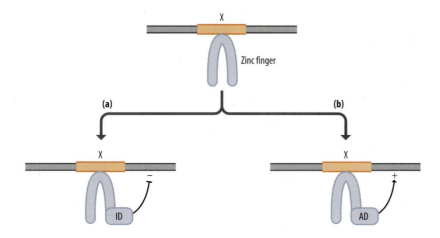

Figure 12.25
It is possible to design zinc fingers which will bind specifically to a DNA sequence (X) in a target gene. Linking this finger to a transcriptional inhibitory domain (ID) can be used to specifically switch off the target gene (a), whereas linkage to a transcriptional activation domain (AD) can specifically switch on the target gene (b).

to the target gene and not to all the other genes in the cell which do not contain its target sequence.

If this designer zinc finger is linked to the inhibitory domain of a transcription factor (see Section 5.3) it will deliver this domain to the target gene and specifically inhibit its transcription (**Figure 12.25a**). This method therefore offers an exciting means of specifically switching off a target gene. It has been used for example, to inhibit the expression of specific viral genes and thereby block infection of cultured cells with viruses causing human disease such as herpes simplex virus or human immunodeficiency virus.

As well as targeting viral gene expression, this method can also be used to target an individual cellular gene. This approach was used, for example, to inhibit the expression of the *CHK2* gene (which is involved in cell proliferation and cancer). Gene chip technology (see Section 1.2) was then used to show that while CHK2 mRNA levels were reduced the levels of all other cellular RNAs were unchanged, indicating the specificity of this method.

Evidently, as well as using this approach to inhibit the expression of a single gene it is also possible to specifically activate an individual gene by linking the designer finger to the activation domain of a transcription factor (see Section 5.2) (**Figure 12.25b**). This approach has been used to activate the gene encoding the **VEGF** growth factor in the intact animal *in vivo*. The enhanced levels of VEGF produced were functional and were able to produce increased blood vessel growth, indicating that this effect could be of therapeutic value in diseases where patients suffer from poor blood supply (**Figure 12.26**).

The effective use of this approach, like gene therapy or small inhibitory RNAs, requires the development of effective means of delivering the designer finger or the DNA encoding it to the patient. Nonetheless, it has considerable therapeutic potential since unlike the other methods we have discussed it can target any gene regardless of its normal method of regulation.

Therapy could be achieved by modulating RNA splicing

So far in this section we have focused on potential therapeutic methods which are designed to modulate transcription either directly or via altering chromatin structure. As noted above, however, it is also possible to manipulate kinase pathways such as the PI3-kinase/Akt/TOR cascade to achieve potentially therapeutic effects on mRNA translation. It is also potentially possible to achieve a therapeutic effect at the post-transcriptional level by modulating RNA splicing. In different cases this could be achieved either by promoting the use of a particular splice site or by inhibiting the use of a particular site.

An example of the potential use of splicing inhibition is provided by the dystrophin gene whose mutation causes muscular dystrophy (see Section 12.3). In one type of this disease, exon 49 of the dystrophin gene is deleted together with its flanking introns, so fusing exons 48 and 50. This changes the reading frame of the resulting mRNA so that translation into protein ceases at an in-frame stop codon in exon 51 (**Figure 12.27**).

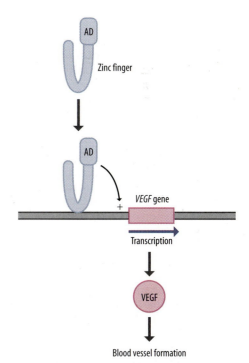

Figure 12.26
An artificial zinc finger linked to an activation domain (AD) binds to the regulatory region of the *VEGF* gene and activates its transcription. In turn, the increased levels of VEGF produced by this means induce enhanced blood vessel formation.

(a) **Normal**

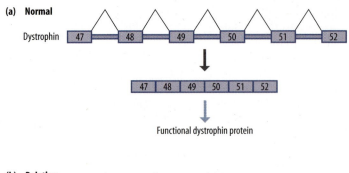

(b) **Deletion**

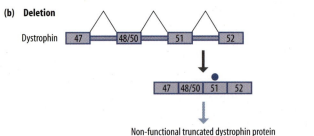

(b) **Therapy**

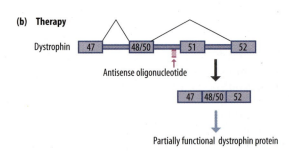

Figure 12.27
Region of the dystrophin gene containing exons 49, 50, 51, and 52. Normally, these exons are spliced together to produce a functional mRNA (a). In some cases of muscular dystrophy exon 49 is deleted, resulting in exons 48 and 50 being fused together. This changes the reading frame of the resulting mRNA producing a premature translational stop codon (circle) in exon 51 and resulting in a non-functional truncated protein (b). A potential therapeutic approach involves using an antisense oligonucleotide to block the inclusion of exon 51 in the mRNA. This removes the premature stop codon and allows all the exons downstream of exon 51 to be translated, so producing a partially functional protein (c).

One potential therapy for this form of the disease would be to inhibit the inclusion of exon 51 in the mRNA. Since the resulting mRNA would lack exon 51, as well as the deleted exon, it would not contain the premature stop codon. Hence, the mRNA would continue to be translated beyond the exon 51 region so producing a less severely damaged protein (see Figure 12.27).

To inhibit the splicing of exon 51, it is possible to use a short antisense oligonucleotide which is complementary to the 3′ splice site of exon 51 and which inhibits the inclusion of exon 51 in the mRNA (see Figure 12.27). This approach has proved effective in animal models of the disease and clinical trials in human patients are now underway.

It is also potentially possible to achieve a therapeutic effect by promoting the inclusion of a particular exon in the mRNA. An example of this is the disease spinal muscular atrophy (SMA) which, as described in Section 12.3, is caused by the deletion of the *SMN1* gene.

All human individuals, including SMA patients, have a second closely related gene, known as *SMN2*. The protein encoded by this gene could potentially substitute for the lack of SMN1 in SMA patients. However, it is expressed at low levels which are insufficient to do this. This is because in all human individuals (including SMA patients), compared to *SMN1*, the *SMN2* gene has a C-to-T base change in an exon splicing enhancer (ESE) located in exon 7 (**Figure 12.28**).

As discussed in Chapter 6 (Section 6.3) ESEs promote the inclusion of the exon in the final mRNA by stimulating the recruitment of SR splicing factors to the RNA. The base change in the exon 7 ESE in SMN2 compared to SMN1 weakens the ESE. This results in exon skipping with exon 7 frequently being omitted from the SMN2 mRNA (see Figure 12.28). In turn, this results in a non-functional SMN2 protein lacking the portion of the protein encoded by exon 7.

One potential therapy which is being developed for SMA would involve preparing an oligonucleotide complementary to the exon 7 ESE in SMN2 and linking it to an SR protein. By recruiting the SR protein to the RNA, this would facilitate the correct splicing of the SMN2 mRNA so as to include exon 7 (see Figure 12.28). In turn, this will result in an increased amount of functional SMN2 protein able to substitute for SMN1.

The use of oligonucleotides to modulate splicing either positively or negatively therefore offers a potential therapeutic opportunity. As with the use of designer zinc fingers or small RNAs, it is necessary to develop methods for the effective delivery of oligonucleotides to human patients. However, recent advances have occurred in this area involving the modification of the oligonucleotide to enhance its stability and its uptake by its target cells, indicating that these delivery problems may be overcome in the near future.

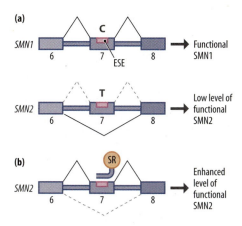

Figure 12.28
Compared to the *SMN1* gene, the *SMN2* gene has a base change (T rather than C) in an exon splicing enhancer (ESE; pink) in exon 7. This results in exon 7 frequently being skipped and excluded from the RNA (a). In SMA patients the SMN1 gene is deleted. However, an oligonucleotide complementary to the ESE could be used to artificially recruit an SR splicing protein to the ESE, so promoting inclusion of exon 7 in the SMN2 mRNA. This would enhance the level of functional SMN2 and have a therapeutic effect by allowing it to substitute for SMN1 (b).

CONCLUSIONS

As discussed in this chapter, many inherited human diseases are caused by alterations in the processes which regulate gene expression at the level of transcription, chromatin structure, or post-transcriptional processes. Similarly, infectious micro-organisms target gene expression so as to enhance the expression of their own genes or to overcome the defense mechanisms of the host organism.

Ultimately, our enhanced understanding of the processes normally regulating gene expression and the way they are altered in disease will lead to a new generation of therapies for human diseases. Currently, many thousands of patients take therapeutic drugs such as salicylate, FK506, or tamoxifen which target transcription factors. However, all these drugs were isolated on the basis of their efficacy and their mechanism of action was only determined subsequently. The identification of small molecules which can block the p53–MDM2 interaction, the production of designer zinc fingers targeted at particular genes, and the development of oligonucleotides which target RNA splicing offer hope of a new generation of therapeutic drugs designed specifically to target particular aspects of gene regulation.

KEY CONCEPTS

- A variety of inherited human diseases involve mutations in genes encoding proteins that regulate gene transcription.

- Genes mutated in this way can encode factors that regulate transcription directly such as DNA-binding transcription factors or co-activators, as well as proteins which regulate chromatin structure; for example, via altered DNA methylation or histone modifications.

- Inherited human diseases can also result from mutations which affect post-transcriptional processes such as RNA splicing or mRNA translation.

- Infectious micro-organisms which cause human diseases can target cellular gene expression either to enhance the expression of their own genes or to overcome host defense mechanisms.

- Gene-specific therapeutic methods can target a specific gene or genes whose expression is altered in a particular disease.

- Such therapeutic methods can involve modulating gene expression at the levels of transcription, chromatin structure, or post-transcriptional processing.

- Zinc fingers designed to have particular DNA-binding specificities can target any gene and increase or decrease its expression to produce a therapeutic benefit.

FURTHER READING

12.1 Transcription and human disease

Latchman DS (1996) Transcription factor mutations and disease. *N. Engl. J. Med.* 334, 28–33.

Riley BE & Orr HT (2006) Polyglutamine neurodegenerative diseases and regulation of transcription: assembling the puzzle. *Genes Dev.* 20, 2183–2192.

Rockman MV & Kruglyak L (2006) Genetics of global gene expression. *Nat. Rev. Genet.* 7, 862–872.

12.2 Chromatin structure and human disease

Bassell GJ & Warren ST (2008) Fragile X syndrome: loss of local mRNA regulation alters synaptic development and function. *Neuron* 60, 201–214.

Chahrour M & Zoghbi HY (2007) The story of Rett syndrome: from clinic to neurobiology. *Neuron* 56, 422–437.

Feinberg AP (2007) Phenotypic plasticity and the epigenetics of human disease. *Nature* 447, 433–440.

Kumari D & Usdin K (2009) Chromatin remodeling in the noncoding repeat expansion diseases. *J. Biol. Chem.* 284, 7413–7417.

12.3 Post-transcriptional processes and human disease

Cooper TA, Wan L & Dreyfuss G (2009) RNA and disease. *Cell* 136, 777–793.

Licatalosi DD & Darnell RB (2006) Splicing regulation in neurologic disease. *Neuron* 52, 93–101.

O'Rourke JR & Swanson MS (2009) Mechanisms of RNA-mediated disease. *J. Biol. Chem.* 284, 7419–7423.

Scheper GC, van der Knaap MS & Proud CG (2007) Translation matters: protein synthesis defects in inherited disease. *Nat. Rev. Genet.* 8, 711–723.

Wang GS & Cooper TA (2007) Splicing in disease: disruption of the splicing code and the decoding machinery. *Nat. Rev. Genet.* 8, 749–761.

Weiner L & Brissette JL (2009) Hair lost in translation. *Nat. Genet.* 41, 141–142.

12.4 Infectious diseases and cellular gene expression

Randall RE & Goodbourn S (2008) Interferons and viruses: an interplay between induction, signalling, antiviral responses and virus countermeasures. *J. Gen. Virol.* 89, 1–47.

von der Haar T (2009) One for all? A viral protein supplants the mRNA cap-binding complex. *EMBO J.* 28, 6–7.

12.5 Gene regulation and therapy of human disease

Bensinger SJ & Tontonoz P (2008) Integration of metabolism and inflammation by lipid-activated nuclear receptors. *Nature* 454, 470–477.

Bonetta L (2009) RNA-based therapeutics: ready for delivery? *Cell* 136, 581–584.

Castanotto D & Rossi JJ (2009) The promises and pitfalls of RNA-interference-based therapeutics. *Nature* 457, 426–433.

Haberland M, Montgomery RL & Olson EN (2009) The many roles of histone deacetylases in development and physiology: implications for disease and therapy. *Nat. Rev. Genet.* 10, 32–42.

Jordan VC (2007) Chemoprevention of breast cancer with selective oestrogen-receptor modulators. *Nat. Rev. Cancer* 7, 46–53.

Kim DH & Rossi JJ (2007) Strategies for silencing human disease using RNA interference. *Nat. Rev. Genet.* 8, 173–184.

Klug A (2005) The discovery of zinc fingers and their development for practical applications in gene regulation. *Proc. Jap. Acad. Ser. B Phys. Biol. Sci.* 81, 87–102.

Latchman DS (2000) Transcription factors as potential targets for therapeutic drugs. *Curr. Pharm. Biotechnol.* 1, 57–61.

Pennisi E (2008) Hopping to a better protein. *Science* 322, 1454–1455.

Pouyssegur J, Dayan F & Mazure NM (2006) Hypoxia signalling in cancer and approaches to enforce tumour regression. *Nature* 441, 437–443.

Smith LT, Otterson GA & Plass C (2007) Unraveling the epigenetic code of cancer for therapy. *Trends Genet.* 23, 449–456.

Vassilev LT (2007) MDM2 inhibitors for cancer therapy. *Trends Mol. Med.* 13, 23–31.

Yuan TL & Cantley LC (2008) PI3K pathway alterations in cancer: variations on a theme. *Oncogene* 27, 5497–5510.

Conclusions and Future Prospects

13

CONCLUSIONS AND FUTURE PROSPECTS

Transcription factors interact with one another to regulate transcription

The extraordinary rate of progress in the study of gene-control processes can be gauged from the fact that by 1980 no transcriptional regulatory protein or its DNA-binding site in a regulated gene promoter had been defined. Over the next decade, however, tremendous advances were made and a relatively clear picture of the action of individual factors which regulate gene expression became available. For example, in the case of gene induction by glucocorticoid or related hormones it was known that these agents act by activating specific receptor proteins (see Section 5.1), that such activated receptors bind to specific DNA sequences upstream of the target gene (see Section 4.3), displacing a nucleosome (see Section 3.5), and that a region in the receptor protein then interacts with a factor bound to the TATA box to activate transcription (see Section 5.2).

Clearly therefore, the process by which a single agent can activate a specific transcription factor and thereby modulate gene expression was reasonably well understood in outline by 1990. However, in subsequent years it became increasingly clear that the activation of a single transcription factor by a single agent cannot be assessed in isolation. Rather, the activity of a specific factor will depend upon its interaction with other transcription factors as well as with other proteins. For example, glucocorticoid hormone activates transcription by promoting the dissociation of the glucocorticoid receptor from the inhibitory HSP90 protein which otherwise anchors it in the cytoplasm and prevents it from moving to the nucleus and activating transcription (see Section 8.1).

A similar example of protein–protein interactions regulating transcription factor activity but in the opposite direction is observed in the Fos and Jun proteins, which regulate gene expression in response to growth factor or phorbol ester treatment (see Section 11.2). As discussed in Chapter 5 (Section 5.1), the Fos protein cannot bind to DNA on its own but can do so only after forming a heterodimer with the Jun protein. This heterodimer then binds to DNA and activates transcription. In this case, the interaction with another protein allows Fos to activate transcription, whereas in the case of the glucocorticoid receptor the interaction with HSP90 has the opposite effect.

Such types of protein–protein interactions regulating transcription factor activation are not confined to the regulation of gene expression by short-term inducers but are also involved in the regulation of cell type and tissue-specific gene expression. For example, as discussed in Chapter 10 (Section 10.1), the MyoD transcription factor plays a key role in regulating muscle-cell-specific gene expression and thereby in producing differentiated skeletal muscle cells. MyoD can bind to the promoters of muscle-specific genes and activate their expression, thereby causing the differentiation of a fibroblast cell line into muscle cells.

The production of differentiated skeletal muscle cells in this situation requires however, that the cells are cultured in the absence of growth factors. In the presence of growth factor, the cells contain high levels of the inhibitory factor Id which heterodimerizes with MyoD and prevents it binding to DNA and activating transcription (see Section 5.3). The production of the skeletal muscle phenotype therefore requires not only high levels of the activating MyoD protein but also a fall in the levels of the inhibitory Id protein which occurs upon growth factor removal.

It is clear therefore that in families of transcription factors which are capable of dimerization by means of specific motifs such as the leucine zipper or helix-loop-helix (see Section 5.1), such dimerization with another member of the family can result in activation (Fos–Jun) or repression (MyoD–Id).

Transcription factors can repress gene expression as well as activating it

The Id example also illustrates another theme which has become of increasing importance in gene regulation, namely the importance of specific inhibition as opposed to activation of gene expression (see Section 5.3). As well as transcription factors such as the Id factor which function only as inhibitors, cases also exist where the same factor can function as a direct activator or repressor depending on the circumstances. This is seen in the case of the thyroid hormone receptor which unlike the glucocorticoid receptor binds to its DNA target site even in the absence of hormone. As discussed in Chapter 5 (Section 5.3), this factor can directly inhibit promoter activity in the absence of thyroid hormone whereas in the presence of thyroid hormone it undergoes a conformational change which allows it to activate gene expression.

DNA-binding transcription factors interact with co-activators/co-repressors and with regulators of chromatin structure

The case of the thyroid hormone receptor illustrates two of the major themes to have emerged in recent years from further study of gene-control processes, namely the role of co-repressor/co-activator molecules (see Chapter 5) and the influence of transcription factors on chromatin structure (see Chapters 2 and 3). As discussed in Chapter 5 (Section 5.3), the inhibitory domain of the thyroid hormone receptor acts indirectly by recruiting an inhibitory molecule known as the nuclear receptor co-repressor which actually produces the inhibitory effect. Although this factor may interact directly with the basal transcriptional complex, it has also been shown to itself recruit another molecule which has the ability to deacetylate histones. As such histone deacetylation is associated with a tightly packed chromatin structure incompatible with transcription (see Section 3.3), the inhibitory effect of the thyroid hormone receptor appears to be mediated, at least in part, by the ability of its co-repressor to recruit a molecule that can organize the chromatin into a non-transcribable form.

As discussed in Chapter 5 (Section 5.3), the addition of thyroid hormone results in a conformational change in the thyroid hormone receptor which causes the release of its co-repressor and allows the receptor to activate transcription. Interestingly, as with transcriptional repression by the receptor, such activation requires a co-factor and involves alterations in histone acetylation. Following exposure to thyroid hormone, the thyroid hormone receptor can bind the CBP co-activator. This factor was originally identified as a co-activator which binds to the CREB transcription factor (see Sections 5.2 and 8.2) but was subsequently shown to be involved in transcriptional activation by nuclear receptors, such as the glucocorticoid receptor and the thyroid hormone receptor.

As discussed in Chapter 5 (Section 5.2), CBP is able to act as a co-activator by interacting with components of the basal transcriptional complex to stimulate its activity. In addition, however, it also has histone acetyltransferase activity (see Sections 2.3 and 3.3), indicating that by acetylating histones it can reverse the effects of the nuclear receptor/co-repressor complex and convert the chromatin into a form where transcription can occur. The addition of thyroid hormone therefore converts the thyroid hormone receptor from a form which can bind an inhibitory co-repressor complex with histone deacetylase activity to a form where it can bind a stimulatory co-activator with histone acetyltransferase activity.

This example therefore indicates the importance of co-activators and co-repressors, as well as providing a critical link between the transcription factors discussed in Chapter 5 and the chromatin structure changes discussed in Chapter 3.

Such an interplay between the control of transcription itself and the regulation of chromatin structure is also well illustrated by an analysis of the processes which regulate gene expression during embryonic development (see Chapter 9). For example, in *Drosophila* many of the factors regulating gene expression have been defined using a genetic approach involving the isolation of mutant flies exhibiting abnormal development and the subsequent characterization of the regulatory protein encoded by each mutated gene (see Section 9.2).

In a number of such cases the proteins identified in this way encode DNA-binding transcription factors such as the homeodomain factors which can bind to DNA and then activate or repress transcription (see Sections 5.1 and 9.2). In contrast, however, other proteins identified in this way encode factors that regulate the chromatin structure of their target genes. For example, polycomb factors act to produce a more tightly packed chromatin structure incompatible with transcription whereas this effect is opposed by trithorax proteins which open up the chromatin structure, allowing transcription to occur.

Interestingly, the chromatin structure and hence the transcription of the genes encoding the homeodomain factors themselves are subject to regulation by the polycomb and trithorax proteins (see Section 5.1). This indicates that specific regulatory proteins can control the expression of other regulatory proteins, as well as regulating target genes encoding proteins with other functions. Moreover, although originally identified in *Drosophila*, both homeodomain proteins and the polycomb/trithorax proteins have subsequently been identified in a wide range of other organisms including mammals where they play a key role in the regulation of gene expression (see, for example, Sections 9.1 and 9.3).

Histone modifications play a central role in the regulation of chromatin structure

The key role played by changes in chromatin structure has led to intensive investigation of the manner in which regulatory proteins can alter the structure of chromatin. It is now clear that the post-translational modification of histones by transcriptional regulators is a key event in the modulation of chromatin structure, leading to transcriptional activation or repression.

As discussed in Chapter 3 (Section 3.3), such histone modifications include not only acetylation but also methylation, phosphorylation, ubiquitination, and sumoylation. For example, the effects of polycomb and trithorax proteins on chromatin structure involve specific modifications of histones with the polycomb complex methylating histones on the lysines at positions 9 and 27 in histone H3, which promotes a more tightly packed chromatin structure. In contrast, the trithorax proteins promote the methylation of histone H3 on lysine 4 which in turn promotes demethylation at positions 9 and 27 (see Section 3.3).

The various histone modifications can interact with one another so that the existence of one modification at a particular amino acid influences positively or negatively the occurrence of another modification at another amino acid. This has led to the idea of a "histone code" in which a particular pattern of modification on the histones in a particular region of DNA promotes or inhibits the changes necessary for transcription to occur by affecting the recruitment of chromatin-remodeling complexes (see Section 3.5) and DNA-binding transcriptional activators or repressors.

Co-activators/co-repressors link together different signaling pathways by interacting with multiple transcription factors

Interestingly, a number of co-activator/co-repressor molecules appear to interact respectively with multiple different activating or inhibitory transcription factors. In the case of the CBP co-activator, which is used by transcription factors activated by different signaling pathways, this has been shown to result in interactions between these pathways. For example, both the glucocorticoid receptor, which is activated by glucocorticoid, and the AP1 (Fos–Jun) complex, which is activated by phorbol esters, require the CBP co-activator to mediate this stimulatory effect on gene transcription (see Section 5.2). Because CBP levels in the cell are relatively low, when both pathways are activated simultaneously they will compete for the limited amount of CBP and neither pathway will be able to activate transcription. Hence, the requirement of both of these pathways for CBP results in a mutual inhibition of gene activation by each of these factors (see Section 8.1).

This example illustrates how the effect of one signaling pathway can be affected by the activation of another signaling pathway. In the absence of glucocorticoid hormone, the glucocorticoid receptor will be anchored in the cytoplasm by HSP90 and the Fos–Jun complex will be able to interact with CBP and activate gene expression in response to growth factors or phorbol esters. However, when the glucocorticoid receptor is freed from HSP90 by the presence of glucocorticoid it can interact with CBP, removing it from Fos–Jun and preventing them activating gene expression.

This is seen in the case of the gene encoding the collagenase enzyme which is activated by the Fos–Jun complex in response to phorbol esters and can produce severe tissue destruction in an inflamed area. As expected on the basis of the above model the activation of this gene by phorbol esters is inhibited by treatment with glucocorticoid hormone accounting for the anti-inflammatory effort of this steroid hormone. Conversely, the activation of glucocorticoid-responsive genes by hormone treatment will be inhibited by the presence of high levels of Fos and Jun induced by growth factor or phorbol ester treatment since they will compete for the limited amount of available CBP (see Section 8.1).

These examples indicate therefore that the effect of a particular factor on gene expression will depend on the nature of the other factors present in the cell and the state of activation of specific signaling pathways. Indeed, as discussed in Chapter 8, the activity of many transcription factors is modulated by specific signaling pathways. Such pathways can enhance or reduce the activity of a particular transcription factor by for example regulating its association with another protein, its post-translational modification or its processing from a precursor protein. Such modifications can affect co-activators as well as DNA-binding transcriptional activators. For example, phosphorylation of the CBP co-activator by the IκB kinase enhances its binding to NFκB while reducing its binding to p53 (see Section 8.2). The balance between different pathways which use the same co-activator can therefore be modulated either by modifying the activator itself or by modifying a co-activator.

It is clear therefore that our simple picture of the glucocorticoid receptor acting in isolation has to be modified to consider its inhibition by interaction with HSP90, as well as its need to recruit other factors in order to stimulate transcription. Such factors include CBP and, as discussed in Chapter 3 (Section 3.5), the SWI–SNF complex which also reorganizes chromatin structure.

Gene regulation is highly complex and involves both transcriptional and post-transcriptional regulation

Over the last few years it has become clear therefore that the regulation of a single gene in an intact cell is vastly more complex than could be predicted from the study of an isolated transcription factor interacting with its binding site. At first sight it may appear that recent progress in this area has merely complicated the issue. However, the process of gene regulation must not only produce all the different types of cells in the body (see Chapter 10) but also ensure that each cell type is produced in the right place and at the right time during development (see Chapter 9).

It is inevitable therefore that gene regulation must be a highly complex process with numerous factors interacting with each other and their respective DNA-binding sites to regulate chromatin structure and/or directly regulate the transcription of specific genes in specific cell types. Moreover, such control of transcription is significantly supplemented by post-transcriptional control processes (see Chapters 6 and 7). These include processes such as RNA splicing or RNA editing which can produce multiple mRNAs encoding related proteins from a single gene. For example, over 90% of human genes which contain multiple introns are subject to alternative splicing. Similarly, transcriptional control is supplemented by other post-transcriptional processes such as the regulation of mRNA stability or of mRNA translation which can produce a rapid change in protein levels in response to changing conditions. Moreover, the processes which control transcriptional initiation and elongation interact with post-transcriptional processes such as capping, splicing, and polyadenylation to ensure the final mRNA is produced in an effective and co-ordinated manner (see Section 6.4).

RNA molecules play a central role in regulating gene expression

It is clear therefore that specific regulatory proteins play key roles in the regulation of gene expression, acting by altering chromatin structure, by controlling transcription itself, or by regulating post-transcriptional processes. However, the last few years have seen the realization that these gene-control processes are also regulated by specific RNA molecules. Small inhibitory RNA molecules such as siRNAs and miRNAs have now been extensively characterized (see Section 1.5) and shown to be involved in, for example, the control of developmentally regulated (see Chapter 9) and cell-type-specific gene expression (see Chapter 10).

Interestingly, small RNAs can repress gene expression both by altering chromatin structure (see Section 3.4) and at the post-transcriptional levels of mRNA degradation and mRNA translation (see Section 7.6). Moreover, as well as regulating expression of structural genes encoding proteins with diverse functions in the cell, they can also regulate the expression of genes which themselves encode regulatory proteins such as transcription factors or alternative splicing factors. In turn, such alteration in the expression of regulatory proteins will obviously have effects on the expression of the downstream target genes for such regulatory proteins. This not only greatly increases the regulatory effects of the small RNAs but allows them to have positive as well as negative effects on gene expression. For example, by inhibiting the expression of an inhibitory transcription factor, a small RNA can neutralize the inhibitory effect of the transcription factor on its target genes (see, for example, Sections 7.6 and 10.2).

As well as such small RNAs, larger non-protein-coding RNA molecules which have an effect on gene expression have been defined. For example, large non-coding RNAs are expressed specifically in undifferentiated embryonic stem (ES) cells and play a role in maintaining them in a proliferating undifferentiated state (see Section 9.1). Similarly, specific RNAs are involved in the processes of gene imprinting, while the XIST RNA, transcription of which from one X chromosome silences all other genes on that chromosome, plays a critical role in X-chromosome inactivation (see Section 3.6).

In contrast to the inhibiting effect of siRNAs/miRNAs and of large RNAs such as XIST, some RNA molecules have a role in activating rather than repressing gene expression. These include the non-coding HSR1 RNA which is involved in the activation of the heat-shock factor so that it can activate transcription of the heat-shock genes following exposure to stress (see

Section 8.1). Similarly, transcription of non-coding RNAs from polycomb/ trithorax DNA-response elements plays a key role in allowing trithorax proteins to produce an open chromatin structure and in counteracting the inhibitory effects of polycomb proteins (see Section 4.4).

It is clear therefore that as well as regulatory proteins, regulatory RNAs are involved in controlling gene expression. Moreover, the increasing number of such regulatory RNAs which are being characterized could explain the findings of the pilot phase of the ENCODE project analyzing the functional elements in the human gene (see the Conclusions section of Chapter 3 and Section 4.3). This project found that although a significant proportion of the human genome is transcribed into RNA, many of these RNAs do not encode proteins and are likely therefore to have a regulatory function.

Like the RNAs derived from polycomb-/trithorax-response elements, many of these RNAs are derived from regulatory DNA sequences adjacent to a transcribed gene but outside the transcribed region. As such they are likely to play a role in controlling the access of DNA-binding proteins to these sequences. However, many other non-coding RNAs which have been identified in the ENCODE project, overlap regions which are transcribed to produce protein-coding RNAs. Moreover, they are transcribed from the opposite strand of the DNA to that producing the protein-coding mRNA. They could therefore potentially regulate expression of the protein-coding RNA, for example by allowing the production of a double-stranded RNA that could generate inhibitory siRNAs (see Section 1.5). Alternatively, the transcription of a non-coding RNA from a particular gene locus may alter its chromatin structure, facilitating or inhibiting the expression of the protein-coding RNA (see Sections 3.6 and 4.4).

Alterations in regulatory RNAs and proteins cause human disease

In view of the key role of regulatory RNAs in gene control, it is not surprising that alterations in, for example, the expression of individual miRNAs have been shown to be involved in human diseases, such as cancer (see Section 11.4). This evidently parallels the involvement in cancer of specific gene-regulatory proteins encoded by oncogenes (see Section 11.2) or anti-oncogenes (see Section 11.3).

As discussed in Chapter 12, alterations in gene-regulatory processes can also produce human diseases other than cancer. For example, the CBP factor, which as discussed above plays a critical role in regulating gene activation in response to specific signaling pathways, also appears to be involved in gene regulation during development and in human disease. As discussed in Chapter 12 (Section 12.1), mutations in the *CBP* gene in humans result in Rubinstein–Taybi syndrome which is characterized by abnormal development leading to mental retardation and physical abnormalities. Interestingly, this disease occurs when only one of the two copies of the *CBP* gene is mutated with the other being capable of producing functional CBP protein; individuals with two mutant *CBP* genes have never been identified. This indicates that CBP is likely to be of such importance that its complete loss by inactivation of both *CBP* genes is incompatible with survival. Moreover, as in the signaling processes described above, the precise level of CBP in a cell is critical for proper development and a single functional gene evidently cannot produce sufficient CBP for normal development to occur.

Regulatory networks control gene expression

Investigation of the processes regulating gene expression both in normal cells and in disease has therefore revealed this to be an enormously complex processes. Studies initiated by the analysis of an individual regulatory factor have led to an understanding of the other proteins and RNAs which interact with it to regulate its activity and/or its expression. In turn, this has led to the ability to progressively build up and analyze the regulatory networks which control gene expression in a particular cell type or in response to a particular signal.

This factor-based approach has been complemented more recently by computational approaches which allow gene networks to be constructed based on a wide variety of data. This includes analysis of the levels of expression of all the genes which are active in a particular cell type (using, for example, DNA microarray analysis; see Section 1.2), analysis of the regulatory factors bound to particular genes (using the ChIP technique; see Section 4.3), and analysis of the effects of artificially reducing or enhancing the expression of genes encoding individual regulatory proteins or RNAs.

Computer analysis of such data produces highly complex networks integrating the regulatory effects of many different proteins and RNAs. Moreover, it is clear that hierarchies of regulatory factors exist in which, for example, one particular factor can modulate the expression and/or activity of other regulatory factors and so on (**Figure 13.1**).

Figure 13.1

Gene-regulatory network controlling the development of endoderm and mesoderm cells in the sea urchin embryo. The products of genes which are active in the mother (gray boxes) (see Chapter 9) activate a cascade of gene expression involving multiple regulatory genes (orange boxes). In turn, this results in the regulation of genes involved in the differentiation of mesoderm (green boxes) or endoderm (blue boxes). Arrows indicate activation of gene expression; lines ending in bars indicate repression of gene expression. From Alberts B, Johnson A, Lewis J et al (2008) Molecular Biology of the Cell, 5th ed. Garland Science/Taylor & Francis LLC.

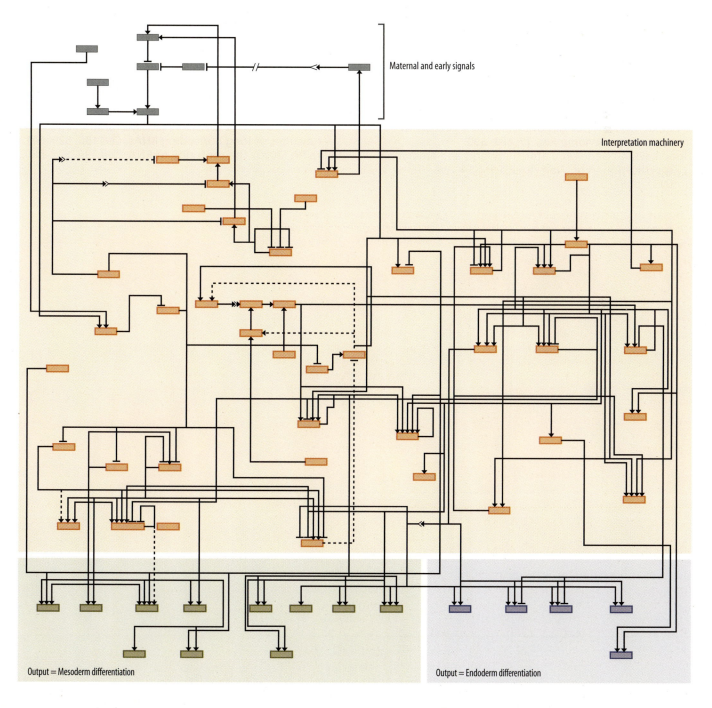

Maternal and early signals

Interpretation machinery

Output = Mesoderm differentiation

Output = Endoderm differentiation

In general, therefore, both analysis based on the interactions of individual factors and computer-based network analysis indicate the enormous complexity of gene-regulatory processes. Nonetheless, the progress in understanding gene regulation in terms of factor–factor interactions, changes in chromatin structure, identification of regulatory RNAs, analysis of gene-regulatory networks, etc., which has taken place in the last few years, indicates that an understanding of the ultimate problem of mammalian development in terms of differential gene expression can eventually be achieved. Similarly, our enhanced understanding of the role of aberrant gene regulation in cancer and other human diseases (see Chapters 11 and 12) offers hope of improved therapies for such diseases based on the artificial manipulation of gene expression in human patients.

FURTHER READING

Carninci P (2009) The long and short of RNAs. *Nature* 457, 974–975.

ENCODE Project Consortium (2007) Identification and analysis of functional elements in 1% of the human genome by the ENCODE pilot project. *Nature* 447, 799–816.

Gingeras TR (2007) Origin of phenotypes: genes and transcripts. *Genome Res.* 17, 682–690.

Hood L (2008) Gene regulatory networks and embryonic specification. *Proc. Natl Acad. Sci. USA* 105, 5951–5952.

Karlebach G & Shamir R (2008) Modelling and analysis of gene regulatory networks. *Nat. Rev. Mol. Cell Biol.* 9, 770–780.

Mercer TR, Dinger ME & Mattick JS (2009) Long non-coding RNAs: insights into functions. *Nat. Rev. Genet.* 10, 155–159.

Ponting CP, Oliver PL & Reik W (2009) Evolution and functions of long noncoding RNAs. *Cell* 136, 629–641.

Sharp PA (2009) The centrality of RNA. *Cell* 136, 577–580.

Glossary

Words in bold type indicate cross references to other glossary entries.

acetylation   Addition of a chemical group derived from acetic acid (CH_3COOH). Acetyl groups are added covalently to some proteins, such as histones, as a **post-translational modification**.

activation domain   Domain of a transcription factor responsible for its ability to activate transcription.

activator (gene-activator protein, transcriptional activator)   Gene-regulatory protein that when bound to its regulatory sequence in DNA activates transcription.

adenomatous polyposis coli (APC) protein   Tumor-suppressor protein that forms part of a protein complex in the **Wnt** signaling pathway. APC recruits free cytoplasmic β-catenin and degrades it.

adenylate cyclase (adenylyl cyclase)   Membrane-bound enzyme that catalyzes the formation of **cyclic AMP** from **ATP**. An important component of some intracellular signaling pathways.

Akt (protein kinase B, PKB)   Serine/threonine protein kinase that acts in the **PI3-kinase**/Akt intracellular signaling pathway involved especially in signaling cells to grow and survive. Also called protein kinase B (PKB).

allele   One of several alternative forms of a gene. In a **diploid** cell each gene will typically have two alleles, occupying the corresponding position (**locus**) on homologous chromosomes.

alpha-helix (α-helix)   Common folding pattern in proteins, in which a linear sequence of amino acids folds into a right-handed helix stabilized by internal hydrogen bonding between backbone atoms.

alternative RNA splicing   Production of different RNAs from the same gene by **splicing** the transcript in different ways.

amino acid   Organic molecule containing both an amino group and a carboxyl group. Those that serve as building blocks of proteins are alpha amino acids, having both the amino and carboxyl groups linked to the same carbon atom.

aminoacyl-tRNA synthetase   Enzyme that attaches the correct amino acid to a **tRNA** molecule to form an aminoacyl-tRNA.

amino-terminus   *see* **N-terminus**

amphipathic   Having both hydrophobic and hydrophilic regions.

Angstrom (Å)   Unit of length used to measure atoms and molecules. Equal to 10^{-10} meter or 0.1 nanometer (nm).

antibody (immunoglobulin, Ig)   Protein produced by **B cells** in response to a foreign molecule or invading microorganism. Binds tightly to the foreign molecule or cell, inactivating it or marking it for destruction. Can be used as a specific reagent to detect the protein it recognizes.

anticodon   Sequence of three nucleotides in a **transfer RNA** (tRNA) molecule that is complementary to a three-nucleotide **codon** in a **messenger RNA** (mRNA) molecule.

antigen   A molecule that can induce an immune response or that can bind to an antibody or T-cell receptor.

anti-oncogene (tumor-suppressor gene)   Gene that appears to help prevent formation of a cancer. Loss-of-function mutations in such genes favor the development of cancer.

anti-oncoprotein   Protein encoded by an **anti-oncogene**.

antiparallel   Describes the relative orientation of the two strands in a DNA double helix or two paired regions of a polypeptide chain; the polarity of one strand is oriented in the opposite direction to that of the other.

antisense RNA   RNA complementary to an RNA transcript of a gene. Can hybridize to the specific RNA and block its function.

AP1 site   DNA-binding site for **Fos** and **Jun** transcription factors. Mediates activation of genes which contain it in response to growth factors or phorbol esters.

APC   *see* **adenomatous polyposis coli protein**

apical   Referring to the tip of a cell, a structure, or an organ. The apical surface of an epithelial cell is the exposed free surface, opposite to the **basal** surface. The basal surface rests on the basal lamina that separates the epithelium from other tissue.

apoptosis   Form of **programmed cell death**, in which a "suicide" program is activated within an animal cell, leading to rapid cell death mediated by intracellular proteolytic enzymes called caspases.

Argonaute   Family of proteins found in the **RISC** and **RITS** complexes which play a key role in the repression of gene expression by **microRNAs** and **siRNAs**.

ATP (adenosine 5′-triphosphate)   Nucleoside triphosphate composed of adenine, ribose, and three phosphate groups. The principal carrier of chemical energy in cells. The terminal phosphate groups are highly reactive in the sense that their hydrolysis, or transfer to another molecule, takes place with the release of a large amount of free energy.

ATPase   Enzyme that catalyzes the hydrolysis of **ATP**. Many proteins have ATPase activity.

autoradiography   Method in which the position of a radioactive probe is detected by exposure to X-ray film. Used in conjunction with **blotting** techniques.

autosome   Any chromosomes other than a sex chromosome.

axon   Long nerve cell projection that can rapidly conduct nerve impulses over long distances so as to deliver signals to other cells.

5-azacytidine   Analog of the C **nucleotide** in DNA which cannot be methylated. Used to probe the consequences of the loss of **DNA methylation** on C residues.

Barr body   Inactive X chromosome whose very condensed chromatin structure makes it visible by microscopy.

basal   Situated near the base of a cell or other structure, opposite to the **apical** surface.

basal transcription complex   Complex of **RNA polymerase** and basal transcription factors which assembles at the **gene promoter** and then initiates transcription.

basal transcription factor   Any of the proteins whose assembly at a promoter is required for the binding and activation of **RNA polymerase** and the initiation of transcription.

base pair   Two nucleotides in an RNA or DNA molecule that are held together by hydrogen bonds: for example, G paired with C, and A paired with T or U.

basic DNA-binding domain   DNA-binding domain rich in basic amino acids which can only bind to DNA as a dimer and is associated with either the **helix-loop-helix** or **leucine-zipper** dimerization motifs.

B cell (**B lymphocyte**)   Type of **lymphocyte** that makes antibodies.

beads on a string   Structure of chromatin in which the DNA is wrapped twice around the **histone octamer** forming a series of **nucleosomes** in a 10 nm fiber.

beta-catenin (β-**catenin**)   Multifunctional cytoplasmic protein involved in cell adhesion which can also act independently as a gene-regulatory protein. Has an important role in animal development as part of the **Wnt** signaling pathway.

beta-sheet (β-**sheet**)   Common structural motif in proteins in which different sections of the polypeptide chain run alongside each other, joined together by hydrogen bonding between atoms of the polypeptide backbone. Also known as a β-pleated sheet.

bicoid   **Homeodomain** transcription factor which plays a key role in specifying the anterior structures in the *Drosophila* embryo. The gene encoding bicoid is an example of an **egg-polarity** gene.

bivalent code   Unusual combination of **histone modifications** found in genes which are inactive in **embryonic stem cells** but which are activated upon differentiation.

blastula   Early stage of an animal embryo, usually consisting of a ball of epithelial cells surrounding a fluid-filled cavity, before **gastrulation** begins.

blotting   Biochemical technique in which macromolecules separated on a gel are transferred to a nylon membrane or sheet of paper, thereby immobilizing them for further analysis. See **Northern**, **Southern**, and **Western** blotting (**immunoblotting**).

B lymphocyte   *see* **B cell**

branch point   Nucleotide in the **lariat** structure of the **intron** with which the 5′ end of the intron forms a 5′-to-2′ bond.

bromodomain   Protein domain found in many proteins which can convert chromatin to a more open structure.

budding yeast   Common name given to the baker's yeast *Saccharomyces cerevisiae*, a model experimental organism, which divides by budding off a smaller cell. *Compare with* **fission yeast**.

Ca²⁺/calmodulin-dependent protein kinase   *see* **CaM kinase**

CaM kinase   Serine/threonine protein kinase that is activated by Ca²⁺/calmodulin. Indirectly mediates the effects of an increase in cytosolic Ca²⁺ by phosphorylating specific target proteins.

cAMP   *see* **cyclic AMP**

cAMP-dependent protein kinase   *see* **protein kinase A**

cancer   Disease featuring abnormal and improperly controlled cell division resulting in invasive growths, or tumors, that may spread throughout the body.

5′ cap   Modified G nucleotide which is added to the 5′ end of the nascent RNA in the process of **capping**.

cap-binding complex   Protein complex which binds to the **cap**.

capping   Addition of a modified G nucleotide to the 5′ end of nascent RNA transcripts by a 5′–5′ triphosphate linkage.

carboxyl-terminus   *see* **C-terminus**

β-**catenin**   *see* **beta-catenin**

CBP   **CREB**-binding protein. Transcriptional co-activator for a variety of transcriptional activators. Originally identified on the basis of its binding to the CREB transcription factor.

CCAAT box   DNA sequence frequently found in many gene **promoters** which binds transcription factors such as CEBP/α and CEBP/β.

Cdk   *see* **cyclin-dependent protein kinase**

cDNA   DNA copy of mRNA.

cell cycle (**cell-division cycle**)   Reproductive cycle of a cell: the orderly sequence of events by which a cell duplicates its chromosomes and, usually the other cell contents and divides into two.

cell fate   In developmental biology, describes what a particular cell at a given stage of development will normally give rise to.

cell line   Population of cells of plant or animal origin capable of dividing indefinitely in culture.

cell memory   Retention by cells and their descendants of persistently altered patterns of gene expression, without any change in DNA sequence. *See also* **epigenetic inheritance**.

cell signaling   The process in which cells are stimulated or inhibited by extracellular signals, usually chemical signals produced by other cells.

centromere   Constricted region of a mitotic chromosome that holds sister **chromatids** together.

CG island Region of DNA with a greater than average density of CG sequences; these regions generally remain unmethylated.

chaperone (**molecular chaperone**) Protein that helps guide the proper folding of other proteins, or helps them avoid misfolding. Includes **heat-shock proteins** (HSPs).

chimera Whole organism formed from an aggregate of two or more genetically different populations of cells (two or more genotypes), originating from different zygotes.

ChIP *see* **chromatin immunoprecipitation**

cholesterol An abundant lipid molecule with a characteristic four-ring steroid structure. An important component of the plasma membranes of animal cells.

chondrocyte (**cartilage cell**) Connective tissue cell that secretes the matrix of cartilage.

chromatid One of the two copies of a duplicated chromosome formed by DNA replication during **S phase**. The two chromatids, called sister chromatids, are joined at the **centromere**.

chromatin Complex of DNA, histones, and non-histone proteins found in the nucleus of a eukaryotic cell. The material of which chromosomes are made.

chromatin immunoprecipitation (**ChIP**) Technique by which chromosomal DNA bound by a particular protein can be isolated and identified, by precipitating it by means of an antibody against the bound protein.

chromatin-remodeling complex Enzyme complex that alters histone-DNA configurations in eukaryotic chromosomes, changing the accessibility of the DNA to other proteins, notably those involved in transcription.

chromatography Broad class of biochemical techniques in which a mixture of substances is separated by charge, size, hydrophobicity, non-covalent binding affinities, or some other property by allowing the mixture to partition between a moving phase and a stationary phase.

chromodomain Protein domain found in many proteins which can convert chromatin to a more tightly packed structure.

chromosome Structure composed of a very long DNA molecule and associated proteins that carries part (or all) of the hereditary information of an organism. Especially evident in plant and animal cells undergoing **mitosis** or **meiosis**, during which each chromosome becomes condensed into a compact rod-like structure visible in the light microscope.

Ci factor Cubitus interruptus, transcription factor which is involved in signaling via the **Hedgehog** pathway. The full size protein acts as a transcriptional activator whereas the proteolytic fragment produced in the absence of Hedgehog signaling acts as a transcriptional repressor.

cis On the same side; used, for example, to indicate that two DNA sequences are linked on the same molecule e.g. *cis*-acting regulatory sequence. *Compare with* **trans**.

clone Population of identical individuals (cells or organisms) formed by repeated (asexual) division from a common ancestor. Also used as a verb: "to clone a gene" means to create multiple copies of a gene by growing a clone of carrier cells (such as *E. coli*) into which the gene has been introduced, and from which it can be recovered, by recombinant DNA techniques.

cloning vector Small DNA molecule, usually derived from a bacteriophage or plasmid, which is used to carry the fragment of DNA to be cloned into the recipient cell, and which enables the DNA fragment to be replicated.

co-activator Protein that does not itself bind DNA but assembles on other DNA-bound gene-regulatory proteins to activate transcription of a gene.

codon Sequence of three nucleotides in a DNA or **mRNA** molecule that represents the instruction for incorporation of a specific amino acid into a growing polypeptide chain.

co-immunoprecipitation (**co-IP**) Method of isolating proteins that form a complex with each other by using an antibody specific for one of the partners.

collagen Fibrous protein rich in glycine and proline that is a major component of the extracellular matrix in animals, conferring tensile strength.

commitment Process by which a cell becomes determined to a particular differentiation pathway. Normally precedes the obvious changes in the cell which occur during **differentiation** itself.

complementary Of nucleic acid sequences: capable of forming a perfect base-paired duplex with each other.

complementary DNA *see* **cDNA**

condensin (**condensin complex**) Complex of proteins involved in chromosome condensation prior to **mitosis**.

consensus sequence Average or most typical form of a sequence that is reproduced with minor variations in a group of related DNA, RNA, or protein sequences. Indicates the nucleotide or amino acid most often found at each position. Preservation of a sequence implies that it is functionally important.

co-repressor Protein that does not itself bind DNA but assembles on other DNA-bound gene-regulatory proteins to inhibit the expression of a gene.

core promoter Region of the promoter close to the transcriptional start site which binds the **basal transcriptional complex** and is essential for transcription.

CPSF Cleavage and polyadenylation specificity factor which binds to the **polyadenylation signal** and plays a key role in the process of **polyadenylation**.

CRE Cyclic AMP-response element. DNA sequence which mediates gene activation in response to **cyclic AMP**.

CREB Cyclic AMP-response-element-binding protein. Transcription factor which binds to the **CRE** and activates transcription in response to **cyclic AMP**.

Cstf Cleavage stimulation factor which binds downstream of the **polyadenylation signal** and plays a key role in the process of **polyadenylation**.

CTD C-terminal domain of **RNA polymerase** II. Contains multiple repeats of the sequence Tyr-Ser-Pro-Thr-Ser-Pro-Ser and is a target for phosphorylation during transcriptional initiation (serine at position 5) and transcriptional elongation (serine at position 2).

C-terminus (carboxyl-terminus)   The end of a polypeptide chain that carries a free carboxyl (-COOH) group.

cyclic AMP (cAMP)   Nucleotide that is generated from **ATP** by **adenylate cyclase** in response to various extracellular signals. It acts as small intracellular signaling molecule, mainly by activating **cAMP-dependent protein kinase** (PKA). It is hydrolyzed to AMP by a phosphodiesterase.

cyclic AMP-dependent protein kinase (protein kinase A, PKA)   Enzyme that phosphorylates target proteins in response to a rise in intracellular **cyclic AMP**.

cyclic GMP (cGMP)   Nucleotide that is generated from GTP by guanylate cyclase in response to various extracellular signals.

cyclin   Protein that periodically rises and falls in concentration in step with the eukaryotic cell cycle. Cyclins activate crucial protein kinases (called **cyclin-dependent protein kinases**, or Cdks) and thereby help control progression from one stage of the cell cycle to the next.

cyclin-dependent protein kinase (Cdk)   Protein kinase that has to be complexed with a **cyclin** protein in order to act. Different Cdk-cyclin complexes trigger different steps in the cell-division cycle by phosphorylating specific target proteins.

cytokine   Extracellular signal protein or peptide that acts as a local mediator in cell–cell communication.

cytokine receptor   Cell-surface receptor that binds a specific cytokine or hormone and acts through the **JAK/STAT** signaling pathway.

cytoplasm   Contents of a cell that are contained within its plasma membrane but in the case of eukaryotic cells outside the nucleus.

Delta   Cell-surface protein which acts as a ligand in the **Notch** signaling pathway.

dendrite   Extension of a nerve cell, often elaborately branched, that receives stimuli from other nerve cells.

deoxyribonucleic acid   *see* **DNA**

deoxyribose   The five-carbon monosaccharide component of DNA. Differs from **ribose** in having H at the 2-carbon position rather than OH.

designer zinc finger   **Zinc finger** which has been artificially synthesized with a particular DNA-binding specificity. Potential therapeutic use to deliver an **activation domain** or an **inhibitory domain** to a particular gene which contains its target DNA sequence.

determination   In developmental biology, an embryonic cell is said to be determined if it has become committed to a particular specialized path of development. Determination reflects a change in the internal character of the cell, and it precedes the much more readily detected process of cell **differentiation**. Also known as **commitment**.

differentiation   Process by which a cell undergoes a change to an overtly specialized cell type.

dimerization domain   Region of a protein which mediates its ability to form a dimer with itself (*see* **homodimer**) or another protein (*see* **heterodimer**).

diploid   Containing a double genome (two sets of homologous chromosomes and hence two copies of each gene or genetic locus). *Compare with* **haploid**.

disulfide bond (-S-S-)   Covalent linkage formed between two sulfhydryl (thiol) groups on cysteines.

DNA (deoxyribonucleic acid)   Polynucleotide formed from covalently linked deoxyribonucleotide units. The store of hereditary information within a cell and the carrier of this information from generation to generation.

DNA affinity chromatography   Technique for purifying sequence-specific DNA-binding proteins by their binding to a matrix to which the appropriate DNA sequence is attached.

DNA-binding domain   Domain of a transcription factor responsible for its ability to bind to DNA.

DNA library   Collection of cloned DNA molecules, representing either an entire genome (genomic library) or complementary DNA copies of the mRNA produced by a cell (cDNA library).

DNA methylation   Addition of **methyl** groups to DNA. Extensive methylation of the cytosine base in CG sequences is used in vertebrates to keep genes in an inactive state.

DNA methyltransferases   Enzyme able to add a **methyl** group to DNA.

DNA microarray   A large array of short DNA molecules (each of known sequence) bound to a glass microscope slide or other suitable support. Used to monitor expression of thousands of genes simultaneously: mRNA isolated from test cells is converted to cDNA, which in turn is hybridized to the microarray.

DNA mobility-shift assay   Technique for detecting proteins bound to a specific DNA sequence by the fact that the bound protein slows down the migration of the DNA fragment through a gel during gel electrophoresis.

DNaseI footprinting assay   Technique for determining the DNA sequence to which a DNA-binding protein binds.

DNaseI hypersensitive sites   Sites within chromatin which are highly sensitive to digestion with DNaseI. Frequently located at target sites for transcription factor binding.

DNaseI sensitivity assay   Technique to probe the structure of chromatin by assaying the rate at which it is digested by the **endonuclease** enzyme DNase I.

DNA tumor virus   General term for a variety of different DNA viruses than can cause tumors.

domain (protein domain)   Portion of a protein that has a tertiary structure of its own. Larger proteins are generally composed of several domains, each connected to the next by short flexible regions of polypeptide chain. Homologous domains are recognized in many different proteins.

domain swap   Technique to test whether a particular region of a transcription factor acts, for example, as an **activation domain**. Involves linking the domain being tested to the **DNA-binding domain** of another transcription factor.

dominant   In genetics, the member of a pair of alleles that is expressed in the phenotype of an organism while the other allele is not, even though both alleles are present. Opposite of **recessive**.

dominant negative mutation Mutation that dominantly affects the phenotype, blocking gene activity and causing a loss-of-function phenotype even in the presence of a normal copy of the gene.

double helix The three-dimensional structure of DNA, in which two antiparallel DNA chains, held together by hydrogen bonding between the bases, are wound into a helix.

double-minute chromosome Small additional chromosomes consisting of amplifed DNA which are observed in cancer cells.

Drosophila melanogaster Species of small fly, commonly called a fruit fly. A model organism in molecular genetics.

duplex DNA Double-stranded DNA.

E12/E47 Constitutively expressed **helix-loop-helix** transcription factors which can heterodimerize with cell type-specific transcription factors such as **MyoD**.

E. coli *see* ***Escherichia coli***

ectoderm Embryonic epithelial tissue that is the precursor of the epidermis and nervous system.

E2F protein Gene regulatory protein that switches on many genes that encode proteins required for entry into the **S phase** of the cell cycle. Interaction with the **retinoblastoma protein** results in inhibition of E2F activity.

egg-polarity genes In *Drosophila* development, a gene whose protein product is expressed in a gradient within the fertilized egg so defining one axis of the embryo.

eIF *see* **eukaryotic initiation factor**

electrophoresis Technique for separating molecules (typically proteins or nucleic acids) on the basis of their speed of migration through a porous medium when subjected to a strong electric field.

elongation factor (EF) Nomenclature used in both transcription and translation. In transcription, elongation factors associate with **RNA polymerase** and allow it to transcribe long stretches of DNA without dissociating. In translation, elongation factors bind to the **ribosome** and, by hybridizing GTP, drive the addition of amino acids to the growing polypeptide chain.

embryoid body Initial product of the differentiation of **embryonic stem cells** contains an outer layer of cells surrounding a fluid-filled cavity and a central mass of undifferentiated cells.

embryonic stem cell (**ES cell**) Cell derived from the inner cell mass of the early mammalian embryo. Capable of giving rise to all the cells in the body.

endoderm Embryonic tissue that is the precursor of the gut and associated organs.

endonuclease Enzyme that cleaves nucleic acids *within* the polynucleotide chain. *Compare with* **exonuclease**.

endoplasmic reticulum (**ER**) Labyrinthine membrane-bounded compartment in the cytoplasm of eukaryotic cells, where lipids are synthesized and membrane-bound proteins and secretory proteins are made.

enhanceosome Multi-protein complex which assembles at an **enhancer**. Acts to open the chromatin structure and activate transcription.

enhancer Regulatory DNA sequence which can increase the rate of transcription of a structural gene that can be many thousands of base pairs away.

enzyme Protein that catalyzes a specific chemical reaction.

epigenetic inheritance Inheritance of phenotypic changes in a cell or organism that do not result from changes in the nucleotide sequence of DNA. Can be due to heritable modifications in chromatin such as **DNA methylation** or **histone modifications**.

ErbA **Oncoprotein** originally identified in avian erythroblastosis virus which is derived from the cellular thyroid hormone receptor.

ERE (estrogen-response element) DNA sequence which binds the estrogen receptor and mediates the activation of genes in response to estrogen.

erythrocyte (**red blood cell**) Small hemoglobin-containing blood cell of vertebrates that transports oxygen to, and carbon dioxide from, tissues. No longer synthesizes globin or other proteins (*see* **reticulocyte**).

ES cell *see* **embryonic stem cell**

Escherichia coli (***E. coli***) Rod-like bacterium normally found in the colon of humans and other mammals and widely used in biomedical research.

ETS family Family of transcription factors including the ETS-1 **proto-oncogene** protein and the TCF-1 transcription factor.

euchromatin Region of an **interphase** chromosome that stains diffusely; "normal" chromatin, as opposed to the more condensed **heterochromatin**.

eukaryote (**eucaryote**) Organism composed of one or more cells that have a distinct nucleus. Member of one of the three main divisions of the living world, the other two being Bacteria and Archaea, both of which are **prokaryotes**.

eukaryotic initiation factor (**eIF**) Protein that promotes the proper association of ribosomes with mRNA and is required for the initiation of protein synthesis.

Eve Even skipped, **homeodomain** transcription factor which is expressed in a striped pattern in the *Drosophila* embryo. The gene encoding eve is an example of a **pair-rule gene**.

exon Segment of a eukaryotic gene that consists of a sequence of nucleotides that will be represented in mRNA or in the final transfer, ribosomal, or other mature RNA molecule. In protein-coding genes, exons encode the amino acids in the protein. Exons are so named because they *ex*it the nucleus. *Compare with* **intron**.

exon junction complex Protein complex which is deposited on the RNA during **RNA splicing** and which promotes the transport of the spliced RNA to the cytoplasm.

exon skipping Process in which an exon is incorrectly spliced out of the mRNA resulting in disease.

exon splicing enhancer Sequences located in the **exons** of the RNA transcript which promote splicing of the RNA.

exonuclease Enzyme that cleaves nucleotides one at a time from the ends of polynucleotides. *Compare with* **endonuclease**.

exosome Multi-protein complex which degrades mRNA.

expression vector   A virus or plasmid that carries a DNA sequence into a suitable host cell and there directs the synthesis of the protein encoded by the sequence.

extracellular signal molecule   Any secreted or cell-surface chemical signal that binds to receptors and regulates the activity of the cell expressing the receptor.

fertilization   Fusion of a male and a female gamete (both **haploid**) to form a **diploid zygote**, which develops into a new individual.

fibroblast   Common cell type found in connective tissue. Secretes an extracellular matrix rich in collagen and other extracellular matrix macromolecules. Migrates and proliferates readily in wounded tissue and in tissue culture.

FISH   *see* **fluorescence *in situ* hybridization**

fission yeast   Common name for the yeast model organism *Schizosaccharomyces pombe*. It divides to give two equal-sized cells. *Compare with* **budding yeast**.

fluorescence *in situ* hybridization (**FISH**)   Technique in which fluorescently labeled nucleic acid probes hybridize to specific DNA or RNA sequences *in situ*.

footprinting   *see* **DNaseI footprinting**

Fos   Oncogenic transcription factor containing **helix-loop-helix** and **basic DNA-binding domains**. Binds to **AP1 site** as a heterodimer with **Jun**.

fusion protein   Protein that combines two or more normally separate polypeptides.

GAGA factor   Transcription factor involved in chromatin remodeling/gene activation. A member of the **trithorax** family.

gain-of-function mutation   Mutation that increases the activity of a gene, or makes it active in inappropriate circumstances. Usually **dominant**.

GAL4   Yeast transcription factor involved in the regulation of gene expression in response to galactose.

gamete   Specialized **haploid** cell (a sperm or egg) formed from primordial germ cells by meiosis and specialized for sexual reproduction. *See* also **germ cell**.

gap gene   In *Drosophila* development, a gene that is expressed in specific broad regions along the anteroposterior axis of the early embryo, and which helps designate the main divisions of the insect body. Mutation results in the absence of several adjacent segments of the fly.

gastrulation   Stage in animal embryogenesis during which the embryo is transformed from a ball of cells to a structure with a gut (a gastrula).

GCN4   Yeast transcription factor involved in the regulation of gene expression in response to amino acid starvation.

GEF   *see* **guanine nucleotide exchange factor**

gene   Region of DNA that is transcribed as a single unit and carries information for a discrete hereditary characteristic, usually corresponding to (1) a single protein (or set of related proteins generated by variant post-transcriptional processing), or (2) a single RNA (or set of closely related RNAs).

gene-activator protein   *see* **activator**

gene-control region   The set of linked DNA sequences regulating expression of a particular gene. Includes **promoter** and regulatory sequences required to initiate transcription of the gene and control the rate of initiation.

gene expression   Production of an observable molecular product (RNA or protein) by a gene.

gene repressor protein   *see* **repressor**

genetic code   Set of rules specifying the correspondence between nucleotide triplets (**codons**) in DNA or RNA and amino acids in proteins.

genetic engineering (**recombinant DNA technology**)   Collection of techniques by which DNA segments from different sources are combined to make a new DNA, often called a recombinant DNA. Recombinant DNAs are widely used in the cloning of genes, in the genetic modification of organisms and in molecular biology generally.

genome   The totality of genetic information belonging to a cell or an organism; in particular, the DNA that carries this information.

genomic DNA   DNA constituting the genome of a cell or an organism. Often used in contrast to **cDNA** (DNA prepared by reverse transcription from mRNA). Genomic DNA clones represent DNA cloned directly from chromosomal DNA, and a collection of such clones from a given genome is a genomic DNA library.

genomic imprinting   Phenomenon in which a gene is either expressed or not expressed in the offspring depending on which parent it is inherited from.

genomics   Study of the DNA sequences and properties of entire genomes.

genotype   Genetic constitution of an individual cell or organism. The particular combination of **alleles** found in a specific individual. *Compare with* **phenotype**.

germ cell   A cell in the germ line of an organism, which includes the **haploid** gametes and their specific **diploid** precursor cells. Germ cells contribute to the formation of a new generation of organisms and are distinct from **somatic cells**, which form the body and leave no descendants.

germ layer   One of the three primary tissue layers (**endoderm**, **mesoderm**, and **ectoderm**) of an animal embryo.

germ line   The cell lineage that consists of the **haploid** gametes and their specific **diploid** precursor cells.

GPCR   *see* **G-protein-coupled receptor**

G_1/S-Cdk   Cyclin–Cdk complex formed in vertebrate cells by a G_1/S-cyclin and the corresponding **cyclin-dependent kinase** (Cdk).

G_1/S-cyclin   **Cyclin** that activates Cdks in late G1 of the eukaryotic cell cycle and thereby helps trigger progression through Start, resulting in a commitment to cell-cycle entry. Its level falls at the start of **S phase**.

G_1 phase   Gap 1 phase of the eukaryotic cell-division cycle, between the end of **mitosis** and the start of DNA synthesis.

G_2 phase   Gap 2 phase of the eukaryotic cell-division cycle, between the end of DNA synthesis and the beginning of **mitosis**.

G protein (**trimeric GTP-binding protein**)   A trimeric GTP-binding protein with intrinsic GTPase activity that couples **GPCRs** to enzymes or ion channels in the plasma membrane.

G-protein-coupled receptor (**GPCR**) A receptor that, when activated by its extracellular ligand, activates a **G protein**.

GRE (**glucocorticoid-response element**) DNA sequence which binds the glucocorticoid receptor and mediates the activation of genes in response to glucocorticoid.

growth factor Extracellular signal protein that can stimulate a cell to grow. They often have other functions as well, including stimulating cells to survive or proliferate. Examples include epidermal growth factor (EGF) and platelet-derived growth factor (PDGF).

GTP (**guanosine 5′-triphosphate**) Nucleoside triphosphate produced by the phosphorylation of guanosine diphosphate (GDP). Like **ATP**, it releases a large amount of free energy on hydrolysis of its terminal phosphate group. Has a role in protein synthesis and cell signaling.

GTPase An enzyme that converts GTP to GDP. GTPases fall into two large families. Large **trimeric G proteins** are composed of three different subunits and mainly couple **GPCRs** to enzymes or ion channels in the plasma membrane. Small monomeric GTP-binding proteins (also called monomeric GTPases) consist of a single subunit and help relay signals from many types of cell-surface receptors and have roles in intracellular signaling pathways. Both trimeric G proteins and monomeric GTPases cycle between an active GTP-bound form and an inactive GDP-bound form and frequently act as molecular switches in intracellular signaling pathways.

GTPase-activating protein (**GAP**) Protein that binds to a **GTPase** and stimulates its GTPase activity, causing the enzyme to hydrolyze its bound GTP to GDP.

GTP-binding protein *see* **GTPase**

guanine nucleotide exchange factor (**GEF**) Protein that binds to a **GTPase** and activates it by stimulating it to release its tightly bound GDP, thereby allowing it to bind GTP in its place.

guanosine 5′-triphosphate *see* **GTP**

haploid Having only a single copy of the genome (one set of chromosomes), as in a sperm cell, unfertilized egg, or bacterium. *Compare with* **diploid**.

haploid insufficiency Disease in which one copy of a gene is mutated and the remaining **wild-type** copy cannot produce sufficient functional protein to prevent disease.

haplotype Haploid genotype.

heat-shock element (**HSE**) DNA sequence in genes encoding **heat-shock proteins** which binds the **HSF** transcription factor. The presence of an HSE allows these genes to be transcriptionally activated in response to heat or other stresses.

heat-shock factor (**HSF**) Transcription factor which is activated by heat or other stresses when activated. It binds to the **HSE** in the genes encoding **heat-shock proteins** and activates their transcription.

heat-shock protein (**HSP, stress-response protein**) One of a large family of highly conserved molecular chaperone

proteins, so named because they are synthesized in increased amounts in response to elevated temperature or other stressful treatment. HSPs have important roles in aiding correct protein folding or refolding. Prominent examples are HSP90 and HSP70.

heavy chain (**H chain**) The larger of the polypeptide chains in an **immunoglobulin** or other protein molecule.

Hedgehog proteins Family of secreted extracellular signal molecules (e.g. Sonic Hedgehog) that have many different roles controlling cell differentiation and gene expression in animal embryos and adult tissues.

α-**helix** *see* **alpha-helix**

helix-loop-helix (**HLH**) Dimerization motif present in many gene-regulatory proteins, consisting of a short α-**helix** connected by a flexible loop to a second longer α-helix. Its structure enables two HLH-containing proteins to dimerize, forming a complex that binds to DNA via the **basic DNA-binding domain**. Distinct from the **helix-turn-helix** motif.

helix-turn-helix DNA-binding structural motif present in many gene-regulatory proteins, consisting of two α helices held at a fixed angle and connected by a short chain of amino acids, constituting the turn. Distinct from the **helix-loop-helix** motif.

heme Cyclic organic molecule containing an iron atom that carries oxygen in hemoglobin.

heterochromatin Region of a chromosome that remains in the form of unusually condensed chromatin; generally transcriptionally inactive. *Compare with* **euchromatin**.

heterodimer Protein complex composed of two different polypeptide chains.

heterothallic Term for yeast strains with separate mating types.

heterozygote **Diploid** cell or individual having two different alleles of one or more specified genes.

HIF-1 (**hypoxia-inducible factor 1**) Transcription factor which mediates gene activation in response to **hypoxia**.

high-performance liquid chromatography (**HPLC**) Type of chromatography that uses columns packed with tiny beads of matrix; the solution to be separated is pushed through under high pressure.

histone One of a group of small abundant proteins, rich in arginine and lysine. Histones form the **nucleosome** cores around which DNA is wrapped in eukaryotic chromosomes.

histone acetyltransferase (**HAT**) Enzyme able to add an acetyl group to histones. *See* **acetylation**.

histone code Combinations of post-translational modifications to histones (e.g. **acetylation**, **methylation**) that are thought to determine how and when the DNA packaged in nucleosomes can be accessed.

histone deacetylase (**HDAC**) Enzyme able to remove acetyl groups from histones.

histone H1 "Linker" (as opposed to "core") histone protein that binds to DNA where it exits from a nucleosome and helps package **nucleosomes** into the 30 nm chromatin fiber.

histone methyltransferase   Enzyme able to add a methyl group to histones. *See also* **methylation**.

histone modification   Post-translational modification of a histone, e.g. by **acetylation**, **methylation**, **phosphorylation**, etc.

histone octamer   Complex containing two molecules of each of the four histones: H2A, H2B, H3, H4.

histone variant   Less-abundant histone forms encoded by different genes to those encoding the more abundant histones (H1, H2A, H2B, H3, and H4).

HIV   Human immunodeficiency virus, the **retrovirus** that is the cause of acquired immune deficiency syndrome (AIDS).

HLH   *see* **helix-loop-helix**

hnRNP protein (**heterogeneous nuclear ribonuclear protein**)   Any of a group of proteins that assemble on newly synthesized RNA, organizing it into a more compact form.

HO   **Endonuclease** enzyme which initiates **mating-type switching** in yeast.

homeodomain   DNA-binding domain that defines a class of gene-regulatory proteins important in animal development.

homeotic mutation   Mutation that causes cells in one region of the body to behave as though they were located in another, causing a bizarre disturbance of the body plan.

homeotic selector gene   In *Drosophila* development, a gene that defines and preserves the differences between body segments.

homodimer   Protein complex composed of two identical polypeptide chains.

homologous   Genes, proteins, or body structures that are similar as a result of a shared evolutionary origin.

homologous chromosomes   The maternal and paternal copies of a particular chromosome in a **diploid** cell.

homothallic   Term for yeast strains which switch **mating type** from **a** to α and vice versa.

homozygote   **Diploid** cell or organism having two identical **alleles** of a specified gene or set of genes.

hormone   Signal molecule secreted into the bloodstream, which can then carry it to distant target cells.

housekeeping gene   Gene serving a function required in all the cell types of an organism, regardless of their specialized role.

***Hox* gene complex**   Cluster of genes coding for gene-regulatory factors, each gene containing a **homeodomain**, and specifying body-region differences. *Hox* mutations typically cause **homeotic** transformations.

HP1   Heterochromatin protein 1. Involved in producing the tightly packed structure of **heterochromatin**.

HPLC   *see* **high-performance liquid chromatography**

HSE   *see* **heat-shock element**

HSF   *see* **heat-shock factor**

HSP   *see* **heat-shock protein**

hybridization   In molecular biology, the process whereby two complementary nucleic acid strands form a base-paired duplex, DNA–DNA, DNA–RNA, or RNA–RNA molecule. Forms the basis of a powerful experimental technique for detecting specific nucleotide sequences.

hypoxia   Low oxygen level.

Id   Inhibitory transcription factor which can heterodimerize with **MyoD** and prevent it activating transcription.

IF   *see* **eukaryotic initiation factor**

Ig   *see* **antibody**

Ig superfamily   Large and diverse family of proteins that contain immunoglobulin domains or immunoglobulin-like domains. Most are involved in cell–cell interactions or antigen recognition.

IκB protein   Inhibitory transcription factor which binds to **NFκB** and prevents it activating transcription. Specific stimuli can activate NFκB by promoting the phosphorylation of IκB, leading to its degradation.

imaginal disc   Group of cells that are set aside, apparently undifferentiated, in the *Drosophila* embryo and which will develop into an adult structure; e.g. eye, leg, wing.

immortalization   Production of a cell line capable of an apparently unlimited number of cell divisions. Can be the result of mutations or viral transformation or of fusion of the original cells with cells of a tumor line.

immune response   Response made by the immune system when a foreign substance or microorganism enters the body.

immune system   System of **lymphocytes** and other cells in the body that provides defense against infection.

immunoblotting   *see* **blotting**

immunoglobulin (**Ig**)   An antibody molecule able to recognize specific proteins.

immunohistochemistry   Use of specific antibody to stain sections of tissue or cells to identify the distribution of a protein.

immunoprecipitation (**IP**)   Use of a specific antibody to draw the corresponding protein antigen out of solution. The technique can identify complexes of interacting proteins in cell extracts by using an antibody specific for one of the proteins to precipitate the complex. *See also* **chromatin immunoprecipitation**, **co-immunoprecipitation**.

imprinting   *see* **genomic imprinting**

inducible promoter   A regulatory DNA sequence that allows expression of an associated gene to be switched on by a particular molecular or physical stimulus (e.g. heat shock).

inhibitory domain   Domain of a transcription factor responsible for its ability to inhibit transcription.

initiator element   DNA sequence close to the transcriptional start site which in some genes serves as an alternative to the **TATA box** for recruitment of the **basal transcriptional complex**.

initiator tRNA   Special **transfer RNA** that initiates translation. It always carries the amino acid methionine, forming the complex Met-tRNA$_i$.

inner cell mass Cluster of undifferentiated cells in the early mammalian embryo from which the whole of the adult body is derived.

inositol Ring-shaped sugar molecule forming part of inositol phospholipids.

in situ **hybridization** Technique in which a single-stranded RNA or DNA probe is used to locate a gene or mRNA molecule in a cell or tissue by **hybridization**.

insulator element DNA sequence that prevents a gene-regulatory protein bound to DNA in the control region of one gene from influencing the transcription of adjacent genes.

insulin Polypeptide hormone that is secreted by β cells in the pancreas to help regulate glucose metabolism in animals.

interferon (**IFN**) Member of a class of cytokines secreted by virus-infected cells and certain types of activated **T cells**. Interferons induce antiviral responses.

interleukin (**IL**) Secreted **cytokine** that mainly mediates local interactions between white blood cells during inflammation and immune responses.

internal ribosome-entry site (**IRES**) Specific site in a eukaryotic mRNA, other than at the 5′ end, at which the ribosome can bind and initiate translation.

interphase Long period of the cell cycle between one mitosis and the next.

intracellular signaling protein Protein involved in a signaling pathway inside the cell. It usually activates the next protein in the pathway or generates a small intracellular mediator.

intron Non-coding region of a eukaryotic gene that is transcribed into an RNA molecule but is then excised by **RNA splicing** during production of the mRNA or other functional RNA. Introns are so named becomes they remain *in* the nucleus. *Compare with* **exon**.

in vitro Taking place in an isolated cell-free extract, as opposed to in a living cell; also sometimes used to distinguish studies in cell cultures from studies in intact organisms (Latin for "in glass").

in vivo In an intact cell or organism (Latin for "in life").

IRES *see* **internal ribosome-entry site**

iron-response element Sequence in an mRNA which regulates its stability or translation in response to iron.

isoelectric point (**pI**) The pH at which a molecule in solution has no net electric charge and therefore does not move in an electric field. Differences in the isoelectric points of different proteins are used to separate proteins in the isoelectric focusing technique. In this technique each protein in a mixture moves in a pH gradient until its charge is neutralized so that proteins with different isoelectric points move to different positions.

isoform One of a set of variant forms of a protein, derived either by **alternative RNA splicing** of a common transcript or as products of different members of a set of closely homologous genes.

JAK/STAT signaling pathway Signaling pathway activated by **cytokines** and some hormones, providing a rapid route from the plasma membrane to the nucleus to alter gene transcription. Involves cytoplasmic Janus kinases (JAKs), and signal transducers and activators of transcription (STATs).

Jun Oncogenic transcription factor containing **helix-loop-helix** and **basic DNA-binding domains**. Binds to **AP1 site** as a homodimer or as a heterodimer with **Fos**.

karyotype Display of the full set of chromosomes of a cell, arranged with respect to size, shape, and number.

kinase Enzyme that catalyzes the addition of phosphate groups to molecules. *See also* **protein kinase**.

knockout An engineered deletion or inactivating mutation of a gene.

Kruppel **Zinc-finger** transcription factor which plays a key role in *Drosophila* development. The gene encoding Kruppel is an example of a **gap** gene.

lampbrush chromosome Huge paired chromosome in **meiosis** in immature amphibian eggs, in which the chromatin forms large stiff loops extending out from the linear axis of the chromosome.

lariat Structure of the **intron** during and after RNA splicing. Contains a bond between the 5′ end of the intron and the 2′ position of the nucleotide at the **branch point**.

lateral inhibition Process in which a differentiating cell prevents adjacent cells from following the same pathway of differentiation.

L chain *see* **light chain**

lethal mutation Mutation that causes the death of the cell or the organism that contains it.

leucine zipper Structural motif seen in many DNA-binding proteins in which two α-**helices** from separate proteins are joined together in a coiled-coil (rather like a zipper), forming a protein dimer.

leukemia Cancer of **white blood cells**.

ligand Any molecule that binds to a specific site on a protein or other molecule. From Latin *ligare*, meaning "to bind".

light chain (**L chain**) One of the smaller polypeptides of a multi-subunit protein such as myosin or immunoglobulin.

lipid Organic molecule that is insoluble in water but tends to dissolve in non-polar organic solvents. A special class, the phospholipids, forms the structural basis of biological membranes.

locus In genetics, the position on a chromosome. For example, in a **diploid** cell different **alleles** of the same gene occupy the same locus.

locus-control region Regulatory element which can determine the chromatin structure of a large region of DNA.

long-term potentiation Long-lasting increase (days to weeks) in the sensitivity of certain synapses in the brain, induced by a short burst of repetitive firing in the presynaptic neurons.

long terminal repeat Repeated sequence present at each

end of the genomes of **retroviruses** which has **promoter/enhancer** activity.

loss-of-function mutation   Mutation which reduces or abolishes the activity of a gene. Usually **recessive**.

LTP   *see* **long-term potentiation**

lymphocyte   **White blood cell** responsible for the specificity of adaptive immune responses. Two main types: **B cells**, which produce antibody, and **T cells**, which interact directly with other effector cells of the immune system and with infected cells. T cells develop in the thymus and are responsible for cell-mediated immunity. B cells develop in the bone marrow in mammals and are responsible for the production of circulating antibodies.

lymphoma   Cancer of lymphocytes, in which the cancer cells are mainly found in lymphoid organs (rather than in the blood, as in **leukemias**).

MADS family   Transcription factor family named for its founder members MCM1, agamous, deficiens, and serum response factor.

maintenance methylase   Enzyme which can recognize and methylate CG sites in DNA which are methylated only on one strand of the DNA. It then fully methylates them.

MALDI   Matrix-assisted laser desorption–ionization method for determining the molecular weight of a peptide. Often combined with **nanospray mass spectrometry**.

malignant   Of tumors and tumor cells: invasive and/or able to undergo **metastasis**. A malignant tumor is a cancer.

MAP kinase (mitogen-activated protein kinase)   Protein kinase at the end of a three-component signaling module involved in relaying signals from the plasma membrane to the nucleus.

MAP-kinase cascade (mitogen-activated protein kinase cascade)   An intracellular signaling cascade composed of three protein kinases, acting in sequence, with **MAP kinase** as the third. Typically activated by a **Ras protein** in response to extracellular signals.

mass spectrometry (MS)   Technique for identifying compounds on the basis of their precise mass-to-charge ratio. Powerful tool for identifying proteins and sequencing polypeptides.

master regulatory transcription factor   Transcription factor which can induce a particular pattern of differentiation. In muscle differentiation, four master regulatory transcription factors have been defined: **MyoD**, Myf5, myogenin, and Mrf4.

mating-type locus (Mat locus)   In budding yeast, the locus that determines the mating type (α or **a**) of the haploid yeast cell.

mating-type switching   Conversion of mating type from **a** to α or vice versa in **homothallic** yeast.

matrix-attachment region   Region of the chromosome which attaches to the **nuclear matrix**. Also known as scaffold-attachment regions.

MDM2   **Oncoprotein** which can inhibit **p53** activity by inducing its degradation.

MDM4   **Oncoprotein** which can inhibit **p53** activity by masking its transcriptional activation domain.

MeCP2   Regulatory protein which binds specifically to methylated DNA.

mediator   Multi-protein complex which links activating transcription factors to the **basal transcriptional complex**.

MEF2 (myocyte enhancer factor 2)   Family of transcription factors. Originally identified on the basis of their key role in muscle cells but now known to regulate transcription in other cell types such as neuronal cells.

meiosis   Special type of cell division that occurs in sexual reproduction. It involves two successive nuclear divisions with only one round of DNA replication, thereby producing **haploid** cells from a **diploid** cell.

mesenchyme   Immature, unspecialized form of connective tissue in animals, consisting of cells embedded in a thin extracellular matrix.

mesoderm   Embryonic tissue that is the precursor to muscle, connective tissue, skeleton, and many of the internal organs.

messenger RNA (mRNA)   RNA molecule that specifies the amino acid sequence of a protein. Produced in eukaryotes by processing of an RNA molecule made by **RNA polymerase** as a complementary copy of DNA. It is translated into protein in a process catalyzed by ribosomes.

metastasis   Spread of cancer cells from their site of origin to other areas in the body.

methyl group (-CH$_3$)   Hydrophobic chemical group derived from methane (CH$_4$).

methylation   Addition of a methyl (-CH$_3$) group to histones or other proteins.

microarray   *see* **DNA microarray**

micron (μm or micrometer)   Unit of measurement equal to 10^{-6} meter or 10^{-3} millimeter.

microRNA (miRNA)   Short (21–26 nucleotide) eukaryotic RNAs, produced by the processing of specialized RNA transcripts coded in the genome, that regulate gene expression through complementary base-pairing with mRNA. Depending on the extent of base pairing miRNAs can lead either to destruction of the mRNA or to a block in its translation.

miRNA   *see* **microRNA**

mitogen   Extracellular signal molecule that stimulates cells to proliferate.

mitogen-activated protein kinase   *see* **MAP kinase**

mitosis   Division of the nucleus of a eukaryotic cell, involving condensation of the DNA into visible chromosomes, and separation of the duplicated chromosomes to form two identical sets. From Greek *mitos*, a thread, referring to the thread-like appearance of the condensed chromosomes.

mitotic chromosome   Highly condensed duplicated chromosome with the two new chromosomes still held together at the **centromere** as sister **chromatids**.

molecular chaperone   *see* **chaperone**

monocistronic   mRNA that encodes only a single protein.

monomeric GTPase   *see* **GTPase**

morphogen Signal molecule that can impose a pattern on a field of cells by causing cells in different places to adopt different fates.

morula Early stage of embryonic development which consists of a ball of 16 cells.

mosaic (genetic mosaic) In developmental biology, an individual organism made of a mixture of cells with different genotypes, but developed from a single zygote. Mosaics can arise naturally, as a result of a mutation in cells that give rise to new tissues, or can be made deliberately to aid genetic analysis. *Compare with* **chimera**.

motif Element of structure or pattern that recurs in many contexts. Specifically, a small structural domain that can be recognized in a variety of proteins.

mRNA *see* **messenger RNA**

muscle cell Cell type specialized for contraction. The three main classes are skeletal, heart, and smooth muscle cells.

mutation Heritable change in the DNA nucleotide sequence of a chromosome.

Myc Gene-regulatory **oncoprotein** that is activated when a cell is stimulated to grow and divide by extracellular signals. It activates the transcription of many genes, including those that stimulate cell growth.

myoblast Mononucleated, undifferentiated muscle precursor cell. A skeletal muscle cell is formed by the fusion of multiple myoblasts.

MyoD Transcription factor expressed in skeletal muscle cells whose over-expression in other cell types can induce skeletal muscle cell differentiation.

myoepithelial cell Type of unstriated muscle cell found in epithelia, e.g. in the iris of the eye and in glandular tissue.

myotome Region of the **mesoderm** which ultimately forms muscle.

myotube Multinucleated muscle cell, formed by fusion of multiple **myoblasts**.

nanometer (nm) Unit of length commonly used to measure molecules and cell organelles. 1 nm=10^{-3} micrometer (μm)=10^{-9} meter.

nanospray mass spectrometry Method for determining amino acid sequence of a peptide. Often combined with the **MALDI** method.

negative feedback Control mechanism whereby the output of a reaction or pathway inhibits an earlier step in the same pathway.

nerve cell *see* **neuron**

neural tube Tube of **ectoderm** that will form the brain and spinal cord in a vertebrate embryo.

neurite Long process growing from a nerve cell in culture. A generic term that does not specify whether the process is an **axon** or a **dendrite**.

neuroblast Embryonic nerve cell precursor.

neuron (nerve cell) Impulse-conducting cell of the nervous system, with extensive processes specialized to receive, conduct, and transmit signals.

NFκB protein Latent transcription factor that is activated by various intracellular signaling pathways when cells are stimulated during immune, inflammatory, or stress responses.

nGRE DNA sequence which binds the glucocorticoid receptor and mediates the repression of genes in response to glucocorticoid.

nm *see* **nanometer**

nonsense-mediated mRNA decay Mechanism for degrading aberrant mRNAs containing in-frame internal **stop codons** before they can be translated into protein.

Northern blotting Technique in which RNA fragments separated by electrophoresis are immobilized on a paper sheet, and a specific RNA is then detected by hybridization with a labeled nucleic acid probe.

Notch Transmembrane receptor protein (and latent gene-regulatory protein) involved in many cell-fate choices in animal development, for example in the specification of nerve cells from ectodermal epithelium. Its ligands are cell-surface proteins such as **Delta**.

notochord Stiff rod of cells defining the central axis of all chordate embryos. In vertebrates becomes incorporated into the vertebral column.

NPC *see* **nuclear pore complex**

N-terminus (amino-terminus) The end of a polypeptide chain that carries a free α-amino group.

nuclear envelope (nuclear membrane) Double membrane (two bilayers) surrounding the nucleus. Consists of an outer and inner membrane and is perforated by nuclear pores. The outer membrane is continuous with the **endoplasmic reticulum**.

nuclear export signal Signal which ensures that the molecules containing it are transported from the nucleus to the cytosol through **nuclear pore complexes**.

nuclear localization signal (NLS) Signal sequence found in proteins destined for the nucleus that enables their selective transport into the nucleus from the cytosol through the **nuclear pore complexes**.

nuclear matrix Network of RNA and protein fibrils extending throughout the nucleus. Also known as nuclear scaffold.

nuclear pore complex (NPC) Large multiprotein structure forming an aqueous channel (the nuclear pore) through the nuclear envelope that allows selected molecules to move between nucleus and cytoplasm.

nuclear receptor superfamily Intracellular receptors for hydrophobic signal molecules such as steroid and thyroid hormones and retinoic acid. The receptor–ligand complex acts as a transcription factor in the nucleus.

nuclear run-on assay Method for measuring the transcription rate of a particular gene by adding radioactive precursor to isolated nuclei and measuring its incorporation into the RNA transcript.

nuclear transplantation Transfer of a nucleus from one cell to another by microinjection.

nuclear transport receptor Protein that escorts macromolecules either into or out of the nucleus; nuclear import receptor or nuclear export receptor.

nuclease   Enzyme that splits nucleic acids by hydrolyzing bonds between nucleotides. *See also* **endonuclease, exonuclease**.

nucleic acid   RNA or DNA, a macromolecule consisting of a chain of nucleotides joined together by phosphodiester bonds.

nucleic acid hybridization   *see* **hybridization**

nucleolus   Structure in the nucleus where **rRNA** is transcribed and ribosomal subunits are assembled.

nucleoside   Purine or pyrimidine base covalently linked to a ribose or deoxyribose sugar.

nucleosome   Bead-like structure in eukaryotic chromatin, composed of a short length of DNA wrapped around a **histone octamer**. The fundamental structural unit of chromatin.

nucleosome remodeling   Process in which **nucleosomes** are either displaced from or along the DNA or are altered structurally so producing altered chromatin structure.

nucleotide   Nucleoside with one or more phosphate groups joined in ester linkages to the sugar moiety. DNA and RNA are polymers of nucleotides.

nucleus   Prominent membrane-bounded organelle in eukaryotic cells, containing DNA organized into chromosomes.

null mutation   Loss-of-function mutation that completely abolishes the activity of a gene.

NURF   Nucleosome-remodeling factor, a protein complex able to produce **nucleosome remodeling**.

octamer motif   DNA sequence ATGCAAAT found in many gene **promoters** which binds **POU** family transcription factors such as Oct1, Oct2, and Oct4.

oligonucleotide   Short DNA sequence.

oncogene   A gene whose protein product can make a cell cancerous when over-expressed or mutated.

oncoprotein   Protein encoded by an **oncogene**.

oocyte   Developing egg, before it has completed **meiosis**.

oogenesis   Formation and maturation of **oocytes** in the ovary.

open reading frame (**ORF**)   A continuous nucleotide sequence free from **stop codons** in at least one of the three reading frames (and thus with the potential to code for protein).

p300   Transcriptional co-activator for a variety of transcriptional activators. Closely related to **CBP**. Named on the basis of its molecular weight of 300 kDa.

p53   **Anti-oncogene** (tumor-suppressor gene) which is mutated in about half of human cancers. Encodes a gene-regulatory protein that is activated by damage to DNA and is involved in blocking further progression through the **cell cycle** or inducing **apoptosis**.

paired domain   DNA-binding domain found in several Pax family transcription factors.

pair-rule gene   In *Drosophila* development, a gene expressed in a series of regular transverse stripes along the body of the embryo and which helps to determine its different segments.

palindromic sequence   Rotationally symmetrical DNA sequence which has the same 5′-to-3′ sequence on both strands of the DNA. Often found as a DNA-binding site for transcription factors which bind as homodimers.

P-body   Cytoplasmic granules in which mRNAs are decapped (see **capping**) and degraded.

PCR (polymerase chain reaction)   Technique for amplifying specific regions of DNA by the use of sequence-specific primers and multiple cycles of DNA synthesis, each cycle being followed by a brief heat treatment to separate complementary strands.

peptide   Short polymer of amino acids.

phenotype   The observable character (including both physical appearance and behavior) of a cell or organism. *Compare with* **genotype**.

phosphatase   Enzyme that catalyzes the hydrolytic removal of phosphate groups from a molecule.

phosphodiester bond   A covalent chemical bond formed when two hydroxyl groups form ester linkages to the same phosphate group, such as between adjacent nucleotides in RNA or DNA.

3-phosphoinositide-dependent kinase I (**PDKI**)   Kinase which is activated by PIP_3 (produced by **PI3-kinase**) and which in turn phosphorylates **Akt**.

phosphoinositide 3-kinase (**PI3-kinase**)   Membrane-bound enzyme that is a component of the PI3-kinase/**Akt** intracellular signaling pathway. It phosphorylates phosphatidylinositol 4,5-bisphosphate (PIP_2) at the 3 position on the **inositol** ring to produce phosphatidylinositol 3,4,5-triphosphate (PIP_3) which then activates 3-phosphoinositide-dependent kinase 1 (PDK1).

phosphorylation   Reaction in which a phosphate group is covalently coupled to another molecule.

phosphorylation cascade   Series of sequential protein phosphorylations mediated by a series of protein kinases, each of which phosphorylates and activates the next kinase in the chain. Such cascades are common in intracellular signaling pathways.

PI3-kinase   *see* **phosphoinositide 3-kinase**

pluripotent   Describes a cell that is able to give rise to several different cell types. *Compare with* **totipotent**.

point mutation   Change of a single nucleotide pair or a very small part of a single gene, in DNA.

polyadenylation   Addition of a long sequence of A nucleotides (the polyA tail) to the 3′ end of a nascent mRNA molecule.

polyadenylation signal   AAUAAA sequence in the RNA which directs its cleavage in the first stage of **polyadenylation**.

polyA tail   *see* **polyadenylation**

polycistronic mRNA   Individual mRNA with several different sites of translational initiation.

polycomb complex   Multi-protein complex which

produces a closed chromatin structure, incompatible with transcription.

polymerase Enzyme that catalyzes polymerization reactions such as the synthesis of DNA and RNA. *See also* **RNA polymerase**.

polymerase chain reaction *see* **PCR**

polymerase pausing Process in which the **RNA polymerase** pauses after initiating transcription and transcribing approximately 30 nucleotides. Subsequently followed by transcriptional elongation phase.

polypeptide Linear polymer of amino acids. Proteins are large polypeptides, and the two terms can be used interchangeably.

polypyrimidine tract Run of multiple pyrimidine (C or T) nucleotides involved in RNA splicing. Located adjacent to the **branch point** and the 3′ **splice site**.

polyribosome (polysome) Messenger RNA molecule to which are attached a number of **ribosomes** engaged in protein synthesis.

polytene chromosome Giant chromosome in which the DNA has undergone repeated replication and the many copies have stayed together.

position effect Difference in gene expression that depends on the position of the gene on the chromosome and probably reflects differences in the state of the chromatin along the chromosome. When an active gene is placed next to **heterochromatin**, the inactivating influence of the heterochromatin can spread to affect the gene to a variable degree giving rise to position effect variegation.

post-transcriptional control Any control of gene expression that is exerted at a stage after transcription has begun.

post-translational modification An enzyme-catalyzed change to a protein made after it is synthesized. Examples are acetylation, cleavage, glycosylation, methylation, and phosphorylation.

POU domain DNA-binding domain named for the transcription factors in which it was originally found (Pit-1, Oct1, Oct2, Unc-86). Consists of a POU-specific domain and a POU-homeodomain related to the classical DNA-binding **homeodomain**.

pre-B cell Immediate precursor of a **B cell**.

pre-mRNA Precursor to messenger RNA. In eukaryotes, includes all intermediate stages of RNA processing.

primary transcript (primary RNA transcript) Newly synthesized transcript, before it has undergone splicing or other modifications.

primer **Oligonucleotide** that pairs with a template DNA or RNA strand and promotes the synthesis of a new complementary strand by a polymerase.

probe Defined fragment of RNA or DNA, radioactively or chemically labeled, used to locate specific nucleic acid sequences by hybridization.

process/discard decision Post-transcriptional control mechanism in which an RNA transcript is either spliced to produce a functional mRNA or is degraded.

programmed cell death A form of cell death in which a cell kills itself by activating an intracellular death program.

prokaryote (procaryote) Single-celled microorganism whose cells lack a well-defined, membrane-enclosed nucleus. Either a bacterium or an archaebacterium. *Compare with* **eukaryote**.

promoter Region of DNA close to the transcriptional start site which directs transcription. *See also* **core promoter**, **upstream promoter elements**.

proneural gene Gene whose expression defines cells with the potential to develop as neural tissue.

proofreading Process by which potential errors in DNA replication, transcription, and translation are detected and corrected.

protease (proteinase, proteolytic enzyme) Enzyme that degrades proteins by hydrolyzing some of the peptide bonds between amino acids.

proteasome Large protein complex in the cytosol with proteolytic activity that is responsible for degrading proteins that have been marked for destruction by ubiquitylation or by some other means.

protein The major macromolecular constituent of cells. A linear polymer of amino acids linked together by peptide bonds in a specific sequence.

protein domain *see* **domain**

protein kinase Enzyme that transfers the terminal phosphate group of ATP to one or more specific amino acids (serine, threonine, or tyrosine) of a target protein.

protein kinase A Protein kinase which is activated by cyclic AMP.

proteolysis Degradation of a protein by hydrolysis at one or more of its peptide bonds.

proteolytic enzyme *see* **protease**

proteomics Study of all the proteins, including all the covalently modified forms of each, produced by a cell, tissue, or organism. Proteomics often investigates changes in this larger set of proteins ("the proteome") caused by changes in the environment or by extracellular signals.

proto-oncogene Normal gene, usually concerned with the regulation of cell proliferation, that can be converted into a cancer-promoting **oncogene** by mutation or overexpression.

pseudogene Nucleotide sequence of DNA that has accumulated multiple mutations that have rendered it inactive and nonfunctional.

PTB **Polypyrimidine tract**-binding protein. The constitutively expressed PTB protein and the neuron-specific nPTB protein act as **alternative RNA splicing** factors regulating specific splicing decisions.

pTEF-b Kinase enzyme which phosphorylates the **CTD** of **RNA polymerase** II on serine 2 and thereby plays a key role in transcriptional elongation.

pulse labeling Process for determining the synthesis rate of RNA (or any other molecule) by adding radioactive precursor to cells for a brief period and measuring its incorporation into the RNA.

purine   Nitrogen-containing base found in DNA and RNA: adenine or guanine.

pyrimidine   Nitrogen-containing base found in DNA and RNA: cytosine, thymine, or uracil.

quenching   Mechanism of transcriptional repression in which an inhibitory factor masks the **activation domain** of an activating transcription factor and prevents it activating transcription.

quiescent   Non-dividing cells.

Ras (Ras protein)   Monomeric **GTPase** of the Ras super-family that helps to relay signals from cell-surface **receptor tyrosine kinases** to the nucleus, frequently in response to signals that stimulate cell division. Named for the *ras* gene, first identified in viruses that cause rat sarcomas.

Ras superfamily   Large superfamily of monomeric **GTPases** (also called small GTP-binding proteins) of which **Ras** is the prototypical member.

Rb   *see* **retinoblastoma protein**

reading frame   Phase in which nucleotides are read in sets of three to encode a protein. An mRNA molecule can be read in any one of three frames, only one of which will give the required protein.

receptor   Any protein that binds a specific signal molecule (**ligand**) and initiates a response in the cell. Some are on the cell surface while others are inside the cell.

receptor serine/threonine kinase   Cell-surface receptor with an extracellular ligand-binding domain and an intracellular kinase domain that phosphorylates signaling proteins on serine or threonine residues in response to ligand binding.

receptor tyrosine kinase (RTK)   Cell-surface receptor with an extracellular ligand-binding domain and an intracellular kinase domain that phosphorylates signaling proteins on tyrosine residues in response to ligand binding.

recessive   In genetics, the member of a pair of alleles that fails to be expressed in the phenotype of the organism when the **dominant** allele is present.

recognition helix   The helix within the DNA-binding **helix-turn-helix motif** which recognizes and binds to a specific DNA sequence.

recombinant DNA   Any DNA molecule formed by joining DNA segments from different sources.

recombinant DNA technology   *see* **genetic engineering**

red blood cell   *see* **erythrocyte**

regulatory gene   Gene encoding a protein that regulates processes such as gene transcription or RNA splicing. *Compare with* **structural gene**.

regulatory sequence   DNA sequence to which a gene-regulatory protein binds to control the rate of assembly of the **basal transcriptional complex** at the **promoter**.

release factor   Protein that enables release of a newly synthesized protein from the ribosome by binding to the ribosome in the place of **transfer RNA** (whose structure it mimics).

reporter gene   Genetic construct, usually artificial, in which a copy of the regulatory DNA of a gene of interest is linked to a sequence coding for a readily quantifiable product. The level of this product (the "reporter protein") in a cell containing the construct indicates the activity of the regulatory region of interest.

repressor (**gene repressor protein, transcriptional repressor**)   Protein that binds to a specific region of DNA to prevent transcription of an adjacent gene.

REST   Repressor element silencing transcription factor. Inhibitory transcription factor which represses neuron-specific gene expression in non-neuronal cells.

restriction fragment   Fragment of DNA generated by the action of **restriction enzyme(s)**.

restriction enzyme (**restriction endonuclease**)   One of a large number of nucleases that can cleave a DNA molecule at any site where a specific short sequence of nucleotides occurs. Extensively used in recombinant DNA technology.

reticulocyte   Precursor to the mature red blood cells (**erythrocytes**) which lacks the nucleus but continues to synthesize globin from long-lived mRNA.

retinoblastoma protein (**Rb**)   **Anti-oncoprotein** (tumor-suppressor protein) involved in the regulation of cell division. Mutated in the cancer retinoblastoma, as well as in many other tumors. Its normal activity is to regulate the eukaryotic cell cycle by binding to and inhibiting the **E2F proteins**, thus blocking progression to DNA replication and cell division.

retrovirus   RNA-containing virus that replicates in a cell by first making an RNA-DNA intermediate and then a double-stranded DNA molecule that becomes integrated into cellular DNA.

reverse transcriptase   Enzyme first discovered in **retroviruses** that makes a double-stranded DNA copy from a single-stranded RNA template molecule.

reverse transcription   Transcription from RNA to DNA. This is in the opposite direction to that prescribed by the central dogma of molecular biology, which holds that DNA is transcribed into RNA and RNA is then translated into protein.

ribonuclease   Enzyme that cuts an RNA molecule by hydrolyzing one or more of its phosphodiester bonds.

ribonucleic acid   *see* **RNA**

ribonucleoprotein   Term for any complex of RNA and protein; *see also* **small nuclear ribonucleoprotein**.

ribose   The five-carbon monosaccharide component of RNA. *Compare with* **deoxyribose**.

ribosomal RNA (**rRNA**)   Any one of a number of specific RNA molecules that form part of the structure of a ribosome and participate in the synthesis of proteins. Often distinguished by their sedimentation coefficient (e.g. 28S rRNA, 5S rRNA).

ribosome   Particle composed of rRNAs and ribosomal proteins that catalyzes the synthesis of protein using information provided by mRNA.

RISC   RNA-induced silencing complex. Protein complex which plays a key role in post-transcriptional repression by **microRNAs** and **siRNAs**.

RITS RNA-induced transcriptional silencing complex. Protein complex which plays a key role in transcriptional repression by **siRNAs**.

RNA (ribonucleic acid) Polymer formed from covalently linked ribonucleotide monomers. *See also* **messenger RNA, ribosomal RNA, transfer RNA.**

RNA editing Type of RNA processing that alters the nucleotide sequence of a pre-mRNA transcript after it is synthesized by inserting, deleting, or altering individual nucleotides.

RNA exporter complex Multi-protein complex which binds to RNAs and promotes their export from the nucleus to the cytoplasm.

RNA interference (RNAi) As originally described, mechanism by which a double-stranded RNA induces sequence-specific destruction of complementary mRNAs. The mechanism, which is highly conserved in eukaryotes, proceeds through short double-stranded **small interfering RNAs** produced by endonucleolytic cleavage. The term RNAi is often used broadly to also include the inhibition of gene expression by **microRNAs**, which are encoded in the cell's own genome. RNA interference is widely used experimentally to study the effects of inactivating specific genes.

RNA polymerase Enzyme that catalyzes the synthesis of an RNA molecule on a DNA template from ribonucleoside triphosphate precursors. RNA polymerase I transcribes the gene encoding the 28S, 18S, and 5.8S **ribosomal RNAs**, RNA polymerase II transcribes protein-coding genes, RNA polymerase III transcribes the genes encoding **transfer RNAs** and the 5S **ribosomal RNA**. RNA polymerases IV and V are found only in plants and are involved in transcriptional repression by **small interfering RNAs**.

RNA polymerase holoenzyme Multi-protein complex of **RNA polymerase** II and basal transcription factors such as TFIIB (but not TFIID).

RNA processing Broad term for the various modifications an RNA transcript undergoes as it reaches its mature form. May include 5′ capping, 3′ polyadenylation, 3′ cleavage, splicing, and editing.

RNA splicing Process in which **intron** sequences are excised from RNA transcripts and **exons** are joined together in the nucleus during formation of messenger and other RNAs.

rRNA *see* **ribosomal RNA**

rRNA gene Gene that specifies a **ribosomal RNA** (rRNA).

RTK *see* **receptor tyrosine kinase**

RT-PCR (reverse transcription-polymerase chain reaction) Technique in which a population of mRNAs is converted into cDNAs via reverse transcription, and the cDNAs are then amplified by **PCR**.

S *see* **S phase**

Saccharomyces Genus of yeasts that reproduce asexually by budding or sexually by conjugation. Economically important in brewing and baking. *Saccharomyces cerevisiae* is widely used as a simple model organism in the study of eukaryotic cell biology.

SAGA complex Multi-protein complex which links activating transcription factors to the **basal transcriptional complex**.

S. cerevisiae *see* ***Saccharomyces***

SDS-PAGE (sodium dodecyl sulfate-polyacrylamide gel electrophoresis) Type of electrophoresis used to separate proteins by size. The protein mixture to be separated is first treated with a powerful negatively charged detergent (SDS) and with a reducing agent (β-mercaptoethanol), before being run through a polyacrylamide gel. The detergent and reducing agent unfold the proteins, free them from association with other molecules and separate the polypeptide subunits. *See also* **electrophoresis**.

second messenger (small intracellular mediator) Small intracellular signaling molecule that is formed or released for action in response to an extracellular signal and helps to relay the signal within the cell. Examples include **cyclic AMP, cyclic GMP,** IP_3, and Ca^{2+}.

segment-polarity gene In *Drosophila* development, a gene involved in specifying the anteroposterior organization of each body segment.

serine/threonine kinase Enzyme that phosphorylates specific proteins on serine or threonine.

sex chromosome Chromosome that may be present or absent, or present in a variable number of copies. determining the sex of the individual; in mammals, the X and Y chromosomes.

β-sheet *see* **beta-sheet**

Shine–Dalgarno sequence Sequence found in prokaryotic mRNAs which is complementary to the 16S ribosomal RNA and therefore promotes binding of the small ribosomal subunit to the mRNA.

signaling cascade Sequence of linked intracellular reactions, typically involving multiple amplification steps in a relay chain, triggered by an activated cell-surface receptor.

signal molecule Extracellular chemical produced by a cell that signals to other cells in the organism to alter cellular behavior.

signal transduction Conversion of a signal from one physical or chemical form to another (e.g. conversion of extracellular signals to intracellular ones).

silencer Regulatory DNA sequence that can decrease the rate of transcription of a structural gene.

siRNA *see* **small interfering RNA**

S6 kinase Enzyme which is phosphorylated by the **TOR** kinase and which in turn phosphorylates the S6 ribosomal protein and translation initiation factors.

Smad protein Latent gene-regulatory protein that is phosphorylated and activated by **receptor serine/threonine kinases** and carries the signal from the cell surface to the nucleus.

small interfering RNA (siRNA) Short (21–26 nucleotides) double-stranded RNAs that inhibit gene expression by directing destruction of complementary mRNAs. Production of siRNAs is triggered by double-stranded RNA.

small intracellular mediator *see* **second messenger**

small nuclear ribonucleoprotein particles (snRNP) Complex of a **small nuclear RNA** with proteins that forms part of a **spliceosome**.

small nuclear RNA (snRNA) Small RNA molecules that are complexed with proteins to form the **small nuclear ribonucleoprotein particles** (snRNPs) involved in **RNA splicing**. Also known as **URNAs**.

snRNA *see* **small nuclear RNA**

solenoid Potential structure of the 30 nm chromatin fiber in which **nucleosomes** are thrown into a single helix. *See also* **zig-zag ribbon**.

somatic cell Any cell of plant or animal other than the **germ cells**.

somite One of a series of paired blocks of **mesoderm** that form during early development and lie on either side of the **notochord** in a vertebrate embryo. They give rise to the segments of the body axis, including the vertebrae, muscles, and associated connective tissue.

Southern blotting Technique in which DNA fragments separated by electrophoresis are immobilized on a paper sheet. Specific fragments are then detected with a labeled nucleic acid probe (named after E.M. Southern, inventor of the technique).

spermatogenesis Development of sperm in the testes.

Sp1 box DNA sequence found in many gene **promoters** which binds the Sp1 transcription factor.

S phase Period of a eukaryotic cell cycle in which DNA is synthesized.

spliceosome Large assembly of RNA and protein molecules that performs pre-mRNA splicing in eukaryotic cells.

splice site **Exon/intron** junctions in an RNA prior to splicing. The junction between the upstream exon and the intron is known as the 5′ splice site whilst the junction between the intron and the downstream exon is known as the 3′ splice site.

splicing Removal of **introns** from a pre-mRNA transcript and joining together of the **exons** that lie on either side of each intron. *See also* **alternative RNA splicing, trans-splicing**.

SR proteins Proteins which are rich in serine (S) and arginine (R). They play a key role in **RNA splicing** binding to **exon splicing enhancers**.

stalled polymerase **RNA polymerase** which has paused after initiating transcription and transcribing approximately 30 nucleotides. This process is known as **polymerase pausing**.

start codon AUG **codon** in the mRNA at which protein synthesis initiates.

STAT (signal transducer and activator of transcription) Latent transcription factor that is activated by phosphorylation by **JAK** kinases and enters the nucleus in response to signaling from receptors of the cytokine receptor family.

stem cell Undifferentiated cell that can continue dividing indefinitely, producing daughter cells that can either commit to differentiation or remain a stem cell (in the process of self-renewal).

steroid Hydrophobic lipid molecule with a characteristic four-ringed structure; derived from cholesterol. Many important hormones, including glucocorticoid, estrogen, and testosterone, are steroids that activate intracellular **nuclear receptors**.

stop codon **Codon** in the mRNA which produces termination of its translation rather than insertion of another amino acid.

structural gene Gene that codes for a protein or for an RNA molecule that forms part of a structure or has an enzymatic function. distinguished from genes that encode proteins that regulate gene expression. *Compare with* **regulatory gene**.

subtractive hybridization **Hybridization** technique in which all mRNAs common to two samples are removed leaving only those specifically expressed in one of the samples which can then be analyzed.

SUMO Small protein related to ubiquitin which, like ubiquitin, can be linked to other proteins. *Compare with* **ubiquitin**.

SV40 Simian virus 40. Small DNA virus which infects monkeys and has been widely used in studies of eukaryotic gene regulation.

SWI–SNF Multi-protein complex which is able to hydrolyze **ATP** and use the energy generated to catalyze **chromatin remodeling**.

synaptic plasticity Increase in synaptic activity following repeated stimulation.

TAFs Factors associated with **TBP** in the TFIID complex which can act as co-activators.

TATA box AT-rich sequence in the **promoter** region of many eukaryotic genes that binds the general transcription factor **TBP** and hence specifies the position at which transcription is initiated.

TBP Transcription factor which plays a key role in transcription by all three **RNA polymerases**. Binds to the **TATA box**.

T cell (T lymphocyte) Type of lymphocyte responsible for T-cell-mediated immune responses.

telomere End of a chromosome, associated with a characteristic DNA sequence that is replicated in a special way. Counteracts the tendency of the chromosome otherwise to shorten with each round of replication. From Greek *telos*, meaning "end".

thalassemia Human disease caused by abnormal synthesis of hemoglobin.

TFI- Component of the **basal transcriptional complex** for **RNA polymerase** I; e.g. TFIA.

TFII- Component of the **basal transcriptional complex** for **RNA polymerase** II, e.g. TFIIB, TFIID.

TFIII- Component of the **basal transcriptional complex** for **RNA polymerase** III, e.g. TFIIIA, TFIIIB, TFIIIC.

T lymphocyte *see* **T cell**

topoisomerase Enzyme that binds to DNA and reversibly breaks a phosphodiester bond in one or both strands.

TOR (target of Rapamycin) Serine/threonine protein

kinase that is a downstream component of the **PI3-kinase/ Akt** signaling pathway.

totipotent   Describes a cell that is able to give rise to all the different cell types in an organism. *Compare with* **pluripotent**.

trans   On the other (far) side, used, for example, to indicate that two molecules are not physically linked to one another; e.g. a *trans*-acting regulatory protein. *Compare with* **cis**.

transcript   RNA product of DNA transcription.

transcription (DNA transcription)   Copying of one strand of DNA into a complementary RNA sequence by the enzyme **RNA polymerase**.

transcription factor   Protein involved in transcriptional initiation or elongation or in their regulation.

transcriptional activator   *see* **activator**

transcriptional repressor   *see* **repressor**

transcriptomics   Study of all the mRNAs of a cell, tissue or organism. *Compare with* **proteomics**.

transdifferentiation   Process in which one differentiated cell type can dedifferentiate, proliferate, and then differentiate into a different cell type.

transfer RNA (tRNA)   Set of small RNA molecules used in protein synthesis as an interface (adaptor) between mRNA and amino acids. Each type of tRNA molecule is covalently linked to a particular amino acid.

transformation   (1) Insertion of new DNA (e.g. a plasmid) into a cell or organism, such as into competent *E. coli*. (2) Conversion of a normal cell into one that behaves in many ways like a cancer cell (i.e. unregulated proliferation, anchorage-independent growth in culture).

transgenic organism   Plant or animal that has stably incorporated one or more genes from another cell or organism (through insertion, deletion, and/or replacement) and can pass them on to successive generations. The gene that has been added is called a transgene.

translation (RNA translation)   Process by which the sequence of nucleotides in an mRNA molecule directs the incorporation of amino acids into protein. Occurs on **ribosomes**.

translocation   Type of mutation in which a portion of one chromosome is broken off and attached to another.

trans-splicing   Type of **RNA splicing** in which **exons** from two separate RNA transcripts are joined together to form an mRNA.

trimeric G protein (trimeric GTP-binding protein)   *see* **G protein**

triplet-repeat disease   Diseases involving abnormal amplification of a 3 bp sequence.

trithorax complex   Multi-protein complex which produces an open chromatin structure, allowing transcription to occur.

tRNA   *see* **transfer RNA**

trophectoderm   Outer cells of the **blastula** surrounding the **inner cell mass**. Trophectoderm cells give rise to the extra-embryonic membranes/placenta.

tubulin   The protein subunit of microtubules.

tumor   Abnormal mass of cells resulting from a defect in cell-proliferation control.

tumor necrosis factor α **(TNFα)**   **Cytokine** that is especially important in inducing inflammatory responses.

tumor virus   Virus that can make the cell it infects cancerous.

tyrosine kinase   Enzyme that phosphorylates specific proteins on tyrosines.

ubiquitin   Small, highly conserved protein present in all eukaryotic cells that becomes covalently attached to lysines of other proteins.

upstream promoter elements   DNA sequences close to and usually upstream of the **core promoter** which increase promoter activity.

URNA   Family of small uridine-rich RNAs. Several such as U1, U2, U4, U5, and U6 are involved in **RNA splicing**. Also known as **snRNAs**.

UTR (untranslated region)   Non-coding region of an mRNA molecule. The 3′ UTR extends from the **stop codon** that terminates protein synthesis to the start of the **polyA tail**. The 5′ UTR extends from the **5′ cap** to the start codon that initiates protein synthesis.

variable region   Region of an immunoglobulin light or heavy chain that differs from molecule to molecule and forms the antigen-binding site.

vascular endothelial growth factor (VEGF)   Secreted protein that stimulates the growth of blood vessels.

V(D)J recombination   Somatic recombination process by which gene segments are brought together to form a functional gene for a polypeptide chain of an immunoglobulin or T-cell receptor.

vector   In cell biology, the DNA of any agent (virus or plasmid) used to transmit genetic material to a cell or organism. *See also* **cloning vector**, **expression vector**.

VEGF   *see* **vascular endothelial growth factor**

VHL   von Hippel Lindau **anti-oncoprotein** which is involved in regulating the stability of other proteins.

virus   Particle consisting of nucleic acid (RNA or DNA) enclosed in a protein coat and capable of replicating only within a host cell and then spreading from cell to cell.

Western blotting (immunoblotting)   Technique by which proteins are separated by electrophoresis and immobilized on a paper sheet and then analyzed, usually by means of a labeled antibody.

white blood cell (leukocyte)   General name for all the nucleated blood cells lacking hemoglobin. Includes lymphocytes, granulocytes, and monocytes.

wild-type   Normal, non-mutant form of a gene or an organism; the form found in nature (in the wild).

Wnt (Wnt protein)   Member of a family of secreted signal proteins that have many different roles in controlling cell differentiation, proliferation, and gene expression in animal embryos and adult tissues.

Wnt signaling pathway Signaling pathway activated by binding of a **Wnt protein** to its cell-surface receptors. Activation causes increased amounts of β-**catenin** to enter the nucleus, where it regulates the transcription of genes controlling cell differentiation and proliferation. Overactivation of the Wnt/β-catenin pathway can lead to cancer.

wobble effect Ability of several different **codons** in the mRNA to bind to the same **anticodon** in a **tRNA** and therefore direct insertion of the same amino acid into the protein.

X chromosome One of the two sex chromosomes of mammals. The cells of women contain two X chromosomes, while those of men contain only one.

Xenopus laevis Species of frog frequently used in studies of early vertebrate development.

XIC *see* **X-inactivation center**

X-inactivation Inactivation of one copy of the X chromosome in the somatic cells of female mammals.

X-inactivation center (**XIC**) Site in an X chromosome at which inactivation is initiated and spreads outwards.

XIST Non-coding RNA whose transcription initiates in the **X-inactivation center** and which is critical for the process of **X-inactivation**.

X-ray crystallography (**X-ray diffraction**) Technique for determining the three-dimensional arrangement of atoms in a molecule such as a protein based on the diffraction pattern of X-rays passing through a crystal of the molecule.

Y chromosome One of the two sex chromosomes of mammals. The cells of men contain one Y and one X chromosome.

yeast Common name for several families of unicellular fungi. Includes species used for brewing and bread-making, as well as pathogenic species. Among the simplest of **eukaryotes**.

zebrafish A fish which is widely used as a model organism for the study of vertebrate development.

zig-zag ribbon Potential structure of the 30 nm chromatin fiber in which **nucleosomes** are thrown into a double helical structure. *See also* **solenoid**.

zinc finger DNA-binding structural motifs present in many gene-regulatory proteins. All zinc finger motifs incorporate one or more zinc atoms that help hold the protein conformation together. In one type of finger, the zinc atom is bound by two cysteine and two histidine amino acids. In the other type of finger it is bound by multiple cysteine amino acids.

zygote **Diploid** cell produced by fusion of a male and female gamete. A fertilized egg.

Index

Page numbers in **boldface** refer to major discussion of a topic. Page numbers followed by F refer to figures, those followed by T refer to tables, and those followed by M refer to method boxes.

Note that prefixes are ignored in the alphabetical order; c-fos gene, for example, appears under the letter 'f'.